Advanced Structured Materials

Volume 132

Common engineering materials reach in many applications their limits and new developments are required to fulfil increasing demands on engineering materials. The performance of materials can be increased by combining different materials to achieve better properties than a single constituent or by shaping the material or constituents in a specific structure. The interaction between material and structure may arise on different length scales, such as micro-, meso- or macroscale, and offers possible applications in quite diverse fields.

This book series addresses the fundamental relationship between materials and their structure on the overall properties (e.g. mechanical, thermal, chemical or magnetic etc.) and applications.

The topics of *Advanced Structured Materials* include but are not limited to

- classical fibre-reinforced composites (e.g. glass, carbon or Aramid reinforced plastics)
- metal matrix composites (MMCs)
- micro porous composites
- micro channel materials
- multilayered materials
- cellular materials (e.g., metallic or polymer foams, sponges, hollow sphere structures)
- porous materials
- truss structures
- nanocomposite materials
- biomaterials
- nanoporous metals
- concrete
- coated materials
- smart materials

Advanced Structured Materials is indexed in Google Scholar and Scopus.

More information about this series at http://www.springer.com/series/8611

Bilen Emek Abali · Ivan Giorgio
Editors

Developments and Novel Approaches in Biomechanics and Metamaterials

Springer

Editors
Bilen Emek Abali
Institute of Mechanics
Technische Universität Berlin
Berlin, Germany

Ivan Giorgio
Department of Mechanical
and Aerospace Engineering
University of Rome La Sapienza
Latina, Italy

ISSN 1869-8433 ISSN 1869-8441 (electronic)
Advanced Structured Materials
ISBN 978-3-030-50466-3 ISBN 978-3-030-50464-9 (eBook)
https://doi.org/10.1007/978-3-030-50464-9

This Springer imprint is published by the registered company Springer Nature Switzerland AG
The registered company address is: Gewerbestrasse 11, 6330 Cham, Switzerland

Preface

The ICoNSOM 2019, International Conference on Nonlinear Solid Mechanics, took place at Palazzo Argiletum, Rome, Italy, from June 16 to June 19, 2019. Over 200 participation from the whole globe, the urge of this proceedings became clear. With the aid of the organizers, Marco Amabili, Francesco dell'Isola, Ivan Giorgio, Nicola Rizzi, and Luca Placidi, the scientific community did show a great interest allowing us to bring together this proceedings collected in two volumes:

- Developments and Novel Approaches in Nonlinear Solid Body Mechanics
- Developments and Novel Approaches in Biomechanics and Metamaterials

ICoNSoM 2019 Conference has been intended to provide an international opportunity for communicating recent developments in various areas of nonlinear solid mechanics. This monograph consists theory, experiments, and applications in mechanics, thermodynamics, and multiphysics simulation in many length scales.

As editors, we intend to thank all authors for their crucial contributions as well as all reviewers for their invaluable time and effort. We delightedly acknowledge Dr. Christoph Baumann (Springer Publisher) for initiating the book project. In addition, we have to thank Dr. Mayra Castro (Senior Editor Applied Sciences; Materials Science; Materials Engineering; Nanotechnology and Nanomedicine) and Mr. Ashok Arumairaj (Production Administrator) giving their support in the process of publication.

Brussels, Rome
May 2020

Bilen Emek Abali
Ivan Giorgio

Contents

List of Contributors

A. M. Abd-Alla
Math. Dept., Faculty of Science, Sohag University, Egypt. - Math. Dept., Faculty of Science, Taif University, Saudi Arabia.
e-mail: mohmrr@yahoo.com

A. N. Abd-alla
Math. Dept., Faculty of Science, Sohag University, Egypt
e-mail: aboelnourabdalla@yahoo.com

S. M. Abo-Dahab
Math. Dept., Faculty of Science, Taif University, Saudi Arabia. - Math. Dept., Faculty of Science, SVU, Qena, Egypt.
e-mail: sdahb@yahoo.com

K. Akhmadeev
Laboratoire des Sciences du Numérique de Nantes, UMR CNRS 6004
1 rue de la Noë, BP 92101 F-44321 Nantes Cedex 3, France
Ecole Centrale de Nantes, Université de Nantes
e-mail: konstantin.akhmadeev@ec-nantes.fr

M. Almasi
Department Of Mechanical Engineering, Isfahan University of Technology, Iran
e-mail: almasi_matin@yahoo.com

N. Almasi
Department of Structural and Geotechnical Engineering,
Medical Laboratory of Isfahan Shariati Hospital (Director), Iran
e-mail: naseralmasi@yahoo.com

M. Angiolilli
University of L'Aquila DICEAA, Via G. Gronchi 18, 67100, L'Aquila, Italy
e-mail: michele.angiolilli@graduate.univaq.it

K. Anselme
Université de Haute-Alsace, CNRS, IS2M, UMR 7361, Mulhouse, France
Université de Strasbourg, Strasbourg, France
e-mail: karine.anselme@uha.fr

C. C. António
Engineering Faculty, University of Porto, Porto, Portugal
Institute of Science and Innovation in Mechanical and Industrial Engineering (LAETA-INEGI), Porto, Portugal
e-mail: cantonio@fe.up.pt

Y. Aoustin
Laboratoire des Sciences du Numérique de Nantes, UMR CNRS 6004
1 rue de la Noë, BP 92101 F-44321 Nantes Cedex 3, France
Ecole Centrale de Nantes, Université de Nantes
e-mail: Yannick.Aoustin@univ-nantes.fr

M. Arricca
Università degli Studi di Brescia, Brescia, 25123, Italy
Department of Mechanical and Industrial Engineering (DIMI)
e-mail: m.arricca@unibs.it

A. Bacigalupo
IMT School for Advanced Studies Lucca, Piazza S. Francesco 19, 55100 Lucca, Italy
e-mail: andrea.bacigalupo@imtlucca.it

E. Barchiesi
Sapienza University of Rome, Roma, Italy
e-mail: barchiesiemilio@gmail.com

M. Bigerelle
Université de Valenciennes et du Hainaut Cambrésis, Laboratoire d'Automatique, de Mécanique et d'Informatique industrielle et Humaine (LAMIH), UMR-CNRS 8201, Le Mont Houy, Valenciennes, France
e-mail: Maxence.Bigerelle@uphf.fr

L. Bouazza
Univ. Lyon, Université Claude Bernard Lyon 1, INSERM, LYOS UMR 1033, 69008 Lyon, France
e-mail: lamia.bouazza@univ-lyon1.fr

M. Brieu
California State University, Los Angeles, ECST College, Dept. Mechanical Engineering, United States
e-mail: mbrieu@calstatela.edu

J-N. Brunet
Inria, Strasbourg, France e-mail: jean-nicolas.brunet@inria.fr

D. I. Caruntu
University of Texas Rio Grande Valley,
Mechanical Engineering Department, Edinburg, Texas 78539, U.S.A.
e-mail: caruntud2@asme.org

E. Le Carpentier
Laboratoire des Sciences du Numérique de Nantes, UMR CNRS 6004
1 rue de la Noë, BP 92101 F-44321 Nantes Cedex 3, France
Ecole Centrale de Nantes, Université de Nantes
e-mail: Eric.Le-Carpentier@ec-nantes.fr

R. P. Carreira
DECATHLON Technical Office, DECATHLON SA, Lille, France
e-mail: ruipedro.carreira@decathlon.com

R. Cartier
Montreal Heart Institute, University of Montreal, Montreal, Quebec, Canada
e-mail: cartierr@icm.umontreal.ca

C. Castoro
University of L'Aquila DICEAA, Via G. Gronchi 18, 67100, L'Aquila, Italy
e-mail: chiara.castoro@graduate.univaq.it

C. F. Castro
Engineering Faculty, University of Porto, Porto, Portugal
Institute of Science and Innovation in Mechanical and Industrial Engineering (LAETA-INEGI), Porto, Portugal
e-mail: ccastro@fe.up.pt

A. Ciallella
Dipartimento di Ingegneria Civile, Edile-Architettura e Ambientale, Università degli Studi dell'Aquila. Via Giovanni Gronchi n. 18, 67100 L'Aquila, Italy
e-mail: alessandro.ciallella@univaq.it

P. Clezardin
Univ. Lyon, Université Claude Bernard Lyon 1, INSERM, LYOS UMR 1033, 69008 Lyon, France
e-mail: philippe.clezardin@inserm.fr

C. Cluzel
Laboratoire de Mécanique et Technologie, Ecole Normale Supérieure Paris Saclay, Cachan, France
Département Science et Genie des Materiaux, Institut Universitaire de Technologie d'Evry Vald'Essonne, Every, France
e-mail: christophe.cluzel@ens-paris-saclay.fr

C. B. Confavreux
Univ. Lyon, Université Claude Bernard Lyon 1, INSERM, LYOS UMR 1033, 69008 Lyon, France
CEMOS, Service de Rhumatologie, Centre hospitalier Lyon Sud, Institut de Cancérologie des Hospices Civils de Lyon, France
e-mail: cyrille.confavreux@chu-lyon.fr

S. Cotin
Inria, Strasbourg, France e-mail: stephane.cotin@inria.fr

V. Damioli
SISSA - International School for Advanced Studies, Trieste, 34136 Italy
mathLab
e-mail: vdamioli@sissa.it

F. Daví
DICEA & ICRYS, Universitá Politecnica delle Marche, via Brecce Bianche, Ancona, Italy
e-mail: davi@univpm.it

M. De Angelo
Civil, Environmental and Architectural Engineering Department, The University of Kansas, 1530 W. 15th Street, Lawrence, KS 66045-7609, USA
Dipartimento di Ingegneria Civile, Edile-Architettura e Ambientale, Università degli Studi dellAquila, Via Giovanni Gronchi 18 - Zona industriale di Pile, 67100 LAquila, Italy
e-mail: michele.deangelo.aq@gmail.com

A. Della Corte
Mathematics Division, School of Sciences and Technology, University of Camerino, Italy
e-mail: alessandro.dellacorte@unicam.it

B. Delpuech
Univ. Lyon, Université Claude Bernard Lyon 1, INSERM, LYOS UMR 1033, 69008 Lyon, France
Univ Lyon, Université Claude Bernard Lyon 1, Univ. Gustave Eiffel, IFSTTAR, LBMC UMR_T 9406, F-69622, Lyon, France
e-mail: benjamin.delpuech@laposte.net

S. Diebels
Applied Mechanics, Saarland University, Campus A4.2, Saarbrücken, Germany
e-mail: s.diebels@mx.uni-saarland.de

S. Dumont
University of Nîmes, IMAG, CNRS UMR 5149 Pl. E. Bataillon, 34095 Montpellier cedex 5, France
e-mail: serge.dumont@unimes.fr

M. Elsagheer
Math. Dept., Faculty of Science, Sohag University, Egypt. - Math. Dept., Faculty of Science, Northern Border University, Arar, Saudi Arabia.
e-mail: elsagheer_mohamed@yahoo.com

H. Follet
Univ. Lyon, Université Claude Bernard Lyon 1, INSERM, LYOS UMR 1033, 69008 Lyon, France
e-mail: helene.follet@inserm.fr

A. Gallard
Laboratoire des Sciences du Numérique de Nantes, UMR CNRS 6004
1 rue de la Noë, BP 92101 F-44321 Nantes Cedex 3, France
Ecole Centrale de Nantes, Université de Nantes
e-mail: albgal42@gmail.com

A. Gregori
University of L'Aquila DICEAA, Via G. Gronchi 18, 67100, L'Aquila, Italy
e-mail: amedeo.gregori@ing.univaq.it

C. P. Grigoropoulos
University of California, Berkeley, 2521 Hearst Ave, 94720-1740, Berkeley, CA, USA
e-mail: cgrigoro@berkeley.edu

E. Grillo
Università degli Studi di Brescia, Brescia, 25123, Italy
Department of Molecular and Translational Medicine (DMMT)
Laboratory for Preventive and Personalized Medicine (MPP Lab)
e-mail: elisabetta.grillo@unibs.it

G. Haïat
CNRS, Univ Paris Est Creteil, Univ Gustave Eiffel, MSME UMR 8208, F-94010 Creteil, France
e-mail: guillaume.haiat@cnrs.fr

V. Hassaine-Daouadji
Univ. Lille, CNRS, Centrale Lille, FRE 2016 - LaMcube - Laboratoire de mécanique multiphysique multiéchelle, F-59000, Lille, France
e-mail: valentin.hassainedaouadji@centralelille.fr

T. Hoellinger
Inria, Strasbourg, France e-mail: tristan.hoelling@inria.fr

A. Jung
Applied Mechanics, Saarland University, Campus A4.2, Saarbrücken, Germany
e-mail: anne.jung@mx.uni-saarland.de

K. Komvopoulos
University of California, Berkeley, 2521 Hearst Ave, 94720-1740, Berkeley, CA, USA
e-mail: kyriakos@me.berkeley.edu

S. Laporte
Arts et Metiers Institute of Technology, Université Sorbonne Paris Nord, IBHGC - Institut de Biomécanique Humaine Georges Charpak, F-75013 Paris, France
e-mail: sebastien.laporte@ensam.eu

Marco Laudato
Dipartimento di Ingegneria e Scienze dell'Informazione e Matematica, Università degli Studi dell'Aquila, Via Vetoio (Coppito 1), 67100 Coppito, L'Aquila, Italy.
e-mail: marco.laudato@graduate.univaq.it

F. Lebon
Aix-Marseille University, CNRS, Centrale Marseille, LMA Impasse Nikola Tesla 4, Marseille, France
e-mail: lebon@lma.cnrs-mrs.fr

T. Lemaire
Univ Paris Est Creteil, Univ Gustave Eiffel, CNRS, MSME UMR 8208, F-94010 Creteil, France
e-mail: thibault.lemaire@univ-paris-est.fr

M. Lepidi
DICCA - Università di Genova, Via Montallegro 1, 16154 Genova, Italy
e-mail: marco.lepidi@unige.it

S. Lurie
Institute of Applied Mechanics of Russian Academy of Sciences, Leningradskiy ave., 7, Moscow, Russia
Moscow Aviation Institute, Moscow, Russia
e-mail: salurie@mail.ru

Luca Manzari
KTH Royal Institute of Technology, Marcus Wallenberg Laboratory for Sound and Vibration Research, SE-100 44, Stockholm, Sweden.
e-mail: manzari@kth.se

M. Martin
Univ Paris Est Creteil, Univ Gustave Eiffel, CNRS, MSME UMR 8208, F-94010 Creteil, France
School of Mechanical, Medical and Process Engineering, Queensland University of Technology, 2 George St, Brisbane, QLD 4000, Australia
e-mail: madge.martin@u-pec.fr

S. McLennan
Department of Biomedical Engineering, McGill University, Montreal, Quebec, Canada
e-mail: stewart.mclennan@mail.mcgill.ca

A. Mendizabal
Inria and University of Strasbourg, Strasbourg, France
e-mail: andrea.mendizabal@inria.fr

M. Mercuri
University of L'Aquila DICEAA, Via G. Gronchi 18, 67100, L'Aquila, Italy
e-mail: micaela.mercuri@graduate.univaq.it

J.-L. Milan
Aix Marseille Univ, CNRS, ISM, Marseille, France
e-mail: jean-louis.milan@univ-amu.fr

A. Misra
Civil, Environmental and Architectural Engineering Department, The University of Kansas, 1530 W. 15th Street, Lawrence, KS 66045-7609, USA
e-mail: amisra@ku.edu

S. Mitola
Università degli Studi di Brescia, Brescia, 25123, Italy
Department of Molecular and Translational Medicine (DMMT)
Laboratory for Preventive and Personalized Medicine (MPP Lab)
e-mail: stefania.mitola@unibs.it

D. Mitton
Univ Lyon, Université Claude Bernard Lyon 1, Univ. Gustave Eiffel, IFSTTAR, LBMC UMR_T 9406, F-69622, Lyon, France
e-mail: david.mitton@ifsttar.fr

R. Mongrain
Department of Biomedical Engineering, McGill University, Montreal, Quebec, Canada
e-mail: rosaire.mongrain@mcgill.ca

U. Mühlich
Universidad Austral de Chile, General Lagos 2086, Campus Miraflores, Valdivia, Chile
e-mail: Uwe.Muehlich@yahoo.com

N. Nejadsadeghi
Mechanical Engineering Department, University of Kansas, 1530 W. 15th Street, Learned Hall, Lawrence, KS 66045-7609, USA.
e-mail: nima.nejadsadeghi@ku.edu

S. Nicolle
Univ Lyon, Université Claude Bernard Lyon 1, Univ. Gustave Eiffel, IFSTTAR, LBMC UMR_T 9406, F-69622, Lyon, France
e-mail: stephane.nicolle@univ-lyon1.fr

S. Nikolaev
Inria, Strasbourg, France e-mail: sergei.nikolaev@inria.fr

M. Nowak
Poznan University of Technology, Chair of Virtual Engineering, Poland, ul. Jana Pawla II 24
60-965 Poznan, Poland
e-mail: Michal.Nowak@put.poznan.pl

L. Pieuchot
Université de Haute-Alsace, CNRS, IS2M, UMR 7361, Mulhouse, France
Université de Strasbourg, Strasbourg, France
e-mail: laurent.pieuchot@uha.fr

S. I. S. Pinto
Engineering Faculty, University of Porto, Porto, Portugal
Institute of Science and Innovation in Mechanical and Industrial Engineering (LAETA-INEGI), Porto, Portugal
e-mail: spinto@fe.up.pt

P. Pivonka
School of Chemistry, Physics and Mechanical Engineering, Queensland University of Technology, 2 George St, Brisbane, QLD 4000, Australia
e-mail: peter.pivonka@qut.edu.au

C. Ravelli
Università degli Studi di Brescia, Brescia, 25123, Italy
Department of Molecular and Translational Medicine (DMMT)
Laboratory for Preventive and Personalized Medicine (MPP Lab)
e-mail: cosetta.ravelli@unibs.it

M. Reis
Applied Mechanics, Saarland University, Campus A4.2, Saarbrücken, Germany
e-mail: martin.reis@uni-saarland.de

C. A. Reyes
University of Texas Rio Grande Valley,
Mechanical Engineering Department, Edinburg, Texas 78539, U.S.A
e-mail: c.reyes47@gmail.com

R. Rizzoni
Department of Engineering, University of Ferrara, Via Saragat 1, 44122, Ferrara, Italy
e-mail: raffaella.rizzoni@unife.it

N. Rogalski
Arts et Metiers Institute of Technology, Université Sorbonne Paris Nord, IBHGC - Institut de Biomécanique Humaine Georges Charpak, F-75013 Paris, France
e-mail: nicolas.rogalski@ensam.eu

E. Romano
Engineering Faculty, University of Porto, Porto, Portugal
Institute of Science and Innovation in Mechanical and Industrial Engineering (LAETA-INEGI), Porto, Portugal
e-mail: up201404532@fe.up.pt

A. Salvadori
Università degli Studi di Brescia, Brescia, 25123, Italy
Department of Mechanical and Industrial Engineering (DIMI)
e-mail: alberto.salvadori@unibs.it

V. Sansalone
Univ Paris Est Creteil, Univ Gustave Eiffel, CNRS, MSME UMR 8208, F-94010 Creteil, France
e-mail: vittorio.sansalone@u-pec.fr

D. Scerrato
International Research Center for the Matematics and Mechanics of Complex Systems, M&MoCS, Università degli Studi dell'Aquila, L'Aquila, Italy.
e-mail: daria.scerrato@gmail.com

M. Serpelloni
Università degli Studi di Brescia, Brescia, 25123, Italy
Department of Mechanical and Industrial Engineering (DIMI)
e-mail: m.serpelloni002@unibs.it

M. Serpilli
Department of Civil and Building Engineering, and Architecture, Università Politecnica delle Marche, Via brecce bianche, 60131 Ancona, Italy
e-mail: m.serpilli@univpm.it

Y. Solyaev
Institute of Applied Mechanics of Russian Academy of Sciences, Leningradskiy ave., 7, Moscow, Russia
Moscow Aviation Institute, Moscow, Russia
e-mail: yos@iam.ras.ru

G. Soulez
Centre Hospitalier de l'Université de Montréal, University of Montreal, Montreal, Quebec, Canada
e-mail: gilles.soulez@umontreal.ca

L. C. Sousa
Engineering Faculty, University of Porto, Porto, Portugal
Institute of Science and Innovation in Mechanical and Industrial Engineering (LAETA-INEGI), Porto, Portugal
e-mail: lcsousa@fe.up.pt

M. Spagnuolo
International Research Center for the Matematics and Mechanics of Complex Systems, M&MoCS, Università degli Studi dell'Aquila, L'Aquila, Italy
e-mail: mario.spagnuolo.memocs@gmail.com

E. Tagliabue
Università degli Studi di Verona e-mail: eleonora.tagliabue@univr.it

A. Ustenko
Institute of Applied Mechanics of Russian Academy of Sciences, Leningradskiy ave., 7, Moscow, Russia
Moscow Aviation Institute, Moscow, Russia
e-mail: audlord@ya.ru

Z. Vangelatos
University of California, Berkeley, 2521 Hearst Ave, 94720-1740, Berkeley, CA, USA
e-mail: zvaggelatos@gmail.com

M. Vassaux
Aix Marseille Univ, CNRS, ISM, Marseille, France
e-mail: m.vassaux@ucl.ac.uk

J. F. Witz
Univ. Lille, CNRS, Centrale Lille, FRE 2016 - LaMcube - Laboratoire de mécanique multiphysique multiéchelle, F-59000, Lille, France
e-mail: jean-francois.witz@centralelille.fr

C. Zikry
Department of Biomedical Engineering, McGill University, Montreal, Quebec, Canada
e-mail: christopher.zikry@mail.mcgill.ca

Chapter 1
Perspectives in Generalized Continua

Marco Laudato & Alessandro Ciallella

Abstract The International Conference on Nonlinear Solid Mechanics (ICoNSoM) 2019, held in Rome from 16th to 19th of June 2019, had as main goal to gather together researchers in the field of nonlinear Solid Mechanics in a stimulating research environment. This work is a rational report of activities of the mini-symposia "Perspectives in Generalized Continua" held during the conference. The main aim is to provide the interesting reader with the main topics treated during the conference and to furnish all the relevant bibliography. Additional information, such as the abstracts of all the talks, can be found at the official web-site of the conference: http://www.memocsevents.eu/iconsom2019/.

Keywords: Continuum mechanics · Metamaterials · Second gradient materials

1.1 Introduction

The multi-scale analysis of mechanical systems with a microstructure is one of the most growing research fields in modern Solid Mechanics. The main reasons are the efficient fast-prototyping techniques developed in the last decades on the one hand, and the enhanced mathematical and numerical understanding of this kind of systems, on the other hand. These results have been exploited by researchers to increase the forecasting power of the mathematical models used to describe the behavior of

M. Laudato
Dipartimento di Ingegneria e Scienze dell'Informazione e Matematica, Università degli Studi dell'Aquila. Via Vetoio, Coppito, 67100 L'Aquila, Italy
e-mail: marco.laudato@graduate.univaq.it

A. Ciallella
Dipartimento di Ingegneria Civile, Edile-Architettura e Ambientale, Università degli Studi dell'Aquila. Via Giovanni Gronchi n. 18, 67100 L'Aquila, Italy
e-mail: alessandro.ciallella@univaq.it

B. E. Abali and I. Giorgio (eds.), *Developments and Novel Approaches in Biomechanics and Metamaterials*, Advanced Structured Materials 132,
https://doi.org/10.1007/978-3-030-50464-9_1

continuous systems and even to tailor the features of the microstructure to obtain a desired macroscopic behavior in the so-called mechanical metamaterials (Barchiesi et al, 2019b; Del Vescovo and Giorgio, 2014; Milton, 2013; Milton et al, 2017). Since a detailed mathematical model of the elements of the microstructure is often very expensive from the numerical point of view, a customary approach is to define, by means of a homogenization procedure, an analogous continuous system (Abdoul-Anziz and Seppecher, 2018; Pideri and Seppecher, 1997; Boutin et al, 2017; Rahali et al, 2015; dell'Isola et al, 2015a; Mandadapu et al, 2018). The dynamics of such a system is defined in terms of an energy functional obtained from the kinematics and dynamics of the microstructure via some limiting procedure (Carcaterra et al, 2015; Turco et al, 2020). When the resulting energy model cannot be framed under the hypothesis of the classical theory of elasticity, the resulting continuous system is called a *generalized continuous system*.

The impact that this idea is having on modern Solid Mechanics in the last years has motivated the creation of a dedicated symposium on new perspectives in the generalized continua framework within the International Conference on Nonlinear Solid Mechanics 2019. Motivated by the need to create a stimulating forum of discussion on the main topics in nonlinear Solid Mechanics, the International Research Center of Mathematics and Mechanics of Complex Systems (M&MoCS) of L'Aquila University, in collaboration with the McGill University of Montreal, the Laboratoire International Associé Coss&Vita, and Roma Tre University have organized the International Conference on Nonlinear Solid Mechanics 2019 (ICoNSoM 2019), held in Rome from 16th to 19th of June 2019.

The topics of the conference have been divided into thematic symposia. In the present work, we focus on the symposium *Perspectives in Generalized Continua*. We present the content of the talks and, to introduce the interested reader into the state-of-art of the treated topics, we furnish the relevant references.

1.2 Perspectives in Generalized Continua

In this section we briefly review the content of the talks presented in the mini symposium *Perspectives in Generalized Continua*. The goal is to furnish the interested reader with the main topics treated and with the relevant bibliography.

The symposium has been animated by ten contributions which have been classified in this work for their experimental or theoretical preeminent nature. We will start by discussing the results in the theoretical framework and then we will conclude by discussing the talks on the experimental results. In the first group we have included the contributions by Eleni Agiasofitou on the J-, M-, and L-integrals, Claude Boutin on highly contrasted plates, Francesco dell'Isola on pantographic metamaterials, Salvatore Federico on convected stress and balance equations, Markus Lazar on singularity-free point of defects, Uwe Mühlich on the influence of randomness on the stiffness of different networks, and Hua Yang on homogenization procedures. In the second group, we have included the contributions by Gregor Ganzosch on experiments on pantographic materials, Marco Laudato on the experimental investigation of the dynamics of pantographic materials, and Mario Spagnuolo on the pivot damage mechanism in pantographic materials.

1.2.1 J-, M-, and L-integrals: from Micromechanics of Dislocations to Body Charges and Forces - Eleni Agiasofitou

Eleni Agiasofitou presented a talk titled "*J-, M-, and L-integrals: from Micromechanics of Dislocations to Body Charges and Body Forces*". The main goal was to provide an exposition of the J-, M-, and L-integrals and their role in field theories (Lazar and Agiasofitou, 2018a). The physical interpretation of these integrals have been discussed by deriving them in different physical scenario, highlighting their fundamental role. In particular, the role of these integrals for a single dislocation (e.g., edge and screw) has been analyzed and the remarkable result is that the M-integral can be used to describe the total energy per unit length of the dislocation which is in turn given by the sum of the self-energy and of the core energy per unit length. In the final part of the talk, electromagnetic systems have been discussed and it has been shown that the J-integral of body charges (in electrostatics) represents the Lorentz force whereas the M-integral between two point sources is proportional to the electrostatic interaction energy. Finally, the results coming from the comparison between these concepts in electromagnetism and elasticity theory have been remarked (Agiasofitou and Lazar, 2017; Lazar and Agiasofitou, 2018b).

1.2.2 Non-conventional Behavior of Highly Contrasted Plates - Theory and Experiments - Claude Boutin

Claude Boutin, in his contribution "*Non-conventional Behavior of Highly Contrasted Plates - Theory and Experiments*", discussed an effective approach to the description of highly contrasted plates exhibiting a non-conventional behavior. Two types of systems have been discussed: stratified plates in which thin inter-layers are softer than the thick layers, and plates characterized by a periodic 1D or 2D array of beams (Fossat et al, 2018; Viverge et al, 2016). A validation of the approach has been discussed in terms of numerical simulations and experiments. In particular, for the stratified plates it has been highlighted the role of inter-layer sliding as an additional kinematical descriptor. Such an enlarged kinematics allows for a tri-Laplacian formulation able to furnish analytical solutions under standard simple loadings. The dynamics of 1D and 2D beams stiffened plates has been analyzed in the final part of the talk. The different modeling approaches for the two cases allow for the investigation of non-typical wave dispersion in presence of high geometrical and mechanical contrast between the structural components (see also Boutin and Viverge, 2016; Boutin et al, 2020; Gazzo et al, 2020).

1.2.3 Pantographic Metamaterials - Francesco dell'Isola

Francesco dell'Isola presented a keynote lecture titled "*Pantographic Metamaterials*". During the lecture, the main features and the state of the art of this mechanical metamaterial (Alibert et al, 2003; dell'Isola et al, 2015b) have been carefully reviewed and discussed (dell'Isola et al, 2019b,a; Placidi et al, 2016; Barchiesi and Placidi, 2017). In particular, it has been stressed the crucial role of the interplay between fast-prototyping techniques such as 3D printing and the mathematical description of this complex system. An extended discussion was made, with the aid of several theoretical (Eremeyev et al, 2018; dell'Isola and Steigmann, 2015; dell'Isola et al, 2016; Steigmann and dell'Isola, 2015), numerical (Turco et al, 2016a; Giorgio, 2016; dell'Isola et al, 2017a; Maurin et al, 2019; Greco et al, 2017; Giorgio et al, 2016, 2017; Turco et al, 2017a; Scerrato et al, 2016; Giorgio et al, 2018a, 2019; Scerrato and Giorgio, 2019) and experimental results, of the static (Turco et al, 2017b; Golaszewski et al, 2019; Barchiesi et al, 2018a, 2019a, 2020b; Nejadsadeghi et al, 2019; Yildizdag et al, 2020) and dynamical Laudato and Barchiesi (2019) behavior of this system. In the final part of the talk, an analysis of the rupture mechanism of the system under quasi-static boundary conditions has been presented, highlighting the possible technological applications of this metamaterial. In particular, it has been shown that pantographic materials show three different deformation regimes. In the first one, the relation between relative displacement of the specimen and the total deformation energy is quadratical and consequently the resultant forces depends linearly on the displacement. In the second regime, the system is still in an elastic regime, although the forces do not depend linearly on the relative displacement. Finally, in the last regime the rupture of the systems starts but the resultant force continues to increase with the displacement up to the final rupture. The final value of the force is several times greater that the maximum obtained in the linear regime. As a consequence, the total energy needed to reach the ultimate failure of a pantographic material is larger that the maximum stored elastic energy. This is a remarkable property reinforced by the advantageous weight/resistance to extension ratio. One of the aspects of the utmost importance to be underlined here is that the pantographic metamaterial is a concrete example of material whose energy must incorporate second gradient terms at a macroscopic level of observation (Mindlin, 1965; dell'Isola et al, 2017b, 2015c).

1.2.4 Convected Stress and Balance Equations - Salvatore Federico

A rigorous mathematical formulation of the spatial equation of balance of linear momentum in field theories has been discussed in the talk "*Convected Stress and Balance Equations*" by Salvatore Federico. In particular, a fully material form of the vectorial equation of balance of linear momentum in terms of the so-called convected stress has been discussed (Alhasadi et al, 2019). By means of the pull-back by the so-called configuration map ϕ, it is indeed possible to define convected accelerations $A = \phi^* a$, convected body forces $B = \phi^* b$ and convected divergence which is computed in terms of the Levi-Civita connection associated with the pull-back via the configuration map of the metric tensor $C = \phi^* g$. Finally, it has been highlighted the relations between the geometrical meaning of convected stress and its relation with the Piola–Kirchhoff stress tensor. In particular, the convected stress is power conjugated to the convected deformation rate, i.e. the Lie derivative of the pulled-back metric C with respect to the convected velocity. Therefore, it is possible to consider the convected stress as a geometrically meaningful alternative to the second Piola–Kirchhoff stress tensor. We refer also to the papers by Federico et al (2019); Epstein (2009); Grillo et al (2016, 2005) for some more details.

1.2.5 Singularity-free Point of Defects: Gradient Elasticity of Bi-Helmholtz Type - Markus Lazar

In the talk by Markus Lazar, "*Singularity-free Point of Defects: Gradient Elasticity of Bi-Helmholtz Type*", a non-singular continuum theory of point defects based on the bi-Helmholtz second strain gradient elasticity theory has been presented. The non-singular expressions of the displacement field, the first displacement gradient, and the second displacement gradient which have been derived in this setting have been discussed. Furthermore, plastic distortion and the gradient of plastic distortion appear to be dependent on a form factor and their expression is non-singular. By means of this approach, the interaction energy and the interaction force between two dilation centers and for one dilation center in the stress field of an edge dislocation have all non singular expression (Lazar et al, 2006; Lazar, 2013). The main advantages have been discussed in the last part of the seminar in which applications on the computation of finite self-energy of a dilation centre have been shown (Lazar, 2019).

1.2.6 Influence of Randomness in Topology and Geometry on the Stiffness of Different Networks Generated from the Same Graph - Uwe Mühlich

Uwe Mühlich presented a talk titled "*Influence of Randomness in Topology and Geometry on the Stiffness of Different Networks Generated from the Same Graph*". The main goal was to present a Ising model aimed at describing the relations between the characteristic of a random network and the effective materials Mühlich et al (2015). During the talk, different networks obtained from the same initial graph have been analyzed when undergoing to an overall strain. In particular, different realizations of the network can be realized by varying the ratio between the numbers of diagonals of different type and a regularity measure. By means of the interpretation of the edges of the network in terms of truss and beams it is possible to study in this way mechanical systems undergoing an overall strain. The correlation between the strain energy density and the matrices describing an effective media theory defined on these networks, such as degree matrix, adjacency matrix and incidence matrix, has been discussed in the last part of the talk. Moreover, for the sake of completeness, some useful concepts can be find in Zybell et al (2014); Mühlich et al (2012, 2009).

1.2.7 Homogenization of Metamaterials with Strain Gradient Effects Based on Asymptotic Analysis - Hua Yang

In the contribution by Hua Yang, "*Homogenization of Metamaterials Based on Asymptotic Analysis by Considering Strain Gradient Effects*", the role of asymptotic analysis in homogenization procedures has been discussed (Abali et al, 2017). It has been shown how to involve, in this approach, higher order terms in the asymptotic expansion of the microscopic deformation energy (Yang et al, 2018). In particular, once a constitutive strain gradient law is defined by assuming the equivalence between strain energy at micro and macro scale over a representative volume element, the deformation energy is developed in an asymptotic series in which the higher order terms are kept in order. Their role is to model the emergent effects of the microstructure at the macroscopic scale. The dependence of this method on the characteristic of the representative volume element has been discussed in the final part of the talk. A useful assortment of papers about the considered topic can be listed as follows (Andreaus et al, 2016; Giorgio et al, 2018b; De Angelo et al, 2019a; Placidi et al, 2017; Turco et al, 2018; Abali et al, 2015, 2019; Yang et al, 2019).

1.2.8 Experiments Performed on Additively Manufactured Pantographic Structures, Validated by Digital Image Correlation - Gregor Ganzosch

Gregor Ganzosch, in his talk "*Experiments Performed on Additively Manufactured Pantographic Structures, Validated by Digital Image Correlation*", discussed the design, additive manufacturing, experimentation, and data analysis of particular examples of pantographic beams and bi-pantographic fabrics. Similarly to pantographic structures, bi-pantographic materials shows wide elastic range during the compliant phase followed by a stiffening when the total strain increases. The mathematical description of the behavior of this kind of system requires therefore higher gradient models. Indeed, 2D pantographic and bi-pantographic materials represent two paradigmatic examples of complete second gradient continua in plane. The results of these investigations have been described, highlighting the presence of buckling phenomena and the high resilience-to-weight ratio exhibited by the various specimens Ganzosch et al (2016). In particular, the presence of non-negligible local buckling phenomena for specimens undergoing shear tests has been discussed. For an overview about the subject, we refer to the papers by Misra et al (2018); Ganzosch et al (2018); Juritza et al (2019); Turco (2019); dell'Isola et al (2020, 2019c); Barchiesi et al (2020a); Tran et al (2020).

1.2.9 Dynamics of 2D Pantographic Metamaterial: Numerical and Experimental Results - Marco Laudato

Marco Laudato, in his contribution "*Dynamics of 2D Pantographic Metamaterial: Numerical and Experimental Results*", has shown the results of his investigation on the dynamical behavior of 2D pantographic fabrics. The comparison between numerical simulations and digital correlation based experiments has been discussed Laudato et al (2018, 2019); Barchiesi et al (2018b). Finally, some information on the spectral properties of this system has been drawn. In particular, the system under investigation was a rectangular pantographic material with one of its short sides clamped to the ground and with the other one undergoing a sinusoidal small oscillation. Under the assumption of linear temporal invariance, the behavior of the system has been analyzed by fitting the displacement vector field in time of the elements of a discrete mesh with a generic sinusoidal function. The proposed linear second gradient model and the experimental observations have been compared by analyzing the distribution of the fit parameters on the geometric domain of the sample (see also Battista et al, 2015, 2017; Eremeyev et al, 2019; Eremeyev and Sharma, 2019; Eremeyev, 2019).

1.2.10 Pivots Damage Mechanism in Metallic Pantographic Structures: An Overview - Mario Spagnuolo

A comprehensive analysis of the behavior of pivot damage in metal pantographic structures has been the main topic of the talk "*Pivots Damage Mechanism in Metallic Pantographic Structures: An Overview*" by Mario Spagnuolo. The relation between the geometrical features of the pivots and the spatial localization of failure signatures of the structure have been discussed (Spagnuolo et al, 2017; De Angelo et al, 2019b). In particular, it has been remarked that in planar rectangular pantographic specimens under bias extension tests the first sign of damage has been always observed in one of the corners of the sheet. By varying the geometric features of the pivots, the localization of such failure points can appear in different places and, more remarkably, can be forecast on the basis of the shear or torsion of the pivots (see, e.g., Spagnuolo et al, 2019; Turco et al, 2016b; Placidi et al, 2018; Placidi and Barchiesi, 2018; Andreaus et al, 2018).

1.3 Conclusions

What emerges from the contributions of the symposium "*Perspectives in Generalized Continua*" is the central role of the generalized continua framework in the modern approach to Solid Mechanics. Indeed, from the discussions outlined in the previous section, the width of applications of this approach is evident. A paradigmatic example is the branch of the theoretical and experimental analysis of the pantographic metamaterial. It has been shown that this system has to be modeled in terms of a generalized theory of elasticity. Of course, this is a powerful tool and it has been already exploited in other areas of investigation. Successful examples are the study of defects and dislocations, multi-physics applications such as the coupling of mechanical systems with electromagnetism, high contrast materials, etc. During and after the talks, several discussions have animated the symposium. Indeed, there are several open problems in this framework (e.g. the role of the inertia) that have to be tackled by the community and that have been outlined during the conference. The importance of events like the International Conference on Nonlinear Solid Mechanics 2019, to this regard, should not be underestimated. They represent the perfect chance for scientists from different fields to confront and to establish new collaborations. The symposium *Perspectives in Generalized Continua* can be considered a successful example of this paradigm. Indeed, thanks to the judicious work made by the chairmen and to the stimulating atmosphere provided by the venue, Palazzo Argiletum in Rome, new collaborations and ideas for future works have been created. In forthcoming papers, the contents of other symposia will be reported. The hope of the authors is to help the spreading of the new interesting ideas that catalyst events like ICoNSoM 2019 help to create.

References

Abali BE, Müller WH, Eremeyev VA (2015) Strain gradient elasticity with geometric nonlinearities and its computational evaluation. Mechanics of Advanced Materials and Modern Processes 1(1):1–11

Abali BE, Müller WH, dell'Isola F (2017) Theory and computation of higher gradient elasticity theories based on action principles. Archive of Applied Mechanics 87(9):1495–1510

Abali BE, Yang H, Papadopoulos P (2019) A computational approach for determination of parameters in generalized mechanics. In: Higher Gradient Materials and Related Generalized Continua, Springer, pp 1–18

Abdoul-Anziz H, Seppecher P (2018) Strain gradient and generalized continua obtained by homogenizing frame lattices. Mathematics and mechanics of complex systems 6(3):213–250

Agiasofitou E, Lazar M (2017) Micromechanics of dislocations in solids: J-, M-, and L-integrals and their fundamental relations. International Journal of Engineering Science 114:16–40

Alhasadi MF, Epstein M, Federico S (2019) Eshelby force and power for uniform bodies. Acta Mechanica 230(5):1663–1684

Alibert JJ, Seppecher P, dell'Isola F (2003) Truss modular beams with deformation energy depending on higher displacement gradients. Mathematics and Mechanics of Solids 8(1):51–73

Andreaus U, dell'Isola F, Giorgio I, Placidi L, Lekszycki T, Rizzi NL (2016) Numerical simulations of classical problems in two-dimensional (non) linear second gradient elasticity. International Journal of Engineering Science 108:34–50

Andreaus U, Spagnuolo M, Lekszycki T, Eugster SR (2018) A Ritz approach for the static analysis of planar pantographic structures modeled with nonlinear Euler–Bernoulli beams. Continuum Mechanics and Thermodynamics 30(5):1103–1123

Barchiesi E, Placidi L (2017) A review on models for the 3D statics and 2D dynamics of pantographic fabrics. In: Wave dynamics and composite mechanics for microstructured materials and metamaterials, Springer, pp 239–258

Barchiesi E, dell'Isola F, Laudato M, Placidi L, Seppecher P (2018a) A 1D Continuum Model for Beams with Pantographic Microstructure: Asymptotic Micro-Macro Identification and Numerical Results, Springer International Publishing, Cham, pp 43–74

Barchiesi E, Laudato M, Di Cosmo F (2018b) Wave dispersion in non-linear pantographic beams. Mechanics Research Communications 94:128–132

Barchiesi E, Ganzosch G, Liebold C, Placidi L, Grygoruk R, Müller WH (2019a) Out-of-plane buckling of pantographic fabrics in displacement-controlled shear tests: experimental results and model validation. Continuum Mechanics and Thermodynamics 31(1):33–45

Barchiesi E, Spagnuolo M, Placidi L (2019b) Mechanical metamaterials: a state of the art. Mathematics and Mechanics of Solids 24(1):212–234

Barchiesi E, dell'Isola F, Hild F, Seppecher P (2020a) Two-dimensional continua capable of large elastic extension in two independent directions: asymptotic homogenization, numerical simulations and experimental evidence. Mechanics Research Communications 103:103,466

Barchiesi E, Eugster SR, dell'Isola F, Hild F (2020b) Large in-plane elastic deformations of bi-pantographic fabrics: asymptotic homogenization and experimental validation. Mathematics and Mechanics of Solids 25(3):739–767

Battista A, Cardillo C, Del Vescovo D, Rizzi NL, Turco E (2015) Frequency shifts induced by large deformations in planar pantographic continua. Nanoscience and Technology: An International Journal 6(2)

Battista A, Del Vescovo D, Rizzi NL, Turco E (2017) Frequency shifts in natural vibrations in pantographic metamaterials under biaxial tests. Technische Mechanik 37(1):1–17

Boutin C, Viverge K (2016) Generalized plate model for highly contrasted laminates. European Journal of Mechanics-A/Solids 55:149–166

Boutin C, dell'Isola F, Giorgio I, Placidi L (2017) Linear pantographic sheets: asymptotic micro-macro models identification. Mathematics and Mechanics of Complex Systems 5(2):127–162

Boutin C, Fossat P, Droz C, Ichchou M (2020) Dynamics of ribbed plates with inner resonance: Analytical homogenized models and experimental validation. European Journal of Mechanics-A/Solids 79:103,838

Carcaterra A, dell'Isola F, Esposito R, Pulvirenti M (2015) Macroscopic description of microscopically strongly inhomogenous systems: A mathematical basis for the synthesis of higher gradients metamaterials. Archive for Rational Mechanics and Analysis 218(3):1239–1262

De Angelo M, Barchiesi E, Giorgio I, Abali BE (2019a) Numerical identification of constitutive parameters in reduced-order bi-dimensional models for pantographic structures: application to out-of-plane buckling. Archive of Applied Mechanics 89(7):1333–1358

De Angelo M, Spagnuolo M, D'annibale F, et al (2019b) The macroscopic behavior of pantographic sheets depends mainly on their microstructure: experimental evidence and qualitative analysis of damage in metallic specimens. Continuum Mechanics and Thermodynamics 31(4):1181–1203

Del Vescovo D, Giorgio I (2014) Dynamic problems for metamaterials: review of existing models and ideas for further research. International Journal of Engineering Science 80:153–172

dell'Isola F, Steigmann D (2015) A two-dimensional gradient-elasticity theory for woven fabrics. Journal of Elasticity 118(1):113–125

dell'Isola F, Andreaus U, Placidi L (2015a) At the origins and in the vanguard of peridynamics, non-local and higher-gradient continuum mechanics: an underestimated and still topical contribution of gabrio piola. Mathematics and Mechanics of Solids 20(8):887–928

dell'Isola F, Lekszycki T, Pawlikowski M, Grygoruk R, Greco L (2015b) Designing a light fabric metamaterial being highly macroscopically tough under directional extension: first experimental evidence. Zeitschrift für angewandte Mathematik und Physik 66(6):3473–3498
dell'Isola F, Seppecher P, Della Corte A (2015c) The postulations á la D'Alembert and á la Cauchy for higher gradient continuum theories are equivalent: a review of existing results. Proceedings of the Royal Society A: Mathematical, Physical and Engineering Sciences 471(2183):20150,415
dell'Isola F, Giorgio I, Pawlikowski M, Rizzi NL (2016) Large deformations of planar extensible beams and pantographic lattices: heuristic homogenization, experimental and numerical examples of equilibrium. Proceedings of the Royal Society A: Mathematical, Physical and Engineering Sciences 472(2185):20150,790
dell'Isola F, Cuomo M, Greco L, Della Corte A (2017a) Bias extension test for pantographic sheets: numerical simulations based on second gradient shear energies. Journal of Engineering Mathematics 103(1):127–157
dell'Isola F, Della Corte A, Giorgio I (2017b) Higher-gradient continua: The legacy of Piola, Mindlin, Sedov and Toupin and some future research perspectives. Mathematics and Mechanics of Solids 22(4):852–872
dell'Isola F, Seppecher P, Alibert JJ, et al (2019a) Pantographic metamaterials: an example of mathematically driven design and of its technological challenges. Continuum Mechanics and Thermodynamics 31(4):851–884
dell'Isola F, Seppecher P, Spagnuolo M, Barchiesi E, Hild F, Lekszycki T, Giorgio I, Placidi L, Andreaus U, Cuomo M, Eugster SR, Pfaff A, Hoschke K, Langkemper R, Turco E, Sarikaya R, Misra A, De Angelo M, D'Annibale F, Bouterf A, Pinelli X, Misra A, Desmorat B, Pawlikowski M, Dupuy C, Scerrato D, Peyre P, Laudato M, Manzari L, Göransson P, Hesch C, Hesch S, Franciosi P, Dirrenberger J, Maurin F, Vangelatos Z, Grigoropoulos C, Melissinaki V, Farsari M, Muller W, Abali BE, Liebold C, Ganzosch G, Harrison P, Drobnicki R, Igumnov L, Alzahrani F, Hayat T (2019b) Advances in pantographic structures: design, manufacturing, models, experiments and image analyses. Continuum Mechanics and Thermodynamics 31(4):1231–1282
dell'Isola F, Turco E, Misra A, Vangelatos Z, Grigoropoulos C, Melissinaki V, Farsari M (2019c) Force–displacement relationship in micro-metric pantographs: Experiments and numerical simulations. Comptes Rendus Mécanique 347(5):397–405
dell'Isola F, Lekszycki T, Spagnuolo M, , et al (2020) Experimental methods in pantographic structures. Discrete and Continuum Models for Complex Metamaterials p 263
Epstein M (2009) The split between remodelling and aging. International Journal of Non-Linear Mechanics 44(6):604–609
Eremeyev VA (2019) On anti-plane surface waves considering highly anisotropic surface elasticity constitutive relations. In: Wave Dynamics, Mechanics and Physics of Microstructured Metamaterials, Springer, pp 1–9
Eremeyev VA, Sharma BL (2019) Anti-plane surface waves in media with surface structure: Discrete vs. continuum model. International Journal of Engineering Science 143:33–38
Eremeyev VA, dell'Isola F, Boutin C, Steigmann D (2018) Linear pantographic sheets: existence and uniqueness of weak solutions. Journal of Elasticity 132(2):175–196
Eremeyev VA, Rosi G, Naili S (2019) Comparison of anti-plane surface waves in strain-gradient materials and materials with surface stresses. Mathematics and mechanics of solids 24(8):2526–2535
Federico S, Alhasadi MF, Grillo A (2019) Eshelby's inclusion theory in light of noether's theorem. Mathematics and Mechanics of Complex Systems 7(3):247–285
Fossat P, Boutin C, Ichchou M (2018) Dynamics of periodic ribbed plates with inner resonance: Analytical homogenized model and dispersion features. International Journal of Solids and Structures 152:85–103
Ganzosch G, dell'Isola F, Turco E, Lekszycki T, Müller WH (2016) Shearing tests applied to pantographic structures. Acta Polytechnica CTU Proceedings 7:1–6
Ganzosch G, Hoschke K, Lekszycki T, Giorgio I, Turco E, Müller W (2018) 3d-measurements of 3d-deformations of pantographic structures. Technische Mechanik 38(3):233–245

Gazzo S, Cuomo M, Boutin C, Contrafatto L (2020) Directional properties of fibre network materials evaluated by means of discrete homogenization. European Journal of Mechanics-A/Solids p 104009
Giorgio I (2016) Numerical identification procedure between a micro-Cauchy model and a macro-second gradient model for planar pantographic structures. Zeitschrift für angewandte Mathematik und Physik 67(4):95
Giorgio I, Della Corte A, dell'Isola F, Steigmann DJ (2016) Buckling modes in pantographic lattices. Comptes rendus Mecanique 344(7):487–501
Giorgio I, Rizzi NL, Turco E (2017) Continuum modelling of pantographic sheets for out-of-plane bifurcation and vibrational analysis. Proceedings of the Royal Society A: Mathematical, Physical and Engineering Sciences 473(2207):20170,636
Giorgio I, dell'Isola F, Steigmann DJ (2018a) Axisymmetric deformations of a 2nd grade elastic cylinder. Mechanics Research Communications 94:45–48
Giorgio I, Harrison P, dell'Isola F, Alsayednoor J, Turco E (2018b) Wrinkling in engineering fabrics: a comparison between two different comprehensive modelling approaches. Proceedings of the Royal Society A: Mathematical, Physical and Engineering Sciences 474(2216):20180,063
Giorgio I, dell'Isola F, Steigmann DJ (2019) Edge effects in hypar nets. Comptes Rendus Mécanique 347(2):114–123
Golaszewski M, Grygoruk R, Giorgio I, Laudato M, Di Cosmo F (2019) Metamaterials with relative displacements in their microstructure: technological challenges in 3d printing, experiments and numerical predictions. Continuum Mechanics and Thermodynamics 31(4):1015–1034
Greco L, Giorgio I, Battista A (2017) In plane shear and bending for first gradient inextensible pantographic sheets: numerical study of deformed shapes and global constraint reactions. Mathematics and Mechanics of Solids 22(10):1950–1975
Grillo A, Zingali G, Federico S, Herzog W, Giaquinta G (2005) The role of material in homogeneities in biological growth. Theoretical and Applied Mechanics 32(1):21–38
Grillo A, Prohl R, Wittum G (2016) A poroplastic model of structural reorganisation in porous media of biomechanical interest. Continuum Mechanics and Thermodynamics 28(1-2):579–601
Juritza A, Yang H, Ganzosch G (2019) Qualitative investigations of experiments performed on 3D-FDM-printed pantographic structures made out of PLA. In: New Achievements in Continuum Mechanics and Thermodynamics, Springer, pp 197–209
Laudato M, Barchiesi E (2019) Non-linear dynamics of pantographic fabrics: Modelling and numerical study. In: Wave Dynamics, Mechanics and Physics of Microstructured Metamaterials, Springer, pp 241–254
Laudato M, Manzari L, Barchiesi E, Di Cosmo F, Göransson P (2018) First experimental observation of the dynamical behavior of a pantographic metamaterial. Mechanics Research Communications 94:125–127
Laudato M, Di Cosmo F, Drobnicki R, Göransson P (2019) Dynamical vector fields on pantographic sheet: Experimental observations. In: New Achievements in Continuum Mechanics and Thermodynamics, Springer, pp 257–269
Lazar M (2013) The fundamentals of non-singular dislocations in the theory of gradient elasticity: Dislocation loops and straight dislocations. International Journal of Solids and Structures 50(2):352–362
Lazar M (2019) A non-singular continuum theory of point defects using gradient elasticity of bi-helmholtz type. Philosophical Magazine 99(13):1563–1601
Lazar M, Agiasofitou E (2018a) Eshelbian dislocation mechanics: J-, M-, and L-integrals of straight dislocations. Mechanics Research Communications 93:89–95
Lazar M, Agiasofitou E (2018b) The J-, M- and L-integrals of body charges and body forces: Maxwell meets Eshelby. Journal of Micromechanics and Molecular Physics 3(03n04):1840,012
Lazar M, Maugin GA, Aifantis EC (2006) Dislocations in second strain gradient elasticity. International Journal of Solids and Structures 43(6):1787–1817
Mandadapu KK, Abali BE, Papadopoulos P (2018) On the polar nature and invariance properties of a thermomechanical theory for continuum-on-continuum homogenization. arXiv preprint arXiv:180802540

Maurin F, Greco F, Desmet W (2019) Isogeometric analysis for nonlinear planar pantographic lattice: discrete and continuum models. Continuum Mechanics and Thermodynamics 31(4):1051–1064

Milton G, Briane M, Harutyunyan D (2017) On the possible effective elasticity tensors of 2-dimensional and 3-dimensional printed materials. Mathematics and Mechanics of Complex Systems 5(1):41–94

Milton GW (2013) Complete characterization of the macroscopic deformations of periodic unimode metamaterials of rigid bars and pivots. Journal of the Mechanics and Physics of Solids 61(7):1543–1560

Mindlin RD (1965) Second gradient of strain and surface-tension in linear elasticity. International Journal of Solids and Structures 1(4):417–438

Misra A, Lekszycki T, Giorgio I, Ganzosch G, Müller WH, dell'Isola F (2018) Pantographic metamaterials show atypical Poynting effect reversal. Mechanics Research Communications 89:6–10

Mühlich U, Zybell L, Kuna M (2009) Micromechanical modelling of size effects in failure of porous elastic solids using first order plane strain gradient elasticity. Computational materials science 46(3):647–653

Mühlich U, Zybell L, Kuna M (2012) Estimation of material properties for linear elastic strain gradient effective media. European Journal of Mechanics-A/Solids 31(1):117–130

Mühlich U, Ballani F, Stoyan D (2015) Influence of randomness in topology, geometry and material properties on the mechanical response of elastic central-force networks. Probabilistic Engineering Mechanics 40:36–41

Nejadsadeghi N, De Angelo M, Drobnicki R, Lekszycki T, dell'Isola F, Misra A (2019) Parametric experimentation on pantographic unit cells reveals local extremum configuration. Experimental Mechanics 59(6):927–939

Pideri C, Seppecher P (1997) A second gradient material resulting from the homogenization of an heterogeneous linear elastic medium. Continuum Mechanics and Thermodynamics 9(5):241–257

Placidi L, Barchiesi E (2018) Energy approach to brittle fracture in strain-gradient modelling. Proceedings of the Royal Society A: Mathematical, Physical and Engineering Sciences 474(2210):20170,878

Placidi L, Barchiesi E, Turco E, Rizzi NL (2016) A review on 2D models for the description of pantographic fabrics. Zeitschrift für angewandte Mathematik und Physik 67(5):121

Placidi L, Andreaus U, Giorgio I (2017) Identification of two-dimensional pantographic structure via a linear d4 orthotropic second gradient elastic model. Journal of Engineering Mathematics 103(1):1–21

Placidi L, Barchiesi E, Misra A (2018) A strain gradient variational approach to damage: a comparison with damage gradient models and numerical results. Mathematics and Mechanics of Complex Systems 6(2):77–100

Rahali Y, Giorgio I, Ganghoffer JF, dell'Isola F (2015) Homogenization à la Piola produces second gradient continuum models for linear pantographic lattices. International Journal of Engineering Science 97:148–172

Scerrato D, Giorgio I (2019) Equilibrium of two-dimensional cycloidal pantographic metamaterials in three-dimensional deformations. Symmetry 11(12):1523

Scerrato D, Zhurba Eremeeva IA, Lekszycki T, Rizzi NL (2016) On the effect of shear stiffness on the plane deformation of linear second gradient pantographic sheets. ZAMM-Zeitschrift für Angewandte Mathematik und Mechanik 96(11):1268–1279

Spagnuolo M, Barcz K, Pfaff A, dell'Isola F, Franciosi P (2017) Qualitative pivot damage analysis in aluminum printed pantographic sheets: numerics and experiments. Mechanics Research Communications 83:47–52

Spagnuolo M, Peyre P, Dupuy C (2019) Phenomenological aspects of quasi-perfect pivots in metallic pantographic structures. Mechanics Research Communications 101:103,415

Steigmann DJ, dell'Isola F (2015) Mechanical response of fabric sheets to three-dimensional bending, twisting, and stretching. Acta Mechanica Sinica 31(3):373–382

Tran CA, Gołaszewski M, Barchiesi E (2020) Symmetric-in-plane compression of polyamide pantographic fabrics—modelling, experiments and numerical exploration. Symmetry 12(5):693
Turco E (2019) How the properties of pantographic elementary lattices determine the properties of pantographic metamaterials. In: New Achievements in Continuum Mechanics and Thermodynamics, Springer, pp 489–506
Turco E, dell'Isola F, Cazzani A, Rizzi NL (2016a) Hencky-type discrete model for pantographic structures: numerical comparison with second gradient continuum models. Zeitschrift für angewandte Mathematik und Physik 67(4):85
Turco E, dell'Isola F, Rizzi NL, Grygoruk R, Müller WH, Liebold C (2016b) Fiber rupture in sheared planar pantographic sheets: numerical and experimental evidence. Mechanics Research Communications 76:86–90
Turco E, Giorgio I, Misra A, dell'Isola F (2017a) King post truss as a motif for internal structure of (meta) material with controlled elastic properties. Royal Society open science 4(10):171,153
Turco E, Golaszewski M, Giorgio I, D'Annibale F (2017b) Pantographic lattices with non-orthogonal fibres: experiments and their numerical simulations. Composites Part B: Engineering 118:1–14
Turco E, Misra A, Pawlikowski M, dell'Isola F, Hild F (2018) Enhanced Piola–Hencky discrete models for pantographic sheets with pivots without deformation energy: numerics and experiments. International Journal of Solids and Structures 147:94–109
Turco E, Barchiesi E, Giorgio I, dell'Isola F (2020) A lagrangian hencky-type non-linear model suitable for metamaterials design of shearable and extensible slender deformable bodies alternative to timoshenko theory. International Journal of Non-Linear Mechanics p 103481
Viverge K, Boutin C, Sallet F (2016) Model of highly contrasted plates versus experiments on laminated glass. International Journal of Solids and Structures 102:238–258
Yang H, Ganzosch G, Giorgio I, Abali BE (2018) Material characterization and computations of a polymeric metamaterial with a pantographic substructure. Zeitschrift für angewandte Mathematik und Physik 69(4):105
Yang H, Abali BE, Timofeev D, Müller WH (2019) Determination of metamaterial parameters by means of a homogenization approach based on asymptotic analysis. Continuum Mechanics and Thermodynamics pp 1–20
Yildizdag ME, Barchiesi E, dell'Isola F (2020) Three-point bending test of pantographic blocks: numerical and experimental investigation. Mathematics and Mechanics of Solids doi: 10.1177/1081286520916911
Zybell L, Hütter G, Linse T, Mühlich U, Kuna M (2014) Size effects in ductile failure of porous materials containing two populations of voids. European Journal of Mechanics-A/Solids 45:8–19

Part I
Biomechanics

Chapter 2
Failure Prediction of Tumoral Bone with Osteolytic Lesion in Mice

Benjamin Delpuech, Stéphane Nicolle, Cyrille B. Confavreux, Lamia Bouazza, Philippe Clezardin, David Mitton, and Hélène Follet

Abstract Pathological fractures due to bone metastases remain difficult to predict. The aim of this study is to assess whether a model taking into account tumor-specific geometry and mechanical properties improves assessment of bone failure, and to determine which criterion has to be taken into account to improve failure detection. To achieve this aim, an osteolytic mice model was considered. Tumoral cells were injected intra-tibially to induce a tumor in the bone. After six weeks, eight mice were sacrificed. Tomographic (μCT) images were obtained to build subject-specific finite element models. A compression test was performed on each tibia and used to assess the finite element models. Implementation of tumor geometry and mechanical properties did not provide better failure prediction in comparison to models based

B. Delpuech
Univ. Lyon, Université Claude Bernard Lyon 1, INSERM, LYOS UMR 1033, 69008 Lyon, France
Univ Lyon, Université Claude Bernard Lyon 1, Univ. Gustave Eiffel, IFSTTAR, LBMC UMR_T 9406, F-69622, Lyon, France
e-mail: benjamin.delpuech@laposte.net

S. Nicolle
Univ Lyon, Université Claude Bernard Lyon 1, Univ. Gustave Eiffel, IFSTTAR, LBMC UMR_T 9406, F-69622, Lyon, France
e-mail: stephane.nicolle@univ-lyon1.fr

C. B. Confavreux
Univ. Lyon, Université Claude Bernard Lyon 1, INSERM, LYOS UMR 1033, 69008 Lyon, France
CEMOS, Service de Rhumatologie, Centre hospitalier Lyon Sud, Institut de Cancérologie des Hospices Civils de Lyon, 69310 Pierre Bénite, France
e-mail: cyrille.confavreux@chu-lyon.fr

L. Bouazza, P. Clezardin, H. Follet
Univ. Lyon, Université Claude Bernard Lyon 1, INSERM, LYOS UMR 1033, 69008 Lyon, France
e-mail: lamia.bouazza@univ-lyon1.fr, philippe.clezardin@inserm.fr, helene.follet@inserm.fr

D. Mitton
Univ Lyon, Université Claude Bernard Lyon 1, Univ. Gustave Eiffel, IFSTTAR, LBMC UMR_T 9406, F-69622, Lyon, France
e-mail: david.mitton@univ-eiffel.fr

B. E. Abali and I. Giorgio (eds.), *Developments and Novel Approaches in Biomechanics and Metamaterials*, Advanced Structured Materials 132,
https://doi.org/10.1007/978-3-030-50464-9_2

on μCT grey levels. The average difference with experiments reached respectively (23 ± 22% and 12 ± 7%). Considering a detection criterion based on the percentage difference between bone global ultimate load and bone local ultimate load (with a region of interest surrounding the tumor) allowed detection of all bones that experienced a partial failure. A next step will be to assess this failure criterion on human bones to help clinicians in decision-making.

Keywords: Bone metastasis · Fracture prediction · Finite Element Analysis · Experimental tests · Mice

2.1 Introduction

The skeleton is the third most common localization of metastases (Du et al., 2010). Metastases usually develop in the vicinity of the bone red marrow in the spine and proximal long bones. Even if radiological aspects of bone metastases are "lytic," "blastic" or "mixed," these features all impair bone strength and increase bone fragility (Wong and Pavlakis, 2011). Pathological fractures are responsible for pain, altered quality of life, comorbidities and are associated with a reduction in overall survival. Nevertheless, pathological fractures remain difficult to predict even for experienced surgeons (Attar et al, 2012). In 1982, Mirels suggested a scoring system to predict failure risk in long bones (Mirels, 2003). Their system was developed based on a retrospective study of 78 metastatic long bone lesions. Mirels's score is based on four criteria: lesion site, size, degree of sclerosis/lysis and pain (Mirels, 2003). To evaluate Mirels's score, Van der Linden et al. led a retrospective study including 102 patients with metastatic femora of whom 14 suffered a fracture during follow-up (Van der Linden et al, 2004). Each patient lesion was rated using Mirels's score. Among the considered cases, 84 (82%) would have been surgically overtreated while 13% of the lesions without fracture were detected at low risk (Van der Linden et al, 2004). Implementing a threshold on metastases with axial cortical involvement reduced overtreatment to 42% (Van der Linden et al, 2004). As surgical overtreatment has a large impact on quality of life for patients, especially when they have limited life expectancy (Van der Linden et al, 2004), a more accurate failure risk predictor would be useful for physicians. Interestingly, most cancer patients undergo regular whole body Computed Tomography (CT) to assess cancer progression and guide oncological treatment. Up to now, such quantitative CT data has not been used to assess fracture risk and the decision relies on surgeon expertise. Nevertheless, CT data may be used to predict bone metastasis strength using Finite Element Analysis (FEA). This strategy has already been tested to assess fracture risk in osteoporosis, and the results were quite interesting (among others, Bessho et al, 2009; Duchemin et al, 2008; Keyak et al, 2001; Kopperdahl et al, 2014; Zysset et al, 2013). More recently FEA was used as a potential tool to improve failure prediction of metastatic bones (Benca et al, 2017, 2019; Derikx et al, 2012; Eggermont et al, 2018; Goodheart et al, 2015; Keyak et al, 2005; Tanck et al, 2009). In an *ex vivo* study with mechanically-induced defects, FEA in this condition outperformed clinical experts (Kendall's tau of 0.78 for FEA against 0.5 $\pm$ 0.03 on average for six clinical experts (Derikx et al, 2012)). The main limitation of this study was the comparison with mechanically-induced defects which are far from real metastases. To overcome this limitation a clinical retrospective study was led by the same team (Eggermont et al, 2018). In that study, 39 patients with non-fractured femoral metastatic lesion were included. Each patient was followed for six months, until femoral fracture or death (whichever occurred first) (Eggermont et al, 2018). Results agreed with previous studies, showing better prediction with FEA (sensitivity (correctly identifying clinical fractures) of 89% versus 0% to 33% for clinical assessments). Even though the difference is incommensurate with previous cited results, it has to be noted that clinicians scored a higher specificity than FE models (specificity (correctly identify cases that do not fracture), which was 79% for the FE models versus 84% to 95% for clinical assessments), leaving room for improvement. However, studies on metastatic bones are difficult to perform, as samples are uncommon. Thus, mice have been widely used as a skeletal model(Fritton et al, 2005)in preclinical studies (Slosky et al, 2015) and even a biomechanical cancer study (Mann et al, 2008) to create tumoral bones. Thus, we decided to consider a mice model to obtain bone with a biologically-induced tumor (lytic). Based on this mice model, our aim was twofold. First, create and assess a FE model taking into account specific tumor properties to predict failure of tumoral bone in a lytic lesion. Second, consider a simpler model (relying only on elastic properties) to evaluate failure criteria taking into account a global and local analysis.

2.2 Material and Methods

The overall process of our investigation is presented in Figure 2.1. Since we used human cell lines, experiments were performed in immunocompromised six-week old BALB/c nude mice (Janvier Laboratories®). Protocols received approval of the University Claude Bernard Lyon I Ethical Committee for Animal Experimentation (DR2015-39). To assess tumoral tissue mechanical properties, subcutaneous tumorigenesis experiments were performed on mice. In addition, a human specimen was collected during surgery of a patient with a myeloma lytic lesion of the hip (authorization number: CT 69HCL19_0266). On different mice, intra-tibial injections were used to create bone metastases in limbs.

2.2.1 *Tumorigenesis*

A first group composed of BABL/c nude female mice ($n = 4$) was created in order to assess tumoral tissue Young's moduli. Mice were anesthetized using an induction box with a 1 L air/min stream, 3% isoflurane concentration, and maintained with a

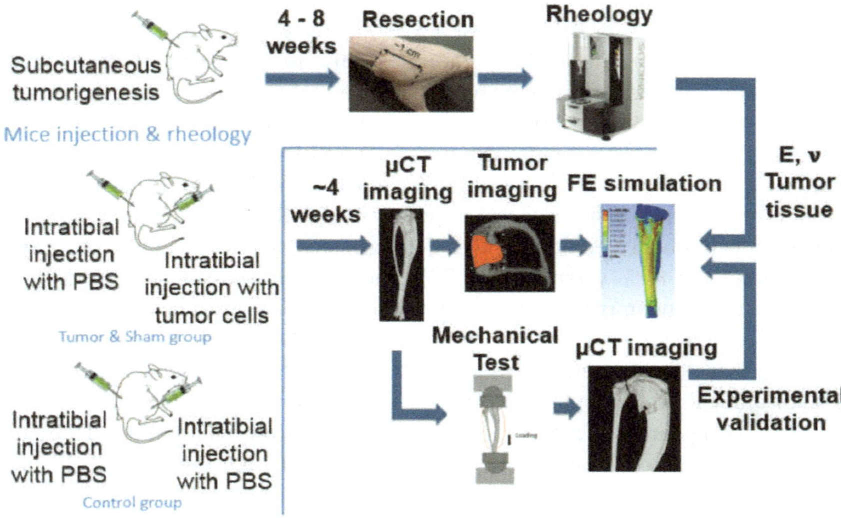

Fig. 2.1 Global Study workflow.

mask at the same stream, 2% isoflurane. The mice were then subcutaneously injected with MDA-B02 human breast cancer cells (106 cells in 100 µL Phosphate Buffered Saline –called PBS– solution using a 1 mL syringe, Myjector 070151, Terumo Japan). After injection, tumor development was followed up through palpation, and as soon as possible, tumors were measured with a caliper. When the tumor reached about 1 cm in diameter without apparent necrosis, animals were anesthetized using once again isoflurane, sacrificed via cervico-dislocation, and the tumor was excised and stored in liquid azote for same-day testing.

2.2.2 Rheological Tests on Tumor and Soft Tissues

Each mice tumor ($\approx$ 1 cm in diameter ball) was sliced into 800 µm thick slices using a home-made double-scalpel. First a B02 tumor allowed for slicing into four slices, while a second provided 6 slices, the third 1 slice, and the fourth 2 slices. Normal tissues (skin, muscle, and fat) were also collected and provided respectively 17, 15, and 7 slices to obtain the rheological properties of normal tissues. The human myeloma sample was also sliced with the same tool and provided 13 slices. Each slice was then placed between two glass slides equipped with a variable gap in order to precisely measure the thickness and the surface of the slice. The sample was then placed in a custom rheometer and subjected to small-displacement harmonical shearing between 0.1 and 10 000 Hz (Nicolle et al, 2005). This test allows the determination of the frequency-dependent shear modulus (dynamic shear modulus) of the tissue. Assuming that the deformation process is quasi-static during the test and that the tumor tissue has a linear elastic behavior in a first approximation (Fung, 2013), a low frequency modulus was chosen to characterize each tumor tissue in our specific FE models. Since Young's modulus is required in FE software when using a linear elastic law, the dynamic shear modulus was thus converted to Young's modulus by assuming that the tumor tissue is quasi-incompressible (Poisson ratio = 0.499) as is usually assumed for other soft biological tissues. In this case, it is demonstrated that Young's modulus was equal to three times the shear modulus.

2.2.3 Intra-tibial Bone Metastases

Six-week-old BALB/c female mice (n=16) were anesthetized as previously described. Mice were injected with tumor cells (n=8) in 15 μL of PBS. The number of tumor cells injected has been previously determined through a pre-study where mice were injected with different concentrations to set a reliable model providing a repeatable lyse at 30 days. Bone metastases development was assessed weekly on plain radiographs. At day 30, mice were sacrificed by cervicodislocation and limbs were excised en bloc, to be stored at -20°C in PBS soaked gauze. Limbs were only unfrozen on the day of the mechanical test. A first group of eight BALB/c nude female mice were injected with human B02 tumor cells in the right limbs and PBS in the left limb (sham limb). A second group, of eight BALB/c nude mice female, were injected in both limbs with PBS to provide sham limbs. For convenience, this second group will be further referred to as "control" limbs, while contralateral PBS injected limbs (first group) will be referred as "sham" limbs. Bone injected with B02 cells will be referred as "tumor" bones (Figure 2.1).

2.2.4 Imaging: μCT Acquisition and Reconstruction

A μ-CT scan was performed prior to and after the mechanical test. μCT imaging were performed on a Bruker Skyscan 1176 (Kontiche, Belgium) with a 10 μm nominal resolution. Each limb was scanned at a 10 μm isotropic resolution (50 kV, 500 μA, 0.5 mm aluminum filter, 0.6° rotation step on 180°). Reconstruction was performed using scan software (Nrecon 1.7.0.4, Bruker, Kontiche, Belgium) with a smoothing, ring artefacts correction and beam-hardening correction (parameters respectively set at 2, 6 and 20%). To ensure density/bone mineral density (BMD) correlation, a classical phantom created by the scanner manufacturer (Skyscan Bruker, Kontiche, Belgium) was used, providing two known densities (0.25 g/mm^3 and 0.75 g/mm^3). All scans were performed with the same parameters and in the same conditions in the PBS soaked gauze.

2.2.5 Compression Tests on Tibia

Before the test, limbs were thawed at an ambient temperature for half an hour. Next, tibia was separated from the femur at the knee junction, and from the foot at the ankle, thus removing all tendons linking tibia muscle to other bones. Finally, the attachment between the tibia and fibula was cut, but soft tissues surrounding the tibia were kept in place in order to limit any possible damage to the tibia or tumoral tissue. To avoid excessive buckling, the tibia was embedded in a 5 mm acorn hex cap nuts stainless steel with methyl methacrylate glue (VariDur ref 10-1027, Buehler, USA). This allowed the bone to stay intact, while limiting its slenderness. The proximal end of the tibia was molded using fast epoxy paste (Pattex, Ref 1875423) and the tibia was imaged imbedded while allowing it to have its exact length once housed along with its orientation. After μCT imaging, two acorn nuts were put in place in a previously countersunk loading plateau, creating a ball joint at each end of the tibia. The mechanical test started with sinusoidal pre-cycling -0.5 N and -2 N for 30 cycles at 0.5 Hz. The destructive test was conducted immediately after pre-cycling by compressing the tibia at a rate of 0.03 mm/s until failure using an electromagnetic testing machine (Access 5500, Bose Corporation, Eden Prairie, USA). Load (with a one-axis load transducer, accuracy 0.03 N) and displacement (3 μm accuracy) were recorded at 60 Hz (WinTest®Digital Control System). A custom Python program was developed to analyze load-displacement curves to determine stiffness and the ultimate load on each bone. Briefly, stiffness was determined by using the derivative of the experimental curve. This was determined using a linear regression on the longest interval where the derivative function variation was under $\pm$5 N/mm. Ultimate load was defined as the maximum load measured by the load cell.

2.2.6 FE Model Creation

Two different types of models were created, one converting grey levels in the images into mechanical properties as proposed in the literature (Keyak et al, 2005), called "heterogeneous," while the other model took into account tumoral tissue mechanical properties (obtained by rheology) called "specific." Results from both mice and patient tumor rheological tests were implemented in these specific models.

"Heterogeneous models" were based on reconstructed and straightened µCT scans, while the volume occupied by each bone was converted to generic surface mesh (STL files) from 8-bit images using the scanner manufacturer's software (CTan 1.16.4.1+, Skyscan Bruker, Kontiche, Belgium) and open source software (3DSlicer 4.8.1, various developers). This surface mesh was then refined (MeshLab ν2016.12, CNR, Roma and ISTI, Pisa, Italy), converted to a volume, and meshed using tetrahedron elements thru FEA software (Ansys 19.0, Ansys inc., Canonsburg, Pennsylvania, USA). Academically-developed software was then used to correlate bone density to the Young's modulus of each element of the model (Bonemat, Istituto

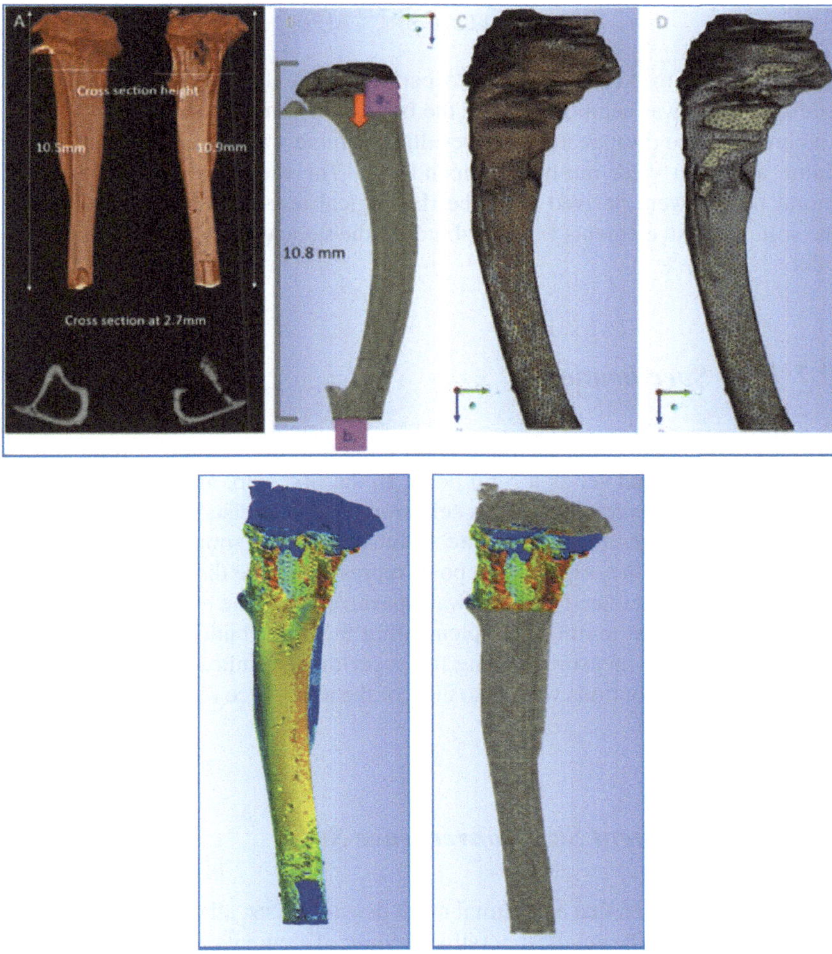

Fig. 2.2 A) Example of tibia μCT reconstruction with cross-sectional view: Sham tibia on left and tumoral tibia on right, B) Boundary condition for FEA, load is applied on top "a" elements, displacement is blocked on bottom "b" elements , C) Heterogeneous model, D) Specific model, tumor can be seen on light grey on the top of bone, E) Global analysis on the entire bone , F) Local analysis focused on the location of the tumor.

Ortopedico Rizzoli, Bologna, Italy) using a published relationship to convert μCT-derived volumetric density to elastic modulus (equation (15.1)) (Easley et al, 2010). A threshold at 0.31 HA g/mm^3 was applied in order to prevent an unrealistic Young's modulus in the shaft due to a beam hardening effect. These heterogeneous models were created for all study groups (control, tumor and sham groups). Finally, models with implemented Young's modulus:

$$E = 11270\,\rho^{1.746}\ \text{MPa}\,, \tag{2.1}$$

were taken back to Ansys, in order to perform FEA. In order to build "specific" models, a tumor was segmented from the bone, assuming that tumoral tissue would be where the bone disappeared. The resulting volume was subtracted from the bone volume and both were meshed independently. The mechanical properties of the tumoral tissues were derived from the rheological tests. Mechanical properties of bone volume mesh elements were attributed in the same way as for the heterogeneous model.

2.2.7 FEA Specifications

In order to mimic the experimental compression test, the first top millimeter of the bone was selected and set as a rigid body. In the first step, an axial force equal to the ultimate load was applied to this selection while the distal bone elements were fixed to represent the embedding. The stiffness from the simulation is assumed to be the slope linking the origin to the point representing the displacement due to the applied load. The simulated stiffness was optimized to fit the experimental stiffness. The applied load was modified, implementing the contribution on X and Y axes. These axes were not measured during the experiments (uniaxial force sensor). The limit for these contributions was set to 20% of the axial force for the sum of the loads on the X and Y axes.

2.2.8 Mesh Element Size Convergence Study

It is generally accepted that an optimal mesh density exists, allowing for the most accurate solution with the smallest possible amount of elements (Burkhart et al, 2013). This is commonly determined through the use of a convergence study (Burkhart et al, 2013). This convergence study was done on the same bone, meshed with a different number of elements. Studied models included 10 k, 100 k, 200 k, 300 k, 400 k, and 500 k elements. To prevent a double variation due to variation in element size between models and variation in mechanical properties, a constant Young's modulus of 20 GPa was applied with a threshold at 0.31 HA g/mm^3. As the failure criteria used is based on strain, the sensitivity study was done with an average strain energy density.

2.2.9 *Failure Criterion Parameters Determination*

As our model is purely elastic, no damage to the elements is recorded and a criterion is needed to assess failure. Pistoia et al (2002) proposed such a criterion on human radii, and defined failure as reached when 2% of the bone meshed volume reaches 0.007 or more in strain. Nyman et al. reworked this in order to adapt it to mice vertebrae (Nyman et al, 2015). To assess the best couple of parameters in the current study, failure volume (FV) was set at 2, 4 and 6%, while failure strain (FailStrain) was set at 0.007, 0.0085, and 0.01. Root Mean Square Error of Estimate (RMSE) was minimized while maximizing R^2 (as close as one). One parameter was added to the comparison from the Nyman et al (2015) study: the slope, which had to be the closest to one as possible in order to avoid under or overestimations of failure load.

2.2.10 *Regions of Interest for Failure Assessment*

Two different analyses of the same simulation were performed in order to test failure detection sensitivity. The first analysis took into account the entire bone volume (except bone embedded in resin) and will be further referred to as "global" analysis (Figure 2.2E). Another analysis took into account only the three millimeters of bone below the load application selection, referred as "local" analysis (Figure 2.2F). This region of interest corresponded to the lyses location of tumoral bone due to the intra-tibial injection.

2.2.11 *Statistical Analysis*

Statistical analysis was performed with R software (RCore TEAM, 2016) using a significance level of 5%. All tests were two-tailed. Results were reported as a scattergram and Bland-Altman's representation. Bland-Altman's differences were calculated as experimental ultimate load-numerical ultimate load values. Dotted lines represented the lower and higher limit of agreement (95% confidence interval (CI) of limits of agreement: average difference ± 1.96 standard deviation of the difference) and the pecked line represented the average difference. Statistical correlations between experimental and FEA ultimate loads were assessed using Spearman's rank correlation (ρ_{spearman}) test. Group comparison was ascertained by Mann-Whitney's unpaired test. The ability of FEA to predict sample ultimate load was ascertained by linear regression to determine the intercept, the slope and the Root Mean Square Error of Estimate (RMSE (Equation (15.2)), representing the square root of the differences between simulated failure load and predicted failure load (i.e., linear regression between experimentally measured peak force and simulated failure load of each tibia, with

$$RMSE = \sqrt{\frac{\sum(y-\bar{y})^2}{n-1-p}} \tag{2.2}$$

2.3 Results

Raw data are presented in Table 2.1.

2.3.1 Rheological Results

Mice harvested tumor tissues showed stronger mechanical properties than other soft tissues. Dynamic modulus of the tissues reached ten times the fat tissue and twice that of the skin and muscle tissues. After conversion, the value of 0.0225 MPa was used for tumoral tissue Young's modulus. The human myeloma sample obtained during surgery had a dynamic modulus 10 times superior to the one obtained on mice samples, given after conversion of a Young's modulus of 0.3 MPa.

2.3.2 Sham and Control Limbs Comparison

No differences were found in the experimental ultimate load between the two groups of sham and control (p=0.77). Therefore, sham and control limbs were pooled for the rest of the analyses.

2.3.3 Mesh Element Size Convergence Study

The 200k element model was chosen because its strain density was close to those of the 500k model (2% on average strain density) and provides a reduced computational cost. Thus, the mesh was composed of elements of 0.0001 mm^3 in volume, a mean surface element of 0.018 mm^2 and an average edge of 0.18 mm.

2.3.4 Results for Heterogeneous Models

The Mouse 8 tumor limb experienced a local fracture before overall failure of the entire structure and was removed from this comparison. The Mouse 9 left control limb was also removed from the analysis due to a fall experienced prior to test. All other limbs were pooled (n=30) for the assessment of the heterogeneous model. FEA

Table 2.1 Raw data of the study.

Group		Injection	Mice number	Tibia Side	Exp. Ult. Load. (N)	FEA Heterogeneous Global Ult. Load. (N)	FEA Heterogeneous Local Ult. Load (N)	FEA specific global Ult. Load (N) E=B02 value	FEA specific global Ult. Load (N) E=Myeloma value
Group 1	CTRL	PBS	9	Left	N/A	N/A	N/A	N/A	N/A
	CTRL	PBS	10	Left	26.6	20.9	20.2	N/A	N/A
	CTRL	PBS	11	Left	21.4	20.0	18.1	N/A	N/A
	CTRL	PBS	12	Left	21.5	23.8	26.2	N/A	N/A
	CTRL	PBS	13	Left	25.2	22.7	20.4	N/A	N/A
	CTRL	PBS	14	Left	26.4	21.8	22.1	N/A	N/A
	CTRL	PBS	15	Left	23.0	21.1	20.3	N/A	N/A
	CTRL	PBS	16	Left	27.5	24.0	23.9	N/A	N/A
	CTRL	PBS	9	Right	24.1	20.3	19.1	N/A	N/A
	CTRL	PBS	10	Right	23.4	23.0	22.7	N/A	N/A
	CTRL	PBS	11	Right	17.8	19.2	17.1	N/A	N/A
	CTRL	PBS	12	Right	21.9	21.5	18.9	N/A	N/A
	CTRL	PBS	13	Right	20.4	20.2	19.3	N/A	N/A
	CTRL	PBS	14	Right	20.7	22.8	25.0	N/A	N/A
	CTRL	PBS	15	Right	19.4	18.9	19.0	N/A	N/A
	CTRL	PBS	16	Right	24.8	23.7	22.8	N/A	N/A
Group 2	B02_SHAM	PBS	1	Left	19.4	17.0	21.1	N/A	N/A
	B02_SHAM	PBS	2	Left	17.8	23.9	25.2	N/A	N/A
	B02_SHAM	PBS	3	Left	18.2	19.8	19.6	N/A	N/A
	B02_SHAM	PBS	4	Left	17.3	20.4	20.3	N/A	N/A
	B02_SHAM	PBS	5	Left	27.8	27.3	27.4	N/A	N/A
	B02_SHAM	PBS	6	Left	29.3	24.4	23.7	N/A	N/A
	B02_SHAM	PBS	7	Left	28.7	22.6	26.0	N/A	N/A
	B02_SHAM	PBS	8	Left	22.5	24.0	26.5	N/A	N/A
	B02_TUMOR	B02	1	Right	20.2	21.0	21.4	21.1	21.2
	B02_TUMOR	B02	2	Right	20.6	22.4	21.1	21.1	22.3
	B02_TUMOR	B02	3	Right	7.7	4.2	2.5	3.7	3.7
	B02_TUMOR	B02	4	Right	13.3	10.6	6.2	13.1	13.1
	B02_TUMOR	B02	5	Right	18.4	22.7	22.3	22.0	22.0
	B02_TUMOR	B02	6	Right	12.5	12.2	6.2	7.4	7.5
	B02_TUMOR	B02	7	Right	11.9	16.5	12.3	17.0	17.0
	B02_TUMOR	B02	8	Right	*7.7*	*18.3*	11.7	17.9	17.9

ultimate load showed good agreement with experimental ultimate load (11±8% difference on average and in absolute value) along with a low RMSE (2.78), an acceptable R^2 (0.65) and slope (0.7) (Figure 2.3A). A significant correlation was found between the numerical ultimate load and experimental load when considering all samples (ρ_{spearman}=0.69, p<0.001) (Figure 2.3A). The mean difference on the Bland-Altman's graph is 0.56. When only the tumoral samples are considered, there is no longer significant correlation (ρ_{spearman}=0.75, p=0.052), even if RMSE=2.99 and R^2=0.85, with a mean difference on the Bland-Altman's graph of -0.71 (Figure 2.3B).

2.3.5 Specific Model Results

The Young's modulus considered for the seven specific models was the one obtained on mice subcutaneous tumors ($E = 0.0225\,\text{MPa}$). The ultimate loads predicted via FEA showed a lower agreement with the experimental results than their heterogeneous counterpart ($23 \pm 22\%$ difference on average and in absolute value). The RMSE remained low (3.66), the R^2 and slope were respectively 0.79 and 1.33. The two methods were significantly linked ($\rho_{\text{spearman}} = 0.77$, $p = 0.041$) (Figure 2.3B). The mean difference on the Bland-Altman's graph is -0.11 (Figure 2.3C). When using the Young's modulus determined on a human myeloma sample ($E = 0.3\,\text{MPa}$) the agreement between the two ultimate loads improved as RMSE stayed at the same value (3.55), R^2 and slope values increased (respectively 0.81 and 1.38) as did the correlation ($\rho_{\text{spearman}} = 0.86$, $p = 0.014$) (Figure 2.3C). The mean difference on the Bland-Altman's graph is -0.31 (Figure 2.3D).

2.3.6 Global and Local Analysis

Local analyses were compared to experimental data in order to provide enlightenment on disruption in load-displacement curves or low fracture (Figure 2.4D). Figure 2.4 shows an example of the different types of experimental curves obtained. For both control and sham limb (without tumoral cells), the experimental behavior corresponded to a classical compressive test on live material (Figure 2.4A)&B)). The same behavior is sometimes also obtained on tumoral limbs (Figure 2.4C), but can also show a total different behavior, which suggest a different analysis (Figure 2.4D, same scales for all graphs) When using the threshold of 20% between local and global ultimate loads, it is possible to detect all bones that experienced a disruption in their loading curves (Figure 2.5), which correspond to five of eight tumoral limbs.

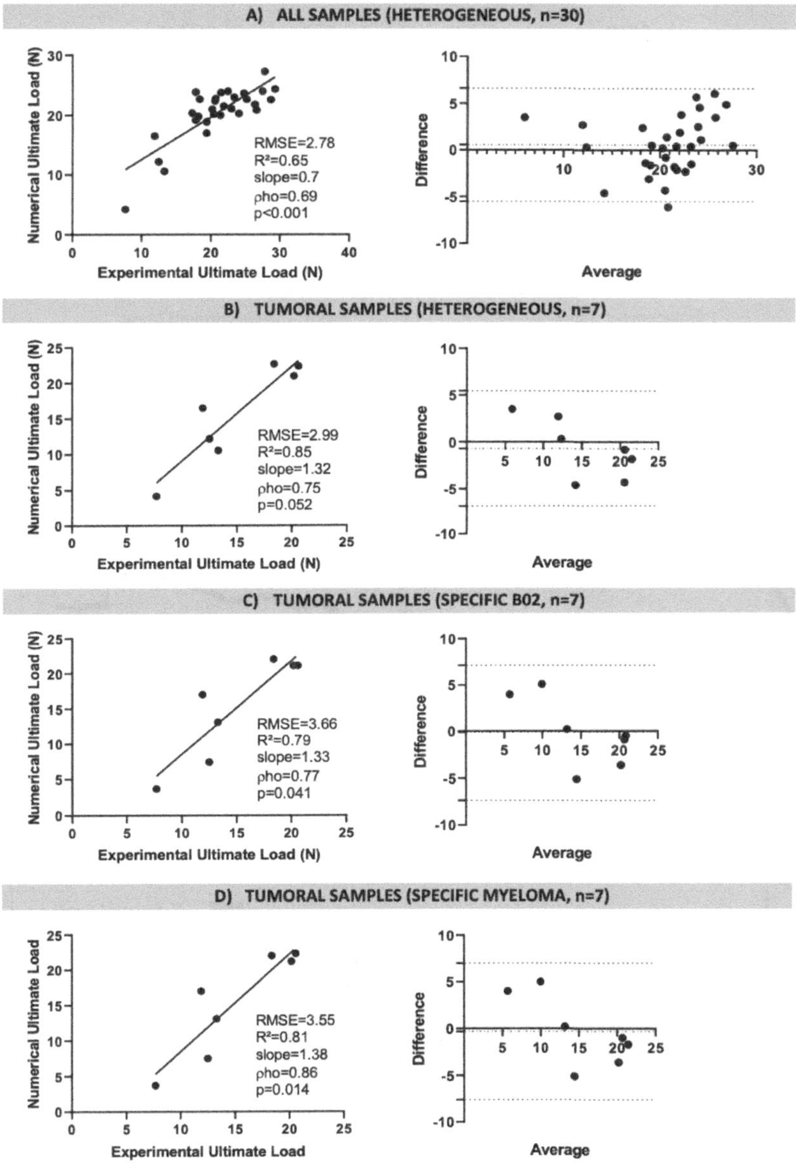

Fig. 2.3 On left, linear regression between experimental ultimate load (N) and Numerical ultimate load and on the right, Bland & Altman graph representing the difference between the numerical ultimate load and the experimental load in function of the mean of the two parameters, for the following A) all sample, using a heterogeneous simulation ($n = 30$), B) tumoral sample using a heterogeneous simulation ($n = 7$), C) tumoral sample using a specific Young's modulus for the simulation ($n = 7$), here breast cancer (B02), D) tumoral sample using a specific Young's modulus for the simulation ($n = 7$), here myeloma cancer (Myeloma).

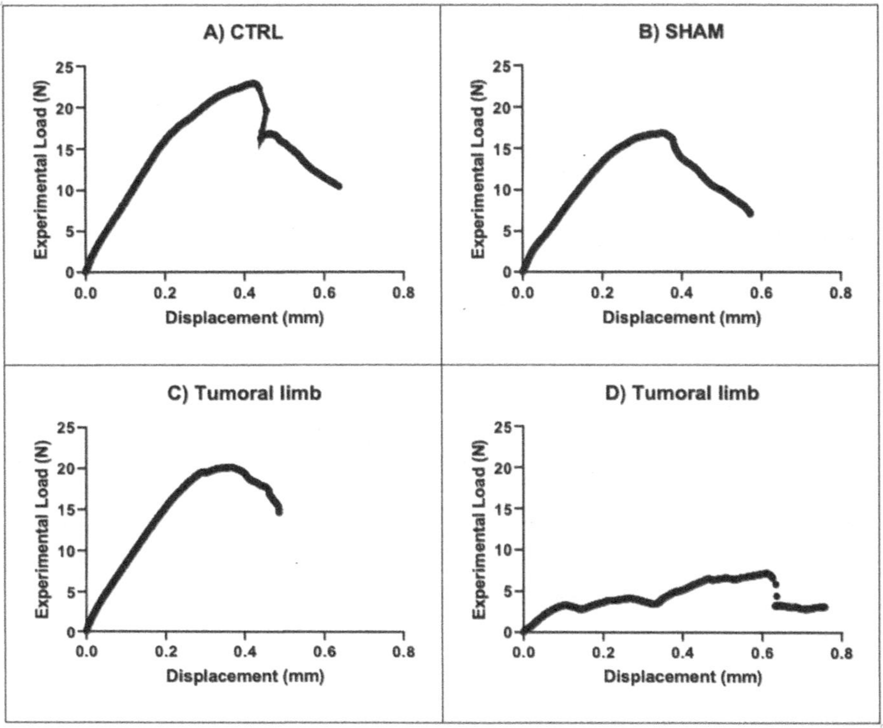

Fig. 2.4 Example of typical experimental curves. A) Control mice, B) Sham limb, C) Tumoral limb, D) Tumoral limb requiring a local analysis. Same scales are used for all graphs.

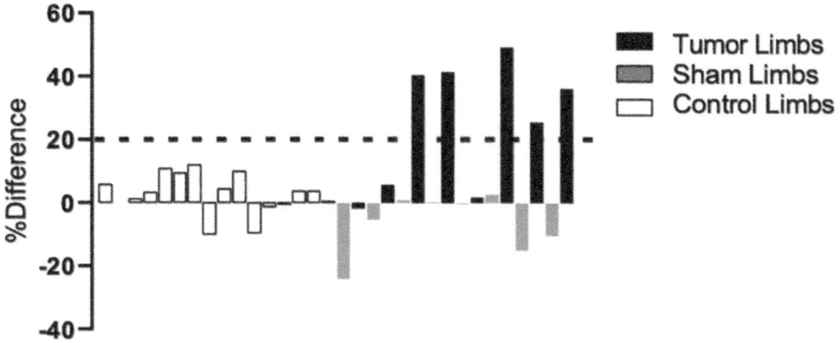

Fig. 2.5 Percentage of the difference between global and local ultimate loads for all limbs, ordered by pairs (left and right tibias, with white bars: control limbs, grey bars: sham limbs and black bar: tumor limbs). The dot line represents a threshold at 20% allows to distinguish limbs with partial failure (five limbs upper the line experienced partial failure).

2.4 Discussion

The aim of this study was to assess FE models taking into account specific tumor properties to predict failure of tumoral bone in lytic lesion in mice. Additionally, a failure criterion was assessed to discriminate a limb with lower strength. The FE model with specific tumor properties did not improve prediction of the experimental failure load compared to a heterogeneous FE model based on CT grey levels. In this first study, the tumoral samples came from subcutaneous tumors. The tumoral properties of bone metastases might be different than these subcutaneous tumors. The mechanical properties of the bone metastases should be quantified in future work. The agreement between the heterogeneous FE model and the experiments obtained in the current study on biologically induced tumor confirms previous results obtained on human bones with mechanically induced defects (Benca et al, 2019) (Tanck et al, 2009). A failure criterion based on the difference between the failure load assessed in the entire tibia and the failure load assessed in a region of interest close to the tumor allowed us to distinguish different strengths (Figure 2.5). The criterion is based on a local region of interest of 3 mm around the tumor. This value is arbitrary. The sensitivity of this value should be assessed in future work, especially when considering other bones in mice or human bones. However, this principle opens new perspectives for FE model prediction. This study has several limitations. The rheological approach was considered to assess Young's modulus of the soft tissues, which induces many assumptions. This limitation was known in the study design, however, this methodology was considered because it is a reference approach (Nicolle et al, 2005) for quantifying soft tissue properties (especially in cases of small samples) and allowed for a comparison between tumors and other soft tissues. The compression test on the tibia had a limitation due to the definition of the loading conditions. The orientation of the tibia according to the loading direction is not known because we decided to keep soft tissues surrounding the tibia so as to not damage the tumor during the dissection. As a consequence, an optimization of the loading conditions was done prior to the simulation of the failure load. A third limitation was the use of an isotropic FE model that is purely linear elastic, whereas bone is in fact an anisotropic and elastoplastic material. Indeed, tumoral area could also be modeled with a mixture theory as done by (Lekszycki and dell'Isola, 2012), (Giorgio et al, 2016),(Madeo et al, 2012). Despite this simplification, the prediction model is promising. These results have to be confirmed on human bones, but the simplicity of the model could be of interest for clinical applications. A fourth limitation is related to the tumoral properties. The mechanical properties of bone metastases were not quantified. No data were available and the human myeloma sample and the subcutaneous tumors in mice might have different properties from the bone metastases. This should be assessed in the future.

2.5 Conclusion

The results obtained from a specific model using the mechanical properties of a tumor (coming from subcutaneous samples or human myeloma) did not improve failure prediction compared to heterogeneous models. Heterogeneous models were more consistent with experimental data than specific models (respective difference with experiments 11% and 23%). Interestingly, we were able to improve the failure assessment of tumoral bone by comparing a global bone analysis with a local analysis (in a region of interest surrounding the tumor). We found that a threshold of a 20% difference between local and global ultimate loads, allowed for the successful differentiation of all tumoral samples. This new prediction approach might be very useful in human bones to help clinicians in their decision-making.

Acknowledgements The authors are deeply grateful to Marc Gardegaront, Jean-Paul Roux, Richard Roussillon and Sandra Geraci for their technical support, and with the use of the platform ALECS. This work was partly funded by LabEx Primes (ANR-11-LABX-0063) and MSDAVENIR Research Grant (CC).

References

Attar S, Steffner RJ, Avedian R, Hussain WM (2012) Surgical intervention of nonvertebral osseous metastasis. Cancer Control 19(2):113–121

Benca E, Reisinger A, Patsch JM, Hirtler L, Synek A, Stenicka S, Windhager R, Mayr W, Pahr DH (2017) Effect of simulated metastatic lesions on the biomechanical behavior of the proximal femur. Journal of Orthopaedic Research 35(11):2407–2414

Benca E, Synek A, Amini M, Kainberger F, Hirtler L, Windhager R, Mayr W, Pahr DH (2019) QCT-based finite element prediction of pathologic fractures in proximal femora with metastatic lesions. Scientific reports 9(1):1–9

Bessho M, Ohnishi I, Matsumoto T, Ohashi S, Matsuyama J, Tobita K, Kaneko M, Nakamura K (2009) Prediction of proximal femur strength using a CT-based nonlinear finite element method: differences in predicted fracture load and site with changing load and boundary conditions. Bone 45(2):226–231

Burkhart TA, Andrews DM, Dunning CE (2013) Finite element modeling mesh quality, energy balance and validation methods: A review with recommendations associated with the modeling of bone tissue. Journal of biomechanics 46(9):1477–1488

Derikx LC, van Aken JB, Janssen D, Snyers A, van der Linden YM, Verdonschot N, Tanck E (2012) The assessment of the risk of fracture in femora with metastatic lesions: comparing case-specific finite element analyses with predictions by clinical experts. The Journal of bone and joint surgery British volume 94(8):1135–1142

Duchemin L, Mitton D, Jolivet E, Bousson V, Laredo J, Skalli W (2008) An anatomical subject-specific FE-model for hip fracture load predictionFE-model for hip fracture load prediction. Computer methods in biomechanics and biomedical engineering 11(2):105–111

Easley SK, Jekir MG, Burghardt AJ, Li M, Keaveny TM (2010) Contribution of the intra-specimen variations in tissue mineralization to pth-and raloxifene-induced changes in stiffness of rat vertebrae. Bone 46(4):1162–1169

Eggermont F, Derikx L, Verdonschot N, Van der Geest I, De Jong M, Snyers A, Van Der Linden Y, Tanck E (2018) Can patient-specific finite element models better predict fractures in metastatic bone disease than experienced clinicians? towards computational modelling in daily clinical practice. Bone & joint research 7(6):430–439

Fritton J, Myers E, Wright T, Van der Meulen M (2005) Loading induces site-specific increases in mineral content assessed by microcomputed tomography of the mouse tibia. Bone 36(6):1030–1038

Fung YC (2013) Biomechanics: mechanical properties of living tissues. Springer Science & Business Media

Giorgio I, Andreaus U, Scerrato D, dell'Isola F (2016) A visco-poroelastic model of functional adaptation in bones reconstructed with bio-resorbable materials. Biomechanics and modeling in mechanobiology 15(5):1325–1343

Goodheart JR, Cleary RJ, Damron TA, Mann KA (2015) Simulating activities of daily living with finite element analysis improves fracture prediction for patients with metastatic femoral lesions. Journal of Orthopaedic Research 33(8):1226–1234

Keyak J, Rossi S, Jones K, Les C, Skinner H (2001) Prediction of fracture location in the proximal femur using finite element models. Medical engineering & physics 23(9):657–664

Keyak JH, Kaneko TS, Rossi SA, Pejcic MR, Tehranzadeh J, Skinner HB (2005) Predicting the strength of femoral shafts with and without metastatic lesions. Clinical Orthopaedics and Related Research (1976-2007) 439:161–170

Kopperdahl DL, Aspelund T, Hoffmann PF, Sigurdsson S, Siggeirsdottir K, Harris TB, Gudnason V, Keaveny TM (2014) Assessment of incident spine and hip fractures in women and men using finite element analysis of ct scans. Journal of Bone and Mineral Research 29(3):570–580

Lekszycki T, dell'Isola F (2012) A mixture model with evolving mass densities for describing synthesis and resorption phenomena in bones reconstructed with bio-resorbable materials. ZAMM-Journal of Applied Mathematics and Mechanics/Zeitschrift für Angewandte Mathematik und Mechanik 92(6):426–444

Van der Linden Y, Dijkstra P, Kroon H, Lok J, Noordijk E, Leer J, Marijnen C (2004) Comparative analysis of risk factors for pathological fracture with femoral metastases: results based on a randomised trial of radiotherapy. The Journal of bone and joint surgery British volume 86(4):566–573

Madeo A, George D, Lekszycki T, Nierenberger M, Rémond Y (2012) A second gradient continuum model accounting for some effects of micro-structure on reconstructed bone remodelling. Comptes Rendus Mécanique 340(8):575–589

Mann KA, Lee J, Arrington SA, Damron TA, Allen MJ (2008) Predicting distal femur bone strength in a murine model of tumor osteolysis. Clinical orthopaedics and related research 466(6):1271–1278

Mirels H (2003) The classic: metastatic disease in long bones a proposed scoring system for diagnosing impending pathologic fractures. Clinical Orthopaedics and Related Research® 415:S4–S13

Nicolle S, Lounis M, Willinger R, Palierne JF (2005) Shear linear behavior of brain tissue over a large frequency range. Biorheology 42(3):209–223

Nyman JS, Uppuganti S, Makowski AJ, Rowland BJ, Merkel AR, Sterling JA, Bredbenner TL, Perrien DS (2015) Predicting mouse vertebra strength with micro-computed tomography-derived finite element analysis. BoneKEy reports 4

Pistoia W, Van Rietbergen B, Lochmüller EM, Lill C, Eckstein F, Rüegsegger P (2002) Estimation of distal radius failure load with micro-finite element analysis models based on three-dimensional peripheral quantitative computed tomography images. Bone 30(6):842–848

RCore TEAM (2016) R: A language and environment for statistical computing. R: Foundation for Statistical Computing, Vienna, Austria

Slosky LM, Largent-Milnes TM, Vanderah TW (2015) Use of animal models in understanding cancer-induced bone pain. Cancer growth and metastasis 8:CGM–S21,215

Tanck E, van Aken JB, van der Linden YM, Schreuder HB, Binkowski M, Huizenga H, Verdonschot N (2009) Pathological fracture prediction in patients with metastatic lesions can be improved with quantitative computed tomography based computer models. Bone 45(4):777–783

Wong M, Pavlakis N (2011) Optimal management of bone metastases in breast cancer patients. Breast Cancer: Targets and Therapy 3:35

Zysset PK, Dall'Ara E, Varga P, Pahr DH (2013) Finite element analysis for prediction of bone strength. BoneKEy reports 2

Chapter 3
Automatic Classification of Intramuscular EMG to Recognize Pathologies

Alban Gallard, Konstantin Akhmadeev, Eric Le Carpentier, Raphaël Gross, Yann Péréon, and Yannick Aoustin

Abstract This paper proposes to assess the relevance of new automated tools for electromyography (EMG) analysis, in order to differentiate neuropathic from myopathic patterns. The challenge is to define the diagnosis with only one iEMG signal per patient. Our proposed method uses the decomposition of the EMG signal to characterize motor unit action potentials (MUAPs). The decomposition of each iEMG signal is carried out with EMGLAB. For each signal, the decomposition provides a code which is used by the automated classification algorithms. We use here the linear Support Vector Machine (SVM) and the Bagging Trees methods. For the learning process we use several EMG signals and in different parts of the muscle. Only one recorded electromyography EMG signal per subject is used for the diagnostic test. We evaluate the $k - fold$ cross-validation and the confusion matrix for both models. The accuracy is 77.3% for the SVM and 68.2% for the Bagging Trees. These are the first developments of this tool to make it useful for clinical practice.

Keywords: Quantitative electromyography · Motor unit action potential · Classification · Support Vector Machine · Bagging Trees

---

A. Gallard, K. Akhmadeev, E. Le Carpentier, Y. Aoustin
Laboratoire des Sciences du Numérique de Nantes, UMR CNRS 6004
1 rue de la Noë, BP 92101 F-44321 Nantes Cedex 3, France
Ecole Centrale de Nantes, Université de Nantes
e-mail: albgal42@gmail.com, konstantin.akhmadeev@ec-nantes.fr,
Eric.Le-Carpentier@ec-nantes.fr, Yannick.Aoustin@univ-nantes.fr

R. Gross
Laboratoire d'Analyse du Mouvement, CHU de Nantes 44000 Nantes
e-mail: Raphael.Gross@chu-nantes.fr

Y. Péréon
Laboratoire d'Explorations Fonctionnelles, Centre de Références Maladies Neuromusculaires Atlantique-Occitanie-Caraïbes (LEF), Université de Nantes,
44000 CHU de Nantes
e-mail: Yann.Pereon@univ-nantes.fr

B. E. Abali and I. Giorgio (eds.), *Developments and Novel Approaches in Biomechanics and Metamaterials*, Advanced Structured Materials 132,
https://doi.org/10.1007/978-3-030-50464-9_3

3.1 Introduction

Electromyography (EMG) is used in routine in clinical practice. Electrodes are inserted into the muscle (intramuscular EMG, iEMG) or placed on the skin (surface EMG, sEMG) to record the muscular electrical activity. The measured EMG signal is a sum of elementary contributions. Each contribution is a wavelet train produced by a motor unit (MU) in the electrode vicinity. A MU corresponds to a spinal motoneuron (MN) and the muscle fibers it innervates (Heckman and Enoka, 2012), and the wavelet is called the Motor Unit Action Potential (MUAP). A variation of muscle activation level produces a variation of the number of active MUs in a process called "spatial recruitment," and a variation of the discharge rate (that is the number of MUAPs per time unit) of the active MUs, called "temporal recrutement."

EMG signals study can be used to detect neuromuscular diseases. It provides information in favor of either muscle lesions (so-called myopathic pattern) or nerve lesions (so-called neuropathic pattern). Nerve and muscle lesions affect both individual MUAP characteristics and recruitment pattern during contraction. In routine, neurologists assess both through a visual and auditory analysis, but differentiating normal recordings from myopathic or neuropathic ones can be difficult with possible misdiagnosis (Pereon, 2015; Harwood et al, 2012).

EMG can also be used to gain knowledge about the neuromuscular mechanisms involved in the force and movement control. For example, in Gross et al (2016), EMG data and capture motion data are used to analyze children walking along a curved path. This study investigated the EMG activity in the lower limbs of typically developing children during turning trajectories with increasing curvature resulting from changes of direction towards targets, and evidenced that differences exist in curved compared with straight walking.

Despite the medical doctors' expertise a purely descriptive approach is not sufficient. An automated analysis can lead to an objective diagnosis minimizing observer bias, (Dorfman and McGILL, 1988). For a few decades now different approaches have been proposed for rendering the EMG examination more quantitative and more automatic to permit precise interpretation of the findings, to minimize observer bias, to facilitate comparative studies over time for different methodologies. An automated analysis relies on a preliminary decomposition of the processed signal, that is the separation of the active MUs contributions together with MUAPs shapes and trains characteristics estimation, to provide EMG features. Pattichis et al (1995) proposed artificial neural networks for the automatic classification of EMG features, which are recorded from normal individuals and patients suffering with neuromuscular diseases. In the field of machine learning, Katsis *et al.* used a Support Vector Machine (SVM) algorithm (Katsis et al, 2006), or a radial basis function network and a decision tree (Katsis et al, 2007) for automated EMG decomposition and MUAPs classification. Let us remark that several methods exist to carry out this automated EMG decomposition: the Convolution Kernel Compensation (CKC) (Holobar and Zazula, 2007; Holobar et al, 2009), Monte Carlo Markov Chain based decomposition (Ge et al, 2011), Bayesian filtering based on a hidden Markov model of the intramuscular EMG (iEMG) (Monsifrot et al, 2014). They allow a full decomposition of single-channel iEMG signals produced during contractions at moderate force levels but presenting superimposed MUAPs. Farina et al (2001) propose a tool to evaluate and compare EMG signal decomposition algorithms.

Note that these previous works to help to diagnosis are about the MUAP classification. Another way can be to make the EMG classification. Pattichis and Schizas (1996) studied a hybrid diagnostic system that combines both neural network and genetics-based machine learning models to diagnose certain neuromuscular diseases from EMG data. 680 MUAP's are collected from twelve healthy subjects, 11 patients with neuropathy and 11 patients with myopathy. The experimental protocol therefore uses 20 MUAPS per patient. Eight people from each group formed the training set, and the other 10 subjects formed the evaluation set. The considered MUAP parameters are duration, spike duration, amplitude, area, spike area, phases, and turns. Each subjects are described mean and the standard deviation of each MUAP parameter in order to characterize the EMG signals and perform the diagnosis. They showed that the diagnostic performance of neural network genetics-based machine learning models is enhanced by the hybrid system. In this paper, a unique EMG signal per subject is used to make a targeted analysis of the MUAP shape, in order to provide a patient diagnosis based on full recorded iEMG signal classification. This paper is also focused on the EMG classification. Sadikoglu et al (2017) analyzes the power spectrum of patients' EMG signals to describe parameters such as mean frequency, median frequency, amplitude in terms of root mean square spectral density, minimum and maximum. The classification results distinguish between healthy EMG signals and EMG signals from neuropathic patients. Yousefi and Hamilton-Wright (2014) provide a critical review of several classification methods including Bayesian techniques, neural networks, multilayer perceptrons, fuzzy approaches, SVM, and neuro-fuzzy systems for EMG signal characterization. They conclude that SVM is a remarkably robust classification method in disease diagnosis. Subasi (2013) proposed to combine the particle swarm optimization (PSO) and SVM to improve the EMG signal classification accuracy. The EMG signals are decomposed into the frequency sub-bands using discrete wavelet transform (DWT). A set of statistical features are extracted from these sub-bands to represent the distribution of wavelet coefficients. The presented results show that the strategy is very efficient. However, the features deduced from the decomposition are not physiologically interpretable.

In this paper, our goal is to mimic the practitioner behavior, that is a combination of overall aspect of the EMG signal with a more targeted analysis of the MUAP shape, in order to provide a patient diagnosis based on full recorded iEMG signal classification. However our method is based on a unique recorded EMG signal per patient. This method necessarily leads to poorer results than those of Pattichis and Schizas (1996) but it is less tedious at the level of the experimental protocol. Since the purpose of this paper is not to compare decomposition tools, we will use a semi-automatic decomposition by means of the EMGLab software with manual corrections (EMG, 2007; Florestal et al, 2006). Each signal will be encoded by the number of active MU, an average MUAP shape and a firing rate for each active MU. Then, we will focus on the use of machine learning tools for the EMG classification stage. Two algorithms are considered: the linear SVM (Breiman, 1995; Yousefi and Hamilton-Wright, 2014) and the Bagging Trees (Breiman, 1996). The bagging method is a suitable mean for increasing efficiency of standard machine learning algorithms, see Machova et al (2006); Gokgoz and Subasi (2015).

This paper is outlined as follows: section 3.2 where presents the decomposition of the EMG signal, the Motor Unit, the general features of the myopathic and neuropathic patterns, the features and the classification of the EMG signal; section 3.3 shows the different numerical results; then section 3.4 discusses about these results and finally the section 3.5 offers our conclusions and perspectives.

3.2 Methods

The method is composed of five steps. First, the decomposition of the EMG signals gives the MUAPs. Secondly, we need to understand how the MUs work. We also need to know how the pathologies modify EMG signals and more particularly the MUAPs. With the second and third steps, the best features to classify the signal can be found. Finally, the computation of the machine learning algorithms can be performed with the features created in the previous step.

3.2.1 Data

Miki Nikolic from the Rigshospitalet of Copenhagen presented in his Ph.D thesis (Nikolic, 2001) an automatic system called EMGPAD to decompose the clinical EMG signal into its constituent motor MUAPs and their corresponding firing patterns (FPs). The signals used for this paper were made available by Miki Nikolic on the website of EMGLab (EMG, 2007; Florestal et al, 2006). Only the Biceps Bracii muscle has been studied to eliminate variations in EMG in the different muscles.

Five healthy people (age rank 21-37 years) were studied, as well as five patients presenting with myopathic lesions (age rank 19-63 years) and five patients with neuropathic lesions (age rank 35-67 years).

3.2.1.1 Preprocessing

The EMG signal was recorded using intramuscular needle electrodes with a cross-section of 0.07 mm^2. A surface electrode has been placed on the limb to allow single pole work. The signal was measured for 11.2 s in each case.

The signals were amplified 4000 times and filtered with a bandpass filter between 2 Hz and 10 kHz. Then, they were sampled and digitized. The sampling frequency was 23437 Hz and the digital conversion is encoded with 16 bits processor.

Several measures were taken in different depths and different locations of the Biceps Bracii muscle to explore it globally.

3.2.1.2 Decomposition

The decomposition of the EMG signals was performed semi-automatically using EMGLab (EMG, 2007).

For simple cases, the automatic decomposition was correct. A manual check was carried out to avoid any errors. For complex cases, the automatic decomposition created too many errors. Automatic decomposition and manual verification take longer than manual decomposition alone.

The first step in manually decomposing a signal is to find isolated MUAPs. It is important to check their repetitions to ensure that they are isolated. Afterwards, each time the MUAPs are found in the EMG signal, a mark is placed. This allows you to decompose overlapping MUAPs and find new ones.

The number of signals decomposed for each patient is detailed in the table 3.1. There are respectively 50 signals for the groups of healthy subjects and patients with myopathy. There are 45 signals for the neuropathic patient group. Thus, 145 signals were decomposed to feed the machine learning algorithms. Each signal is represented by the number of active MU's and, for each active MU, by the firing rate and the average MUAP.

Table 3.1 Number of signals for each healthy subject and patient.

	Subject numbers	Number of signals	Total
Healthy Subjects	1	17	50
	2	12	
	3	4	
	4	14	
	5	3	
Myopathic Subjects	1	15	50
	2	8	
	3	6	
	4	9	
	5	12	
Neuropathic Subjects	1	12	45
	2	9	
	3	8	
	4	5	
	5	11	

3.2.2 *Motor unit (MU)*

Studying the creation and propagation of the electrical signal is important to understand the differences between the pathologies. The central nervous system composed of the brain and the spinal cord sends electrical messages to the muscles (Heckman and Enoka, 2012). These messages are sent by using the motoneurons (MNs) (Silani et al, 1999), which are located in the brain stem and in the anterior part of the grey matter of the spinal cord, through the motor axon along the peripheral nerves to the muscles fibers. The muscle fibers innervated by a motoneuron are called muscle unit. The motor unit (MU) defines the association of the motoneuron and the muscles fibers it innervates. The MUs transform the message of the central nervous system to contract a muscle and to extend the opposite one. In that way, a synergistic effect and an antagonistic force are produced (Farina et al, 2016).

A MUAP found in the EMG is the sum of the action potentials of all the muscle fibers innervated by a single motoneuron. The electrode location is important because an EMG measures a part of the electrical signal of the muscle fibers of the MUs. Moreover, a fiber closer to the electrodes gives a more important signal.

Waveform characteristics are different for all MUAPs, for example the duration or the amplitude. The number of phases, which is the number of sign changes of the signal and also changes; such as the number of peaks.

The rate at which the MUAPs are discharged is the firing rate. In other words, it is the number of MUAPs per unit of time. It increases with strength.

The number of muscle fibers per MU is different for all muscles. That is why, it is important to always use the same muscle to compare the EMG.

3.2.3 *Pathologies*

To find the best features for machine learning it is essential to know how the pathologies affect the MUs and the EMG signals. Typically, in case of nerve or root lesions, a lower number of MUs are recruited (poor recruitment because fewer motor neurons are available), and at a higher rate. The chronic dennervation-reinnervation process is responsible for the increase in size and duration of MUAPs due to distal reinnervation: this defines the EMG neuropathic pattern. In the case of muscular dystrophy, a lower number of muscle fibres per MU is present, providing MUAPs of low amplitude, short duration, multiple turns, and recruitment is abnormally increased (rich recruitment) in order to provide as much as possible efficient strength: this induces the EMG myopathic pattern. Figure 3.1 shows this phenomenon by simplifying it (Kubis and Lozeron, 2014).

Table 3.2 summarizes the consequences of the disease on the MUAPs. The healthy case is the standard, which is why there is the sign $=$ at every lines. The $+$ sign means higher and the $-$ sign means lower.

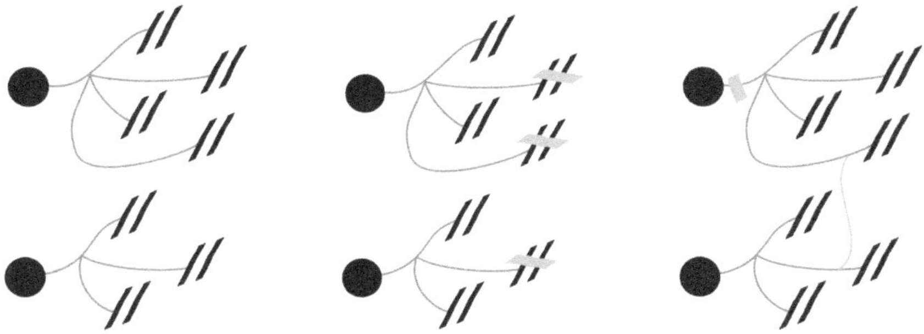

Fig. 3.1 Diagram of the MUs: Healthy (left), Myopathy (center), Neuropathy (right). The black circles represent the motoneurons and the pairs of black lines are the muscle fibers. The grey lines represent the connection between a motoneuron and the muscle fibers.

Table 3.2 Consequences of the pathologies (# means number).

	Healthy	Myopathy	Neuropathy
# active MUs	=	+	-
MU firing rate	=	+	-
MUAP duration	=	-	+
MUAP amplitude	=	-	+
MUAP # phases	=	+	+
MUAP # turns	=	+	-

3.2.4 Features

The features of the Table 3.2 can be used to differentiate the pathologies. The number of active MUs is the first feature of the signal, but the others are defined for each MU. To obtain a unique length of the feature vector, the minimum, the mean and the maximum values are used.

The vector of features for a signal, considering the set of MUAPs, has 16 elements: the number of MUAPs and the minimum, mean and maximum of the duration, amplitude, firing (discharge) rates, number of phases and number of turns.

In this vector, some of the features can be useless to differentiate the pathologies. A T-test is used to find them. A T-Test rejects or not the null hypothesis: there is no significant statistical difference between the two distributions. In that case, the two distributions are one feature for a pathology and the same feature with another pathology (Rice, 2006).

The T-test is composed of three variables. The first one is the significant value α that determines the percentage of rejection of the hypothesis. The two others are the TValue TV and the degree of freedom of the T-test df. Let us consider two distributions Healthy *Subjects−Myoapthic Subjects, Healthy Subjects−Neuropathic Subjects, or Myoapathic Subjects−Neuropathic−Subjects* . The values of T-test and *df* are calculated with the mean of the distribution $\bar{x}$ and $\bar{y}$, their standard deviation σ_x and σ_y, their number of data n and m:

$$TV = \frac{|\bar{x} - \bar{y}|}{\sqrt{\frac{\sigma_x^2}{n} + \frac{\sigma_y^2}{m}}} \tag{3.1}$$

$$df = n + m - 2 \tag{3.2}$$

TV is after compared with the result found with *df* and the significant value α in the table of T-Test. The value of α is chosen equal to 0.05. If it is higher, the null hypothesis will be rejected and the features must be kept. Otherwise, it will not be rejected and the features can be deleted.

The T-test shows that the number of active MUs is not significant.

The correlation analysis indicates the strength of the relationship between two features. The correlation coefficient $\rho_{A,B}$ determines the correlation between the two distributions A and B. In our case, A and B are two different features with all pathologies. The following equation gives $\rho_{A,B}$:

$$\rho_{A,B} = \frac{1}{N-1} \sum_{i=1}^{N} \left(\frac{A_i - \bar{a}}{\sigma_A} \right) \left(\frac{B_i - \bar{b}}{\sigma_b} \right) \tag{3.3}$$

Where $\bar{a}$ and $\bar{b}$ are the means of the distributions, σ_A and σ_B their standard deviations and N their number of data.

This analysis shows the mean of the duration, the mean of the number of phases and the mean of the number of turns are highly correlated to the other features.

The 12 features remaining are:

- Minimum and maximum of the duration.
- Minimum, mean and maximum of the amplitude.
- Minimum and maximum of the number of phases.
- Minimum and maximum of the number of turns.
- Minimum, mean and maximum of the firing rate.

3.2.5 Classification

With the features found in the previous step, the machine learning models can be studied. The two models used are the Support Vector Machine (SVM) and the Bagging Trees. The first one has been chosen for its simplicity and the second one for its speed and its good performance.

3.2.5.1 Linear SVM

The SVM separates the data in several parts. For the linear SVM, the separation is an hyperplane. Its equation is:

$$h(\mathbf{x}) = w_0 + \mathbf{x}^\top \mathbf{w} \tag{3.4}$$

The SVM uses margins that have to be the largest to decrease the error. Usually, the data can not be separated. To have less errors, the SVM attributes a slack variable ξ_i to each point. Its value depends on the classification. It will be null, if it is correct; less than 1, if it is in the margin; and more than 1, if it is false. The problem to find the parameters $\mathbf{w}$, w_0 and ξ_i is solved thanks the following minimization:

$$\min_{w_0,\mathbf{w}} \frac{1}{2}||\mathbf{w}||^2 + C\sum_{i=1}^{n} \xi_i \tag{3.5}$$

With the conditions:

$$\xi_i \geq 0 \quad \forall i \tag{3.6}$$

$$y_i(w_0 + \mathbf{x}_i^\top \mathbf{w}) \geq 1 - \xi_i \quad \forall i \tag{3.7}$$

The value of the parameter C is chosen depending on the importance of the slack variables that we want. This problem can be solved with the Lagrangian function where the Lagrange multipliers are maximized.

3.2.5.2 Bagging Trees

The Bagging Trees is a model that uses several Trees. A Tree is a succession of choices. A tree takes one of the features and a value for it. Then a tree compares the data with the value and separates it in two parts. For each part, the Tree does the same thing until a shutoff parameter like a number of points is reached. The Bagging Trees does not use all data for each Tree, but only a part that is different for all Trees.

When the Bagging Trees tests a new data, it finds the number $P_{d,t}$ of all pathologies d for each Tree t. Let N_t be the number of trees. After, the Bagging Trees assembles the N_t trees:

- Major voting.

$$\underset{d}{\text{argmax}} \sum_{t} (d = \underset{d}{\text{argmax}} P_{d,t}) \tag{3.8}$$

- Sum rule.

$$\underset{d}{\text{argmax}} \sum_{t} P_{d,t} \tag{3.9}$$

- Mean rule.

$$\underset{d}{\text{argmax}} \frac{\sum_t P_{d,t}}{N_t} \tag{3.10}$$

- Product rule.

$$\underset{d}{\operatorname{argmax}} \prod_{t} P_{d,t} \tag{3.11}$$

We will use the major voting is usually used for the classification case Hakim et al (2017) with 30 trees and 100 possible choices.

3.3 Results

The quality of the decomposition can be defined with the signal to interference ratio (SIR) and the standard deviation (SD). These criteria are described for the healthy subjects, the neuropathic patients, and the myopathic patients in Table 3.3.

Table 3.3 Decomposition: Quality indices for the healthy subjects, the neuropathic patients, and the myopathic patients.

	Healthy	Myopathy	Neuropathy
SIR	8.57 dB	7.53 dB	9.33 dB
SD	3.37	4.04	3.82

We can remark that the SIR is lower for myopathy because the magnitude signals are lower. Conversely, SIR is higher for neuropathy because the magnitude of the signals are higher.

The data are split up in two sets. The first, called a training set contains 85% of the data and it is used to train the model. The other one contains 15% of the data and is called the test set. After the training, the model uses the data in the test to verify the number of misclassified signals. The elements of both parts are randomly selected.

k-fold cross-validation method is a reliable method for predicting the error rate of a classification technique (Sandvig et al, 2008; Gokgoz and Subasi, 2015). k-fold cross-validation arbitrarily divides the data in a given number of subsets, the "folds." For the number of employed folds $k = 10$ we have respectively for each classifier:

- for SVM classifier, Cross-validation accuracy: 68.92%, with STD=13.5.
- for Bagging Trees with 30 trees and and 100 split nodes, Cross-validation accuracy: 84.55%, with STD=8.3.

The Confusion Matrix gives the errors between each pathology for the prediction and the actual pathologies according to expert. The Confusion Matrix for the tests of the SVM and Bagging Trees is represented in Tables 3.4 and 3.5. To understand the gathered data let us consider of the healthy subjects in Table 3.4. The numbers in blue bold illustrate the agreement between the prediction and reality. There are 8 healthy subjects and 8 predicted healthy subjects. The percentage of good prediction is thus 100%. The first line indicates that the classifier predicts $8 + 4 + 1$ healthy subjects but 4 are myopathic patient and 1 is a neuropathic patient. Percentage of good prediction is thus 61.4%.

Table 3.4 Confusion Matrix for one test of the SVM (Pred = Prediction, Heal = Healthy, Myo = Myopathy, Neuro = Neuropathy, Per = Percentage of good prediction)

		Truth			Per
		Heal	Myo	Neuro	
Pred	Heal	8	4	1	61.4%
	Myo	0	6	0	100%
	Neuro	0	0	3	100%
Per		100%	60%	75%	77.27%

Table 3.5 Confusion Matrix for one test of the Bagging Trees (Pred = Prediction, Heal = Healthy, Myo = Myopathy, Neuro = Neuropathy, Per = Percentage of good prediction)

		Truth			Per
		Heal	Myo	Neuro	
Pred	Heal	5	3	1	55.56%
	Myo	0	7	0	100%
	Neuro	3	0	3	50%
Per		62.5%	70%	75%	68.18%

In the Confusion Matrix, the columns are the real pathologies and the rows are the predictions. The last column gives the percentage of good predictions for a pathology on all predictions of this pathology. For the last row, this is the percentage of good predictions for a pathology on all real signal for this pathology. The number in the last cell is the percentage of the total of good predictions on all signals. The accuracy is 77.3% for the SVM and 68.2% for the Bagging Trees.

3.4 Discussion

The percentage of errors for the two models is not very high, 23% for the linear SVM and 32% for the Bagging Trees. For our study, the SVM seems a better model. The errors of the linear SVM come from the mixed data between the pathologies. For the Bagging Trees, the errors can come from overfitting, but as for the linear SVM, it can also be the mixed signals.

The SVM classifier has better test results than validation from the study of k-fold cross-validation. The opposite is true for Bagging Trees. We have to be careful with these results because it depends on the training set and the test set.

The Confusion Matrix of the Tables 3.4 and 3.5 gives information about the localization of the errors. The models usually make the difference between the myopathic and the neuropathic signals. The most important part of the errors is between the healthy and the myopathy. They are more present between the healthy subjects and the myopathy patients than between the healthy subjects and neuropathy patients.

The last column of the Confusion Matrix shows the good prediction for the myopathic and neuropathic signals. It is possible to trust them, but it is more difficult to believe the models when they say the signal is from a healthy subject.

The last line indicates a patient with neuropathy will usually have the right diagnostic. It is not the case for the healthy subjects and myopathic patients.

Both models can have better results. For example, with a higher number of signals, it will be possible to have models whose the train has signals with a lot of differences to take into account all cases and to avoid errors in the test.

Another way to improve the classification is to use more features. It can be by finding new temporal features or by using frequency ones. The force can also be used. Indeed, the measure can be done with a specific strength to have every features without depending of the force. Moreover, the signal can be measured to different values of the strength to take into consideration the evolution of the features depending to the force.

The SVM can be improved by using a non-linear method. For example, it can be a parabolic, a cubic or a Gaussian SVM.

Bagging Trees can have better results with a larger number of trees. The trees can also have a different stop parameter, but it is important to deal with the overfitting.

3.5 Conclusion

This new methodology automatically classifies the intramuscular EMG signal into three classes: healthy, myopathy and neuropathy. Two machine learning algorithms, the linear SVM and the Bagging Trees, have been studied. For our study SVM classifier has better performance. The classifier tools can give information to the doctor to help for the diagnosis of the pathology. The methodology needs a decomposition of the signal. The obtained MUAPs give features, that must be normalized to be used by the machine learning models. In this work the decomposition was semi-automatic. In future with an automatic decomposition, this methodology can be integrated on-line in a decision support system to help the doctors.

Acknowledgements This work was supported University of Nantes, Transdisciplinary project: **M**odélisation et **D**écomposition des **EMG** pour la **R**econnaissance et l'**A**ssistance aux **T**roubles neur**O**musculaires (Moderato)).

References

(2007) Emglab: A forum for sharing software, data, and information related to emg decomposition (http://www.emglab.net). EMC-Neurologie

Breiman L (1995) Support vector networks. Machine learning 20(3):273–297

Breiman L (1996) Bagging predictors. Machine learning 24(2):123–140

Dorfman LJ, McGILL KC (1988) Aaee minimonograph #29: Automatic quantitative electromyography. Muscle & Nerve 11:804–818

Farina D, Colombo R, Merletti R, Olsen HB (2001) Evaluation of intra-muscular emg signal decomposition algorithms. Journal of Electromyography and kinesiology 11:175–187

Farina D, Negro F, Muceli S, Enoka RM (2016) Principles of motor unit physiology evolve with advances in technology. Physiology 31(2):83–94

Florestal JR, Mathieu PA, Malanda A (2006) Automated decomposition of intramuscular electromyographic signals. IEEE Transactions on Biomedical Engineering 53(5):832–839

Ge D, Le Carpentier E, Farina D, Idier J (2011) Spike Sorting by Stochastic Simulation. IEEE Transactions on Neural Systems and Rehabilitation Engineering 19(3):249–259

Gokgoz E, Subasi A (2015) Comparaison of decision tree algorithms for emg signal classification using dwt. Biomedical signal processing and control 18(2):138–144

Gross R, Leboeuf F, Lempereur M, Michel T, Perrouin-Verbe B, Vieilledent S, Rémy-Noris O (2016) Modulation of lower limb muscle activity induced by curved walking in typically developing children. Gait & Posture 50:34–41

Hakim L, Sartono B, Saefuddin A (2017) Bagging based ensemble classification method on imbalance datasets. International Journal of Computer Science and Network 6(6):670–676

Harwood CA, McDermott CJ, Shaw PJ (2012) Clinical aspect of motor neurone desease. Medicine 40(10):540–545

Heckman C, Enoka RM (2012) Motor unit. Comprehensive Physiology 2:2629–2682

Holobar A, Zazula D (2007) Multichannel blind source separation using convolution kernel compensation. IEEE Trans Signal Processing 55(9):4487–4496

Holobar A, Farina D, Gazzoni M, Merletti R, Zazula D (2009) Estimating motor unit discharge patterns from high-density surface electromyogram. Clinical Neurophysiology 120:551–562

Katsis CD, Goletsis Y, Likas A, Fotiadis DI, Sarmas I (2006) A novel method for automated emg decomposition and muap classification. Artificial Intelligence in Medicine 37(1):55–64

Katsis CD, Exarchos TP, Papaloukas C, Goletsis Y, Fotiadis DI, Sarmas I (2007) A two-stage method for muap classification based on emg decomposition. Computers in Biology and Medicine 37(9):1232–1240

Kubis N, Lozeron P (2014) Vitesses de conduction nerveuse. EMC-Neurologie 12(3):1–13

Machova K, Barcak F, Bednar P (2006) A bagging method using decision trees in the role of base classifiers. Acta Polytechnica hungarica 3(2):121–132

Monsifrot J, Carpentier EL, Aoustin Y, Farina D (2014) Sequential decoding of intramuscular emg signals via estimation of a markov model. IEEE Trans on Neural Systems and Rehabilitation Engineering 22(5):1030–1040

Nikolic M (2001) Detailed analysis of clinical electromyography signals emg decomposition, findings and firing pattern analysis in controls and patients with myopathy and amytrophic lateral sclerosis. PhD thesis, PhD Thesis, Faculty of Health Science, University of Copenhagen

Pattichis CS, Schizas CN (1996) Genetics-based machine learning for the assessment of certain neuromuscular disorders. Journal of the Neurological Sciences 7(2):427–439

Pattichis CS, Schizas CN, Middleton LT (1995) Neural network models in emg diagnosis. IEEE Transactions on Biomedical Engineering 42(5):486–496

Pereon Y (2015) Neurogène ou myogène. Revue Neurologique 171S 1189

Rice JA (2006) Mathematical Statistics and Data Analysis (3rd ed.). Duxbury Advanced

Sadikoglu F, Kavalcioglu C, Dagman B (2017) Electromyogram (EMG) signal detection, classification of EMG signals and diagnosis of neuropathy muscle disease. In: Procedia Computer Science, Budapest, Hungary, 120, pp 422–429

Sandvig JJ, Mobasher B, Burke R (2008) A survey of collaborative recommendation and the robustness of model-based algorithms. IEEE Comput Soc Tech Comm Data Eng 31(2):3–13

Silani V, Ciammola A, Pizzuti A, Cardin V, Scarlato G (1999) Motor neuronne metabolism. Journal of the Neurological Sciences 169:161–169

Subasi A (2013) Classification of emg signals using pso optimized svm for diagnosis of neuromuscular disorders. Computers in Biology and Medicine 43:576–586

Yousefi J, Hamilton-Wright A (2014) Characterizing emg data using machine-learning tools. Computers in Biology and Medicine 51:1–13

Chapter 4
Algorithmic Formulation of Bone Fabric Evolution Based on the Dissipation Principle: A 2D Finite-Element Study

Madge Martin, Peter Pivonka, Guillaume Haïat, Thibault Lemaire, and Vittorio Sansalone

Abstract Bone remodeling is a highly complex biological process allowing bone tissue to evolve throughout life. While this continuous change involves various events at the microscopic scale, macroscopic observations show a correlation between bone architecture and *in vivo* loading conditions. In particular, the orientation of bone microstructure, also referred as bone fabric, relates to principal stresses induced by *in vivo* loading conditions. We utilize here a thermodynamically-consistent formulation based on generalized continuum mechanics to describe the evolution of bone fabric. Our macroscopic model accounts at the continuum level for the reorientation of bone orthotropic elasticity axes, referred to as *rotary remodeling*. We use our model to study the stress-induced rotary remodeling in two benchmark problems. To this end, we developed an algorithm combining finite-element analysis (FEA) and fabric tensor evolution. This iterative routine allows one to compute and update the orthotropic elastic properties of bone material in each element of the FE mesh. Stresses and strains are computed via FEA and fed forward into the rotary remodeling algorithm, resulting in the evolution of bone tissue material axes. In the

M. Martin
Univ Paris Est Creteil, Univ Gustave Eiffel, CNRS, MSME UMR 8208, F-94010 Creteil, France
School of Mechanical, Medical and Process Engineering, Queensland University of Technology, 2 George St, Brisbane, QLD 4000, Australia
e-mail: madge.martin@u-pec.fr

P. Pivonka
School of Mechanical, Medical and Process Engineering, Queensland University of Technology, 2 George St, Brisbane, QLD 4000, Australia
e-mail: peter.pivonka@qut.edu.au

G. Haïat
CNRS, Univ Paris Est Creteil, Univ Gustave Eiffel, MSME UMR 8208, F-94010 Creteil, France
e-mail: guillaume.haiat@cnrs.fr

T. Lemaire, V. Sansalone
Univ Paris Est Creteil, Univ Gustave Eiffel, CNRS, MSME UMR 8208, F-94010 Creteil, France
e-mail: thibault.lemaire@univ-paris-est.fr, vittorio.sansalone@u-pec.fr

B. E. Abali and I. Giorgio (eds.), *Developments and Novel Approaches in Biomechanics and Metamaterials*, Advanced Structured Materials 132,
https://doi.org/10.1007/978-3-030-50464-9_4

first benchmark, using a simple toy problem, we found that the meshing strategy and time discretization choices can significantly impact the rotary evolution of bone microstructure. In the second benchmark, depicting bone rotary remodeling in the proximal femur, we recognized that the material axes tend to align with the typical directions of the trabeculae, while noting the influence of intrinsic bone tissue properties and geometry. This work provides a numerical framework for potential clinical applications, in particular studying the influence of mechanical and biochemical parameters on bone remodeling at the organ scale.

Keywords: Generalized continuum thermodynamics · Bone remodeling · Trabecular adaptation · Finite element analysis · Convergence

4.1 Introduction

The close link between mechanics and bone architecture has been noted since the 19th century and the early observations of Culmann, Von Meyer, Wolff and Roux (Culmann, 1866; Roux, 1881; von Meyer, 1867; Wolff, 1870). Georg Hermann Von Meyer was a German professor of anatomy with a strong interest in the human skeleton. Benefitting from his interactions with the German civil engineering professor Culmann, he noticed the arched patterns of bone inner porous structure (trabecular bone). Culmann noted that the struts appeared to be aligned along principal stress directions. The result of this collaboration deeply influenced the work of Julius Wolff, a German surgeon convinced that Culmann and von Meyer's findings could not be a coincidence (Culmann, 1866; von Meyer, 1867). At the end of the 19th century, Wolff (1870) introduced a theory on bone remodeling, nowadays referred to as "Wolff's law," stating that the mechanical function of bone drives the evolution of its architecture.

Several macroscopic theories were introduced over the past decades to model how bone may adapt its microstructural orientation to mechanical loads (Cowin, 1986; Cowin et al, 1992; Doblaré and García, 2002; Miller et al, 2002; Pettermann et al, 1997; Watzky and Naili, 2004). One may stress the particularly important work of Cowin on that matter, who introduced the *fabric tensor* (a measure of tissue anisotropy, Cowin, 1986) and largely investigated the optimal micro-orientations of bone under varying loading configurations (Cowin, 1986, 1995). In his 1986 work, Cowin found that, at remodeling equilibrium, the eigenspaces of the stress and strain tensors are aligned. This sound observation was also derived mathematically later on by Sansalone et al (2011) as a result of the dissipation principle.

As Skedros and Baucom (2007) note in their study of trabecular "trajectories," a popular example for cancellous bone functional adaptation is the human femur, even though the first experiment showing the close correspondence between arched trabecular patterns and orientations of principal strain was performed on sheep calcanei (Lanyon, 1974) which undergo simpler loading conditions . Hence, several numerical studies attempted to recover the identified femoral neck patterns (see the illustration of the variety in the representation of trabecular and stress patterns in Figure 4.1). These works rely on optimization theories (Fernandes et al, 1999; Jacobs et al, 1997; Luo and Gea, 1998) or on bone remodeling algorithms based on *ad hoc* expressions of directional elastic moduli evolution laws (Geraldes et al, 2016; Miller et al, 2002; Shang et al, 2007) or continuum damage mechanics (Doblaré and García, 2001).

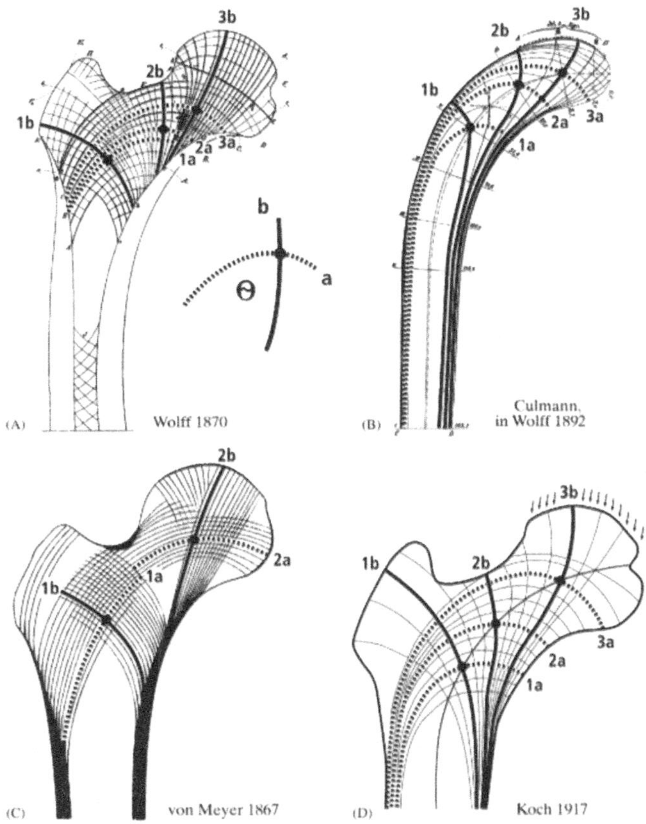

Fig. 4.1 Examples of the diversity in femoral trabecular trajectories representations: (A) Wolff's femur (Wolff, 1870); Culmann's crane (Wolff, 1892); (C) von Meyer's femur (von Meyer, 1867); (D) Koch's drawing (Koch, 1917). Labeled trabecular tracts mark main trajectories. From Skedros and Baucom (2007), with permission.

On the one hand, optimization theories rely on the definition and minimization of an objective function $\mathcal{Q}$ of the microstructural orientation (θ) under a certain load σ, that can also depend on other variables such as density (ρ):

$$\mathcal{Q}(\rho, \hat{\theta}, \sigma) = \min_{\theta} \mathcal{Q}(\rho, \theta, \sigma), \tag{4.1}$$

where $\hat{\theta}$ is the optimal orientation. Optimization theories usually also include constraints, generally on mass. For instance, Fyhrie and Carter (1986) developed an optimization theory using a cost function based on the strain energy, which is interpreted as a measure of bone's structural integrity depending both on density and orientation. This concept was used later on by Luo and An (1998) who also formulated two constraints, being the assumption of a stationary total mass and an upper-limit value to density (saturated cancellous bone density). In the same vein, Jacobs et al (1997) built a model minimizing the dissipation, therefore measuring the efficiency of the translation of external loads into a variation of the internal energy. In their model based on Lagrangian optimization, the imposed constraint is a criterion to remodeling, ensuring that net bone remodeling only takes place when the criterion is violated. Building from this work, Fernandes et al (1999) proposed an approach accounting for the metabolism needed to maintain tissue: their model relies on a minimization of the sum of the compliance and the metabolic cost (assumed to be proportional to tissue mass), while respecting the principle of virtual power. On the other hand, several models rely on the mathematical expression of the evolution of directional elastic moduli. In these approaches, the evolution of the elastic moduli are dependent on directional measures of strain (Geraldes and Phillips, 2014; Geraldes et al, 2016) or stress (Miller et al, 2002; Shang et al, 2007). A different approach was proposed by Doblaré and García (2001) who described bone remodeling in a continuum damage mechanics framework, including the simultaneous evolution of the density and the directions of the material axes. In their theory, the remodeling tensor is considered as an internal variable and its evolution follows from the definition of suitable damage criteria which depend on the local principal directions of anisotropy and the principal directions of equivalent strains.

The present work introduces a 2D finite-element study of bone rotary remodeling based on a previously-introduced theoretical framework (DiCarlo et al, 2006; Martin et al, 2019). In this formulation, bone remodeling is envisaged in an enriched continuum mechanics framework with the introduction of a state variable $\boldsymbol{R}$ that accounts for the micro-rotation of bone tissue microstructure. The spatial organization of bone microstructural elements (osteons in cortical tissue and trabeculae in trabecular tissue) is supposed to lead to an orthotropic symmetry at the continuum scale. Thus, bone is assumed to behave as a linearly elastic, orthotropic material and $\boldsymbol{R}$ is meant to describe the orientation of the orthotropy axes. (Note that $\boldsymbol{R}$ cannot be considered, in general, as a surrogate of the main trabecular orientation.) Hence, DiCarlo et al proposed a constitutive equation defining the free energy as a function of the visible (macroscopic) deformation $\boldsymbol{E}$ and the micro-rotation $\boldsymbol{R}$: $\psi = \frac{1}{2}\ (\mathfrak{C} : \boldsymbol{E}) : \boldsymbol{E}$ where $\mathfrak{C}$ is the time-dependent 4th-order elastic tensor. The elastic stiffness $\mathfrak{C}$ represents the prototype elastic tensor $\mathfrak{C}_0$ (which itself is time-independent) rotated by the action of a rotation tensor $\boldsymbol{R}$, namely

$$\forall \boldsymbol{E},\ \mathfrak{C}(\boldsymbol{E}) = \boldsymbol{R}\,\mathfrak{C}_0(\boldsymbol{R}^{\mathsf{T}}\boldsymbol{E}\,\boldsymbol{R})\,\boldsymbol{R}^{\mathsf{T}}\,. \tag{4.2}$$

Additionally, DiCarlo et al postulated a passive remodeling: the outer remodeling couple (a skew-symmetric couple) is assumed to be null, meaning that no external action targeted at bone rotary remodeling is accounted for. In this scope, bone remodeling is driven by mechanical forces. Moreover, they assumed classically that the deformation $\boldsymbol{E}$ does not induce dissipation, which means that the dissipation would only come from the remodeling process.

In this context, they obtained an expression of the intrinsic dissipation $\mathcal{D}_{int}$ due to the remodeling process, reading

$$\mathcal{D}_{int} = \overset{+}{\boldsymbol{T}}(\dot{\boldsymbol{R}}\boldsymbol{R}^{\mathsf{T}}) \geq 0\,, \tag{4.3}$$

where $\overset{+}{\boldsymbol{T}} = [\boldsymbol{S}, \boldsymbol{E}]$ denotes the dissipation couple related to remodeling, $\boldsymbol{S}$ is the stress tensor, $\boldsymbol{E}$ is the small strain tensor and the brackets denote the commutator operator: $\forall\ (\boldsymbol{A},\ \boldsymbol{B}) \in\ \boldsymbol{Lin},\ [\boldsymbol{A}, \boldsymbol{B}] = \boldsymbol{AB} - \boldsymbol{BA} = 2\,\mathsf{skw}(\boldsymbol{AB})$. Assuming $\overset{+}{\boldsymbol{T}}$ to be linearly related to the remodeling rate: $\overset{+}{\boldsymbol{T}} = \mathfrak{D}\,(\dot{\boldsymbol{R}}\boldsymbol{R}^{\mathsf{T}})$, one ends up with the evolution law of the micro-rotation tensor, reading

$$\mathfrak{D}\,(\dot{\boldsymbol{R}}\boldsymbol{R}^{\mathsf{T}}) = [\boldsymbol{S}, \boldsymbol{E}]\,, \tag{4.4}$$

where $\mathfrak{D}$ represents the 4th-order dissipation tensor which can be interpreted as a resistance to remodeling.

We focus here on two benchmark problems. The first benchmark is a square with cortical bone properties on which a tensile load is applied (Figure 4.2). This example aims to investigate algorithmic convergence and dependency on several solver parameters (iteration method, meshing method and time step). The second example illustrates the proposed algorithm using a simplified 2D femur geometry and boundary conditions as per Beaupré et al (1990), see Figure 4.3. The focus is set on the rotary remodeling of the trabecular tissue. Note that the micro-orientation of the trabecular tissue, *i.e.* the orientation of the material axes of the trabecular tissue, is related to the distribution of the trabeculae but cannot be considered as a surrogate of the main trabecular orientations. In this scope, several computational questions, such as the influences of mesh, time step, approximation method, geometry and material properties, are investigated.

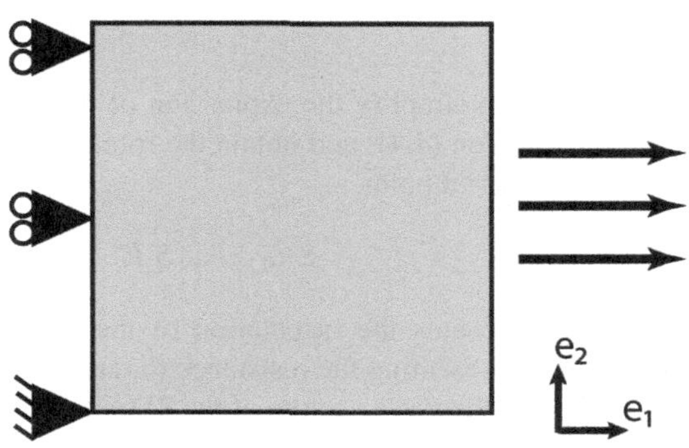

Fig. 4.2 Benchmark 1. Model of a 10 mm x 10 mm square under boundary traction: the load is distributed over a 2 mm region at the center of the right side of the square. Plane stress conditions are assumed.

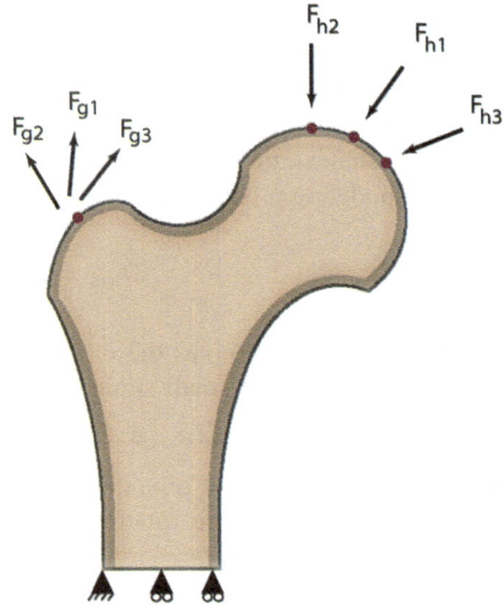

Fig. 4.3 Benchmark 2. Simplified femur model: geometry and boundary conditions. (See text for more details.)

4.2 Materials and Methods

4.2.1 Kinematics and Remodeling Law

As noted earlier, a 2D rotary remodeling law based on the energetic considerations developed by DiCarlo et al (2006) is used. Hence, in this framework of generalized continuum mechanics, the state of each material point $\boldsymbol{X}$ at time t is described by

(i) the position $\boldsymbol{x} = \widehat{\boldsymbol{x}}(\boldsymbol{X}, t)$,
(ii) the local orientation of the microstructure $\boldsymbol{R} = \widehat{\boldsymbol{R}}(\boldsymbol{X}, t)$.

Additionally, the material behavior is assumed to be linearly elastic. The elastic tensor $\mathfrak{C}$ can evolve with time, as it is dependent on the tissue orientation $\boldsymbol{R}$ (see Equation (4.2)). Thus, the stress-strain relationship reads

$$\boldsymbol{S} = \mathfrak{C} : \boldsymbol{E} \,. \tag{4.5}$$

In 2D, one can simplify the expression of the dissipation derived by DiCarlo et al (2006) (Equation (4.4)) and obtain the rotation rate as a function of the stress/strain state of a material point:

$$2\, d_0\, \dot{\alpha} = (\boldsymbol{S}\boldsymbol{E} - \boldsymbol{E}\boldsymbol{S}) : \boldsymbol{W}_{\mathrm{z}}, \tag{4.6}$$

where α represents the orientation of the material axes and d_0 is a dissipation coefficient representing the resistance to remodeling, $\boldsymbol{W}_{\mathrm{z}} := \boldsymbol{e}_2 \otimes \boldsymbol{e}_1 - \boldsymbol{e}_1 \otimes \boldsymbol{e}_2$, and $\boldsymbol{e}_1$ and $\boldsymbol{e}_2$ are the basis vectors of the 2D Euclidean space.

4.2.2 Numerical Solution

The algorithm developed to study bone remodeling relies on an iterative method depicted in Figure 4.4. At each iteration, two main steps are carried out: first, solving a boundary value problem (by means of Finite Element Analysis, FEA) while keeping fixed the material properties; then, updating the orientation of the material microstructure—to be called micro-orientation for short—and therefore the material properties.

The algorithm is initialized by defining the geometry, the boundary conditions and the initial value of the micro-orientation field. Moreover, the FE mesh and the remodeling grid are constructed. After assigning the initial orientation on the nodes of the remodeling grid, the iterative loop starts. At each step n, current material properties are calculated based on the values of the micro-orientation on the remodeling grid points $(\alpha_i^n)_{i\in\mathfrak{I}}$, the index $i \in \mathfrak{I}$ corresponding to a remodeling grid point and $\mathfrak{I}$ being the set of points of the remodeling grid. Micro-orientations are interpolated over the FE domain, delivering a continuous field of micro-orientations. The latter is then used to calculate a $\mathcal{C}^0$ continuous elastic stiffness matrix. This matrix is then read into the nodal points of the FE grid and used to compute the distribution of stress and strain fields with FEA. In the next step of the algorithm, stress and strain fields are projected onto the nodes of the remodeling grid to update the micro-orientation therein using the remodeling law (Equation (4.6)). Eventually, if the exit criterion is not met, the new values of micro-orientation are eventually calculated and made available to the FE model to restart the loop.

FE Mesh and Remodeling Grid

As depicted in Figure 4.5, the nodes of the FE mesh (black lines) may be different from those of the remodeling grid (red dots). If the nodes of the remodeling grid coincide with the FE mesh (case (a) in Figure 4.5), there is a one-to-one correspondence between the node-wise results of the FEA and the nodes where the micro-orientation is computed. Hence, the remodeling law is directly computed at the mesh nodes as a result of the FEA. Otherwise, the FE mesh can be finer than the remodeling grid (case (b) in Figure 4.5). In the latter case, the results of the FEA will be interpolated to evaluate the rotation rate at the remodeling grid points.

Boundary Value Problem

This step is carried out using a commercial FE software (Comsol Multiphysics® 5.3a) (COMSOL AB, 2017). Explicit formulas giving the relationship between each elastic modulus and micro-orientation are implemented in the finite-element model: $\mathfrak{C}(x, y) = \widehat{\mathfrak{C}}(\alpha(x, y))$, where x and y denote the 2D spatial coordinates. Micro-orientations $\alpha(x, y)$ are interpolated from the values at remodeling grid points $(\alpha_i^n)_{i\in\mathfrak{I}}$, which are updated at each step in an external file from the remodeling

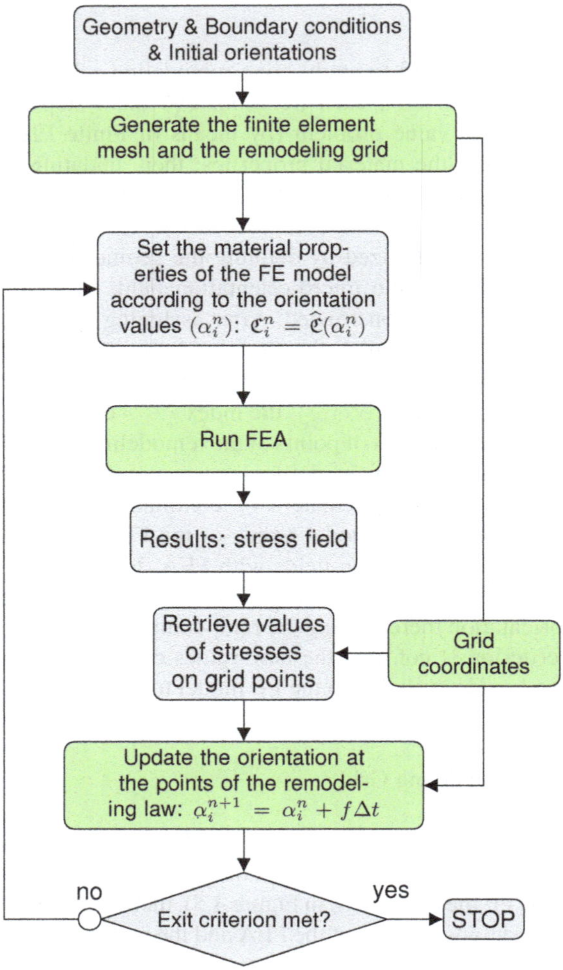

Fig. 4.4 Algorithm structure. Block colors identify computational calculation (green) or data processing (gray). "FEA" refers to finite-element analysis.

law calculation. Thus, the material properties at Gauss points $\mathfrak{C}(x_{\mathsf{G}}, y_{\mathsf{G}})$ are defined. FEA is performed using the plane stress approximation and quadratic elements.

Updating Micro-Orientation

This step is carried out using the Livelink™ for MATLAB® (Matlab Inc., 2017) module of Comsol Multiphysics® 5.3a. More precisely, at the n-th iteration ($n \in \mathfrak{N}$), the discrete version of the remodeling law Equation (4.6) in a remodeling grid point $i \in \mathfrak{I}$ reads

$$\alpha_i^{n+1} = \alpha_i^n + f\,\Delta t \qquad \forall n \in \mathfrak{N},\ \forall i \in \mathfrak{I}\,. \tag{4.7}$$

The accuracy of the above formula depends on the choice of the time step Δt and of the function f. The latter represents a discrete approximation of $\dot{\alpha}$. Two approximation methods are considered here: Euler–Cauchy (EC) and fourth-order Runge–Kutta (RK). Both of them are one-step explicit methods. Therefore, the function f depends only on the current values of the micro-orientation and strain (or, equivalently, stress), reading

$$f = f(\alpha_i^n, \boldsymbol{E}_i^n)\,, \tag{4.8}$$

where $\boldsymbol{E}_i^n$ is interpolated from the FEA strain field.

In the case of Euler–Cauchy approximation, the function f is the current value of $\dot{\alpha}$, reading

$$f_{EC}(\alpha_i^n, \boldsymbol{E}_i^n) = \dot{\alpha}_i^n = \frac{1}{2\,d_0}\left(\boldsymbol{S}_i^n\,\boldsymbol{E}_i^n - \boldsymbol{E}_i^n\,\boldsymbol{S}_i^n\right) : \boldsymbol{W}_{\mathbf{z}}\,, \tag{4.9}$$

where $\boldsymbol{S}_i^n$ is computed using Equation (4.5): $\boldsymbol{S}_i^n = \mathfrak{C}(\alpha_i^n) : \boldsymbol{E}_i^n$.

The 4th-order Runge–Kutta approximation incorporates the approximation of the micro-orientation at intermediate time steps,

$$f_{RK}(\alpha_i^n, \boldsymbol{E}_i^n) = \frac{1}{6}\left(k_1 + 2\,k_2 + 2\,k_3 + k_4\right)\,, \tag{4.10}$$

where

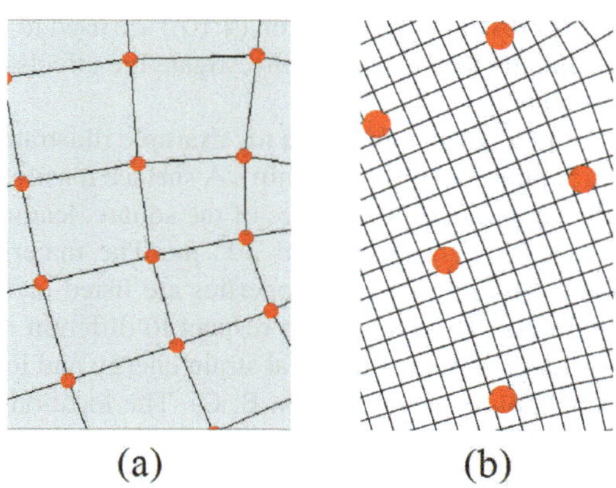

Fig. 4.5 Meshing options: (a) One grid strategy; (b) Two-grid strategy.

$$
\begin{aligned}
k_1 &= f_{EC}(\alpha_i^n, \boldsymbol{E}_i^n)\,,\\
k_2 &= f_{EC}(\alpha_i^n + \frac{k_1}{2}\Delta t, \boldsymbol{E}_i^n)\,,\\
k_3 &= f_{EC}(\alpha_i^n + \frac{k_2}{2}\Delta t, \boldsymbol{E}_i^n)\,,\\
k_4 &= f_{EC}(\alpha_i^n + k_3\Delta t, \boldsymbol{E}_i^n)\,.
\end{aligned}
\tag{4.11}
$$

Exit Criterion

Several exit criteria have been considered based on local or global convergence of mechanical quantities such as the strain energy of the system or the micro-orientation. All these criteria require to define a threshold which can be hardly identified based on biological arguments. Therefore, a simpler criterion was set, by fixing the duration of the simulation so as to observe a qualitative stabilization of the remodeling process. Therefore, the simulated physical time ranged from several months to several years according to the specific application.

4.3 Results and Discussion

In this section, the influence of several solver and modeling parameters is investigated using the two benchmarks in Figures 4.2 and 4.3. Concerning the solver parameters, the focus is set on the meshing methods and on the two key parameters used to update the micro-orientation (Equation (4.7)), namely the time step Δt and the approximation function f. If not stated differently, the FE mesh coincides with the remodeling grid (case (a) in Figure 4.5) and a $\Delta t = 1$ day time step and the Runge–Kutta approximation (Equation (4.10)) are used to update the micro-orientation. The femur model is also used to investigate the effects of the material parameters and of the geometry.

The first benchmark is the toy example illustrated in Figure 4.2, a square piece of cortical bone (10 mm × 10 mm). A surface tensile load $\Sigma_{app} = 100$ MPa is applied on the central part of the edge of the square, leading to strains $\epsilon_{\max} = \max(|\epsilon_i|) < 4000\ \mu\epsilon, \overline{\epsilon} = \mathrm{mean}(|\epsilon_i|) \simeq 10^3\ \mu\epsilon$. The material is assumed to be orthotropic and the relevant material properties are listed in Table 4.1. The dependency of the evolution of the system with respect to different meshing and iteration methods is analyzed globally via the total strain energy and locally via the micro-orientation at selected points of interest (A, B, C). The location of the latter can be visualized in Figure 4.6.

The second benchmark represents a simplified model of the human proximal femur. Geometry and boundary conditions are depicted in Figure 4.3. Arrows represent point loads resulting from joint reaction forces (F_{hi}) and hip abductor forces (F_{gi}) (see Beaupré et al (1990)). Information on the material properties and applied forces are provided in Table 4.2 and Table 4.3, respectively. The properties of the

Table 4.1 Benchmark 1. Material properties for the toy problem in Figure 4.2. (Kelvin notation is used for the elastic coefficients.)

Elastic coefficients	GPa
C_{11}	30
C_{22}	20.85
C_{12}	11.49
C_{66}	13.2
Dissipation coefficient	kPa day
d_0	555

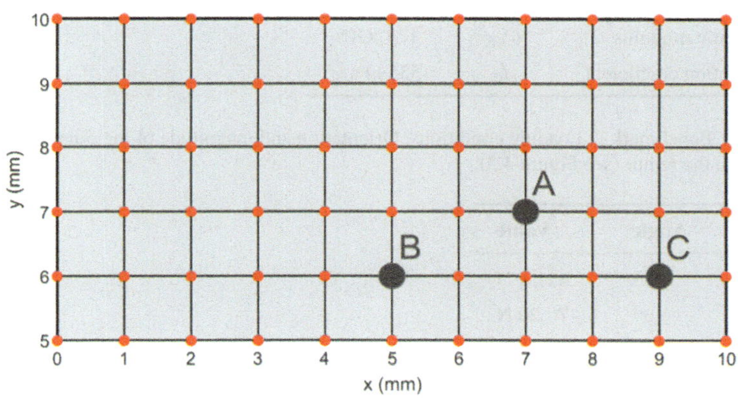

Fig. 4.6 Benchmark 1. 10x10 remodeling grid of the square under boundary traction (see Figure 4.2, only the upper half is represented here) and location of points of interest (A, B, C).

outer shell of the femur (brown region, in Figure 4.3) were linearly interpolated between their values at the external boundary (cortical properties as per Table 4.2, not evolving with time) and the properties at the inner boundary (elastic properties of the trabecular bone as per Table 4.2, evolving with time), unless otherwise specified.

4.3.1 Meshing Strategy

In this subsection, different approaches are investigated for the evaluation of the strain/stress at the remodeling grid points. In the first case (named 1M in the figures), the remodeling grid coincides with the FE nodes (Figure 4.5(a)). Thus, stress/strain are directly transferred from the FE model to the remodeling grid. In a second strategy (named 2M in the figures), stress/strain in the nodes of the remodeling grid

Table 4.2 Benchmark 2. Material properties of cortical and trabecular bone (plane stress). (Kelvin notation is used for the elastic coefficients.)

Material property	Symbol	Value
Cortical bone		
Young modulus	E_c	14 GPa
Poisson's ratio	ν_c	0.3
Trabecular bone		
Elastic modulus	C_{11}	2.50 GPa
Elastic modulus	C_{22}	1.42 GPa
Elastic modulus	C_{12}	636 MPa
Elastic modulus	C_{66}	1.20 GPa
Dissipation coefficient	$\tilde{d}_0$	555 kPa τ

Table 4.3 Benchmark 2. Loading conditions: Orientation and magnitude of the point forces applied on the femur (see Figure 4.3).

Force	Angle	Value
$\mathbf{F_{g1}}$	28°	421.2 N
$\mathbf{F_{g2}}$	−8°	70.20 N
$\mathbf{F_{g3}}$	35°	93.60 N
$\mathbf{F_{h1}}$	24°	−1390 N
$\mathbf{F_{h2}}$	−15°	−213.6 N
$\mathbf{F_{h3}}$	56°	−309.6 N

points (the coarse grid, corresponding to the red dots in Figure 4.5(b)) are obtained by interpolating the stress/strain fields obtained from the FEA. (Runge–Kutta method and a time step $\Delta t = 1$ day are used to update the micro-orientation.)

In Figure 4.7, the effect of the above meshing strategies are evaluated with respect to Benchmark 1. The 1M and 2M meshing strategies are compared, including several 2M FE mesh options of increasing densities, from a mesh 2.5 times finer than the remodeling grid to 10 times finer. Results obtained with the 1M strategy are represented with solid lines in Figure 4.7 whereas the results of the 2M strategy are represented with dotted (2.5 x), dash-dotted (5 x), and dashed (10 x) lines.

The meshing strategy affects the evolution of the system to different extents. The effect is globally small with respect to the local micro-orientation in points A, B, and C (Figure 4.7(a)). However, larger differences exist when looking at the total strain energy (Figure 4.7(b)). These differences are mainly due to the mesh dependency of the FE solution in the singular points of the system (data not shown).

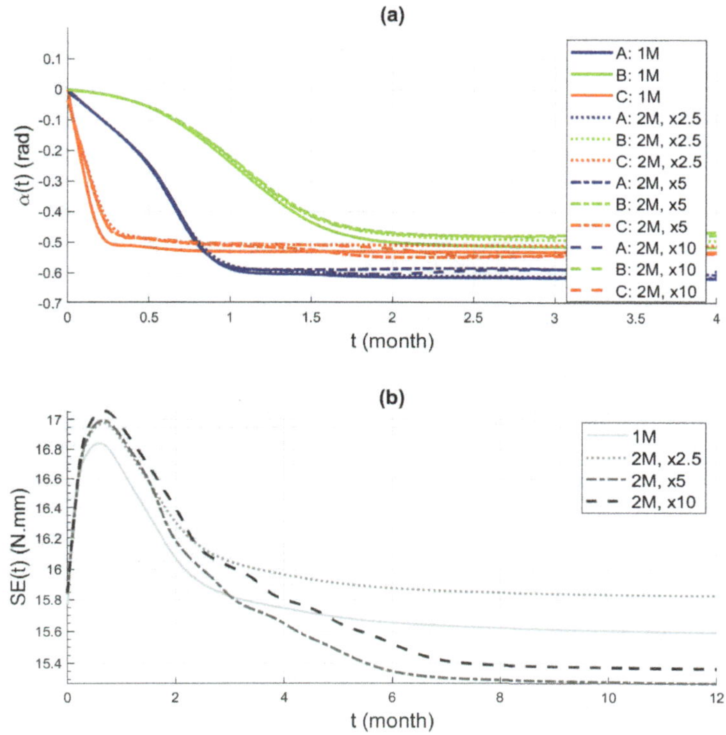

Fig. 4.7 Benchmark 1. Evolutions of (a) micro-orientations in points A, B and C and (b) total strain energy (SE) with respect to time, for varying mesh strategies (one mesh (1M) and two meshes (2M)).

4.3.2 Quadrature of Micro-Rotation Rate

This section assesses the role of the approximation scheme (function f) and time discretization (time step Δt) in the quadrature of the micro-rotation rate, Equation (4.7). (The 1M meshing strategy is used throughout this subsection.)

Figure 4.8 shows the results of a sensitivity analysis with respect to Benchmark 1. Simulations were undertaken with time steps Δt ranging from 0.5 day to 2 days. In parallel, two different approximation schemes were used, namely fourth-order Runge–Kutta (RK, (a)) and Euler–Cauchy (EC, (b)). The evolution of the micro-orientation at the points of interest A, B and C (blue, green, and red lines, respectively) is shown for three different time steps $\Delta t \in \{0.5, 1, 2\}$ day (dotted, dashed and solid lines, respectively) and the RK and EC quadrature schemes ((a) and (b), respectively). As expected, the RK scheme performs better than the EC scheme.

The former renders a stable evolution of the micro-orientation even for large time steps and the behavior of the system is approximately the same, irrespective of the time step (see the dashed lines in any of the points of interest). By contrast, the EC scheme leads to numerical instabilities for large time steps (see the bright red solid line in Figure 4.8 corresponding to point C and a time step $\Delta t = 2$ day).

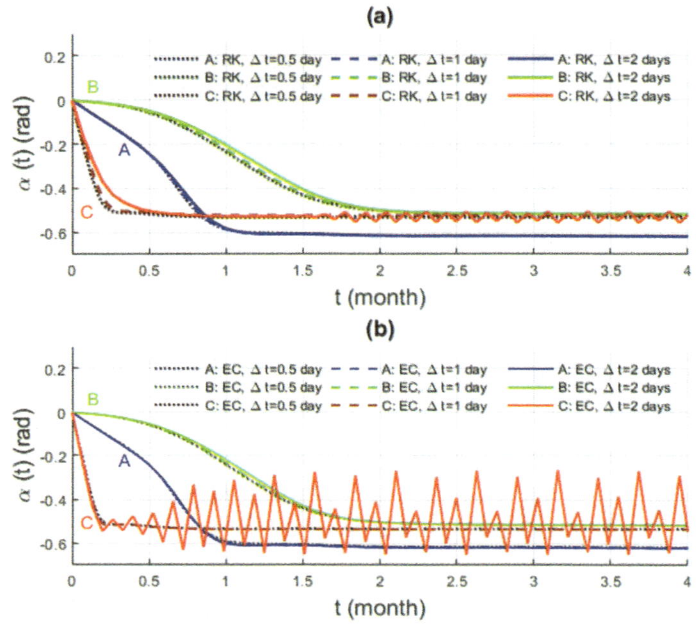

Fig. 4.8 Benchmark 1. Evolutions of the micro-orientation in points A, B and C for varying time steps and approximation methods (EC and RK). Dotted, dashed and solid lines correspond to time steps $\Delta t = 0.5,\ 1,\ 2$ day, respectively. Subfigure (a) corresponds to the Runge–Kutta approximation (RK) and subfigure (b) to the Euler–Cauchy approximation (EC).

A similar analysis was performed on the femur model of Benchmark 2. The time step was varied between from 0.5 to 20 units time (τ) and both EC and RK quadrature schemes were used.

The evolution of the system did not vary significantly when imposing different approximation and time discretization methods from 0.5 to 20 time units. The results of the remodeling simulations are displayed in Figure 4.9, which illustrates the finite-element mesh (a), initial micro-orientation field (b), and micro-orientation fields after 5000 (c) and 10000 (d) time units. Note that, as expected, the strain energy map changes with time (Figure 4.9(b-d)).

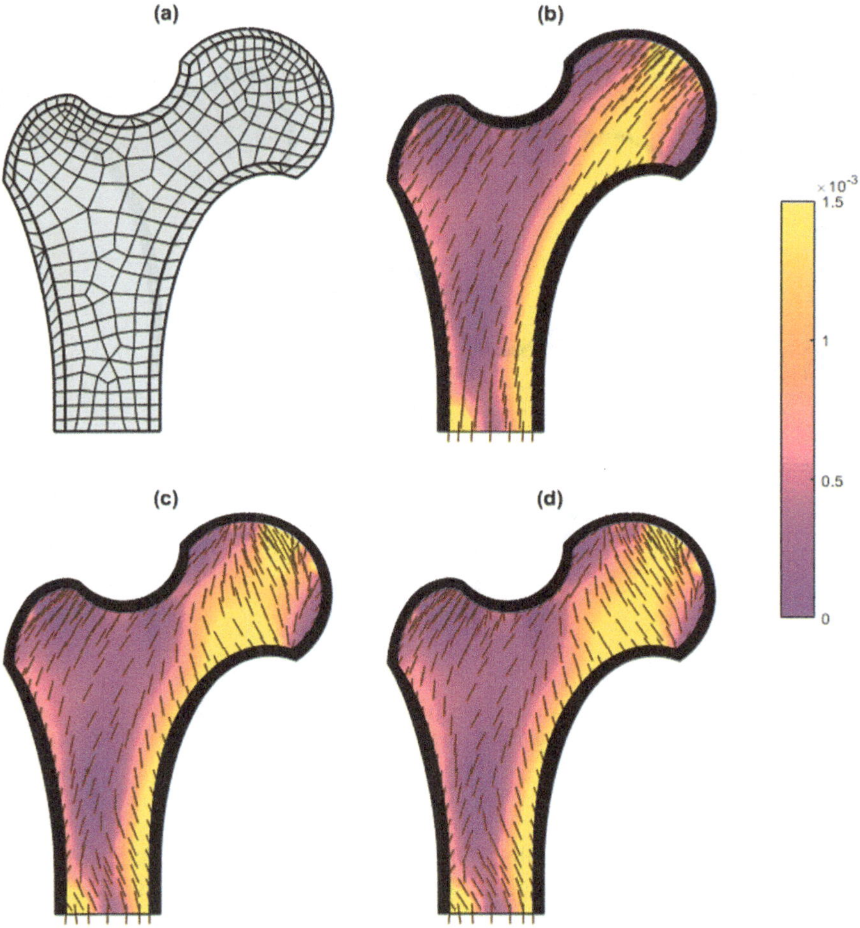

Fig. 4.9 Evolution of the micro-orientation (brown sticks) in the femur model: (a) finite-element mesh; (b) initial micro-orientation field; (c) micro-orientation field after 5000 time units; (d) micro-orientation field after 10000 time units. Color map: strain energy density (MPa).

4.3.3 Geometry

In this subsection, the effect of the presence of the marrow cavity on the remodeling process in the proximal femur is investigated. Therefore, an elliptical cavity was added in the diaphysis of the original femur model to account for the marrow cavity. In one case (Figure 4.10(b)), the cavity was assumed empty. In the other case (Figure 4.10(c)), the marrow cavity was filled with a linearly elastic, almost incompressible isotropic material (E_m = 10 kPa, ν_m = 0.499) mimicking the presence of the marrow.

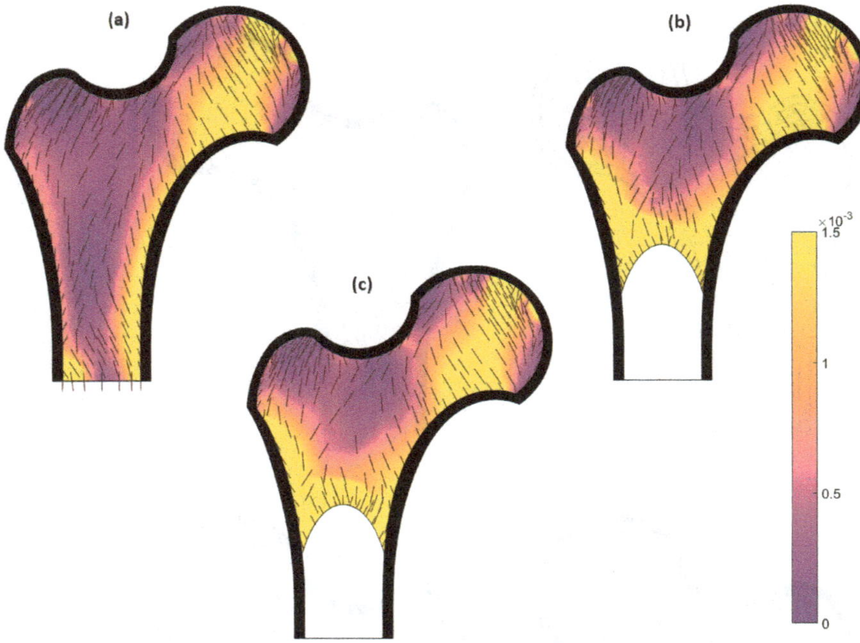

Fig. 4.10 Micro-orientations (brown sticks) in the femur model after 10000 units time for varying geometries: (a) original model; (b) empty marrow cavity; (c) filled marrow cavity. Color map: strain energy density (MPa).

Figure 4.10 displays the results of the previously-mentioned simulations as well as the original femur model (Figure 4.10(a)). These simulations were performed with a time step of 1 time unit and with the 1M method. In all cases, the initial micro-orientation field was assumed as in Figure 4.9(b). While there is a significant discrepancy in the results when introducing the marrow cavity in the geometry, the specific values of the material properties of the marrow space do not seem to notably impact the micro-orientation field in the femur (comparison of Figures 4.10(b) and (c)).

4.3.4 Material Properties

Material properties of bone tissue play a substantial role in remodeling. In particular, the modification of the shear modulus of trabecular bone may change the remodeling response dramatically. Hence, one can also infer that a change in mechanical loading is not the only trigger to a change in bone remodeling activity: changes in bone structure and composition also prompt a different bone cell activity, as the mechanical environment changes. Here, this issue is addressed by considering a modification of the shear modulus C_{66} which is set to twice its original value: $C_{66} = 2.4\,\mathrm{GPa}$.

A different shear modulus can modify the material response to loading in terms of micro-rotation. The influence of material properties on rotary remodeling has been investigated by Martin et al (2019) considering a uniaxial, stationary load. In that scope, the remodeling response has been shown to strongly depend on the sign of the function f_3 defined hereafter:

$$f_3(\{C_{ij}\}) = 2\,C_{44}C_{55}(\sqrt{2}C_{13}^2C_{22} + \sqrt{2}C_{12}^2C_{33} + \sqrt{2}C_{11}(C_{23}^2 - C_{22}C_{33}) \\ -C_{23}^2C_{66} + C_{12}C_{33}C_{66} + C_{22}C_{33}C_{66} - C_{13}C_{23}(2\sqrt{2}C_{12} + C_{66})).$$

The function f_3 governs the tendency of the microstructure to align with the principal stress directions. In particular, it has been shown by Martin et al (2019) that a necessary condition for the micro-orientation to align with the 1D stress direction is the function f_3 to take positive value.

The analytical results of Martin et al (2019) cannot be applied straightforwardly to the femur model since the stress state is more complex and evolves in time. However, it is reasonable to expect the shear modulus to affect the remodeling process by favoring the alignment of the material axes with the principal stress directions when f_3 is positive. Actually, the function f_3 takes negative value when using the elastic coefficients in Table 4.2. However, f_3 takes positive value assuming $C_{66} = 2.4\,\mathrm{GPa}$.

Remodeling of the trabecular tissue can be also affected by the elasticity of the surrounding cortical shell. A linear transition between the trabecular and cortical material properties reported in Table 4.2 was postulated through the outer shell of the femur. This research hypothesis was investigated by smoothing the trabecular-to-cortical transition. To this end, a parabolic variation of the elastic moduli through the cortical shell was also considered. The elastic moduli variations are displayed in Figure 4.11.

Figure 4.12 depicts the material orientations after 10000 time units, using a time step of 1 time unit, a 1M meshing method and varying material properties. The material properties of the original model are shown in the subplot (a). Subplot (b) shows the results obtained when the shear modulus of the trabecular is increased ($C_{66} = 2.4\,\mathrm{GPa}$). One can note the significant changes in the micro-orientation field, in particular at the femoral head. Subplot (c) shows the results obtained by considering a quadratic variation of the elastic moduli through the cortical shell. It is apparent that there are no major changes to the orientations associated with this choice.

4.4 Conclusion

This study on numerical solutions of benchmark problems shows the potential of the proposed algorithm and the relevance of the questions of mesh, time discretization, geometry and material properties. These questions are closely related to the problem

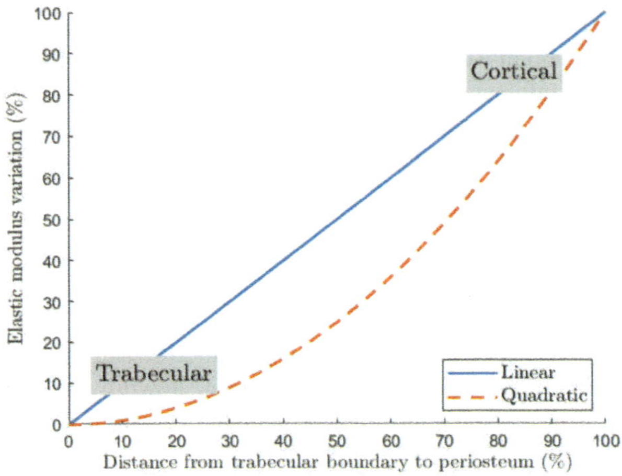

Fig. 4.11 Evolution of the elastic moduli: trabecular-to-cortical transition in the linear and quadratic interpolation cases.

studied, in particular when dealing with complex *in vivo* geometries, where one must find a balance between model convergence and physical meaning. One of the findings of this study is the importance of the definition of material properties, in accordance with earlier theoretical developments (Martin et al, 2019).

The results of this study will be useful in tackling the next developments of our remodeling model. A first development consists in coupling the (macroscopic) remodeling model with a (microscopic) mechanobiology model. This is an ongoing project which introduces new challenges on both modeling and numerics. From this latter point of view, numerical issues are related to the sensitivity and stability of the mechanobiology model. Further developments concern the clinical application of our model, *e.g.* to understand bone remodeling around the implants or in bone pathologies such as adolescent idiopathic scoliosis (AIS) or osteoporosis. This will introduce additional challenges such as dealing with patient-specific geometries. Besides the computational cost and the possible convergence issues related to 3D numerical simulations of real systems, the proposed algorithm provides a useful numerical framework to tackle these studies as it allows to couple FEA and bone remodeling laws in a straightforward way.

To conclude, it should be noted that in this work we considered only *passive* rotary remodeling, *i.e.* a rotation of the material axes triggered by a local mechanical stimulus. Recent works pointed out the relevance of including non-local mechanical and biological stimuli in bone remodeling (Lekszycki and dell'Isola, 2012; Giorgio et al, 2016; Spingarn et al, 2017; Bagherian et al, 2019; George et al, 2019; Giorgio et al, 2019; Kazempour et al, 2019; Sheidaei et al, 2019). Investigating the effects of non-local multi-physics stimuli is an interesting topic that will be tackled in the future.

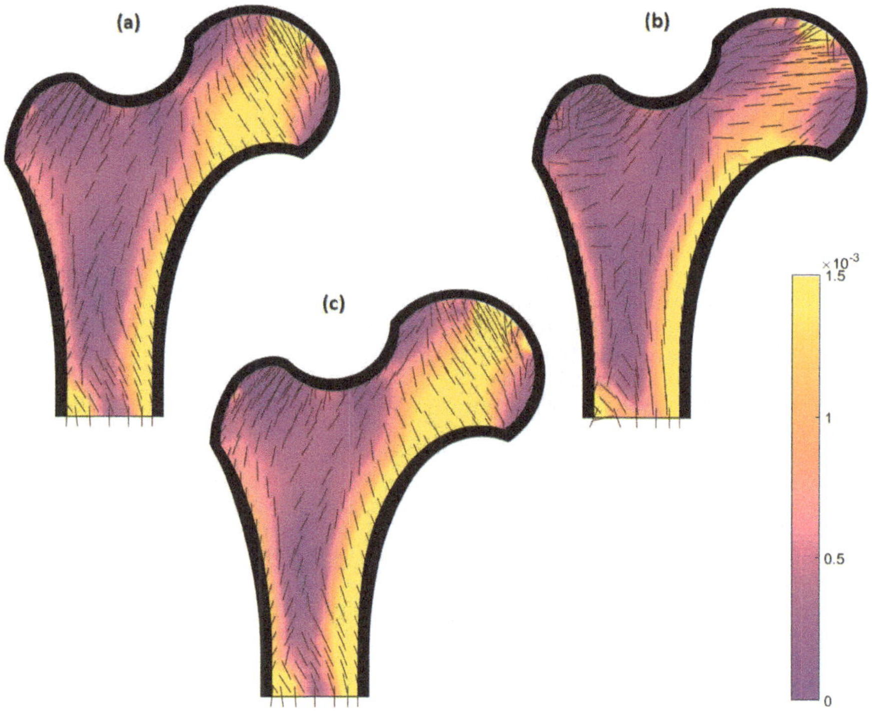

Fig. 4.12 Micro-orientations (brown sticks) in the femur model after 10000 units time for varying material properties: (a) original model; (b) modified trabecular elastic modulus ($C_{66} = 2.4$ GPa); (c) quadratic interpolation between the boundaries of the cortical shell. Color map: strain energy density (MPa).

Acknowledgements This work was partially supported by the IRP CNRS Coss&Vita and from the European Research Council (ERC) under the European Union's Horizon 2020 research and innovation program (grant agreement No 682001, project ERC Consolidator Grant 2015 BoneImplant).

References

Bagherian A, Baghani M, George D, Rémond Y, Chappard C, Patlazhan S, Baniassadi M (2019) A novel numerical model for the prediction of patient dependent bone density loss in microgravity based on micro-CT images. Continuum Mechanics and Thermodynamics

Beaupré GS, Orr TE, Carter DR (1990) An approach for time-dependent bone modeling and remodeling-application: A preliminary remodeling simulation. Journal of Orthopaedic Research 8(5):662–670

COMSOL AB (2017) Comsol Multiphysics® (5.3a). www.comsol.com

Cowin SC (1986) Wolff's Law of Trabecular Architecture at Remodeling Equilibrium. Journal of Biomechanical Engineering 108(1):83

Cowin SC (1995) On the minimization and maximization of the strain energy density in cortical bone tissue. Journal of Biomechanics 28(4):445–447

Cowin SC, Sadegh AM, Luo GM (1992) An Evolutionary Wolff's Law for Trabecular Architecture. Journal of Biomechanical Engineering 114(1):129

Culmann K (1866) Die graphische Statik. Meyer & Zeller (A. Reimann), Zurich

DiCarlo A, Naili S, Quiligotti S (2006) Sur le remodelage des tissus osseux anisotropes. Comptes Rendus Mécanique 334(11):651–661

Doblaré M, García JM (2001) Application of an anisotropic bone-remodelling model based on a damage-repair theory to the analysis of the proximal femur before and after total hip replacement. Journal of Biomechanics 34(9):1157–1170

Doblaré M, García JM (2002) Anisotropic bone remodelling model based on a continuum damage-repair theory. Journal of Biomechanics 35(1):1–17

Fernandes P, Rodrigues H, Jacobs CR (1999) A Model of Bone Adaptation Using a Global Optimisation Criterion Based on the Trajectorial Theory of Wolff. Computer Methods in Biomechanics and Biomedical Engineering 2(2):125–138

Fyhrie DP, Carter DR (1986) A unifying principle relating stress to trabecular bone morphology. Journal of Orthopaedic Research 4(3):304–317

George D, Allena R, Rémond Y (2019) Integrating molecular and cellular kinetics into a coupled continuum mechanobiological stimulus for bone reconstruction. Continuum Mechanics and Thermodynamics 31(3):725–740

Geraldes DM, Phillips ATM (2014) A comparative study of orthotropic and isotropic bone adaptation in the femur. International Journal for Numerical Methods in Biomedical Engineering 30(9):873–889

Geraldes DM, Modenese L, Phillips AT (2016) Consideration of multiple load cases is critical in modelling orthotropic bone adaptation in the femur. Biomechanics and Modeling in Mechanobiology 15(5):1029–1042

Giorgio I, Andreaus U, Scerrato D, dell'Isola F (2016) A visco-poroelastic model of functional adaptation in bones reconstructed with bio-resorbable materials. Biomechanics and Modeling in Mechanobiology 15(5):1325–1343

Giorgio I, dell'Isola F, Andreaus U, Alzahrani F, Hayat T, Lekszycki T (2019) On mechanically driven biological stimulus for bone remodeling as a diffusive phenomenon. Biomechanics and Modeling in Mechanobiology 18:1639–1663

Jacobs CR, Simo JC, Beaupré GS, Carter DR (1997) Adaptive bone remodeling incorporating simultaneous density and anisotropy considerations. Journal of Biomechanics 30(6):603–613

Kazempour M, Bagherian A, Sheidaei A, Baniassadi M, Baghani M, Rémond Y, George D (2019) Numerical Simulation of Osteoporosis Degradation at Local Scale: A Preliminary Study on the Kinematic Loss of Mechanical Bone Stiffness and Microstructure. Biomedical and Health Research 79:86–93

Koch JC (1917) The laws of bone architecture. American Journal of Anatomy 21(2):177–298

Lanyon LE (1974) Experimental support for the trajectorial theory of bone structure. Journal of Bone and Joint Surgery - Series B 56(1):160–166

Lekszycki T, dell'Isola F (2012) A mixture model with evolving mass densities for describing synthesis and resorption phenomena in bones reconstructed with bio-resorbable materials. ZAMM-Zeitschrift für Angewandte Mathematik und Mechanik 92(6):426–444

Luo JH, Gea HC (1998) Optimal orientation of orthotropic materials using an energy based method. Structural Optimization 15(3-4):230–236

Luo ZP, An KN (1998) A theoretical model to predict distribution of the fabric tensor and apparent density in cancellous bone. Journal of Mathematical Biology 36(6):557–568

Martin M, Pivonka P, Haiat G, Lemaire T, Sansalone V (2019) Bone orthotropic remodeling as a thermodynamically-driven evolution. Journal of Mechanics in Medicine and Biology *(accepted)*

Matlab Inc (2017) MATLAB® (R2017b). www.mathworks.com

von Meyer GH (1867) Die Architektur der Spongiosa. (Zehnter Beitrag zur Mechanik des menschlichen Knochgerüstes). Archiv fur Anatomie, Physiologie, und wissenschaftliche Medizin 6:615–628

Miller Z, Fuchs MB, Arcan M (2002) Trabecular bone adaptation with an orthotropic material model. Journal of Biomechanics 35(2):247–256

Pettermann HE, Reiter TJ, Rammerstorfer FG (1997) Computational simulation of internal bone remodeling. Archives of Computational Methods in Engineering 4(4):295–323

Roux W (1881) Der Kampf der Teile des Organismus. Engelmann, Leipzig

Sansalone V, Naili S, Di Carlo A (2011) On the rotary remodelling equilibrium of bone. Computer Methods in Biomechanics and Biomedical Engineering 14(sup1):203–204

Shang Y, Peng L, Bai J (2007) A Three-dimensional Orthotropic Adaptation Algorithm for Femur Remodeling. In: Magjarevic R, Nagel JH (eds) World Congress on Medical Physics and Biomedical Engineering 2006, Springer Berlin Heidelberg, Berlin, Heidelberg, pp 3038–3040

Sheidaei A, Kazempour M, Hasanabadi A, Nosouhi F, Pithioux M, Baniassadi M, Rémond Y, George D (2019) Influence of bone microstructure distribution on developed mechanical energy for bone remodeling using a statistical reconstruction method. Mathematics and Mechanics of Solids 24(10):3027–3041

Skedros JG, Baucom SL (2007) Mathematical analysis of trabecular 'trajectories' in apparent trajectorial structures: The unfortunate historical emphasis on the human proximal femur. Journal of Theoretical Biology 244(1):15–45

Spingarn C, Wagner D, Rémond Y, George D (2017) Multiphysics of bone remodeling: a 2D mesoscale activation simulation. Bio-medical Materials and Engineering 28(s1):S153–S158

Watzky A, Naili S (2004) Orthotropic bone remodeling: case of plane stresses. Mechanics Research Communications 31(5):617–625

Wolff J (1870) Ueber die innere Architectur der Knochen und ihre Bedeutung für die Frage vom Knochenwachsthum. Archiv für Pathologische Anatomie und Physiologie und für Klinische Medizin 50(3):389–450

Wolff J (1892) Das Gesetz der Transformation der Knochen. Berlin, Germany

Chapter 5
Data-Driven Simulation for Augmented Surgery

Andrea Mendizabal, Eleonora Tagliabue, Tristan Hoellinger, Jean-Nicolas Brunet, Sergei Nikolaev, and Stéphane Cotin

Abstract To build an augmented view of an organ during surgery, it is essential to have a biomechanical model with appropriate material parameters and boundary conditions, able to match patient specific properties. Adaptation to the patient's anatomy is obtained by exploiting the image-rich context specific to our application domain. While information about the organ shape, for instance, can be obtained preoperatively, other patient-specific parameters can only be determined intraoperatively. To this end, we are developing data-driven simulations, which exploit information extracted from a stream of medical images. Such simulations need to run in real-time. To this end we have developed dedicated numerical methods, which allow for real-time computation of finite element simulations.

The general principle consists in combining finite element approaches with Bayesian methods or deep learning techniques, that allow to keep control over the underlying computational model while allowing for inputs from the real world. Based on *a priori* knowledge of the mechanical behavior of the considered organ, we select a constitutive law to model its deformations. The predictive power of such constitutive law highly depends on the knowledge of the material parameters and the boundary conditions. In our first approach, material properties are modeled as stochastic parameters whose probability distributions will be modified in real-time using Kalman filters, given observations extracted from intraoperative data. The second option we propose is to directly learn material parameters and boundary conditions from

---

A. Mendizabal
Inria and University of Strasbourg, Strasbourg, France
e-mail: andrea.mendizabal@inria.fr

E. Tagliabue
Università degli Studi di Verona
e-mail: eleonora.tagliabue@univr.it

T. Hoellinger, J-N. Brunet, S. Nikolaev, S. Cotin
Inria, Strasbourg, France
e-mail: tristan.hoelling@inria.fr, jean-nicolas.brunet@inria.fr,
sergei.nikolaev@inria.fr, stephane.cotin@inria.fr

B. E. Abali and I. Giorgio (eds.), *Developments and Novel Approaches in Biomechanics and Metamaterials*, Advanced Structured Materials 132,
https://doi.org/10.1007/978-3-030-50464-9_5

patient-specific data using deep neural networks. This has been applied to the modeling of liver biomechanics, its real-time simulation, and parametrization to achieve patient-specific augmented reality during

Keywords: Data-driven simulation · Bayesian filtering · Deep neural networks · Finite element method · Augmented reality

5.1 Introduction

In computer-aided intervention, the correct alignment of the preoperative images to the intraoperative ones remains a real challenge especially when large deformations are involved. In the context of hepatic surgery for instance, the objective is to accurately locate the internal structures such as tumors and blood vessels (that need to be preserved for the post-operative regeneration of the liver tissue). While the initial position of these structures is known from the preoperative images, their actual position during surgery is often hidden or uncertain. To guide the surgeon, augmented reality techniques are used to enrich visual information through fusion of intraoperative images and a preoperative 3D model of the patient's anatomy. This is usually done by overlaying a virtual representation of the liver built from preoperative images over intraoperative images or through augmented reality glasses. However, surgical manipulations and interactions with the surrounding anatomy can induce significant deformations to the patient's liver. As a consequence, the virtual model of the liver has to account for non-rigid transformations and produce its deformed state in real-time, which is difficult given the complexity of the physical systems needed for accurate biomechanical modeling.

Existing works in this area rely on patient-specific biomechanical models that can provide in-depth motion given surface deformation (Clements et al, 2008; Haouchine et al, 2013; Suwelack et al, 2014; Alvarez et al, 2018). The Finite Element (FE) method is the preferred one due to its ability to numerically solve the complex partial differential equations that come into play. However, the demanding accuracy of medical applications (e.g. registration of internal structures below 5 mm (Ruiter et al, 2006)) raise several challenges that are far from being solved.

In augmented surgery, the computational efficiency of the FE method becomes crucial. In the case of augmented hepatic surgery, intraoperative images are acquired at about 20 Hz leading to update times of less than 50 ms. During this small amount of time, acquisition and processing of the images as well as model update need to take place. As a result, FE computation times should require less than 30 ms. If only small deformations take place, achieving such computation times is feasible (Meier et al, 2005). However, if large non-linear deformations happen, computation times become incompatible with such time constraints. A solution might be the use of the co-rotational FE method, where geometrical non-linearities can be handled in real-time (Haouchine et al, 2013; Petit and Cotin, 2018). Nevertheless, when more complex biomechanical models need to be used these optimizations no longer hold.

Alternative solutions have been proposed with different trade-offs regarding the ratio between computation time and model accuracy (Peterlík et al, 2012; Suwelack et al, 2014; Modrzejewski et al, 2018; Niroomandi et al, 2008; Johnsen et al, 2015; Allard et al, 2007). Marchesseau et al (2010) proposed the Multiplicative Jacobian Energy Decomposition (MJED) that allows for fast and realistic liver deformations including hyperelasticity, porosity and viscosity. Also, Miller et al (2007) introduced Total Lagrangian explicit dynamics (TLED) which can achieve real-time performances when coupled with explicit time integration and GPU-based solvers (Joldes et al, 2010).

More recently, another class of methods made use of machine learning (ML) algorithms to solve the deformed state of a model (Lorente et al, 2017; Roewer-Despres et al, 2018; Tonutti et al, 2017; Pellicer-Valero et al, 2020). Such ML models are often trained with synthetic data generated by the FE method.While the offline training phase can be computationally expensive, the online predictions satisfy real-time compliance and can provide very accurate estimations of the displacement.

Furthermore, to guarantee the high level of precision required, accurate modeling adapted to the patient anatomy needs to be pursued. The first step towards patient-specific modeling is the patient-specific geometry of the organ. Generally, the 3D anatomical model of the organ is constructed from preoperative volumetric medical images such as CT scan or MRI without too much difficulty. Moreover, boundary conditions (BCs) are essential for the FE method to produce accurate results (Bosman et al, 2014). The location and the elastic properties of the BCs are also patient-specific but are not visible on preoperative images. The partial intraoperative images give inaccurate information about the BCs that are often out of the field of view of the laparoscopic camera. In addition, the correct identification of the patient-specific elasticity parameters is essential for an accurate estimation of the deformation of the considered tissue. Note that the values of these parameters are intrinsic to the choice of the constitutive law.

To ensure the aforementioned requirements in terms of model parameterization and computational efficiency, we combine FE approaches with either Bayesian methods or deep learning techniques, in order to keep control over the underlying computational model while allowing for intraoperative inputs. We first propose an image-driven stochastic assimilation method to identify the BCs on the one hand, and the elasticity parameters on the other hand. Second, we present a data-driven deep neural network that learns the desired biomechanical model including its BCs and material parameters, to predict complex non-linear deformations in real-time. This chapter is divided in three main segments. First of all, the biomechanical formulation of the general problem we want to solve is presented, with an emphasis on the role of each parameter in the predictive power of such models. Next, we look at the importance of the correct estimation of patient specific boundary conditions and material parameters, which are identified based on real observations using Kalman filtering. In the third section, we go a step further by directly learning from data the mechanical behavior of a liver through deep neural networks. Lastly, the learned model is adapted to patient specific properties through transfer learning.

5.2 Numerical Simulation of Hyperelasticity Problems

5.2.1 Hyperelasticity of Soft Tissues

The deformable behavior of soft tissues can be described following the laws of continuum mechanics. Hyperelastic formulations are usually exploited to characterize biological materials undergoing large deformations, which cannot be accurately handled by linear models (Delingette and Ayache, 2004). Our reference problem corresponds to the boundary value problem of computing the deformation of a hyperelastic material under both Dirichlet and Neumann boundary conditions. The solid occupies a volume Ω whose boundary is Γ. We assume the Dirichlet conditions on Γ_D, a subset of Γ, known *a priori*, while Neumann boundary conditions on Γ_N can vary at any time step. Relying on the Lagrangian formulation, the relationship between the deformed $\mathbf{x}$ and undeformed state $\mathbf{X}$ of each point of the solid reads as

$$\mathbf{x} = \mathbf{X} + \mathbf{u} \tag{5.1}$$

where $\mathbf{u}$ is the displacement field. Throughout this chapter, we describe material behavior with the Saint-Venant–Kirchhoff constitutive model, which is the simplest generalization of the linear model for large displacements. The Green–Lagrange strain tensor $\mathbf{E} \in \mathbf{R}^{3\times3}$ is computed as a non-linear (quadratic) function of the deformation gradient $\mathbf{F} = \mathbf{I} + \nabla_{\mathbf{X}}\mathbf{u}$,

$$\mathbf{E} = \frac{1}{2}(\mathbf{F}^T\mathbf{F} - \mathbf{I}) \tag{5.2}$$

where $\mathbf{I} \in \mathbf{R}^{3\times3}$ is the identity matrix. The strain-energy density function $\mathbf{W}$ for a St. Venant–Kirchhoff material is obtained according to the following equation:

$$\mathbf{W}(\mathbf{E}) = \frac{\lambda}{2}[\mathrm{tr}(\mathbf{E})]^2 + \mu\,\mathrm{tr}(\mathbf{E}^2) \tag{5.3}$$

where λ and μ are the material parameters called Lame's constants, derived from the Young's modulus Y and the Poisson's ratio ν such that

$$\begin{aligned} \lambda &= \frac{Y\nu}{(1+\nu)(1-2\nu)}, \\ \mu &= \frac{Y}{2(1+\nu)}. \end{aligned} \tag{5.4}$$

The constitutive law is then obtained by differentiating $\mathbf{W}$ with respect to $\mathbf{E}$,

$$\mathbf{S} = \frac{\partial \mathbf{W}(\mathbf{E})}{\partial \mathbf{E}} = [\lambda \mathrm{tr}(\mathbf{E})\mathbf{I} + 2\mu\mathbf{E}] : \mathbf{E} \tag{5.5}$$

where $\mathbf{S}$ is the second Piola–Kirchhoff stress. $\mathbf{S}$ is related to the first Piola–Kirchhoff stress tensor $\mathbf{P}$ by $\mathbf{P} = \mathbf{FS}$.

Ignoring time-dependent terms, the boundary value problem formulated in material coordinates (i.e., considering kinematic quantities with respect to the undeformed geometry) is then given by

$$\begin{cases} \nabla(\mathbf{FS}) = \mathbf{b} \ \text{ in } \Omega \\ \mathbf{u}(\mathbf{X}) = 0 \ \text{ on } \Gamma_D \\ (\mathbf{FS})\mathbf{n} = \mathbf{t} \ \text{ on } \Gamma_N \end{cases} \tag{5.6}$$

where **b** gathers the external body forces, **n** is the unit normal to Γ_N and **t** is the traction applied to the boundary Γ_N. The weak form of (5.6), obtained from the principle of virtual work, brings forward the boundary term and reads as

$$\int_{\Omega} \mathbf{S} : \delta\mathbf{E} \, d\Omega = \int_{\Omega} \mathbf{b}\,\eta \, d\Omega + \int_{\Gamma_N} \mathbf{t}\,\eta \, d\Gamma \tag{5.7}$$

where $\delta\mathbf{E} = \frac{1}{2}(\mathbf{F}^T\nabla\eta + \nabla^T\eta\mathbf{F})$ is the variation of the strain, and $\eta = \{\eta \in H^1(\Omega) \mid \eta = 0 \text{ on } \Gamma_D\}$ is any vector-valued test function ($H^1(\Omega)$ being a Hilbert space). The left side of equation (5.7) denotes the internal virtual work, and the right side, the virtual work from the applied external load.

5.2.2 Finite Element Method

A typical approach to find a numerical solution to equation (5.7) is the finite element method (FEM). FEM relies on a discretization of the domain into a finite number of elements, usually hexahedral (H8) or tetrahedral (T4). The displacement of each point in the volume is represented as a function of the displacement values at the element nodes. The methods we propose in the following rely on either H8 or T4 elements. H8 elements are known to have better convergence and stability, but it is difficult to use them to describe irregular shapes (Shepherd and Johnson, 2008). On the contrary, T4 elements can fit complex geometries, but can be highly inaccurate in the computation of stresses and strains (Benzley et al, 1995). Therefore, tetrahedral meshes are the main approach used for solid organs.

Due to the non-linearity of equation (5.2), the unknown displacements are obtained as the solution of a non-linear system of equations. Using an iterative Newton–Raphson method, from an initial displacement $\mathbf{u}^0$, we try to find a correction δ_u^n after n iterations that balances the linearized set of equations:

$$\dot{\mathbf{K}}^{n-1}\delta_u^n = \mathbf{r}(\mathbf{u}^0 + \delta_u^{n-1}) + \mathbf{b} \tag{5.8}$$

where $\dot{\mathbf{K}}$ is the tangent stiffness matrix and **r** is the internal elastic force vector. In order to solve the linearized system, both the matrix $\dot{\mathbf{K}}$ and the vector **r** need to be re-computed at each iteration. A more detailed description of the solution process is described in (Bro-Nielsen and Cotin, 1996).

5.3 Stochastic Identification of Patient-Specific Properties

Our finite element simulations depend on parameters of the constitutive model such as Young's modulus, Poisson's ratio and the boundary conditions (BCs). Since the liver may be represented with an incompressible material, Poisson's ratio can be safely set to a value close to 0.5. However, the value of Young's modulus E is more difficult to estimate as it varies with the age of the patient or even with pathology. For instance, a cirrhotic liver is significantly stiffer than the average. Therefore, values from the literature do not directly match each patient. Besides that, the location and the elastic properties of the attachments of the organ play a major role in the accurate approximation of the displacement field. Such BCs are not visible in the preoperative images and it is difficult to estimate them intraoperatively as they are often out of the field of view of the surgery.

The elastic properties of materials can be identified by solving inverse problems (Lu and Zhao, 2009; Zhao et al, 2009; Gee et al, 2010; Sinkus et al, 2010) or using elastography techniques (Sarvazyan et al, 1998; Muthupillai et al, 1995; Xu et al, 2007) initially developed for diagnosis purposes. Some works have focused on the estimation of BCs intraoperatively such as (Peterlik et al, 2014; Plantefève et al, 2016; Johnsen et al, 2015) but these methods are difficult to use in practice as either additional intraoperative scanning is required or they are sensitive to anatomical variations. Moreover, when acquiring information intraoperatively, observational errors may occur, thus introducing uncertainty to the system. Alternative solutions accounting for such uncertainty rise from the use of Bayesian methods. For instance, authors in (Mendizabal et al, 2019a), employed the reduced-order unscented Kalman filter (ROUKF) to estimate Young's modulus of a porcine sclera based on observations extracted from optical coherence tomography images. Also works in (Nikolaev et al, 2018; Peterlik et al, 2017), employed the ROUKF to model the BCs of a liver as stochastic parameters, leading to more accurate simulations of the deformations of the organ.

Similarly to works in (Peterlik et al, 2017; Mendizabal et al, 2019a), we propose to use the ROUKF to estimate the value of Young's modulus and the BCs of a liver using observations of the target model. To this end, each sought parameter p is described as a stochastic parameter associated to a Gaussian probability density function (PDF). Initially $p \sim \mathcal{N}(\mu_0, \sigma_0)$ with μ_0 the mean value of p reported in the literature and σ_0 its standard deviation. The aim of the assimilation process is to iteratively reduce the standard deviation σ of p in order to find the most likely value for μ. To this end, the PDF of the parameter p is transformed based on observations. The transformation of the PDF is modelled using a ROUKF which can handle non-linear processes, and is computationally efficient (Moireau and Chapelle, 2011).

In this section we provide a brief description of the ROUKF algorithm, that is first used to estimate Young's modulus of a synthetic liver, and in a second time used to estimate the boundary conditions of an *in vivo* liver.

5.3.1 ROUKF: Overview of the Algorithm

Once the FE model of the organ is built (from a preoperative CT scan for instance), the constraints imposed on the surface Γ_D need to be identified in order to generate a deformation. In other words, the organ attachments representing Dirichlet boundary conditions and the traction or displacement imposed on the free part of the boundary need to be identified. In this work, the former can be either fixed (Sect. 5.3.2) or set as a set stochastic of parameters (Sect. 5.3.3) and the latter can be determined intraoperatively. During the intervention, points in the surface of the organ can be tracked in each video frame. Such points are called *features* and are separated into *control features* and *observation features*. The control features govern the deformation of the liver model (imposed displacement on Γ_D or traction if a force sensor is available) and the observation features correspond to ground truth data (used in the filter *correction* phase to compute the Kalman gain). The control features can be selected close to the surgical tool and be used to prescribe displacements in the mechanical model.

An efficient implementation of a Bayesian inference method able to process nonlinear systems like our models is the Unscented Kalman Filter (UKF) (Julier et al, 1995). Compared to an Extended Kalman Filter, it does not require to compute the Jacobian of the system, which would be prohibitive given the size of our problem. The unknown data to be estimated (the *stochastic state* of the system) is described as a Gaussian distribution, which transformation through the nonlinear system is performed using an *unscented transformation* (see Julier et al (1995) for details). The main idea is to parameterize the Gaussian distribution using a set of *sigma points*, which hold the mean and covariance information, but are easier to transfer through a nonlinear function. The general algorithm is described in Alg. 1. It consists of a loop that contains two main steps. During the *prediction* step we form the new hypothesis about the estimated state, while during the *correction* step we correct it by comparing the predicted measurements with (noisy and partial) observations.

The prediction step can be very costly when using a model with many degrees of freedom, as it is the case when using a FEM method. Using the simplex method to generate the sigma points would require $m + 1$ simulations if m is the number of elements in the stochastic state vector (line 9 of the algorithm). With a mesh of n nodes and k stiffness parameters, this would mean $3n + k + 1$ simulations. A simple FEM mesh of only a few hundred nodes would be too time-consuming for a clinical application, as it would require more than 300 simulations for each step of the assimilation process. To solve this issue, we use a Reduced Order Kalman Filter (ROUKF) instead of the UKF. This method significantly reduces the computation cost since only $k + 1$ simulations (in the best case) are required. This approach was proposed in Peterlik et al (2017).

Let us assume there are k unknown parameters in our model, so k different parameters to estimate that can be either the elasticity of the material or the elasticity of the organ attachments. Since we are using the simplex version of the ROUKF, there are $k+1$ sigma points meaning that $k+1$ evaluations of the model are performed in each prediction step of the assimilation process. At each step of the assimilation, the

Algorithm 1: Main steps of unscented Kalman filter

Initialize data:
set $\mathbf{x}_1$ - *model positions and unknown parameters*
set $T = T(\mathbf{x}_1)$ - *finite element model*
set $\mathcal{I}, \mathbf{P}_1, \mathbf{Q}, \mathbf{W}$ - *initial filter parameters*
for *each simulation step* i **do**
  Compute prediction phase:
  $\mathbf{x}_i^{\sigma_*} = \mathbf{x}_i + \sqrt{\mathbf{P}_i}\mathcal{I}$ - *generate sigma points*
  for *each sigma point* k **do**
    $\widetilde{\mathbf{x}}_{i+1}^{\sigma_k} = T(\mathbf{x}_i^{\sigma_k})$ *get result from deformation step*
  end for
  $\widetilde{\mathbf{x}}_{i+1} = \mathcal{E}(\widetilde{\mathbf{x}}_{i+1}^{\sigma_*})$ - *compute predicted state as mean of sigma points*
  $\widetilde{\mathbf{P}}_{i+1} = (\widetilde{\mathbf{x}}_{i+1}^{\sigma_*} - \widetilde{\mathbf{x}}_{i+1})(\widetilde{\mathbf{x}}_{i+1}^{\sigma_*} - \widetilde{\mathbf{x}}_{i+1})^T + \mathbf{Q}$ - *compute predicted covariance*
  Compute correction phase:
  get $\mathbf{q}_{i+1}^{(o)}$ - *observation features*
  for *each sigma point* k **do**
    $\widetilde{\mathbf{q}_{i+1}^{(o)}}^{\sigma_k} = H(\widetilde{\mathbf{x}}_{i+1}^{\sigma_k})$ - *get predicted observation*
  end for
  $\mathbf{P}_{\mathbf{xq}^{(o)}} = (\widetilde{\mathbf{x}}_{i+1}^{\sigma_*} - \widetilde{\mathbf{x}}_{i+1})(\widetilde{\mathbf{q}_{i+1}^{(o)}}^{\sigma_k} - \mathcal{E}(\widetilde{\mathbf{q}_{i+1}^{(o)}}^{\sigma_*}))^T$ - *compute cross covariance*
  $\mathbf{P}_{\mathbf{q}^{(o)}} = (\widetilde{\mathbf{q}_{i+1}^{(o)}}^{\sigma_k} - \mathcal{E}(\widetilde{\mathbf{q}_{i+1}^{(o)}}^{\sigma_*}))(\widetilde{\mathbf{q}_{i+1}^{(o)}}^{\sigma_k} - \mathcal{E}(\widetilde{\mathbf{q}_{i+1}^{(o)}}^{\sigma_*}))^T + \mathbf{W}$ - *comp. obs. cov.*
  $\mathcal{K}_{i+1} = \mathbf{P}_{\mathbf{xq}^{(o)}}\mathbf{P}_{\mathbf{q}^{(o)}}^{-1}$ - *compute Kalman gain*
  $\mathbf{x}_{i+1} = \widetilde{\mathbf{x}}_{i+1} + \mathcal{K}_{i+1}(\mathbf{q}_{i+1}^{(o)} - \mathcal{E}(\widetilde{\mathbf{q}_{i+1}^{(o)}}^{\sigma_*}))$ - *compute corrected state*
  $\mathbf{P}_{i+1} = \widetilde{\mathbf{P}}_{i+1} - \mathbf{P}_{\mathbf{xq}^{(o)}}\mathbf{P}_{\mathbf{q}^{(o)}}^{-1}\mathbf{P}_{\mathbf{xq}^{(o)}}^T$ - *compute corrected covariance*
end for

control features are extracted from the actual video frame and mapped onto the FE model through barycentric coordinates, in order to prescribe displacements. At the first step, μ and σ are initialized to μ_0 and σ_0 for each parameter. Then, $k+1$ vectors of parameters are sampled and $k+1$ simulations are performed. Each simulation corresponds to one of the sampled $k+1$ values of the parameter and they can be done in parallel as they are independent. After the simulations for all sigma points are performed, the *a priori* expected value and covariance matrix are updated. This is called the prediction phase. Later, in a correction phase, the extracted observation features are compared to the model predicted positions to compute the innovation that is used to compute the Kalman gain. The *a posteriori* expected values and the covariance matrix are computed based on the Kalman gain.

5.3.2 Estimation of the Young's Modulus Using Kalman Filters

In this paragraph, we aim at estimating the value of the Young's modulus of the liver using the ROUKF. This estimation is done using synthetic data, but a similar process can be followed for real data. We build a biomechanical model of the liver,

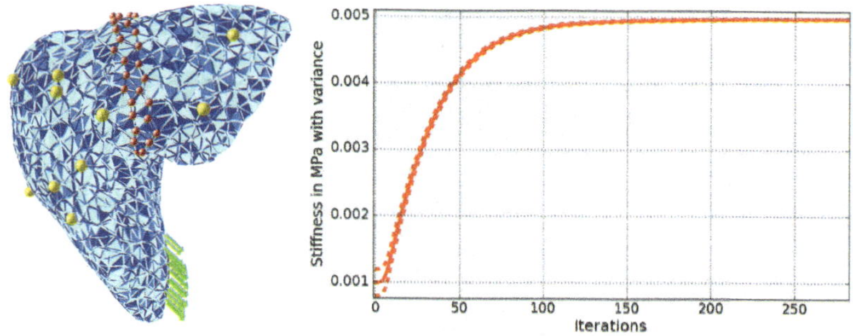

Fig. 5.1 Liver simulation mesh made of $11,000$ tetrahedral elements (left). The red points highlight the fixed points (Dirichlet boundary conditions), the yellow points correspond to the observed features, and the green arrows illustrate the direction of the applied forces. Variation of μ and σ for Young's modulus estimation using the ROUKF (right). The value of the parameter converges to $4992 \pm 15Pa$ in 500 seconds.

with fixed boundary conditions (red points in Figure 5.1(left)) to mimic the effect of the falciform ligament and of the vascular tree. A force of fixed magnitude and varying amplitude is continuously applied to one of the liver lobes to generate observations (yellow points in Figure 5.1(left)). The amplitude of such force follows the sinusoidal function $\frac{1}{2} \times (1 - \cos(2 \times \pi \times \tau))$ where τ is a period. In this case, the control features defined in section 5.3.1 correspond to the force applied (that is known). The Young's modulus is set to $5,000$ Pa in the reference simulation.

For the initialization of the ROUKF, we set μ_0 to $1,000$ Pa and σ_0 to 200 Pa. The state vector contains all the degrees of freedom of the mesh and the parameters to estimate (one parameter in our case). Hence, there are only 2 sigma points which allows a very fast assimilation process to take place as only two evaluations of the model need to be performed at each prediction phase. As depicted in Figure 5.1(right), the value of the Young's modulus reaches rapidly a value close to the ground truth (at iteration 150, $\mu = 4948$ and $\sigma = 97$). The value of the parameter converges to 4992 ± 15 Pa in 631 iterations (that is 500 seconds). Such assimilation process could take place before the surgery starts. Note that if the assimilation needs to be done in real-time, the simulations could be parallelized and simplified (we chose here a relatively high mesh resolution).

5.3.3 Estimation of Boundary Conditions Using Kalman Filters

Apart from material properties, the same approach can be employed to estimate the unknown attachments of the liver. We consider a scenario where an *in vivo* porcine liver is deformed with laparoscopic pincers. A video sequence of 7 s was acquired with a laparoscopic monocular camera inserted in the porcine abdomen inflated

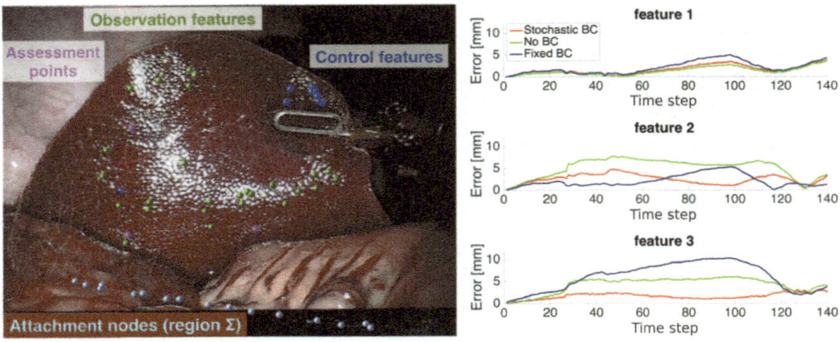

Fig. 5.2 The first frame of the video sequence with features (left). Temporal evolution of the assessment error computed for each assessment point (right).

with gas. We assume there is a region Σ on which hidden boundary conditions are applied.

Before the intervention, a CT scan was collected from the organ's geometry and a FE model was built following the pipeline described in section 5.2.2. A tetrahedral mesh having 315 nodes was generated. The obtained model is fixed with elastic springs in a region Σ that is hidden to the laparoscopic camera view. There are 35 nodes in Σ meaning that 35 nodes were attached with springs to mimic the boundary conditions at these specific locations. The elasticity parameters of such springs are modelled as stochastic parameters (see section 5.3.1). The elasticity values can range from 0 (no attachment) to high values (stiff attachment). A different elasticity is associated with each spring.

Known surface displacements are prescribed on a small area of the visible surface to mimic the effect of the surgical tool based on the control features. The considered scenario falls within the category of displacement-zero traction problems, where the relation between surface and volume displacements is independent of the Young's modulus, for homogeneous materials (Miller and Lu, 2013). As a consequence, without lack of generality we set Y to the fixed value of $5,000$ Pa, which is the average stiffness value for a liver.

For the initialization step of the ROUKF, $\mu = 0$ and $\sigma = 0.01$ for each parameter. Three assessment points are placed on the surface of the liver mesh in order to compute the prediction error between the observed data and the model. We ensure that the assessment points are different from the control and the observation features (see Fig. 5.2(left)). In Fig. 5.2(right) is shown the temporal evolution of the prediction error computed over the three assessment points. The error achieved with the stochastic simulation is compared to the error obtained with either fixed BCs (e.g. stiff attachment) or without BCs (e.g. spring stiffness set to zero). Results show that the stochastic simulation leads to smaller errors than the deterministic simulation.

5.4 Deep Neural Networks for Data-Driven Simulations of a Liver

Methods described in Section 5.3 allow to estimate the boundary conditions and the material parameters of the anatomy, assuming that the constitutive model is known. In practice, the choice of the constitutive relation is usually based on previous works and/or phantom tests, and it is not possible to guarantee that it is the most representative of the real clinical scenario. Within this context, it seems natural to try to learn the complete biomechanical behavior of the organ directly from real data acquired intraoperatively (including its constitutive equation).

In recent years, machine learning (ML) started to revolutionize several fields (vision, language processing, image recognition, genomics). With sufficient ground truth data, machine learning algorithms can map the input of a function to its output without any mathematical formulation of the problem, thus actuating like a black box. The high inferring speed of these methods makes it useful for many applications where the prediction speed is of critical importance. Due to this characteristic and the fact that they are driven directly by data, these methods seem promising for the learning of the entire mechanical behavior of the anatomy without relying on prior models.

Some first attempts that exploit learning methods to estimate the deformation of biological tissues have recently been made. By implicitly encoding soft tissue mechanical behavior in the trained ML models, they proved successful to predict the entire 3D organ deformation starting either by applied surface forces (Morooka et al, 2008; Tonutti et al, 2017; Rechowicz and McKenzie, 2013; Mendizabal et al, 2020) or by surface displacements (Pfeiffer et al, 2019; Brunet et al, 2019; Lorente et al, 2017; Mendizabal et al, 2019b). However, the accuracy of a ML model highly depends on the quality and on the amount of data used to train it. In an ideal scenario, such a model would be trained with an infinite amount of real patient-specific noise-free data, which is in practice not possible. As a matter of fact, acquiring large number of volumetric deformations of an organ is a challenging problem. Moreover, the constraints applied should be precisely controlled which is in practice very complicated to guarantee. Within this framework, FE simulations can be exploited to generate synthetic data that is highly representative of the reality, to be used as training samples.

Among the various ML techniques, the use of neural networks (NN) has considerably increased. This is due to the fact that they are the building blocks of deep learning, a class of methods which is able to learn data representations and has demonstrated strong abilities at extracting high-level representations of complex processes. For example, neural networks are used by Tonutti and Rechowicz to predict the displacement of brain tumors and of the rib cage surface respectively, starting from the acting forces (Tonutti et al, 2017; Rechowicz and McKenzie, 2013). However, both these works do not predict whole volume deformation but only surface displacements. Neural network based methods have been also used to predict liver deformation in augmented surgery. Morooka et al. trained a NN to predict liver deformations for a given input force. They use their model together with PCA to compress the size of the output deformation modes, and thus reduce the training time. Although the model proved able to learn the deformation modes, it was only applied to simulated data and not to real cases (Morooka et al, 2008). An additional example is proposed by Pfeiffer et al., who used a deep NN to estimate liver deformations from the known displacement of partial liver surface. The innovative aspect of this work is that the proposed network is able to provide an accurate prediction on a liver mesh even though the synthetic data used for the training were generated from a set of random meshes. Similarly, Pellicer-Valero et al. trained a NN on various liver geometries by registering them to an average liver geometry. However, the authors rely on the assumption that both the boundary conditions and the elastic properties of the object are known, and they did not test their network performance on real data acquired during surgery (Pfeiffer et al, 2019; Pellicer-Valero et al, 2020).

From all these works it emerges that the main advantage of using neural networks to predict anatomical deformations is that the prediction speed is in the order of few milliseconds and is not affected by the complexity of the model used to generate the training dataset. In this section, we propose a method that, similarly to the approaches described above, allows for extremely fast and accurate simulations by using an artificial neural network that learns the stress-strain relationship directly from data, without any *a priori* mathematical formulation of the problem. Such a network can not only learn the desired biomechanical model, but also the desired boundary conditions and material properties; and predict deformations at haptic feedback rates with very good accuracy. This section is divided in three main segments. First of all, we present the selected network based on a U-Net architecture and the strategy adopted for the generation of training data. Afterwards, we report some representative results, obtained when using the proposed U-Net both in simulated and real scenarios. In a further section, we explain how we employed transfer learning methods to make the U-Net able to generalize to new unseen boundary conditions and elastic parameters.

5.4.1 Method

In this section we propose to use a deep neural network that learns the relation function between surface constraints and volumetric deformation accounting for the specificity of each patient. The general procedure, named the *U-Mesh framework*, consists in training a 3D U-Net architecture (Ronneberger et al, 2015) with synthetic deformed meshes generated with the FEM described in section 5.2.2.

5.4.1.1 The U-Mesh Framework

Formally, our network h is a parameterized function that accepts a $3 \times n_x \times n_y \times n_z$ tensor of input constraints $\mathbf{C}$ and produces a tensor of volume displacements $\mathbf{U_v}$ of the same size as output. The computational domain Ω is sampled by a 3-dimensional grid of resolution $n_x \times n_y \times n_z$ (see Fig 5.5(left)). The tensor $\mathbf{C}$ represents the constraints applied over the surface boundary Γ of the domain.

Our problem consists of finding the function h that produces the best estimations of the displacement field given prescribed constraints $\mathbf{C}$. This is performed by minimizing the expected error over the training set $\{(\mathbf{C}_n, \mathbf{U}_{\mathbf{v}n})\}_{n=1}^{N}$ of N samples:

$$\min_{\theta} \frac{1}{N} \sum_{n=1}^{N} \|h(\mathbf{C}_n) - \mathbf{U}_{\mathbf{v}n}\|_2^2. \tag{5.9}$$

To characterize our network h, we choose a 3D U-Net (Ronneberger et al, 2015) architecture for its similarities with model order reduction techniques from the mechanics community. It is a modified fully convolutional network with an encoding path that transforms the input into a low-dimensional space and a decoding path that expands it back to the original size (see Fig. 5.3). Additional skip connections transfer detailed information along matching levels from the encoding path to the decoding path. For a more detailed explanation of the U-Net, the reader may refer to our previous publication on the subject (Mendizabal et al, 2020).

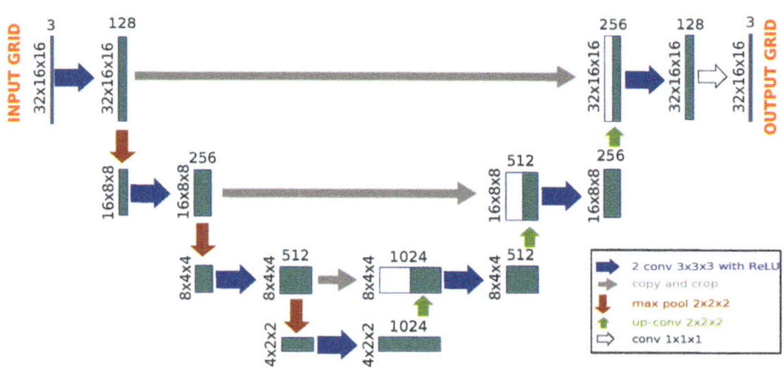

Fig. 5.3 Network architecture for an initial grid resolution of size $20 \times 16 \times 15$, padded to $32 \times 16 \times 16$, 128 channels in the first layer and 3 steps

5.4.1.2 Synthetic Data Generation for U-Net Training

Training data for our network are made of pairs of $(\mathbf{C}, \mathbf{U}_\mathbf{v})$ which are obtained from the previously explained FEM. We perform multiple simulations by imposing random constraints on the boundary Γ. At the end of each simulation, the pair of imposed constraints and obtained volumetric deformation is stored as an element of the data set.

For a correct spatial understanding, the U-Net requires regular grids as input, meaning that the displacements of the considered object need to be encoded in a regular grid. To this end, we propose to mesh the domain Ω using tetrahedral elements to compute the FE deformations and then, map a 3-dimensional regular grid onto the tetrahedral mesh to follow its deformation (see Fig 5.5(left)). This mapping introduces an approximation error that is reduced when the grid resolution is increased. Random forces are directly applied to the surface of the object and then mapped to the nodes of the grid and to the T4 elements. At this stage, we can store the imposed constraints $\mathbf{C}$ (applied traction on grid nodes or grid surface displacement) and the resulting volumetric deformation $\mathbf{U_v}$ as a point of the data set. In total N training samples and M testing samples are generated.

The generated N training samples are used to train the network by minimizing Eq. 5.9 with the Adam optimizer (Kingma and Ba, 2014). All our experiments are performed in a GeForce 1080 Ti using a batch size of 4 and $100,000$ iterations for training. We use a PyTorch implementation of the U-Net. We recall that the batch size is the number of samples that are given to the network at each iteration of the minimization process.

5.4.1.3 Validation Metrics

To assess the efficiency of our method, we perform a statistical analysis of the error over the testing data set $\{(\mathbf{C}_m, \mathbf{U_v}_m)\}_{m=1}^{M}$. Let $\mathbf{U_v}_m$ be the ground truth displacement tensor for sample m generated using the FEM described in section 5.2.2 and $h(\mathbf{C}_m)$ the U-Mesh prediction. We define the *mean Euclidean error* e between $\mathbf{U_v}_m$ and $h(\mathbf{C}_m)$ for sample m as:

$$e(\mathbf{U_v}_m, h(\mathbf{C}_m)) = \frac{1}{n}\sum_{i=1}^{n} \left\|\mathbf{U_v}_m^i - h(\mathbf{C}_m)^i\right\|_2 \tag{5.10}$$

where n is the number of nodes of the mesh. We compute the average $\overline{e}$ and standard deviation $\sigma(e)$ of such norm over the testing data set. The *mean Euclidean error* represents the intuitive nodal distance, averaged over all the nodes of the mesh.

5.4.2 Predicting the Deformation of the Liver

In this section we will show the performance of the U-Mesh in predicting the deformations of a liver. To start with, the U-Net learns to predict the displacement field of a virtual liver given an input contact force. In a follow-up, the U-Net is used in an augmented reality context, where the full volumetric displacement field needs to be estimated from a partial surface deformation. In both cases the network is trained with FEM-generated data since for the moment, we do not know how to collect a sufficient amount of real volumetric information of a liver using current imaging techniques.

5.4.2.1 U-Mesh on a Synthetic Liver

A surface mesh is obtained from a pre-operative CT scan of a human liver. The liver volume is meshed with 4859 tetrahedral elements and Dirichlet boundary conditions are used (67 nodes between the two lobes were fixed) to mimic the effect of the vascular tree and of the falciform ligament (Abdel-Misih and Bloomston, 2010). The length of the liver is 0.2 m. The Young's modulus Y is set to $5,000$ Pa and the Poisson's ratio to 0.48. A regular grid of resolution $21 \times 23 \times 21$ is mapped onto the tetrahedral mesh to encode the displacement fields and the forces in a "U-Net interpretative manner".

Normal forces of random magnitudes are computed on the liver surface. Only one force is applied at each time step on a small surface area. To fit the time requirements of a clinical routine (e.g. a few hours between the pre-operative CT scans and the surgery), we decided to limit the size of the data set to $2,000$ samples (generated in 180 min). The data set consists of $2,000$ pairs of input forces and corresponding volume deformation. $N = 1,600$ samples are used to train the network in 327 min and $M = 400$ samples are left for validation.

Metrics obtained on the validation set are reported in Table 5.1. The maximal *mean Euclidian error* over the testing data set is of only $6.8e-4$ m (see Fig 5.5(right)) and the maximal deformation is 0.08 m. The most impacting result is the small prediction time: outputs are predicted in only 3.47 ± 0.60 ms. In Fig. 5.4 are shown some samples of U-Mesh-deformed livers with the corresponding reference solutions.

These results show the potential of the U-Mesh in applications requiring both high accuracy and speed. In the following, we will show its performance on an augmented reality scenario where the available information can be noisy and sparse.

Table 5.1 Error measures on a liver of length 0.2 m immersed in a $21 \times 23 \times 21$ grid. The maximal e is $6.8e-4$ m and the maximal deformation is 0.08 m.

$\bar{e}$ in m	$\sigma(e)$ in m	prediction time in ms	training time in min
$6.94e-5$	$7.81e-5$	3.47 ± 0.60	327

Fig. 5.4 Various liver samples from the testing data set. The reference solution appears in red and the U-Mesh prediction is in green. The rest shape of the liver is shown in gray.

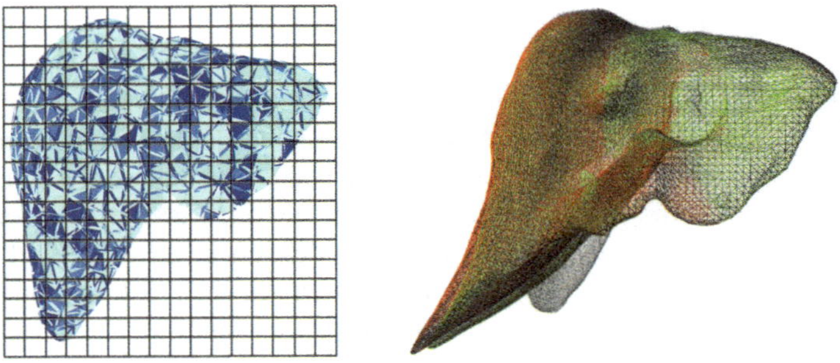

Fig. 5.5 The computational domain Ω is meshed with T4 elements and a 3-dimensional grid of resolution $n_x \times n_y \times n_z$ is mapped onto the tetrahedral mesh (left). Sample with maximal nodal error 0.00068 m (right). The reference solution is shown in red and the U-Mesh prediction is in green. The rest shape is shown in gray.

5.4.2.2 U-Mesh for Augmented Hepatic Surgery

To build an augmented view of a liver during surgery we need to perform an elastic registration of the preoperative model to the intraoperative images acquired with a 3D imaging device (see Fig 5.6). While in minimally invasive surgeries a laparoscopic camera can be used to acquire a video of the abdominal cavity, in open surgeries and RGB-D sensor can capture the surface deformation of the tissues. From such images, a partial point cloud of the liver surface can be extracted using one of the methods listed in Petit and Cotin (2018). To perform the elastic alignment between the preoperative internal structures and the surgical live images, the preferred method is the co-rotational FEM as it can provide a real-time estimation of the displacement field (Petit and Cotin, 2018). However, the complex deformations happening during surgical manipulations may not be correctly taken into account by such a simple model. To overcome this issue we propose to replace the FEM step with the U-Mesh trained on a more sophisticated FEM model (typically not capable of achieving real-time computations).

The U-Mesh needs to learn to predict full volumetric displacements from partial surface point clouds that give information about the position of some points of the surface of the liver. These positions can be translated as prescribed constraints. As explained in section 5.2.2, using H8 elements lead to better approximation of the stress and the strain. Therefore, to generate the data sets, we choose to use a FEM combined with an immersed boundary method (Düster et al, 2008) as it allows for the use of regular hexahedral meshes to compute accurate deformations of the liver. It is worth noting that in this scenario the FEM mesh directly matches the input to the U-Net, thus avoiding the approximation error introduced by the mapping in the previous section.

We can assume that during surgery, half of the surface of the liver is visible to the camera. As depicted in Figure 5.7, 100 points are uniformly sampled in the visible part of the surface to mimic a point cloud. Then, 100 simultaneous forces of random magnitude and direction are applied to these points in order to generate nearly random displacements. The training data set consists of pairs of $(\mathbf{U_s}, \mathbf{U_v})$ where the input to the network $\mathbf{U_s}$ corresponds to the surface point cloud mapped onto the regular grid. For the same reasons stated in previous section, we limited the size of the data set to $2,000$ samples ($N = 1,600$ for training and $M = 400$ for testing)(see Fig. 5.8 for examples of the generated deformations). It is worth mentioning that no patient-specific parameterization of the biomechanical model is required since for homogeneous materials, the relation between the surface and the volumetric displacements is independent of the stiffness of the object (Miller and Lu, 2013), and only depends on the Poisson's ratio (set to 0.49 as soft tissues can generally be described as incompressible).

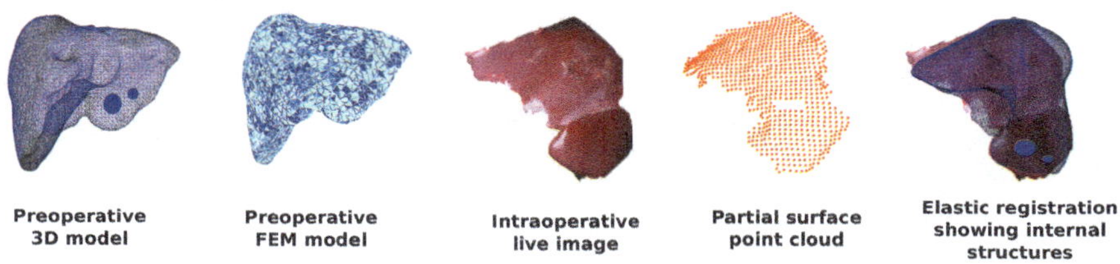

Fig. 5.6 Augmented reality pipeline: preoperative internal structures are mapped in real-time onto the live image of the organ using a FEM model.

Once the network is trained, we assess our approach on *ex vivo* human liver data, on which ten markers are embedded to compute target registration errors (TRE). During the experiments, surface data is obtained with an RGB-D sensor and ground truth data acquired at different stages of deformation using a CT scan. The RGB-D point clouds can be interpolated onto the regular grid to obtain per-node displacements on the surface and can be given as input to the network that in turn predicts the volumetric displacement fields. Each new RGB-D point cloud can be fed to the network, thus generating a continuous visualization of the internal structures of the organ.

The marker predicted positions are compared to our *ex vivo* ground truth by computing TREs (see Fig 5.9). The average TRE at the 10 markers is of only 2.92 mm with a maximal value of 5.3 mm. The same scenario, but this time using a co-rotational FE method, leads to an average TRE of 3.79 mm and is computed at about 25 ms. The solution of the Saint-Venant–Kirchhoff model, used to train our model, gives nearly the same error (which was expected) but for a computation time of 1550 ms.

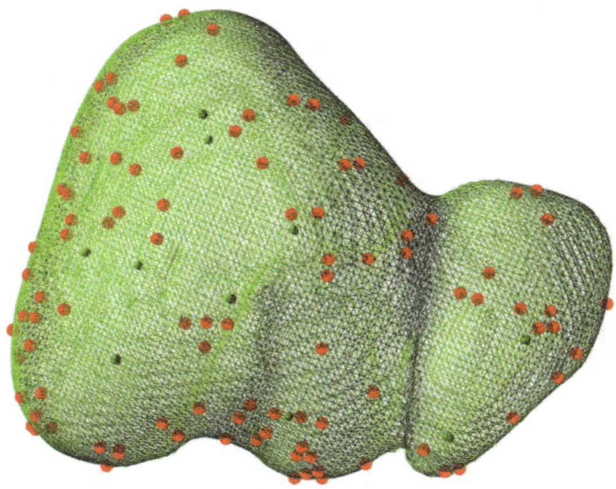

Fig. 5.7 Virtual point cloud on the visible surface of the liver to generate random displacements.

Fig. 5.8 Examples of generated deformations to train the network.

	FEM co-rotational model	FEM StVK model	U-Mesh
Computation time (per time step)	25 ms	1550 ms	3 ms
Accuracy (mean TRE)	3.8 mm	2.9 mm	2.9 mm

Fig. 5.9 Comparison between the reference co-rotational FEM, the Saint-Venant–Kirchoff model used to train the network and the U-Mesh.

5.5 Updating the Trained Model Through Transfer Learning

As mentioned above, there exists a correlation between our method and model reduction techniques. There is an important body of work in this area, with a well-established understanding of the process linking the fast (macro) model to the full (micro) model (De Angelo et al, 2019). Such theory-driven approaches define how to generate reduced models with adapted parameters that characterize the full (micro-scale) model (Boutin et al, 2017). Our Deep Learning approach does something similar by learning the key characteristics (deformation and parameters) of the full-scale model, but using a data-driven approach for this.

When applied in the context of surgery, both approaches share the same limitation. The full model (micro-model) cannot always be correctly parametrized until the surgery has started, as some model parameters are not measurable in pre-operative images. In this case, the use of transfer learning methods can offer a natural, data-driven solution for adapting the neural network to a particular patient. For methods based on reduced models, Bayesian approaches are probably a good alternative, as they can estimate material properties from a probability distribution and a priori knowledge of the parameter value.

As mentioned in the section 5.3, boundary conditions have a significant impact on the accuracy of the predictions computed by biomechanical models. However, since they are hard to identify, we want to ensure the robustness of the U-Mesh to the variability of the BCs. We will show that small amount of data is required to learn patient-specific BCs, when refining a network pre-trained with variable BCs from an appropriate distribution. This could help to significantly reduce the expensive cost of the offline data generation phase. Lastly, since real data can be sparse and noisy, we explore the behaviour of the U-Net when the input tensor $\mathbf{C}$ is highly sparse, and the effect of noise on the quality of the predictions.

5.5.1 Beam with Hidden Fixed Dirichlet BCs

In this section, we compare the accuracy of the U-Mesh either when trained from scratch with up to $16,128$ samples, or when pre-trained on various BCs and refined on the target BCs.

We consider a deformable beam (size 4x1x1 m^3, $E = 300$ Pa, $\nu = 0.4$, 500 regular H8 elements) subject to fixed boundaries on a rectangular cuboid of its bottom part (see Figure 5.10). The beam follows the Saint-Venant–Kirchhoff behavior described in section 5.4.1.2. We generate three different training data sets ($N_1 = 16,128$; $N_2 = 1,209$; $N_3 = 100$) and one testing data set ($M = 4,032$), all drawn from the same distribution. We performed 10 trainings to compare 7 different strategies, summarised in Table 5.2. In strategies 1, 2, and 4, the U-Net is trained from scratch whereas in strategies 3, 5, 6 and 7, the U-Net is refined starting from a network pre-trained with 16,128 data with different boundary conditions (see Table 5.2).

In Table 5.3 are reported the validation metrics computed for each strategy on the same testing dataset (M = 4,032), as well as the index of the best iteration over 200,000 (with a saving step of 5,000). We see that strategy 3 performs better than strategy 2. More impressive yet are the strategies 5, 6 and 7 (especially 5, which, by refining, led to errors comparable to the one obtained with 12x more data without refining). Furthermore, they are substantially better than strategy 4 where no refinement was done. This is an example of a scenario where the U-Net cannot

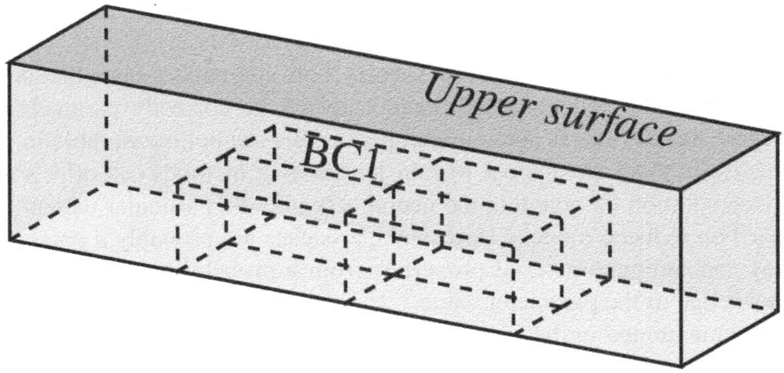

Fig. 5.10 Cuboid-like boundary conditions on which the U-Net is pre-trained in strategies 3 and 5. In strategy 6, there are four more cuboids so that the lower part of the beam is fully covered. In strategy 7, the U-Net is pre-trained on BC1.

Table 5.2 Summary of the 7 strategies of interest. "BC4" stands for 4 adjacent cuboids in the middle of the hidden part of the beam. "BC8" stands for 8 adjacent cuboids fully covering the hidden part of the beam (see figure 5.10).

Strategy ID	1	2	3	4	5	6	7
Training data set #	N_1	N_2		N_3			
Pre training data set	–	–	BC4	–	BC4	BC8	BC1

accurately learn a deformation model from scratch with very few (100) data, whereas it does learn an accurate model in a few thousands iterations using transfer learning.

The mild differences between the metrics obtained for strategies 5, 6 and 7 show that the data generated for pre-training must be reasonably distributed. Indeed, even though the network benefits from the diversity of BCs encountered in the pre-training stage, it is more efficient when these BCs are close enough to the target boundary conditions. Hence the need for a reasonable distribution.

So far we have seen that refining from an average model significantly reduces the quantity of data required to learn a deformation model. Results in Table 5.4 highlight the fact that it also speeds up the model convergence. Computing more metrics, we found that a good accuracy is reached approximately 20x faster when refining with 100 data than when starting from scratch with either 16,128 or 1,209 samples.

For completeness, we also investigated the case where the constitutive law changed between pre-training and refining stages. For pre-training, we modelled a beam with the linear Hooke's law, and for refining, we chose the Saint-Venant–Kirchoff constitutive equation to model the deformations of the beam. In this scenario as well,

Table 5.3 Error measures over all seven scenarios. Best iterations are given in thousands. Transfer learning situations are highlighted in red, first and second best results in green and blue.

	e in mm			# training dataset	best iteration
	avg	max	$\sigma(e)$		
1	**0.45**	2.48	0.29	16,128	200
2	**0.71**	2.96	0.47	1,209	180
3	**0.52**	2.66	0.32	1,209	40
4	**3.49**	32.0	3.59	100	200
5	**0.80**	4.85	0.61	100	15
6	**1.11**	5.89	0.80	100	5
7	**0.98**	8.88	0.82	100	5

Table 5.4 Error measures at iteration 5000. We relaunched the training of 6 and 7 with a step of 50 iterations to ensure there was no significant overfitting before iteration 5,000, hence the values "1.5" and "1.15" (as a matter of fact, iterations 1,500 and 1,150 were actually slightly better than 5,000). Transfer learning situations are highlighted in red.

	e in mm			# training dataset	iteration
	avg	max	σe		
1	**5.74**	27.4	3.73	16,128	5
2	**7.47**	42.1	5.36	1,209	5
3	**0.55**	2.91	0.35	1,209	5
4	**5.25**	44.2	4.71	100	5
5	**0.81**	4.1	0.58	100	5
6	**1.09**	5.37	0.7	100	1.5
7	**0.95**	8.66	0.84	100	1.15

we observed that transfer learning reduces the amount of data required to reach a given accuracy even when the base equations are complexified.

5.5.2 New Boundary Conditions and Sparse Data

As mentioned in previous paragraphs, real intra-operative data can be sparse and noisy. In this section, we show that the U-Net can still learn models when the training input tensors only contain a sparse view of the displacement $\mathbf{U_s}$ imposed on the upper surface Γ_D. In return, the accuracy is reduced and we show that transfer learning is barely relevant in such an adverse scenario.

We consider the same beam as described in 5.5.1, except that the beam is supported on both ends (fixed beam). In order to train the network, we built 2 data sets of sizes $N_1 = 10,080$ and $N_2 = 1,008$ (see 5.5.1). Here, every tensor $\mathbf{C}$ contains the values of an imposed surface displacement, on a randomly selected subdomain of the upper surface (in between 13 and 114 non-zero displacements in the testing data set, 67 in average) - see figure 5.11. We trained the network either directly with N_1 or N_2 samples (strategies 1 and 2), or with N_2 data after a pre-training on a stiffer beam fixed at one end (Young's modulus of 500 Pa). The pre-training was done either with sparse data (strategy 4, same distribution as the refining data set), or dense data (strategy 3, full view of the imposed upper surface displacement $\mathbf{U_s}$).

Fig. 5.11 Randomly visible sub-domains of the upper face of the beam (in yellow).

In Table 5.5 are reported the validation metrics at best iteration. The average error with $N_1 = 10,080$ is of only 3.03 mm and the maximal error is 57.1 mm - meaning less than 1.5 % of the length of the beam as maximal error. This shows that even though U-Net may learn much more accurate models with dense data, it still deals pretty well with sparse data when provided with a large enough training dataset. We should mention that we obtained very similar results by applying an additive white Gaussian noise of variance $N = 10^{-3}$ m on the testing dataset. With a variance $N = 10^{-2}$ m, the average mean Euclidean error $\overline{e}$ barely exceeds one centimeter. On another note, we see that there is no meaningful difference between the validation metrics of strategy 2 (1,008 data without refining) and strategy 4 (1,008 data with refining). Eventually, except when the data set is very small, we found that refining doesn't enhance accuracy in such a scenario. What is more, these results highlight the importance of pre-training the network with sparse data whenever the refining data is sparse. We further investigated the case where only very few data ($N_3 = 100$) are available, and found that it was not sufficient (with or without pre-training), although the refined model was more accurate. What remains valid is that in any scenario, the U-Mesh maintained a better accuracy with transfer learning in the first thousands of iterations. Reiterating these tests with sparse data without modifying the Young's modulus between pre-training and training stage corroborated these results.

5.6 Conclusion

In this work we have proposed a method fulfilling the real-time and precision requirements of patient-specific augmented reality. Based on *a priori* knowledge of the biomechanics of the organ, we select a constitutive model describing the relation between stresses and strains. Such relation is heavily affected by patient-specific properties such as boundary conditions and material characteristics. While obtaining

Table 5.5 Error measures at best iteration. Transfer learning situations are highlighted in red.

	e in mm			# training dataset	best iteration
	avg	max	$\sigma(e)$		
1	**3.03**	57.1	3.57	10,080	185
2	**5.69**	101	7.11	1,008	115
3	**7.73**	106	9.02	1,008	145
4	**5.79**	88.3	6.85	1,008	95

these properties preoperatively may be troublesome, having information about them intraoperatively can be straightforward. In our approach, the parameters of the preoperative finite element or deep learning models are updated based on intraoperative observations. These data-driven simulations are obtained by exploiting Bayesian filtering to update the parameters of the finite element model and by employing transfer learning to update a deep learning model. In the former, each parameter can only be modified individually. Indeed, in our pipeline using the Kalman filters, to estimate the stiffness of boundary conditions, the Young's modulus of the material needs to be fixed (and *vice versa*). A simultaneous estimation of both sets of parameters would be more complicated (yet possible), less precise (variance of the stochastic parameters will remain high) and would require very tedious fine-tuning of the filter. In the latter, not only the elasticity parameters and the boundary conditions can be changed simultaneously, but also the constitutive model itself. An interesting point of using deep neural networks is that the parameters do not need to be explicitly identified. They are encoded in the data. Hence, the network builds its own representation and through transfer learning, the weights of the network can be modified to match the targeted function.

In the second part of this work, we have presented a method that can approximate complex elastic deformations of a real liver and generate a deformed state from an RGB-D point cloud. The obtained accuracy is comparable to the one obtained with the finite element model used to train it while being about 500 times faster. Whenever the preoperative model differs from the intraoperative one, we can use transfer learning to exploit the already learned knowledge in a fast and efficient way. We have reasons to believe that the U-Net learns local correlations in the displacement field rather than an overall model only. As a consequence, if the pre-trained model represents an average liver, transfer learning should not break the constitutive model learned previously. Note that the variability between livers can be high but it will always vary in a bounded range. For this reason, we believe that transfer learning is the key to an accurate and fast simulation of the deformations of a liver. However, the results dedicated to transfer learning are only a proof of concept as for now, we do not know how to collect intraopertaive volumetric data needed to refine our network.

References

Abdel-Misih SR, Bloomston M (2010) Liver anatomy. Surgical Clinics 90(4):643–653

Allard J, Cotin S, Faure F, Bensoussan PJ, Poyer F, Duriez C, Delingette H, Grisoni L (2007) Sofa-an open source framework for medical simulation. In: MMVR 15-Medicine Meets Virtual Reality, vol 125, pp 13–18

Alvarez P, Chabanas M, Rouzé S, Castro M, Payan Y, Dillenseger JL (2018) Lung deformation between preoperative CT and intraoperative CBCT for thoracoscopic surgery: a case study. In: Medical Imaging 2018: Image-Guided Procedures, Robotic Interventions, and Modeling, International Society for Optics and Photonics, vol 10576, p 105761D

Benzley SE, Perry E, Merkley K, Clark B, Sjaardama G (1995) A comparison of all hexagonal and all tetrahedral finite element meshes for elastic and elasto-plastic analysis. In: Proceedings, 4th international meshing roundtable, Sandia National Laboratories Albuquerque, NM, vol 17, pp 179–191

Bosman J, Haouchine N, Dequidt J, Peterlik I, Cotin S, Duriez C (2014) The role of ligaments: Patient-specific or scenario-specific? In: International Symposium on Biomedical Simulation, Springer, pp 228–232

Boutin C, dell'Isola F, Giorgio I, Placidi L (2017) Linear pantographic sheets: asymptotic micro-macro models identification. Mathematics and Mechanics of Complex Systems 5(2):127–162

Bro-Nielsen M, Cotin S (1996) Real-time volumetric deformable models for surgery simulation using finite elements and condensation. In: Computer graphics forum, Wiley Online Library, vol 15, pp 57–66

Brunet JN, Mendizabal A, Petit A, Golse N, Vibert E, Cotin S (2019) Physics-based deep neural network for augmented reality during liver surgery. In: International Conference on Medical Image Computing and Computer-Assisted Intervention, Springer, pp 137–145

Clements LW, Chapman WC, Dawant BM, Galloway Jr RL, Miga MI (2008) Robust surface registration using salient anatomical features for image-guided liver surgery: algorithm and validation. Medical physics 35(6Part1):2528–2540

De Angelo M, Barchiesi E, Giorgio I, Abali BE (2019) Numerical identification of constitutive parameters in reduced-order bi-dimensional models for pantographic structures: application to out-of-plane buckling. Archive of Applied Mechanics 89(7):1333–1358

Delingette H, Ayache N (2004) Soft tissue modeling for surgery simulation. Handbook of Numerical Analysis 12:453–550

Düster A, Parvizian J, Yang Z, Rank E (2008) The finite cell method for three-dimensional problems of solid mechanics. Computer methods in applied mechanics and engineering 197(45-48):3768–3782

Gee MW, Förster C, Wall W (2010) A computational strategy for prestressing patient-specific biomechanical problems under finite deformation. International Journal for Numerical Methods in Biomedical Engineering 26(1):52–72

Haouchine N, Dequidt J, Berger MO, Cotin S (2013) Deformation-based augmented reality for hepatic surgery. Studies in health technology and informatics 184

Johnsen SF, Taylor ZA, Clarkson MJ, Hipwell J, Modat M, Eiben B, Han L, Hu Y, Mertzanidou T, Hawkes DJ, et al (2015) Niftysim: A GPU-based nonlinear finite element package for simulation of soft tissue biomechanics. International journal of computer assisted radiology and surgery 10(7):1077–1095

Joldes GR, Wittek A, Miller K (2010) Real-time nonlinear finite element computations on GPU–Application to neurosurgical simulation. Computer methods in applied mechanics and engineering 199(49-52):3305–3314

Julier SJ, Uhlmann JK, Durrant-Whyte HF (1995) A new approach for filtering nonlinear systems. In: Proceedings of 1995 American Control Conference-ACC'95, IEEE, vol 3, pp 1628–1632

Kingma DP, Ba J (2014) Adam: A method for stochastic optimization. arXiv preprint arXiv:14126980

Lorente D, Martínez-Martínez F, Rupérez MJ, Lago M, Martínez-Sober M, Escandell-Montero P, Martínez-Martínez JM, Martínez-Sanchis S, Serrano-López AJ, Monserrat C, et al (2017) A framework for modelling the biomechanical behaviour of the human liver during breathing in real time using machine learning. Expert Systems with Applications 71:342–357

Lu J, Zhao X (2009) Pointwise identification of elastic properties in nonlinear hyperelastic membranes—part I: theoretical and computational developments. Journal of applied mechanics 76(6)

Marchesseau S, Heimann T, Chatelin S, Willinger R, Delingette H (2010) Multiplicative jacobian energy decomposition method for fast porous visco-hyperelastic soft tissue model. In: International Conference on Medical Image Computing and Computer-Assisted Intervention, Springer, pp 235–242

Meier U, López O, Monserrat C, Juan MC, Alcaniz M (2005) Real-time deformable models for surgery simulation: a survey. Computer methods and programs in biomedicine 77(3):183–197

Mendizabal A, Sznitman R, Cotin S (2019a) Force classification during robotic interventions through simulation-trained neural networks. International journal of computer assisted radiology and surgery 14(9):1601–1610

Mendizabal A, Tagliabue E, Brunet JN, Dall'alba D, Fiorini P, Cotin S (2019b) Physics-based deep neural network for real-time lesion tracking in ultrasound-guided breast biopsy

Mendizabal A, Márquez-Neila P, Cotin S (2020) Simulation of hyperelastic materials in real-time using deep learning. Medical image analysis 59:101,569

Miller K, Lu J (2013) On the prospect of patient-specific biomechanics without patient-specific properties of tissues. Journal of the mechanical behavior of biomedical materials 27:154–166

Miller K, Joldes G, Lance D, Wittek A (2007) Total Lagrangian explicit dynamics finite element algorithm for computing soft tissue deformation. Communications in numerical methods in engineering 23(2):121–134

Modrzejewski R, Collins T, Bartoli A, Hostettler A, Marescaux J (2018) Soft-body registration of pre-operative 3d models to intra-operative RGBD partial body scans. In: International Conference on Medical Image Computing and Computer-Assisted Intervention, Springer, pp 39–46

Moireau P, Chapelle D (2011) Reduced-order unscented Kalman filtering with application to parameter identification in large-dimensional systems. ESAIM: Control, Optimisation and Calculus of Variations 17(2):380–405

Morooka K, Chen X, Kurazume R, Uchida S, Hara K, Iwashita Y, Hashizume M (2008) Real-time nonlinear FEM with neural network for simulating soft organ model deformation. In: International Conference on Medical Image Computing and Computer-Assisted Intervention, Springer, pp 742–749

Muthupillai R, Lomas D, Rossman P, Greenleaf JF, Manduca A, Ehman RL (1995) Magnetic resonance elastography by direct visualization of propagating acoustic strain waves. science 269(5232):1854–1857

Nikolaev S, Peterlik I, Cotin S (2018) Stochastic correction of boundary conditions during liver surgery. In: 2018 Colour and Visual Computing Symposium (CVCS), IEEE, pp 1–4

Niroomandi S, Alfaro I, Cueto E, Chinesta F (2008) Real-time deformable models of non-linear tissues by model reduction techniques. Computer methods and programs in biomedicine 91(3):223–231

Pellicer-Valero OJ, Rupérez MJ, Martínez-Sanchis S, Martín-Guerrero JD (2020) Real-time biomechanical modeling of the liver using machine learning models trained on finite element method simulations. Expert Systems with Applications 143:113,083

Peterlík I, Duriez C, Cotin S (2012) Modeling and real-time simulation of a vascularized liver tissue. In: International Conference on Medical Image Computing and Computer-Assisted Intervention, Springer, pp 50–57

Peterlik I, Courtecuisse H, Duriez C, Cotin S (2014) Model-based identification of anatomical boundary conditions in living tissues. In: International Conference on Information Processing in Computer-Assisted Interventions, Springer, pp 196–205

Peterlik I, Haouchine N, Ručka L, Cotin S (2017) Image-driven stochastic identification of boundary conditions for predictive simulation. In: International Conference on Medical Image Computing and Computer-Assisted Intervention, Springer, pp 548–556

Petit A, Cotin S (2018) Environment-aware non-rigid registration in surgery using physics-based simulation. In: ACCV - 14th Asian Conference on Computer Vision

Pfeiffer M, Riediger C, Weitz J, Speidel S (2019) Learning soft tissue behavior of organs for surgical navigation with convolutional neural networks. International journal of computer assisted radiology and surgery 14(7):1147–1155

Plantefève R, Peterlik I, Haouchine N, Cotin S (2016) Patient-specific biomechanical modeling for guidance during minimally-invasive hepatic surgery. Annals of biomedical engineering 44(1):139–153

Rechowicz KJ, McKenzie FD (2013) Development and validation methodology of the Nuss procedure surgical planner. Simulation 89(12):1474–1488

Roewer-Despres F, Khan N, Stavness I (2018) Towards finite element simulation using deep learning. In: 15th International Symposium on Computer Methods in Biomechanics and Biomedical Engineering

Ronneberger O, Fischer P, Brox T (2015) U-net: Convolutional networks for biomedical image segmentation. In: International Conference on Medical image computing and computer-assisted intervention, Springer, pp 234–241

Ruiter NV, Stotzka R, Muller TO, Gemmeke H, Reichenbach JR, Kaiser WA (2006) Model-based registration of X-ray mammograms and MR images of the female breast. IEEE Transactions on Nuclear Science 53(1):204–211

Sarvazyan AP, Rudenko OV, Swanson SD, Fowlkes JB, Emelianov SY (1998) Shear wave elasticity imaging: a new ultrasonic technology of medical diagnostics. Ultrasound in medicine & biology 24(9):1419–1435

Shepherd JF, Johnson CR (2008) Hexahedral mesh generation constraints. Engineering with Computers 24(3):195–213

Sinkus R, Daire JL, Van Beers BE, Vilgrain V (2010) Elasticity reconstruction: Beyond the assumption of local homogeneity. Comptes Rendus Mécanique 338(7-8):474–479

Suwelack S, Röhl S, Bodenstedt S, Reichard D, Dillmann R, dos Santos T, Maier-Hein L, Wagner M, Wünscher J, Kenngott H, et al (2014) Physics-based shape matching for intraoperative image guidance. Medical physics 41(11):111,901

Tonutti M, Gras G, Yang GZ (2017) A machine learning approach for real-time modelling of tissue deformation in image-guided neurosurgery. Artificial intelligence in medicine 80:39–47

Xu L, Lin Y, Han J, Xi Z, Shen H, Gao P (2007) Magnetic resonance elastography of brain tumors: preliminary results. Acta radiologica 48(3):327–330

Zhao X, Chen X, Lu J (2009) Pointwise identification of elastic properties in nonlinear hyperelastic membranes—part II: experimental validation. Journal of applied mechanics 76(6)

Chapter 6
New Aspects of the Trabecular Bone Remodeling Regulatory Model—Two Postulates Based on Shape Optimization Studies

Michał Nowak

Abstract In the paper the results in the area of shape optimization but related to the simulation models of the trabecular bone remodeling are presented. The interpretation of the results in the context of the trabecular bone adaptation leads to two postulates concerning the simulation of the trabecular remodeling phenomenon.

Keywords: Trabecular bone remodeling · Shape optimization

6.1 Introduction

The regulatory model of the trabecular bone remodeling phenomenon has been described in many papers and books (Huiskes, 2000; Huiskes et al, 2000; Van Oers et al, 2008). The first observation of the amazing behavior of the trabecular bone tissue was made by Wolff and published in 1892 Wolff (1892). The observation proposed by Julius Wolff—called the Wolff's law—can be described as a structural adaptation of the bone to the external forces. Since then, scientific research on the description and development of numerical models of the phenomenon of bone remodeling is still ongoing. As the process of bone remodeling is extremely complex, taking into account the dependence on external loads and various biological aspects, very different numerical modeling approaches are considered for this phenomenon. One of the most interesting of them is the coupled model based on global stimulus of the bone remodeling. The global stimulus approach allows to combine in one model different factors related to the remodeling sequence. The factors are related both to mechanical stimulation but also to other specific biological aspects of the remodeling process like cellular migration and differentiation or nutriments supply (George

M. Nowak
Poznan University of Technology, Chair of Virtual Engineering, Poland, ul. Jana Pawla II 24
60-965 Poznan, Poland
e-mail: Michal.Nowak@put.poznan.pl

B. E. Abali and I. Giorgio (eds.), *Developments and Novel Approaches in Biomechanics and Metamaterials*, Advanced Structured Materials 132,
https://doi.org/10.1007/978-3-030-50464-9_6

et al, 2018, 2019). Other models consider the relationship between mechanical and biological stimulation using the concept of mechanical and biological models coupling to describe the phenomenon of bone remodeling (Giorgio et al, 2016, 2019). From a medical point of view it is also an important issue to include in the models both bone tissue and the tissue reconstructed or reinforced by the addition of an artificial resorbable material (Lekszycki and dell'Isola, 2012). Regardless of the approach used, however, the modeling of structural adaptation to variable load becomes a central research topic.

The same structural adaptation problem is a subject of the shape optimization studies, especially in the area of the stiffest design research (Haftka and Gürdal, 2012; Pedersen, 2003; Plotnikov and Sokołowski, 2012). In this paper, I attempt to present results of our team in the area of shape optimization related to the simulation models of the trabecular bone remodeling phenomenon. The interpretation of the results in the context of the trabecular bone adaptation is based on our previous papers (Nowak et al, 2018, 2020).

6.2 The Stiffest Design Problem

To begin with, the problem of how to obtain the stiffest design and its setting could be presented as follows. In order to maximize the stiffness of a structure (the stiffest design goal), one needs to minimize the functional:

$$J(\Omega) = \int_{\Gamma_1} \boldsymbol{t} \cdot \boldsymbol{u}\, ds \tag{6.1}$$

The constraint is the given volume:

$$\int_{\Omega} dx - V_0 = 0 \tag{6.2}$$

The state equations are as follows:

$$\begin{aligned} \text{div}\boldsymbol{\sigma}(\boldsymbol{u}) &= 0 \quad \text{in} \quad \Omega \\ \boldsymbol{\sigma}(\boldsymbol{u}) \cdot \boldsymbol{n} &= \boldsymbol{t} \quad \text{on} \quad \Gamma_1 \\ \boldsymbol{\sigma}(\boldsymbol{u}) \cdot \boldsymbol{n} &= 0 \quad \text{on} \quad \Gamma_v \\ \boldsymbol{u} &= 0 \quad \text{on} \quad \Gamma_0 \end{aligned} \tag{6.3}$$

Herein above, Ω represents domain of the elasticity system, u the displacement, Ω_0 a given volume, Γ_0 part of the boundary with Dirichlet condition, Γ_1 part of the boundary loaded by traction forces t, Γ_v part of the boundary subject to modification. The domain of the elasticity system and the featured parts are presented in Fig. Nowak-fig1.eps. By analyzing the picture above and construing the presented parts of the domain as the elements of the trabecular bone structure—despite the fact that a single trabecula size ranges from tens to hundreds of micrometres (Hamed et al,

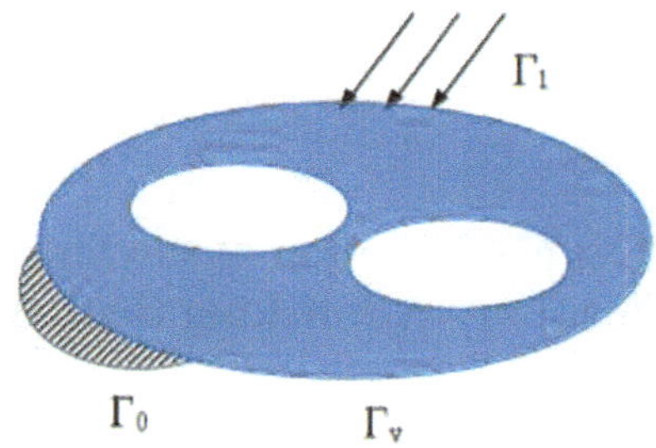

Fig. 6.1 The domain of the elasticity system Ω, Γ_0 part of the boundary with Dirichlet condition, Γ_1 part of the boundary loaded by traction forces, Γ_v part of the boundary subject to modification.

2012), one may imagine that the trabecular structure is the domain of the elasticity system. Subsequently, one can use their result from the area of shape optimization, which concerns Γ_v—a part of the boundary subject to modification.

After defining the Lagrangian for the problem under considerations, where Ω_t is an image of Ω in transformation T_t,

$$L(\Omega_t, \lambda) = \int_{\Gamma_1} \boldsymbol{t} \cdot \boldsymbol{u}_t \, ds + \lambda \left[\int_{\Omega_t} dx - V_0 \right] \tag{6.4}$$

The shape derivative Sokołowski and Zolesio (1992) of the state equation in the weak form gives the formula:

$$\int_{\Gamma_1} \boldsymbol{t} \cdot \boldsymbol{u}' \, ds = - \int_{\Gamma_v} \boldsymbol{\sigma}(\boldsymbol{u}) : \boldsymbol{\varepsilon}(\boldsymbol{u}) \boldsymbol{V} \cdot \boldsymbol{n} \, ds \tag{6.5}$$

Then the shape derivative of Lagrange function using speed method Sokołowski and Zolesio (1992) is taken

$$[L(\Omega_t, \lambda)]' = \int_{\Gamma_1} \boldsymbol{t} \cdot \boldsymbol{u}' \, ds + \lambda \int_{\Gamma_v} \boldsymbol{V} \cdot \boldsymbol{n} \, ds = 0 \tag{6.6}$$

and using the formula (6.5) the derivative of Lagrange function gives

$$\int_{\Gamma_v} \left[\lambda - \boldsymbol{\sigma}(\boldsymbol{u}) : \boldsymbol{\varepsilon}(\boldsymbol{u}) \right] \boldsymbol{V} \cdot \boldsymbol{n} \, ds = 0 \tag{6.7}$$

The derivative at the stationary point should vanish—for details see Nowak et al (2018)—and it can be concluded that

$$\boldsymbol{\sigma}(\boldsymbol{u}) : \boldsymbol{\varepsilon}(\boldsymbol{u}) = \lambda = \text{const.} \tag{6.8}$$

It means that for the stiffest design, the strain energy density on the part of the boundary subject to modification Γ_v must be constant. By referring this result to the structure of the trabecular bone, it can be concluded that if the strain energy density on the structural surface of the trabecular tissue is constant—from the shape optimization point of view—the stiffest design state is obtained.

6.3 The Modification of the Structural Surface Position—Mimicking the Remodeling Process

Going back to the numerical models developed by Huiskes and others Huiskes (2000); Huiskes et al (2000), the regulatory model is based on the strain energy density U as the mechanical stimulation measure. The regulatory model presented in Fig. 6.2. is described by the following equations:

$$\frac{dE}{dt} = \begin{cases} U > U_u & : C_e(U - U_u) \\ U_l \leq U \leq U_u & : 0 \\ U < U_l & : C_e(U - U_l) \end{cases} \tag{6.9}$$

where U_h is the strain energy density value corresponding to homeostasis of bone loss and gain, C_e is a constant value, $2s$ is the size of the "lazy zone", $U_l = U_h(1-s)$, $U_u = U_h(1+s)$ are the values corresponding to lower and upper strain energy level inside the lazy zone, and E denotes Young's modulus of the tissue. The model assumes that the Young's modulus (as a local elastic modulus) of the tissue may vary depending on the level of the mechanical stimulation (strain energy density). The existence of homeostatic value of strain energy density surrounded by the "lazy zone" is assumed. In this way, the bone (trabecular tissue together with marrow) adapts its density value. The "lazy zone" concept was originally proposed by Carter Carter (1984).The approach is similar to this used in topology optimization SIMP method (Solid Isotropic Microstructure with Penalization Bendsoe and Kikuchi (1988)), and

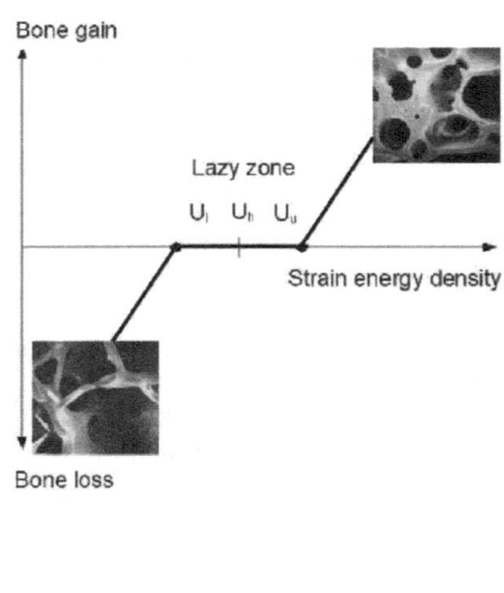

Fig. 6.2 The regulatory model and the "lazy zone" concept.

the Young modulus of the material varies from a value close to zero to the real value, characterized by the Young modulus value for the used material.

To stay closer to the real trabecular remodeling phenomenon, one should take a look at the BMU (Basic Multicellular Unit) Van Oers et al (2008), presented in Fig.6.3, i.e. the area where the remodeling process takes place. The BMU-s are located on the surface of the trabecular tissue to allow cells migration from the marrow. Accordingly, the remodeling process occurs on the surface rather than in the whole trabecular area. Subsequently, one can formulate the first postulate regarding the modification of the trabecular bone remodeling regulatory model.

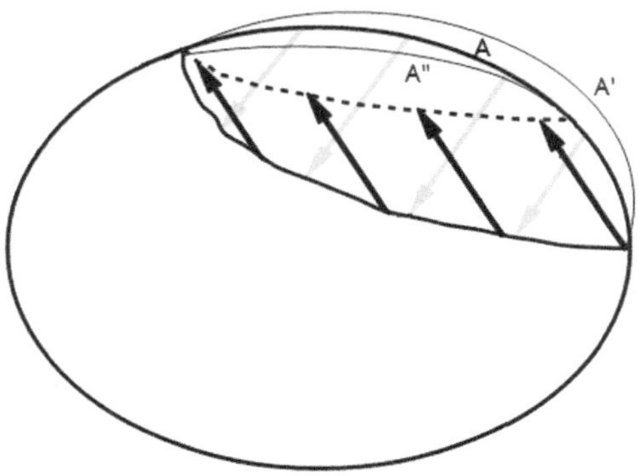

Fig. 6.3 The cross-section of the single trabecula—a scheme of Basic Multicellular Unit (BMU) the bright arrows depict bone loss while the dark arrows represent new tissue formation. A: an initial position of the trabecular tissue surface. A': a new position of the surface for a higher value of strain energy density than the "lazy zone" borders. A": a new position of the surface for a lower value of strain energy density value than "lazy zone" borders.

Postulate 1.: during the remodeling process, the trabecular bone tends to maximize the stiffness of a structure (i.e. to find the stiffest design) by the strain energy density equalization on the structural surface of the trabecular tissue. According to formula (6.8), the strain energy density on the part of the boundary subject to modification (in this case the trabecular bone tissue surrounded by the marrow) must be constant. The local change on the structural surface leads to global minimization of the strain energy for the whole structure. Importantly, the condition from formula (6.8) concerns the structural surface rather than the whole structural volume. However, the fulfillment of this surface condition is tantamount to reaching the goal in the form of minimizing the energy of the entire structure in the domain, i.e. minimization of the functional (6.1). In fact, assuming the possibility of moving the surface of the trabecular tissue in the virtual space, we can modify its position depending on the local value of strain energy density, as presented in Fig. 6.3. If the local value of the strain energy density (on the BMU surface) is higher than the "lazy zone" borders, the surface moves from position A to A'. If, in turn, the local value of the strain energy density (on the BMU surface) is lower than the "lazy zone" borders, the surface moves from position A to A". There is no need for the evolution of Young modulus in such a model, and the obtained structural configurations will reflect the real structure of the trabecular net.

6.4 The Problem of Multiple-Load Conditions

In the regulatory model, the strain energy density is a measure of the mechanical stimulation of the bone tissue. According to formula (6.8), the strain energy density on the structural surface of the trabecular tissue must be constant. However, the loads of the whole bone are very different in time and the distribution of the strain energy density along the structural surface of the trabecular tissue will also be different for various load conditions. Assuming that the trabecular bone can form an optimal structure for multiple loads too, one may consider two different loads, i.e. t_1 and t_2 acting on the same Γ_1 boundary, see Fig. 6.1. Two compliances correspond to these load cases:

$$\boldsymbol{C}_1(\Omega; \boldsymbol{t}_1), \qquad \boldsymbol{C}_2(\Omega; \boldsymbol{t}_2) \tag{6.10}$$

In order to maximize the stiffness of a structure (for multiple loads), one needs to minimize the functional, similar to formula (6.1), with the same constraint (volume) (6.2),

$$J(\Omega) = \alpha_1 \boldsymbol{C}_1(\Omega; \boldsymbol{t}_1) + \alpha_2 \boldsymbol{C}_2(\Omega; \boldsymbol{t}_2) \tag{6.11}$$

where $\alpha_1 + \alpha_2 = 1$, $\alpha_i \geq 0$.

After defining Lagrangian for the problem under considerations, the state equation in the weak form can be rewritten, and then the shape derivative of the Lagrange function can be taken. The derivative at the stationary point should vanish (for details see Nowak et al (2020)), and one may conclude that

$$\alpha_1 \boldsymbol{\sigma}(\boldsymbol{u}_1) : \boldsymbol{\varepsilon}(\boldsymbol{u}_1) + \alpha_2 \boldsymbol{\sigma}(\boldsymbol{u}_2) : \boldsymbol{\varepsilon}(\boldsymbol{u}_2) = \lambda = \text{const.} \tag{6.12}$$

Since $\alpha_1 + \alpha_2 = 1$, one can analyze different load cases and, in order to treat the value of λ as homeostatic value of the strain energy density on the structural surface in the modified regulatory model, formula (6.12) can be extended to other linear combinations of load cases. From the shape optimization point of view, also the result is very interesting. The method was tested for the common multiple load benchmark example (the analytical solution exists), see Nowak et al (2020).

Thereafter, the "lazy zone" concept is coming back to show its necessary presence in the model. If there was no "lazy zone" and different load cases were analyzed separately, the solution (the structural form of the trabecular net) would oscillate between configurations corresponding to each load case. When, for example, the load is applied only in the vertical plane, the resulting structure will also be arranged only in such a plane. If the direction of the load is changed to horizontal, the same solution will be obtained, only rotated by 90 degrees. This will be the case where the "lazy zone" is not taken into consideration but the λ value is assumed. Fig. 6.4.

depicts the Boolean sum of separated solutions of the cantilever beam bending on the left side (vertical and horizontal bending force). Nevertheless, the regulatory

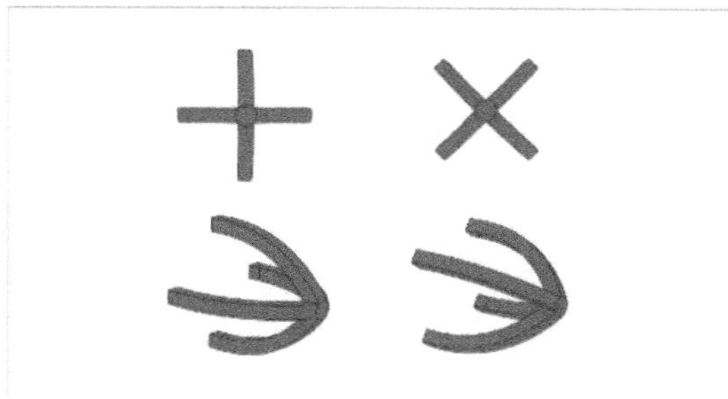

Fig. 6.4 Multiple load problem. *Left*: the Boolean sum of two solution (bending of cantilever beam) for vertical and horizontal bending force. *Right*: the solution of multiple load problem obtained with use of the regulatory model wit assumed "lazy zone".

model is used, the problem can be solved for both load cases according to formula (6.12). The "lazy zone", i.e. insensitivity zone is necessary to avoid oscillations and to consider the impact of all load cases.

Afterwards, the second postulate regarding the modification of the trabecular bone remodeling regulatory model can be formulated. If the existence of the "lazy zone" is not taken into account, then - for the load in the horizontal direction - the term in formula (6.12) corresponding to the load in vertical direction disappears (neutral axis of bending). By the same token, if the direction of force is changed to vertical, the term in formula (6.12) corresponding to the load in horizontal direction will disappear. The solution will oscillate between these two configurations corresponding to each load case. However, if the existence of the "lazy zone" is assumed, the situation will be completely different and one solution for different load cases will be obtained. The solution is presented on the right side in Fig. 6.4. Please note that in this case there is no material in the vertical and horizontal plane. Hence the solution obtained has no common areas with any of the single load bending solutions. Furthermore, it should be emphasized that according to formula (6.12), the same λ value was assumed regardless of the load case. Please note also, that both structures (on the left and right site in Fig. 6.4.) has the same "density," i.e. the same rate of real material and voids in a domain. Both structures have the same volume, but the total structural displacement of the structure on the right side in Fig. 6.4 is lower by 9.8%, while the Huber–Mises stress is lower by 7.9% for each of single load cases.

Finally, it is possible to formulate the second postulate regarding the modification of the regulatory model.

Postulate 2.: the regulatory model of the trabecular bone remodeling can be applied to the multiple load problem maintaining its character and the basic assumptions about the existence of the homeostatic value of the strain energy density, according to formula (6.12). The "lazy zone" is an important element of the model which provides the opportunity to find a solution for many load cases. The local change (in this case related to different loads) on the structural surface leads to global minimization of the strain energy for the whole structure.

6.5 Conclusions

The presented postulates regarding the modification of the trabecular bone remodeling regulatory model require the preparation of appropriate numerical models. Since the local change on the structural surface leads to global minimization of the strain energy for the whole structure, the fulfillment of both postulates requires energy distribution analysis on the structural surface.

To conclude, the need for geometric modeling of structural surface of the trabecular tissue is evident. In addition, the estimation of the λ value and the size of the "lazy zone" seem to be very promising research areas.

Acknowledgements This work was supported by a statutory activity subsidy from the Polish Ministry of Science and Higher Education for the Faculty of Transport Engineering of Poznan University of Technology - grant no. 0554/SBAD/6493 and the computational grant no. 362 Biomimetic approach to compliance optimization and multiple load cases of Poznan Supercomputing and Networking Center (PCSS), Poznan, Poland.

References

Bendsoe MP, Kikuchi N (1988) Generating optimal topologies in structural design using a homogenization method. Computer Methods in Applied Mechanics and Engineering 71(2):197–224

Carter DR (1984) Mechanical loading histories and cortical bone remodeling. Calcified Tissue International 36(1):S19–S24

George D, Allena R, Remond Y (2018) A multiphysics stimulus for continuum mechanics bone remodeling. Mathematics and Mechanics of Complex Systems 6(4):307–319

George D, Allena R, Remond Y (2019) Integrating molecular and cellular kinetics into a coupled continuum mechanobiological stimulus for bone reconstruction. Continuum Mechanics and Thermodynamics 31(3):725–740

Giorgio I, Andreaus U, Scerrato D, dell'Isola F (2016) A visco-poroelastic model of functional adaptation in bones reconstructed with bio-resorbable materials. Biomechanics and Modeling in Mechanobiology 15(5):1325–1343

Giorgio I, dell'Isola F, Andreaus U, Alzahrani F, Hayat T, Lekszycki T (2019) On mechanically driven biological stimulus for bone remodeling as a diffusive phenomenon. Biomechanics and Modeling in Mechanobiology 18(6):1639–1663

Haftka RT, Gürdal Z (2012) Elements of structural optimization, vol 11. Springer Science & Business Media

Hamed E, Jasiuk I, Yoo A, Lee Y, Liszka T (2012) Multi-scale modelling of elastic moduli of trabecular bone. Journal of The Royal Society Interface 9(72):1654–1673

Huiskes R (2000) If bone is the answer, then what is the question? The Journal of Anatomy 197(2):145–156

Huiskes R, Ruimerman R, Van Lenthe GH, Janssen JD (2000) Effects of mechanical forces on maintenance and adaptation of form in trabecular bone. Nature 405(6787):704–706

Lekszycki T, dell'Isola F (2012) A mixture model with evolving mass densities for describing synthesis and resorption phenomena in bones reconstructed with bio-resorbable materials. ZAMM-Journal of Applied Mathematics and Mechanics/Zeitschrift für Angewandte Mathematik und Mechanik 92(6):426–444

Nowak M, Sokołowski J, Zochowski A (2018) Justification of a certain algorithm for shape optimization in 3D elasticity. Structural and Multidisciplinary Optimization 57(2):721–734

Nowak M, Sokołowski J, Zochowski A (2020) Biomimetic approach to compliance optimization and multiple load cases. Journal of Optimization Theory and Applications 184(1):210–225

Pedersen P (2003) Optimal designs–structures and materials-problems and tools, scientific report

Plotnikov P, Sokołowski J (2012) Compressible Navier-Stokes equations: theory and shape optimization, vol 73. Springer Science & Business Media

Sokołowski J, Zolesio JP (1992) Introduction to Shape Optimization: Shape Sensitivity Analysis, Springer Series in Computational Mathematics, vol 16. Springer-Verlag, Heidelberg

Van Oers RF, Ruimerman R, Tanck E, Hilbers PA, Huiskes R (2008) A unified theory for osteonal and hemi-osteonal remodeling. Bone 42(2):250–259

Wolff J (1892) Das Gesetz der Transformation. Transformation der Knochen

Chapter 7
Bone Mechanics and Cell Populations: Mathematical Description and Parametric Study of the Model

Alessio Ciro Rapisarda, Matin Almasi, Naser Almasi, Emilio Barchiesi, Alessandro Della Corte, and Daria Scerrato

Abstract In this paper we study a recently proposed mathematical model for the description of the mechanics of bone as well as bone remodeling processes and bone cell populations dynamics. We investigate the biological meaning and a suitable value for the numerical parameters of the model. To do so, we compare biological data with a systematic numerical investigation of the model. We also propose some corrections to the original model aimed at better describing the observed phenomena.

Keywords: Bone mechanics, Bone cells, Estimate of bone-related parameters, Cell populations dynamics.

A. C. Rapisarda
University of Naples Federico II, Italy
e-mail: a.rapisarda@studenti.unina.it

M. Almasi
Department Of Mechanical Engineering, Isfahan University of Technology, Iran
e-mail: almasi_matin@yahoo.com

N. Almasi
Department of Structural and Geotechnical Engineering,
Medical Laboratory of Isfahan Shariati Hospital (Director), Iran
e-mail: naseralmasi@yahoo.com

E. Barchiesi
Sapienza University of Rome, Roma, Italy
e-mail: barchiesiemilio@gmail.com

A. Della Corte
Mathematics Division, School of Sciences and Technology, University of Camerino, Italy
e-mail: alessandro.dellacorte@unicam.it

D. Scerrato
M&MoCS, University of L'Aquila, Italy
e-mail: daria.scerrato@gmail.com

B. E. Abali and I. Giorgio (eds.), *Developments and Novel Approaches in Biomechanics and Metamaterials*, Advanced Structured Materials 132,
https://doi.org/10.1007/978-3-030-50464-9_7

7.1 Introduction

In the recent past there has been an increasing interest in the development of mathematical descriptions of bone-related phenomena. In fact the subject has quite a long history (for general reference works see e.g. Currey (2012); Pivonka and Komarova (2010); Martin et al (1998)), but in the last years the access to larger amounts of experimental data and also the developments in the mechanics of micromorphic continua (see e.g. Eremeyev and Pietraszkiewicz (2016); Altenbach and Eremeyev (2015); Engelbrecht and Berezovski (2015); dell'Isola et al (2018); Andreaus et al (2018); Franciosi et al (2018) for recent interesting results and Goda et al (2012); Madeo et al (2012); Goda et al (2014); Misra et al (2013, 2015); Sheidaei et al (2019) for applications to bone modeling) have led to a sharp increase in the number of studies. Different models have been proposed, covering almost all the relevant aspects of bone biology, from remodeling (Madeo et al (2012); Lekszycki (2001); Pivonka et al (2008); Scala et al (2017); Giorgio et al (2019)) to functional adaptation (Prendergast and Taylor (1994); Lekszycki (2005, 2002)) to fracture (Lu and Lekszycki (2015); Bednarczyk et al (2018); Lu and Lekszycki (2016); Doblaré et al (2004); Giorgio et al (2017b)) to interaction with artificial grafts (Giorgio et al (2017a); Lu and Lekszycki (2018); Andreaus et al (2008, 2015); Giorgio et al (2016a); George et al (2019)) to cell populations dynamics (Hambli (2014); Komarova et al (2003)). A major problem in order to pass from promising theoretical researches to reliable and predictive models is an accurate parametrization of the model itself. Indeed, an evaluation of the parameters involved in the mathematical description of bone has been attempted several times, but in the opinion of the authors there has been generally an insufficient interaction between mathematical and mechanical literature on one side and biological researches on the other.

In this paper we propose an estimate for the parameters used in the model introduced (in a simplified form) in Della Corte (2017), and described in Rapisarda et al (2018). The model was mainly aimed at the description of bone mechanics and bone cell populations dynamics. Although our investigation will be based on an inspection of recent and authoritative biological literature on bone and bone cells, this has to be considered as a preliminary work mainly aimed at evaluating the order of magnitude of the parameters, although in some cases we are confident to have established a reasonably accurate estimate. Rather than accuracy, however, the principal difficulty in performing the analysis is the intrinsic variability of the involved parameters. In fact, the next step towards a predictive model will probably require the introduction of a stochastic model, in which at least some of the involved quantities are treated as random variables.

7.2 The Model

The mathematical model studied herein is based on the enrichment of the one described in Lekszycki and dell'Isola (2012); Giorgio et al (2016b) with bone cell-populations dynamics. The main novelty of the employed cell populations dynamics model (presented in Rapisarda et al (2018)) is the introduction of the osteocytes density with the relative differentiation terms.

In this section we want to quickly recall the proposed model and the ideas behind it. For a more accurate description the reader is referenced to the works mentioned before.

The model is based on the following assumptions:

- The elastic properties of the bone tissue are described by means of an isotropic Cauchy continuum in small deformations regime.
- The communication between the sensor and actor cells is modeled by means of a scalar field called *Stimulus* and denoted by *S*, defined at every point of the bone and depending on the mechanical load and on the density of sensor cells.
- The status of the system at a given time is given by the macromechanical variables as well as the *internal* state i.e. the scalar fields representing the cells densities;
- The precursor cells are indefinitely available everywhere.

The system of ODEs describing the evolution of cell densities and bone tissue density is

$$\frac{\partial x_k}{\partial t} = -\beta_k X_k + \gamma_{bk} x_b \mathcal{K}(\varphi), \tag{7.1}$$

$$\frac{\partial x_b}{\partial t} = -\beta_b X_b - \gamma_{bk} x_b \mathcal{K}(\varphi) + \alpha_b S^+ x_k, \tag{7.2}$$

$$\frac{\partial x_c}{\partial t} = -\beta_c X_c + \gamma_c x_c \mathcal{K}(\varphi) + \alpha_c S^- x_k, \tag{7.3}$$

$$\frac{\partial \rho}{\partial t} = (a x_b - b x_c) H(\varphi), \tag{7.4}$$

Where x_k, x_b, x_c are respectively the density of osteocytes, osteoblasts and osteoclasts and ρ is the density of the bone tissue. The terms X_k, X_b, X_c are threshold functions defined by

$$X_i = \begin{cases} x_i, & \text{if } x_i > \tilde{x}_i \\ 0, & \text{if } x_i \leqslant \tilde{x}_i \end{cases} \qquad i = k, b, c$$

The coefficients $-\beta_k$, $-\beta_b$, $-\beta_c$ indicate the removal rates of the cells.

The terms $-\gamma_{bk} x_b \mathcal{K}(\varphi)$ and $+\gamma_{bk} x_b \mathcal{K}(\varphi)$ are related to the differentiation from osteoblasts to osteocytes, while $\mathcal{K}$ is a function of bone tissue porosity.

The term $+\gamma_c \mathcal{K}(\varphi)$ models birth and activation of osteoclasts. The terms $+\alpha_b S^+ x_k$ and $+\alpha_c S^- x_k$ model respectively the creation of osteoblasts and osteoclasts due to the stimulus and the osteocytes density: S^+ is the positive part of the stimulus and promotes the birth of osteoblasts, and S^- means the negative part of the stimulus and promotes the birth of osteoclasts.

The parameters a and b are the synthesis and resorption rate of bone tissue (or osteoid) for a single osteoblast or osteoclast. The variable φ indicates the porosity defined like in Lekszycki and dell'Isola (2012); Andreaus et al (2014) as:

$$\varphi = 1 - \theta \frac{\rho}{\rho_{max}} \quad 0 < \theta \leqslant 1 \tag{7.5}$$

where we chose the simplest case $\theta = 1$ and set $\rho_{max} = 0.002\ \frac{\mathrm{g}}{\mathrm{mm}^3}$. (Bochud et al (2017)). The function $H(\varphi)$ used in Rapisarda et al (2018) was a parabola (7.6), and was used in order to account for the influence of effective porosity on the biological activity of actor cells: when effective porosity is too large there is not enough material which actor cells can adhere to, when it is too small there is not enough free space in the pores to allow their mobility and adhesion. In formulas (setting $\tilde{\rho} := \frac{\rho}{\rho_{\max}}$):

$$H(\varphi) = H(1 - \tilde{\rho}) = 4(1 - \tilde{\rho})\tilde{\rho} \tag{7.6}$$

The function $\mathcal{K}(\varphi)$ used in Rapisarda et al (2018) was

$$\mathcal{K}(\varphi) = \mathcal{K}(1 - \tilde{\rho}) = 4(1 - \tilde{\rho})\tilde{\rho} \tag{7.7}$$

In this paper, however, since we are mainly interested in modeling cortical bone, the functions H and K play a different (and less relevant) role, as porosity is always quite small. Thus we assumed for simplicity $H \equiv 1$ and $K \equiv 1$.

We will denote the initial data of (1-4) by x_{k0}, x_{b0},x_{c0} and ρ_0. In a stationary state, they are the physiological values respectively for the osteocytes, osteoblasts, osteoclasts and bone tissue density in a healthy bone sample.

7.2.1 The Stimulus Function

The stimulus function is defined as:

$$S(x,t) = \left(\frac{\int_{\mathcal{B}} U(y,t)\eta x_k(y,t) e^{-\frac{\|x-y\|^2}{D^2}}\, dy}{\int_{\mathcal{B}} e^{-\frac{\|x-y\|^2}{D^2}}\, dy} \right) - S_0(x,t). \tag{7.8}$$

Here $\mathcal{B}$ is the reference configuration, U is the deformation energy density, η measures how much the stimulus is affected by the density of osteocytes and D the range of action of sensor cells. The density of osteocytes appears here because it has been established that osteocytes *in vivo* can amplify the mechanical stimuli through the mechanotrasduction (Bonewald and Johnson (2008); Schaffler and Kennedy (2012); George et al (2018)). The denominator is a normalization factor introduced in order to avoid edge effects (Kumar et al (2011)), while S_0 is a positive function that defines a reference value of stimulus, associated with a physiological amount of loading, which entails a biological equilibrium state where the effect of resorption and synthesis are balanced. Herein for simplicity we will assume S_0 as constant and uniform.

7.2.2 The Mechanical Equations

The Deformation gradient ($\boldsymbol{F}$), its determinant (J), and the Green-Saint-Venant strain tensor ($\boldsymbol{G}$) are defined as usual:

$$\boldsymbol{F} = \nabla\chi, \qquad J = \det \boldsymbol{F}, \qquad 2\boldsymbol{G} = \boldsymbol{F}^T\boldsymbol{F} - \boldsymbol{I}, \tag{7.9}$$

where $\chi : \mathcal{B} \to \mathbb{R}^3$ is the placement function. The strain energy is defined as:

$$U(\boldsymbol{G}, \rho, x) = \mu \mathrm{tr}(\boldsymbol{G}^2) + \frac{\lambda}{2}(\mathrm{tr}(\boldsymbol{G}))^2. \tag{7.10}$$

where μ and λ are the Lamé parameters. Since the material is not homogeneous, and its density evolves with the time, μ and λ are assumed as functions depending on t and x:

$$\mu = \hat{\mu}(\rho(t), x), \quad \lambda = \hat{\lambda}(\rho(t), x). \tag{7.11}$$

Young's modulus and Poisson ratio are related to Lamé parameters by the well-known relations:

$$\lambda = \frac{Y\nu}{(1+\nu)(1-2\nu)}, \tag{7.12}$$

$$\mu = \frac{Y}{2(1+\nu)} \tag{7.13}$$

where Y is the Young modulus, ν is the Poisson ratio, λ and μ are the Lamé coefficients. The Young modulus for the bone tissue is assumed to be (Carter and Hayes (1977))

$$Y = Y_{mb}\left(\frac{\rho}{\rho_{\max}}\right)^{\omega_b}, \tag{7.14}$$

where $\omega_b = 2$, Y_{mb} is the maximum theoretical value for the Young modulus. A physiological value for the Young modulus of a cortical bone is 18 GPa (Cowin (2001)). Assuming 0.2 as a typical value for the porosity of a cortical bone Giorgio et al (2016b), this leads to $Y_{mb} \approx 28$ GPa. The Poisson ratio was set as $\nu = 0.14$, adapting to the isotropic case and averaging the experimental results given in Shahar et al (2007) (see Cluzel and Allena (2018); Allena and Cluzel (2018) for a better insight on mechanical properties of bone). Notice that these results concern cortical bone. In our numerical results we thus expect values of the density and the porosity which are consistent with this fact.

The equilibrium equations are

$$\mathrm{Div}\boldsymbol{T} = \mathrm{Div}\left(\boldsymbol{F}\cdot\frac{\partial U}{\partial \boldsymbol{G}}\right) = -\boldsymbol{b}^{ext}, \tag{7.15}$$

$$\boldsymbol{T}[\boldsymbol{N}] = \boldsymbol{F}\cdot\frac{\partial U}{\partial \boldsymbol{G}}\cdot\boldsymbol{N} = \boldsymbol{f}^{ext}, \tag{7.16}$$

where $\boldsymbol{T}$ is the first Piola stress tensor, $\boldsymbol{F}$ is the deformation gradient and $\boldsymbol{G}$ the strain tensor. The (7.15) gives the force acting on an RVE of the considered body's volume ($-\boldsymbol{b}^{ext}$) instead the (7.16) gives the force acting on the unity surfaces of the considered body ($\boldsymbol{f}^{ext}$) (with outward unit normal $\boldsymbol{N}$).

Finally, we point out that we will interpret our 2D model as a thin layer of (cortical) bone tissue having the thickness of an average *lamella*, i.e. $\approx 10\mu$m (Pazzaglia et al (2012)). We remark that this choice corresponds to the average thickness of the layer of osteoid deposited by an osteoblast or of bone tissue resorbed by an osteoclast (Dumitrescu et al (2007)).

7.3 Estimate of the Model's Parameters

In this section we start the systematic study of the parameters of the model, so as to give realistic values for them (in Rapisarda et al (2018) all of them were non dimensional and assessed simply by means of agreement between numerical results and observations). The parameters that we will consider are x_{k0}, x_{b0}, x_{c0}, a, b, D, β_b, β_c, β_k, γ_{bk}, γ_c, α_b and α_c. Our chosen time unit is one day. Since the stress response of the cells takes place in minutes (Schaffler and Kennedy (2012)), with this choice we are able to treat the stimulus function as a quantity which is averaged over a suitably long time interval.

7.3.1 Estimate of Initial Data and Removal/Production Rates of the Cells

Initial Cell Populations

We are interested in reproducing the physiological behavior of a healthy bone sample in a stationary state. Therefore, we wanted to use, as initial data for cell densities, average physiological values. It is not easy to assess these values, because they have significant intrinsic variability (being strongly influenced by age, sex and pathological conditions (Dalzell et al (2009); Parfitt et al (1997)). We used percentage values taken from Schaffler and Kennedy (2012) and Pawlina and Ross (2006):

$$x_{k0} = 0.94N \quad x_{b0} = 0.05N \quad x_{c0} = 0.01N \tag{7.17}$$

where N is the total number of bone cells in a unit volume. Averaging the results from 12 human samples in Bromage et al (2016), where the number of osteocytes has been evaluated counting the *lacunae*, we obtain

$$x_k = 316/\text{mm}^2 \tag{7.18}$$

whence, using (7.17) and approximating to the closer integer

$$x_{b0} = 17/\text{mm}^2$$
$$x_{c0} = 3/\text{mm}^2$$

The Value of D

The parameter D represents the range of action of sensor cells. In Ruimerman et al (2005) the proposed value for D was 100 μm. However, in Burra et al (2010) the length of dendritic processes of osteocytes is estimated at 10 μm. The last article supports the claim with confocal imaging, and therefore will be considered reliable herein. However there is experimental evidence that dendritic processes can elongate up to more than 30 μm *in vitro* if suitably stimulated (Zhang et al (2006)). Moreover it is known that osetoecytes can respond to and transmit signals over long distances while embedded apart from each other in a calcified matrix (Takano-Yamamoto (2014); Ishihara et al (2008, 2013)). It is very difficult to estimate this distances, so we chose to use for D a value which is one order of magnitude larger than the average length of non-stimulated dendritic processes. Therefore we set

$$D = 100\ \mu\text{m}$$

We observe that the term $e^{-\frac{\|x-y\|^2}{D^2}}$ approaches very quickly zero when $||x - y||$ becomes larger than D.

Removal/Production Rates of the Cells

The parameters β_b, β_c, β_k are the removal rates (meanly due to apoptosis) of the corresponding cell densities and are assumed to be constant. We obtained an estimate for these parameters starting from the lifespan of the different cell populations (data from Manolagas (2000); Rosenberg et al (2012)). In Manolagas (2000) birth and death of osteoblasts and osteoclasts are studied *in vitro*, and therefore no osteoblasts differentiation into osteocytes has to be taken into account (we recall that osteoblasts do not undergo mitosis see e.g. Manolagas (2000)). This allows us to estimate the apoptosis rates for this two cell populations applying stationarity and neglecting the term $-\gamma_{bk} x_b \mathcal{K}(\varphi)$ in (7.2). Furthermore, we are in absence of external load and of osteocytes, so the term $\alpha_b S^+ x_k$ can be neglected too. Finally, the threshold $\tilde{x}_b$ describes a feedback due to bio-chemical communications among the different kinds of cells, and therefore is neglected too, so that we study directly the variable x_b.

Using 45 days for the half-life of osteoblasts we obtain β_b, as follows:

$$x_b(45) = \frac{x_{b0}}{2} = x_{b0} e^{-45\beta_b} \tag{7.19}$$

whence

$$\beta_b = 0.015 \tag{7.20}$$

With a similar procedure we can obtain β_c. We neglect the term $\gamma_c x_c \mathcal{K}(\varphi)$ in the equation (7.3), and precisely as before also the term $\alpha_c S^- x_k$ can be neglected. Using 7 days for the half-life of osteoclasts (Manolagas (2000)) we obtain β_b (again we consider the variable x_c instead of X_c for similar reasons as before)

$$x_c(7) = \frac{x_{c0}}{2} = x_{c0} e^{-7\beta_c}$$

$$\beta_c = 0.099$$

As for β_k, let us start observing that the life-span of osteocytes is 10-20 years Pawlina and Ross (2006), but osteocytes death rate is strongly influenced by the evolution of the bone tissue. Indeed, it has been recently reported that the bone loss observed in Crohn's disease is associated with increased osteocyte death and decreased bone remodeling (Dallas et al (2013); Dallas and Bonewald (2010)) and furthermore the death of osteocytes is connected with bone tissue resorption (Dallas et al (2013); Noble (2008); Zarrinkalam et al (2012)). In addition, dying osteocytes can undergo a process of self-preservation called autophagy to preserve themselves until favorable conditions return (Bonewald (2011)). Because of this evidence, a more accurate form of the term $\beta_k X_k$ is

$$-\lambda \beta^* x_c H(\varphi) x_k - \beta_k X_k$$

where $\lambda \beta^* x_c H(\varphi) x_k$ represents the osteocytes dead as a consequence of bone resorption. The factor λ is a simple threshold function which lets this term activate only above a certain density threshold (which constitutes another parameter of the model):

$$\lambda = \begin{cases} \bar{\lambda}, & \text{if } \rho > \rho_{critical} \\ 0, & \text{if } \rho \leqslant \rho_{critical} \end{cases}$$

In the present contribution we do not want to enrich the model in this way, so we simply estimated β_k with the same procedure of β_b and β_c considering a much longer half-life, namely 7.5 years $\approx$ 2700 days. In this way we obtain the estimate:

$$\beta_k = 2.5 \cdot 10^{-4}$$

The Values of Cells Populations Thresholds

Measuring directly $\tilde{x}_k$, $\tilde{x}_b$ and $\tilde{x}_c$ is not a simple task. These thresholds are introduced to model what in reality is a very complex feedback mechanism due to bio-chemical communications between different kinds of cells. Herein we chose to have very low thresholds, so as to keep the apoptotic process activated unless there is a sharp lack in a specific cell population. This is in agreement with the fact that generally stationary behaviors in cell populations dynamics are a consequence of an equilibrium between birth and death of cells, in which apoptotic processes are normally activated (Lawen

(2003)). Therefore we set

$$\tilde{x}_k = x_{k0}/10$$
$$\tilde{x}_b = x_{b0}/10$$
$$\tilde{x}_c = x_{c0}/10$$

7.3.2 Estimate of the Differentiation Rates and Coupling Coefficients

The Value of γ_{bk}

The parameter γ_{bk} can be seen (in the simplest case $\mathcal{K}(\varphi) = 1$) as the probability for an osteoblast to become an osteocyte in a given day. It is known that about 10-20% of osteoblasts experience differentiation into osteocytes in their lifetime Noble (2008); Pawlina and Ross (2006). Using 15% and supposing an initial population with uniformly distributed ages we get

$$(1 - \gamma_{bk})^{90} = 0.85$$

whence

$$\gamma_{bk} = 1.8 \cdot 10^{-3}$$

The Value of γ_c

The term $\gamma_c \mathcal{K}(\varphi)$ models the birth of new osteoclasts. Since in the proposed model (differently from, e.g. Komarova et al (2003)) we introduced another source term ($\alpha_c S^- x_k$) depending on the interaction with other cells and on the stimulus, the parameter γ_c has the only potential role of describing osteoclasts proliferation. Since mature osteoclasts are terminally differentiated cells that do not undergo mitosis (see e.g. Li et al (2015)), we can set

$$\gamma_c = 0$$

The Value of a and b

The term a represents the synthesis rate of a single osteoblast in a time unit, while b represents the resorption rate for a single osteoclast in a time unit. Imposing stationarity to Eq.(7.4) we get

$$\frac{a}{b} = \frac{x_{c0}}{x_{b0}} \approx 0.18 \tag{7.21}$$

A direct estimate of the bone production rate per osteoblast is in Gruber et al (1986), where it is obtained from a sample of women suffering of postmenopausal osteoporosis

$$\text{production rate per osteoblast} = 2.9 \cdot 10^{-5} \frac{\text{mm}^2}{\text{day}}$$

In Bose and Bandyopadhyay (2016) it is reported that osteoporotic patients have a normal rate of osteoid production per cell. Therefore we will use the value for a estimated in Gruber et al (1986). To transform it in grams per day we have to consider that the thickness of a lamella is ≈ 10 µm (Pazzaglia et al (2012)) and recall that a physiological value for cortical bone density is $1.6 \cdot 10^{-3}$ g/mm^3. This leads to the estimate:

$$a = 4.6 \cdot 10^{-10} \frac{\text{g}}{\text{day}}$$

Using Eq.(7.21) we get

$$b = 2.6 \cdot 10^{-9} \frac{\text{g}}{\text{day}}$$

As an internal consistency check we can compute b from data given in Gruber et al (1986), as done for a. The proposed value is

$$\text{resorption rate per osteoclast nucleus} = 1.6 \cdot 10^{-5} \frac{\text{mm}^2}{\text{day}}$$

Using again 10µm for the thickness and $1.6 \cdot 10^{-3}$ g/mm^3 for the density, we get $2.56 \cdot 10^{-9} \frac{g}{\text{day}}$. This is in agreement with the value obtained imposing stationarity in our model[1].

The Value of α_b and α_c

The terms α_b and α_c model respectively the differentiation rate of osteoblasts and osteoclasts from progenitor cells due to the stimulus and osteocytes density. In fact the presence of these two parameters allows us to set $\eta = 1$ (more precisely, to move it outside the integral and suitably redefine α_b, α_c and S_0). Imposing stationarity to Eqs.(7.2)-(7.3), we get

$$\alpha_b S^+ = (\beta_b + \gamma_{bk}) \frac{x_{b0}}{x_{k0}}$$
$$\alpha_c S^- = (\beta_c + \gamma_c) \frac{x_{c0}}{x_{k0}}$$

[1] We remark that in this case we used the lowest extremum of the range of values for osteoclasts resorption rate given in Gruber et al (1986). However, even using the other extremum we have agreement for what concerns the order of magnitude

This leaves us with three final degrees of freedom, i.e. S_0, α_b and α_c, to be adjusted using numerical simulations.

7.4 Numerical Simulations

The model has been simulated using Comsol Multiphysics (Weak Form PDEs and ODE solver). Finite Element Method has been employed to solve the static elastic problems. In particular, for the mechanical computations Hermite elements with third order polynomials have been used, whereas the ODE system has been solved using the Backward Differentiation Formula, with a variable order depending on the desired accuracy. Our sample is a rectangle sized 1 mm $\times$ 0.2 mm, meant to model a portion of the thin bone layer (*lamella*) deposited by osteoblasts. For the evaluation of magnitudes distributed per unit volume, we therefore treated the sample as a thin parallelepiped having the typical thickness ($\approx$ 10µm) of a *lamella*.

Our first goal was to estimate the parameters S_0, α_b and α_c so as to obtain a reasonable stationary state when the initial values for bone tissue density and cell populations were physiological. We point out that in fact these parameters could not have been estimated from purely biological data as the other ones, as they describe the way external loads affect cell populations dynamics. We considered a compressive load applied to the short side (while the opposite side is clamped) whose time history has a normal distribution around the mean value $L_m = 2$ N/mm with standard deviation of 0.07 N/mm. A 120-days simulation using the initial values given in Table 7.1 is shown in Fig.7.1. The figure refers to cell concentrations and

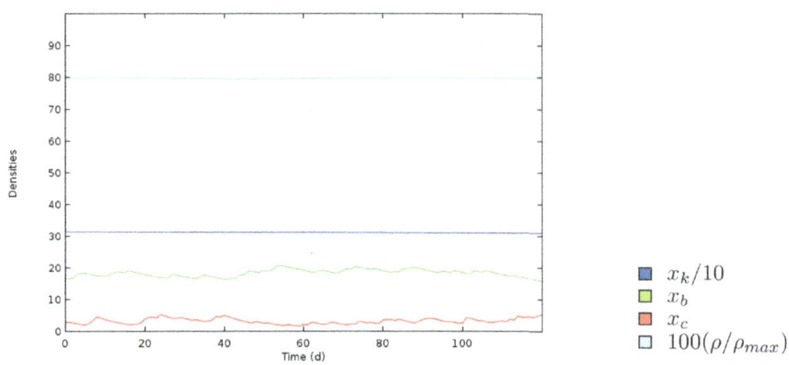

Fig. 7.1 Cell concentrations and bone tissue density at the central point of the sample in a compressive test.

bone tissue density at the central point of the sample. In the simulations the employed values are $S_0 = 0.0344$, $\alpha_b = 1$ days^{-1}, $\alpha_c = 1.2$ days^{-1}. With these values, it can be seen that the initial physiological state is maintained, so that the system can

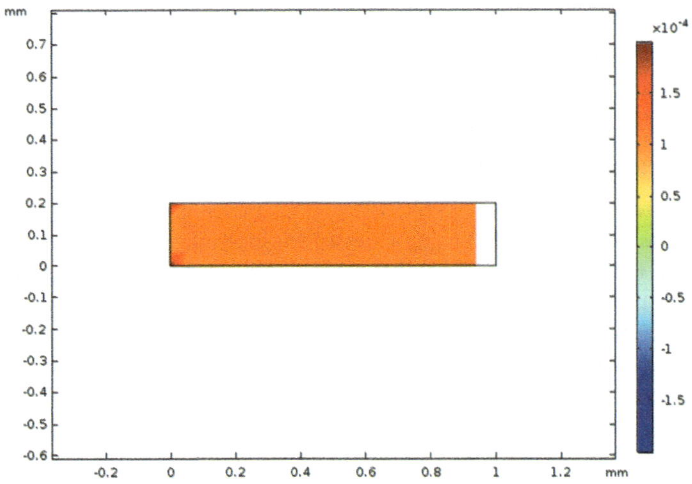

Fig. 7.2 Deformation energy at day 120.

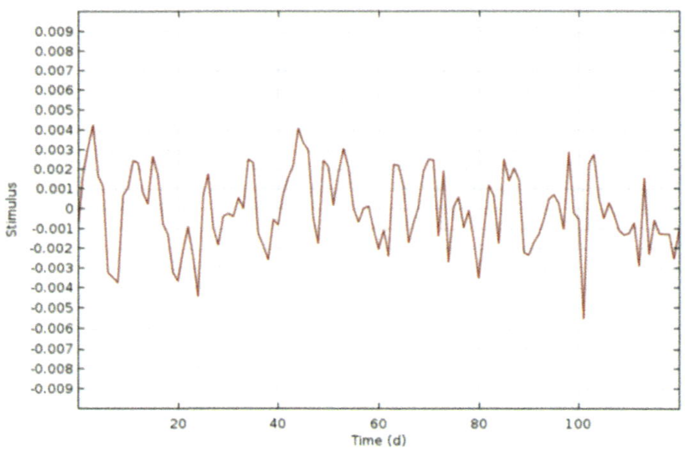

Fig. 7.3 Stimulus evaluated at the central point of the sample.

be considered in a stationary state. The final distribution of deformation energy and the stimulus function $S(t)$ are shown in Figs. 7.2-7.3.

We remark that every time we refer to values evaluated in the central point of the sample; no significant differences are visible considering other points. Next we checked what happens if we start with values of the initial cell populations densities which are far from the normal physiological values. In Fig.7.4, we start with an osteoblasts population which is everywhere less than in the previous case, i.e. $x_{b0} = 6$ instead of $x_{b0} = 17$. We can see that in presence of normal production

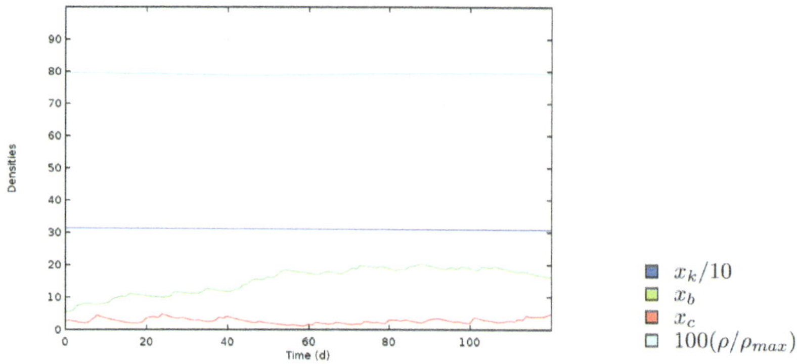

Fig. 7.4 Cell concentrations and bone tissue density at the central point of the sample in a compressive test with initial low value for osteoblasts population.

rates, the system adjusts itself to normal physiological values in the considered time period of 120 days. A similar result is obtained if we start with a higher value for the initial osteoclasts population, i.e. $x_{c0} = 9$ instead of $x_{c0} = 3$, as shown in Fig. 7.5. It can be seen that the system lets the osteoblasts population increase beyond normal

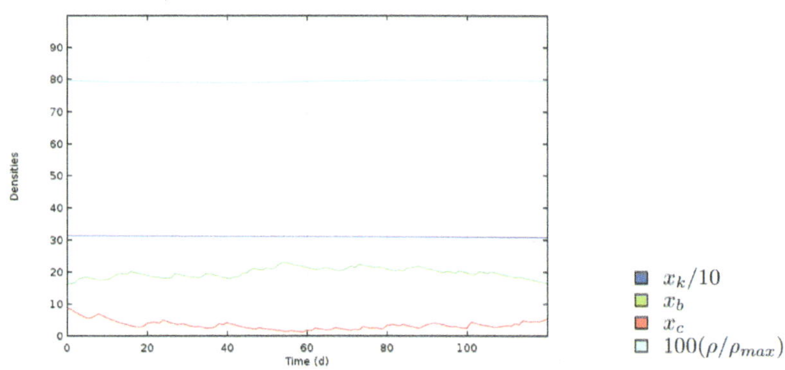

Fig. 7.5 Cell concentrations and bone tissue density at the central point of the sample in a compressive test with initial high value for osteoclasts population.

values during the phase in which the osteoclasts are more numerous. In this way, the bone tissue density can remain more or less stable throughout. In the following simulation we checked what happens if the external load is significantly lower, i.e. $L_m = 1.5$ N/mm instead of $L_m = 2$ N/mm. The result is shown in Fig. 7.6. It is interesting to observe that, after an initial phase in which the osteoclasts increase much, the system reaches a new stationary state with a different (lower) value of bone tissue density. It is also noticeable that the stimulus function has a negative

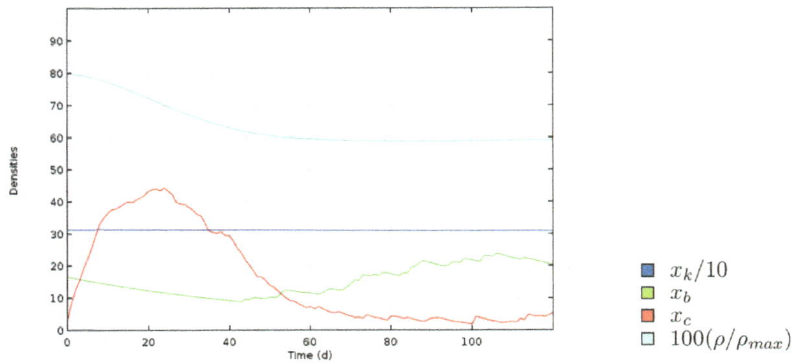

Fig. 7.6 Cell concentrations and bone tissue density at the central point of the sample in a compressive test with a lower value (-25%) of the load.

average at the beginning, but after some time it adjusts to the usual fluctuations around zero, even if the average load per day remains lower than in the previous simulation (see Fig. 7.7). The numerical simulations presented in this section show

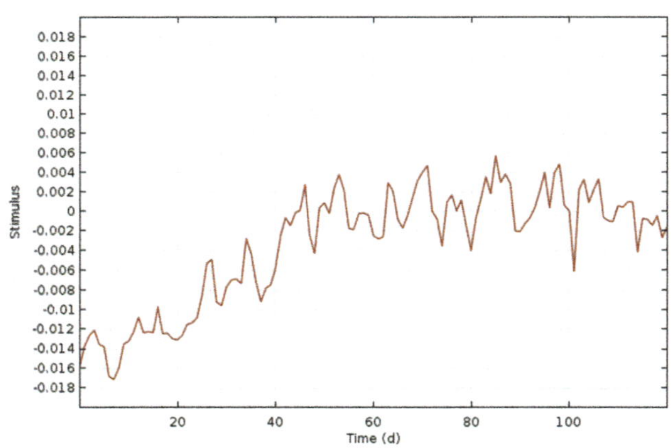

Fig. 7.7 Stimulus evaluated at the central point of the sample with a lower (-25%) external load.

that the parametrization of the model was effective, in that the results are consistent and realistic.

7.5 Summary of the Results

Values and short definitions of the parameters proposed in the paper are summarized in Table 7.1.

Table 7.1 Legend of symbols and their initial values.

ρ_0	physiological cortical bone density	g/mm^3	1.6×10^{-3}
ρ_{max}	maximum cortical bone density	g/mm^3	2×10^{-3}
Y_p	physiological cortical bone Young modulus	GPa	18.6
Y_{mb}	maximum cortical bone Young modulus	GPa	28
ν	cortical bone Poisson coefficient	-	0.14
D	osteocytes range of influence	mm	0.1
x_{k0}	physiological density of osteocytes	1/mm^2	316
x_{b0}	physiological density of osteoblasts	1/mm^2	17
x_{c0}	physiological density of osteoclasts	1/mm^2	3
$\tilde{x}_k$	lower threshold for osteocytes density	1/mm^2	31.6
$\tilde{x}_b$	lower threshold for osteoblasts density	1/mm^2	1.7
$\tilde{x}_c$	lower threshold for osteoclasts density	1/mm^2	0.3
a	rate of bone synthesis for one osteoblast	g/day	4.6×10^{-10}
b	rate of bone resorption for one osteoclast	g/day	2.6×10^{-9}
β_k	death rate of osteocytes	1/day	2.5×10^{-4}
β_b	death rate of osteoblasts	1/day	0.015
β_c	death rate of osteoclasts	1/day	0.099
γ_{bk}	rate of differentiation from osteoblasts to osteocytes	1/day	1.8×10^{-3}
γ_c	proliferation rate of osteoclasts	1/day	0
α_b	production rate of osteoblasts	1/day	1
α_c	production rate of osteoclasts	1/day	1.2
S_0	reference value of the stimulus	-	0.0344
L	longitudinal length of the sample	mm	1
l	width of the sample	mm	0.2
h	thickness of the sample	mm	0.01
L_m	average load	N/mm	2

7.6 Conclusions

In this work we addressed the problem of determining, with reasonable accuracy, the values of the parameters of a mathematical model for bone mechanics and bone cell populations dynamics (described in Rapisarda et al (2018)). The parameters were determined using biological data from recent literature, with the exception of the coefficients α_b and α_c measuring how much the mechanical stimulation affects the production/removal rates of osteoblasts and osteoclasts, and the normal physiological value for the stimulus (see in that regard Abali (2016)). The numerical simulations showed that in this way the model is capable of producing consistent and realistic results. Further investigations are required to improve the reliability of the model, and probably the introduction of stochastic ingredients, capable of accounting for the intrinsic variability of individual biology, have to be taken into account. Moreover, in order to describe trabecular bone, in which microstructure plays an even more relevant role, the assumption of a classical Cauchy continuum can be relaxed in favor of more general continuum theories (see e.g. Placidi et al (2018b); Alibert et al (2003); dell'Isola et al (2009); Giorgio (2016); Placidi and Barchiesi (2018); Placidi et al (2018a); Abali et al (2012); dell'Isola et al (2015); Placidi and Hutter (2005)).

References

Abali BE (2016) Computational Reality: Solving Nonlinear and Coupled Problems in Continuum Mechanics, vol 55. Springer

Abali BE, Völlmecke C, Woodward B, Kashtalyan M, Guz I, Müller WH (2012) Numerical modeling of functionally graded materials using a variational formulation. Continuum Mechanics and Thermodynamics 24(4-6):377–390

Alibert JJ, Seppecher P, dell'Isola F (2003) Truss modular beams with deformation energy depending on higher displacement gradients. Mathematics and Mechanics of Solids 8(1):51–73

Allena R, Cluzel C (2018) Heterogeneous directions of orthotropy in three-dimensional structures: finite element description based on diffusion equations. Mathematics and Mechanics of Complex Systems 6(4):339–351

Altenbach H, Eremeyev V (2015) On the constitutive equations of viscoelastic micropolar plates and shells of differential type. Mathematics and Mechanics of Complex Systems 3(3):273–283

Andreaus U, Giorgio I, Lekszycki T (2014) A 2-D continuum model of a mixture of bone tissue and bio-resorbable material for simulating mass density redistribution under load slowly variable in time. ZAMM-Journal of Applied Mathematics and Mechanics/Zeitschrift für Angewandte Mathematik und Mechanik 94(12):978–1000

Andreaus U, Giorgio I, Madeo A (2015) Modeling of the interaction between bone tissue and resorbable biomaterial as linear elastic materials with voids. Zeitschrift für angewandte Mathematik und Physik 66(1):209–237

Andreaus U, Spagnuolo M, Lekszycki T, Eugster SR (2018) A Ritz approach for the static analysis of planar pantographic structures modeled with nonlinear euler–bernoulli beams. Continuum Mechanics and Thermodynamics pp 1–21

Andreaus UA, Colloca M, Toscano A (2008) Mechanical behaviour of a prosthesized human femur: a comparative analysis between walking and stair climbing by using the finite element method. Biophysics and Bioengineering Letters 1(3)

Bednarczyk EI, Lekszycki T, Glinkowski W (2018) Effect of micro-cracks on the angiogenesis and osteophyte development during degenerative joint disease. Computer Assisted Methods in Engineering and Science 24(3):149–156

Bochud N, Vallet Q, Minonzio JG, Laugier P (2017) Predicting bone strength with ultrasonic guided waves. Scientific reports 7:43,628

Bonewald LF (2011) The amazing osteocyte. Journal of bone and mineral research 26(2):229–238

Bonewald LF, Johnson ML (2008) Osteocytes, mechanosensing and Wnt signaling. Bone 42(4):606–615

Bose S, Bandyopadhyay A (2016) Materials and Devices for Bone Disorders. Academic Press

Bromage TG, Juwayeyi YM, Katris JA, Gomez S, Ovsiy O, Goldstein J, Janal MN, Hu B, Schrenk F (2016) The scaling of human osteocyte lacuna density with body size and metabolism. Comptes Rendus Palevol 15(1):32–39

Burra S, Nicolella DP, Francis WL, Freitas CJ, Mueschke NJ, Poole K, Jiang JX (2010) Dendritic processes of osteocytes are mechanotransducers that induce the opening of hemichannels. Proceedings of the National Academy of Sciences 107(31):13,648–13,653

Carter DR, Hayes WC (1977) Compact bone fatigue damage: a microscopic examination. Clinical Orthopaedics and Related Research 127:265–274

Cluzel C, Allena R (2018) A general method for the determination of the local orthotropic directions of heterogeneous materials: application to bone structures using μCT images. Mathematics and Mechanics of Complex Systems 6(4):353–367

Cowin SC (ed) (2001) Bone Mechanics Handbook, 2nd edn. CRC Press, Boca Raton

Currey JD (2012) The structure and mechanics of bone. Journal of Materials Science 47(1):41–54

Dallas SL, Bonewald LF (2010) Dynamics of the transition from osteoblast to osteocyte. Annals of the New York Academy of Sciences 1192(1):437–443

Dallas SL, Prideaux M, Bonewald LF (2013) The osteocyte: an endocrine cell? and more. Endocrine reviews 34(5):658–690

Dalzell N, Kaptoge S, Morris N, Berthier A, Koller B, Braak L, Van Rietbergen B, Reeve J (2009) Bone micro-architecture and determinants of strength in the radius and tibia: age-related changes in a population-based study of normal adults measured with high-resolution pqct. Osteoporosis international 20(10):1683–1694

Della Corte A (2017) Modeling synthesis and resorption phenomena in bones by means of mixture models enhanced with computational population dynamics. Part 2: models of bone cells population dynamics. In: France-Italy Workshop Bone biomechanics: multiscale and multiphysical aspects, Giuliano di Roma, Italy

dell'Isola F, Sciarra G, Vidoli S (2009) Generalized hooke's law for isotropic second gradient materials. In: Proceedings of the Royal Society of London A: Mathematical, Physical and Engineering Sciences, The Royal Society, pp rspa–2008

dell'Isola F, Andreaus U, Placidi L (2015) At the origins and in the vanguard of peridynamics, non-local and higher-gradient continuum mechanics: An underestimated and still topical contribution of Gabrio Piola. Mathematics and Mechanics of Solids 20(8):887–928

dell'Isola F, Seppecher P, et al (2018) Pantographic metamaterials: an example of mathematically driven design and of its technological challenges. Continuum Mechanics and Thermodynamics doi: 10.1007/s00161-018-0689-8:1–34

Doblaré M, Garcıa J, Gómez M (2004) Modelling bone tissue fracture and healing: a review. Engineering Fracture Mechanics 71(13-14):1809–1840

Dumitrescu G, Ştef L, Drinceanu D (2007) Studies regarding the identification of some histo-morphological parameters taht could be used as markers for bone mineralization apprecisal in pigs–scientific support in osteoporosis study. Scientific Papers Animal Science and Biotechnologies 40(1):340–347

Engelbrecht J, Berezovski A (2015) Reflections on mathematical models of deformation waves in elastic microstructured solids. Mathematics and Mechanics of Complex Systems 3(1):43–82

Eremeyev VA, Pietraszkiewicz W (2016) Material symmetry group and constitutive equations of micropolar anisotropic elastic solids. Mathematics and Mechanics of Solids 21(2):210–221

Franciosi P, Spagnuolo M, Salman OU (2018) Mean Green operators of deformable fiber networks embedded in a compliant matrix and property estimates. Continuum Mechanics and Thermodynamics pp 1–32 (doi: 10.1007/s00,161–018–0668–0)

George D, Allena R, Remond Y (2018) A multiphysics stimulus for continuum mechanics bone remodeling. Mathematics and Mechanics of Complex Systems 6(4):307–319

George D, Allena R, Remond Y (2019) Integrating molecular and cellular kinetics into a coupled continuum mechanobiological stimulus for bone reconstruction. Continuum Mechanics and Thermodynamics 31(3):725–740

Giorgio I (2016) Numerical identification procedure between a micro-cauchy model and a macro-second gradient model for planar pantographic structures. Zeitschrift für angewandte Mathematik und Physik 67(4):95

Giorgio I, Andreaus U, Madeo A (2016a) The influence of different loads on the remodeling process of a bone and bioresorbable material mixture with voids. Continuum Mechanics and Thermodynamics 28(1-2):21–40

Giorgio I, Andreaus U, Scerrato D, dell'Isola F (2016b) A visco-poroelastic model of functional adaptation in bones reconstructed with bio-resorbable materials. Biomechanics and Modeling in Mechanobiology 15(5):1325–1343

Giorgio I, Andreaus U, dell'Isola F, Lekszycki T (2017a) Viscous second gradient porous materials for bones reconstructed with bio-resorbable grafts. Extreme Mechanics Letters 13:141–147

Giorgio I, Andreaus U, Scerrato D, Braidotti P (2017b) Modeling of a non-local stimulus for bone remodeling process under cyclic load: Application to a dental implant using a bioresorbable porous material. Mathematics and Mechanics of Solids 22(9):1790–1805

Giorgio I, dell'Isola F, Andreaus U, Alzahrani F, Hayat T, Lekszycki T (2019) On mechanically driven biological stimulus for bone remodeling as a diffusive phenomenon. Biomechanics and Modeling in Mechanobiology 18(6):1639–1663

Goda I, Assidi M, Belouettar S, Ganghoffer J (2012) A micropolar anisotropic constitutive model of cancellous bone from discrete homogenization. Journal of the mechanical behavior of biomedical materials 16:87–108

Goda I, Assidi M, Ganghoffer JF (2014) A 3D elastic micropolar model of vertebral trabecular bone from lattice homogenization of the bone microstructure. Biomechanics and modeling in mechanobiology 13(1):53–83

Gruber H, Ivey J, Thompson E, Baylink D, et al (1986) Osteoblast and osteoclast cell number and cell activity in postmenopausal osteoporosis. Mineral and electrolyte metabolism 12(4):246–254

Hambli R (2014) Connecting mechanics and bone cell activities in the bone remodeling process: an integrated finite element modeling. Frontiers in bioengineering and biotechnology 2:6

Ishihara Y, Kamioka H, Honjo T, Ueda H, Takano-Yamamoto T, Yamashiro T (2008) Hormonal, ph, and calcium regulation of connexin 43–mediated dye transfer in osteocytes in chick calvaria. Journal of Bone and Mineral Research 23(3):350–360

Ishihara Y, Sugawara Y, Kamioka H, Kawanabe N, Hayano S, Balam TA, Naruse K, Yamashiro T (2013) Ex vivo real-time observation of ca2+ signaling in living bone in response to shear stress applied on the bone surface. Bone 53(1):204–215

Komarova SV, Smith RJ, Dixon SJ, Sims SM, Wahl LM (2003) Mathematical model predicts a critical role for osteoclast autocrine regulation in the control of bone remodeling. Bone 33(2):206–215

Kumar C, Jasiuk I, Dantzig J (2011) Dissipation energy as a stimulus for cortical bone adaptation. Journal of Mechanics of Materials and Structures 6(1):303–319

Lawen A (2003) Apoptosis - an introduction. Bioessays 25(9):888–896

Lekszycki T (2001) On a class of bone cell-based remodeling laws with spatial fading influence of stimuli. Engineering Transactions 49(2-3):155–164

Lekszycki T (2002) Modelling of bone adaptation based on an optimal response hypothesis. Meccanica 37(4-5):343–354

Lekszycki T (2005) Functional adaptation of bone as an optimal control problem. Journal of Theoretical and Applied Mechanics 43(3):555–574

Lekszycki T, dell'Isola F (2012) A mixture model with evolving mass densities for describing synthesis and resorption phenomena in bones reconstructed with bio-resorbable materials. ZAMM - Zeitschrift für Angewandte Mathematik und Mechanik 92(6):426–444

Li F, Sun X, Zhao B, Ma J, Zhang Y, Li S, Li Y, Ma X (2015) Effects of cyclic tension stress on the apoptosis of osteoclasts in vitro. Experimental and therapeutic medicine 9(5):1955–1961

Lu Y, Lekszycki T (2015) Modeling of an initial stage of bone fracture healing. Continuum Mechanics and Thermodynamics 27(4-5):851–859

Lu Y, Lekszycki T (2016) A novel coupled system of non-local integro-differential equations modelling young's modulus evolution, nutrients-supply and consumption during bone fracture healing. Zeitschrift für angewandte Mathematik und Physik 67(5):111

Lu Y, Lekszycki T (2018) New description of gradual substitution of graft by bone tissue including biomechanical and structural effects, nutrients supply and consumption. Continuum Mechanics and Thermodynamics pp 1–15

Madeo A, George D, Lekszycki T, Nierenberger M, Remond Y (2012) A second gradient continuum model accounting for some effects of micro-structure on reconstructed bone remodelling. Comptes Rendus Mécanique 340(8):575–589

Manolagas SC (2000) Birth and death of bone cells: basic regulatory mechanisms and implications for the pathogenesis and treatment of osteoporosis. Endocrine reviews 21(2):115–137

Martin RB, Burr DB, Sharkey NA, et al (1998) Skeletal tissue mechanics, vol 190. Springer

Misra A, Marangos O, Parthasarathy R, Spencer P (2013) Micro-scale analysis of compositional and mechanical properties of dentin using homotopic measurements. In: Biomedical Imaging and Computational Modeling in Biomechanics, Springer, pp 131–141

Misra A, Parthasarathy R, Singh V, Spencer P (2015) Micro-poromechanics model of fluid-saturated chemically active fibrous media. ZAMM-Zeitschrift für Angewandte Mathematik und Mechanik 95(2):215–234

Noble BS (2008) The osteocyte lineage. Archives of biochemistry and biophysics 473(2):106–111

Parfitt A, Han ZH, Palnitkar S, Rao DS, Shih MS, Nelson D (1997) Effects of ethnicity and age or menopause on osteoblast function, bone mineralization, and osteoid accumulation in iliac bone. Journal of Bone and Mineral Research 12(11):1864–1873

Pawlina W, Ross MH (2006) Histology: a text and atlas: with correlated cell and molecular biology. Lippincott Wiliams & Wilkins Philadelphia, PA

Pazzaglia UE, Congiu T, Marchese M, Spagnuolo F, Quacci D (2012) Morphometry and patterns of lamellar bone in human haversian systems. The Anatomical Record: Advances in Integrative Anatomy and Evolutionary Biology 295(9):1421–1429

Pivonka P, Komarova SV (2010) Mathematical modeling in bone biology: From intracellular signaling to tissue mechanics. Bone 47(2):181–189

Pivonka P, Zimak J, Smith DW, Gardiner BS, Dunstan CR, Sims NA, Martin TJ, Mundy GR (2008) Model structure and control of bone remodeling: a theoretical study. Bone 43(2):249–263

Placidi L, Barchiesi E (2018) Energy approach to brittle fracture in strain-gradient modelling. Proc R Soc A 474(2210):20170,878

Placidi L, Hutter K (2005) An anisotropic flow law for incompressible polycrystalline materials. Zeitschrift für angewandte Mathematik und Physik ZAMP 57(1):160–181

Placidi L, Barchiesi E, Misra A (2018a) A strain gradient variational approach to damage: a comparison with damage gradient models and numerical results. Mathematics and Mechanics of Complex Systems 6(2):77–100

Placidi L, Misra A, Barchiesi E (2018b) Two-dimensional strain gradient damage modeling: a variational approach. Zeitschrift für angewandte Mathematik und Physik 69(3):56

Prendergast PJ, Taylor D (1994) Prediction of bone adaptation using damage accumulation. Journal of biomechanics 27(8):1067–1076

Rapisarda A, Della Corte A, Drobnicki R, Di Cosmo F, Rosa L (2018) A model for bone mechanics and remodelling including cell populations dynamics. Submitted

Rosenberg N, Rosenberg O, Soudry M (2012) Osteoblasts in bone physiology - mini review. Rambam Maimonides medical journal 3(2)

Ruimerman R, Hilbers P, Van Rietbergen B, Huiskes R (2005) A theoretical framework for strain-related trabecular bone maintenance and adaptation. Journal of biomechanics 38(4):931–941

Scala I, Spingarn C, Remond Y, Madeo A, George D (2017) Mechanically-driven bone remodeling simulation: Application to LIPUS treated rat calvarial defects. Mathematics and Mechanics of Solids 22(10):1976–1988

Schaffler MB, Kennedy OD (2012) Osteocyte signaling in bone. Current osteoporosis reports 10(2):118–125

Shahar R, Zaslansky P, Barak M, Friesem A, Currey J, Weiner S (2007) Anisotropic poisson's ratio and compression modulus of cortical bone determined by speckle interferometry. Journal of biomechanics 40(2):252–264

Sheidaei A, Kazempour M, Hasanabadi A, Nosouhi F, Pithioux M, Baniassadi M, Remond Y, George D (2019) Influence of bone microstructure distribution on developed mechanical energy for bone remodeling using a statistical reconstruction method. Mathematics and Mechanics of Solids p 1081286519828418

Takano-Yamamoto T (2014) Osteocyte function under compressive mechanical force. Japanese Dental Science Review 50(2):29–39

Zarrinkalam M, Mulaibrahimovic A, Atkins G, Moore R (2012) Changes in osteocyte density correspond with changes in osteoblast and osteoclast activity in an osteoporotic sheep model. Osteoporosis International 23(4):1329–1336

Zhang K, Barragan-Adjemian C, Ye L, Kotha S, Dallas M, Lu Y, Zhao S, Harris M, Harris SE, Feng JQ, et al (2006) E11/gp38 selective expression in osteocytes: regulation by mechanical strain and role in dendrite elongation. Molecular and cellular biology 26(12):4539–4552

Chapter 8
Non-Linear or Quasi-Linear Viscoelastic Property of Blood for Hemodynamic Simulations

Ernesto Romano, Luísa C. Sousa, Carlos C. António, Catarina F. Castro, and Sónia Isabel Silva Pinto

Abstract Hemodynamic simulations with the complex rheology of blood is still a challenge. They can be used to obtain an auxiliary clinical tool, as close as possible to reality, with great potential for the development of preventive measures, diagnosis and treatment of cardiovascular diseases. A wide range of models defining the rheological behavior of blood, ranging from the Newtonian to the purely shear-thinning non-Newtonian models have been used by many authors. However, in vessels, such as carotid or coronary arteries, the validity of such simplified models for blood is not completely clear, mainly in stenotic or aneurysm cases - regions of high velocity gradients. It is well-known, from literature, that blood has complex rheology, behaving as a viscoelastic non-Newtonian fluid due to the storage and release of elastic energy from red blood cells aggregates. Therefore, authors of the present work implemented the viscoelastic property of blood, in UDFs of Ansys® software, in order to simulate the most accurate hemodynamics. Afterwards, the velocity contours, in the middle plane of a 3D idealized coronary artery, were obtained considering the purely shear-thinning model, Carreau model, and two viscoelastic non-Newtonian models. Using the Generalized Oldroyd-B, a quasi-linear model, the viscoelastic effects are not highlighted. Comparing results taking into account the multi-mode Giesekus, a non-linear model, and Carreau model, differences are significant and equal to 0.20 m/s under a maximum velocity of 1.40 m/s (14.3%). Using the multi-mode Giesekus model, the viscoelastic effects are pronounced in addition to the shear-thinning, mainly in regions with high velocity gradients as the stenotic region.

E. Romano, L. C. Sousa, C. C. António, C. F. Castro, S. I. S. Pinto
Engineering Faculty, University of Porto, Porto, Portugal
Institute of Science and Innovation in Mechanical and Industrial Engineering (LAETA-INEGI), Porto, Portugal
e-mail: up201404532@fe.up.pt, lcsousa@fe.up.pt, cantonio@fe.up.pt, ccastro@fe.up.pt, spinto@fe.up.pt

B. E. Abali and I. Giorgio (eds.), *Developments and Novel Approaches in Biomechanics and Metamaterials*, Advanced Structured Materials 132,
https://doi.org/10.1007/978-3-030-50464-9_8

Keywords: Viscoelasticity · Non-linear models · Quasi-linear models · Blood rheology · Hemodynamics

8.1 Introduction

Hemodynamic simulations have proven to be an auxiliary clinical tool with great potential for the development of preventive measures, diagnosis and treatment of cardiovascular diseases. However, the numerical tool should mimic physiological conditions and blood properties as close as possible to reality. There are several models in the literature that can simulate the behavior of blood. Auffray et al (2015) formulated a description for second gradient continua in order to mimic capillary fluids, i.e, fluids for which the deformation energy depends on the second gradient of placement. A Lagrangian action was introduced in both the material and spatial descriptions. The corresponding Euler-Lagrange equations and boundary conditions were found. These conditions were formulated in terms of an objective deformation energy. Eremeyev and Altenbach (2014) have discussed the equilibrium equations and natural boundary conditions also for a second-gradient fluid interacting with a nonlinear elastic solid under finite deformations. They have also taken into account the surface stresses acting at the surface of the solid according to the model. They applied the variational approach based on the energy functional. Rickert et al (2019) have described the flow of fluids with internal rotational degrees of freedom, for example a blood plasma carrying red blood cells (RBC). This blood behavior can be described by the theory of Eringen. Eringen's approach, also known as the micropolar theory of fluids, relies on a consistent use of the complete spin balance and the concept of the conservation of microinertia. They studied such fluids not only from the mechanical point of view, i.e., determining the linear and angular velocities, but also from a thermodynamic one, such as studying the generation of a temperature field during the flow due to internal dissipation. Thus, this requires the balance of momentum, spin and internal energy in combination.

Many authors specialized, concretely, in cardiovascular engineering field have used a wide range of models defining the rheological behavior of blood, ranging from the Newtonian to the purely shear-thinning non-Newtonian models. The particles are oriented randomly in the minimum energy states and the RBC in plasma undergo reversible aggregation, the rouleaux (Thanapong Chaichana, Zhonghua Sun, 2012; De Santis et al, 2013; Lee et al, 2008; Morbiducci et al, 2011; Van Canneyt et al, 2013). However, in vessels, such as carotid or coronary arteries, the validity of the Newtonian and the purely shear-thinning non-Newtonian hypotheses is not completely clear, mainly in stenotic or aneurysm cases – regions of high velocity gradients. It is well-known from literature that blood has a viscoelastic non-Newtonian behavior (Baskurt and Meiselman, 2003; Bodnár et al, 2011; Campo-Deaño et al, 2013, 2015) due to the storage and release of elastic energy from RBC aggregates.

Baskurt and Meiselman (2003) have described the way in which blood viscosity is affected by hematocrit, shear rate and red blood cells aggregation. They also emphasize the importance of red blood cell deformability and list factors which affect the cellular mechanical property. Campo-Deaño et al (2013) achieved, experimentally, several parameters, namely the mobility factor and the extensibility coefficient, for viscoelastic non-Newtonian models of blood at 37° C –the multi-mode Giesekus and simplified Phan-Thien-Tanner (sPTT) models. Later, Campo-Deaño et al (2015) presented a state-of-the-art review of the different models used in the hemodynamics, focusing on modeling blood as a viscoelastic non-Newtonian fluid, in order to understand the role of the complex rheology of blood upon the dynamics in aneurysms.

Nevertheless, few authors have considered the viscoelastic property of blood in numerical simulations. Bodnár et al (2011) demonstrated and quantified the most relevant non-Newtonian characteristics of blood flow in vessels, namely its shear-thinning and viscoelastic behavior. Numerical simulations, through a finite-volume method, were performed in a 3D idealized stenosed vessel, with nominal vessel diameter equal to 6.2 mm. Four models for blood were taken into account: the Newtonian (NS) and the Generalized Newtonian (GNS) models; and the Oldroyd-B (OB) and the Generalized Oldroyd-B (GOB) models. The NS model assumes constant viscosity of blood at infinite shear rate, the GNS considers fluid with variable shear-thinning viscosity, OB takes into account the elastic property of blood and constant viscosity at infinite shear rate and GOB assumes the elastic property of blood with variable shear-thinning viscosity. At constant flow rate, the impact of non-Newtonian effects was observed and viscoelasticity of blood was highlighted. Therefore, simulations considering the complex rheology of blood, viscoelasticity, are of great interest since the most accurate hemodynamic is essential for clinical practice. There is a need for the use of models depicting this behavior.

Thus, authors of the present paper want to take a step forward in the numerical hemodynamic simulations through the implementation and validation of a more accurate rheological model for blood in *User-Defined Functions* (UDF) associated to the *Ansys*® software package. *Ansys*® software was chosen since it is a user-friendly software, widely used by other authors. So that, the UDFs implemented by authors of the present paper can be, in the future, easily used by other authors.

In the present work, a 3D idealized geometry of a stenosed bifurcation, mimicking a right coronary artery (RCA) bifurcation, was chosen to show the accuracy of using the implemented viscoelastic models. Two different viscoelastic non-Newtonian models also able to predict shear-thinning behavior - a Generalized Oldroyd-B model, a quasi-linear model (Bird et al, 1987), and a multi-mode Giesekus model, a non-linear model (Larson, 1988) - were compared with a simpler Generalized Newtonian model – Carreau Model. For all models, time-dependent velocity and pressure profiles of pulsatile flow and pressure waveforms, characteristics of a right coronary artery, were imposed as boundary conditions for numerical simulations.

8.2 Materials and Methods

8.2.1 Mathematical Models for Blood Rheology

The governing equations, taking into account the principles of mass conservation and linear momentum conservation for an incompressible fluid, used in blood flow dynamics, can be defined by:

$$\begin{aligned} \nabla \cdot \mathbf{u} &= 0 \\ \rho\left(\frac{\partial \mathbf{u}}{\partial t} + \mathbf{u} \cdot \nabla \mathbf{u}\right) &= -\nabla p + \nabla \cdot \boldsymbol{\tau} \end{aligned} \tag{8.1}$$

where $\mathbf{u}$ is the velocity vector, ρ the blood density, p the pressure, t the instant time and $\boldsymbol{\tau}$ the extra stress tensor. These governing equations can also describe fluids with viscoelastic non-Newtonian behavior using a constitutive equation defining $\boldsymbol{\tau}$.

Generally, the total stress $\boldsymbol{\tau}$ is expressed by the sum of the solvent part $\boldsymbol{\tau}_s$ and the elastic part $\boldsymbol{\tau}_e$:

$$\boldsymbol{\tau} = \boldsymbol{\tau}_s + \boldsymbol{\tau}_e \tag{8.2}$$

where $\boldsymbol{\tau}_s$ is equal to:

$$\boldsymbol{\tau}_s = 2\mu_s \mathbf{D} \tag{8.3}$$

depending on the viscosity of the solvent part (μ_s) and the strain rate tensor ($\mathbf{D}$).

The elastic stress, $\boldsymbol{\tau}_e$, satisfies the following equations:

$$\begin{aligned} f(\boldsymbol{\tau}_e)\boldsymbol{\tau}_e + \lambda \overset{\nabla}{\boldsymbol{\tau}}_e + \alpha \frac{\lambda}{\mu_e}(\boldsymbol{\tau}_e \cdot \boldsymbol{\tau}_e) &= 2\mu_e \mathbf{D} \\ f(\boldsymbol{\tau}_e) &= 1 + \frac{\lambda \varepsilon}{\mu_e} tr(\boldsymbol{\tau}_e) \end{aligned} \tag{8.4}$$

where μ_e is the viscosity related to the elastic part of the fluid, α is the mobility factor, ε the extensibility coefficient and $\overset{\nabla}{\boldsymbol{\tau}}_e$ is the upper-convected derivative in the elastic contribution of the extra stress tensor.

Since blood has complex rheology, three models were considered in order to observe the importance of considering the viscoelasticity of blood in the hemodynamics. The simplest model chosen is a Generalized Newtonian Model, purely shear-thinning model (without viscoelasticity), through Carreau Model:

$$\mu_s(\dot{\gamma}) = \mu_\infty + (\mu_0 - \mu_\infty) \times \left[1 + (\lambda\dot{\gamma})^2\right]^{\frac{n-1}{2}} \tag{8.5}$$

In this model, $\boldsymbol{\tau}$ is equal to $\boldsymbol{\tau}_s$ and $\boldsymbol{\tau}_e$ is equal to 0. μ_s is the viscosity of the solvent part and $\dot{\gamma}$ the shear rate. For blood at 37°C, the viscosity at infinite shear rate (μ_∞) is equal to 0.00345 Pa s, the viscosity at zero shear rate (μ_0) equal to 0.056 Pa s, the relaxation time (λ) is 3.313 s and the power index (n) equal to 0.3568 (Johnston et al, 2004).

Two different viscoelastic non-Newtonian models were used to also predict the shear-thinning behavior of blood: the Generalized Oldroyd-B model and the multi-mode Giesekus model.

The Generalized Oldroyd-B considers both the mobility factor (α) and the extensibility coefficient (ε) of Equation (8.4) equal to 0 – a quasi-linear model (Bird et al, 1987). Thus, the constitutive equation becomes:

$$\boldsymbol{\tau}_e + \lambda \overset{\triangledown}{\boldsymbol{\tau}}_e = 2\mu_e \mathbf{D} \tag{8.6}$$

The viscosity related to the elastic part (μ_e) is equal to 4.0×10^{-6} Pa s and the relaxation time (λ) is 0.06 s, for blood (Bodnár et al, 2011). The shear-thinning viscosity (μ_s) was defined through Carreau Model represented by Equation (9.8) and parameters for blood are the same as defined previously (Johnston et al, 2004).

The Giesekus model, defining viscoelasticity and shear-thinning, was used in multi-mode form. Each mode number is defined by k. The viscoelastic multi-mode Giesekus model does not take into account the extensibility coefficient (ε) of Equation (8.4) ($\varepsilon = 0$). However, the model considers the mobility factor (α). Therefore, the viscoelastic multi-mode Giesekus model is a non-linear model (Larson, 1988) represented by:

$$\boldsymbol{\tau}_{e_k} + \lambda_k \overset{\triangledown}{\boldsymbol{\tau}}_{e_k} + \frac{\alpha_k \lambda_k}{\mu_{e_k}} (\boldsymbol{\tau}_{e_k} \cdot \boldsymbol{\tau}_{e_k}) = 2\mu_{e_k} \mathbf{D} \tag{8.7}$$

The total elastic stress ($\boldsymbol{\tau}_e$) is the sum of the elastic stress of each k mode ($\boldsymbol{\tau}_{e_k}$) in the total of m modes.

$$\boldsymbol{\tau}_e = \sum_{k=1}^{m} \boldsymbol{\tau}_{e_k} \tag{8.8}$$

Parameters of the multi-mode Giesekus for whole human blood were obtained experimentally by Campo-Deaño et al (2013) and can be shown in Table 9.1.

In addition to the shear-thinning and viscoelastic property of blood, blood was also considered as isotropic, incompressible and homogeneous fluid with constant density (ρ = 1060 kg/m^3).

Table 8.1 Parameters of the multi-mode Giesekus model for human blood (Campo-Deaño et al, 2013)

Mode	μ_{e_k} $[Pa \cdot s]$	λ_k $[s]$	α_k
1	0.05	7	0.06
2	0.001	0.4	0.001
3	0.001	0.04	0.001
4	0.0016	0.006	0.001
Solvent		$\mu_s = 0.0012\,\mathrm{Pa\,s}$	

8.2.2 Implementation of the Viscoelastic Models

The previous viscoelastic models are not included in *Ansys® Fluent* package. However, they can be implemented through *user-defined-functions* (UDFs). UDFs are functions or subroutines programmed in a modified C language which are loaded in *Ansys® Fluent*. This software was used in the present work, to implement the viscoelastic models for blood and further hemodynamic simulations, since it is a user-friendly software widely used by other authors. Therefore, the UDFs implemented by authors of the present paper can be, in the future, easily manipulated by other authors.

The Einstein notation was used in order to compact extensive equations. Einstein notation implies the sum of a set of indexed terms in a formula. In the current case, the subscript *n* must be replaced for a sum of the different Cartesian components, i.e, *x*, *y* and *z*. Thus, the upper-convected derivative equation becomes:

$$\overset{\triangledown}{\tau}_{ij_k} = \frac{\partial \tau_{ij_k}}{\partial t} + u_n \frac{\partial \tau_{ij_k}}{\partial x_n} - \tau_{nj_k} \frac{\partial u_i}{\partial x_n} - \tau_{in_k} \frac{\partial u_j}{\partial x_n} \tag{8.9}$$

Adding Eq. (8.9) to Eq. (9.6), the equation with the upper convected derivative terms on the left side was obtained:

$$\begin{aligned} \frac{\partial \tau_{ij_k}}{\partial t} + u_n \frac{\partial \tau_{ij_k}}{\partial x_n} - \tau_{nj_k} \frac{\partial u_i}{\partial x_n} - \tau_{in_k} \frac{\partial u_j}{\partial x_n} = \frac{2\mu_{e_k} D_{ij}}{\lambda_k} - \\ \frac{1}{\lambda_k} f(\tau_{ij_k}) \tau_{ij_k} - \frac{\alpha_k}{\mu_{e_k}} (\tau_{in_k} \cdot \tau_{nj_k}) \end{aligned} \tag{8.10}$$

Eq. (8.10) can be simplified as:

$$\frac{\partial \tau_{ij_k}}{\partial t} + u_n \frac{\partial \tau_{ij_k}}{\partial x_n} = S_{\tau_{ij_k}} \tag{8.11}$$

where $S_{\tau_{ij_k}}$ are the source terms for each stress component, and for each mode, defined as:

$$S_{\tau_{ij_k}} = \frac{2\mu_{e_k} D_{ij}}{\lambda_k} - \frac{1}{\lambda_k} f(\tau_{ij_k})\tau_{ij_k} - \frac{\alpha_k}{\mu_{e_k}}(\tau_{in_k} \cdot \tau_{nj_k}) + \tau_{nj_k}\frac{\partial u_i}{\partial x_n} + \tau_{in_k}\frac{\partial u_j}{\partial x_n} \tag{8.12}$$

The last step of the implementation was the analysis of the Navier–Stokes equations. The basic conversion of the moment equation must be modified in order to include the decomposition of the stress tensor (Eq. (8.10)). As such, there is a need to take into account the elastic parts of the stress, which are calculated as scalars. This was achieved through the addition of the divergence of the extra stress tensor, $\boldsymbol{\tau}_e$, as sources to the momentum equations, known in *Ansys*® *Fluent* as momentum sources.

$$\begin{aligned} S_{M_x} &= \sum_{k=1}^{m} \frac{\partial \tau_{xn_k}}{\partial x_n} \\ S_{M_y} &= \sum_{k=1}^{m} \frac{\partial \tau_{yn_k}}{\partial x_n} \\ S_{M_z} &= \sum_{k=1}^{m} \frac{\partial \tau_{zn_k}}{\partial x_n} \end{aligned} \tag{8.13}$$

8.2.3 3D Geometry and Computational Mesh

A 3D idealized geometry representing a bifurcation of a RCA was constructed in *Solidworks*® (Fig. 9.1a). The main branch representing a RCA starts with 3 mm diameter and after bifurcation decreases to 2.5 mm. The side-branch, with much lower diameter than the main branch, was considered to have a 1.5 mm diameter. A 40% lumen stenosis in the main branch just after the bifurcation was also designed in order to observe the viscoelastic effects in these regions of flow acceleration and recirculation.

The 3D idealized geometry with the inlet and outlet boundaries perpendicular to the blood flow, and the axis defined at the inlet, was imported to *Meshing Ansys*® to construct the computational mesh.

A tetrahedral mesh was defined in all the domain of the artery (Fig. 9.1b). The *Path Independent Method of Meshing Ansys*® was selected in order to uniform the elements and to obtain an accurate mesh (Ansys, 2013). So, the statistical parameter *Skewness* was used to verify the precision of the mesh. A *Maximum Skewness* of 0 indicates the best case scenario, equilateral cells, while a *Maximum Skewness* equal of 1 indicates the worst case scenario, completely degenerated cells. Following the tutorial guide of *Ansys*®, the mesh is accurate when the *Maximum Skewness* is lower than 0.95 (Ansys, 2013). The computational mesh of this work has a *Maximum*

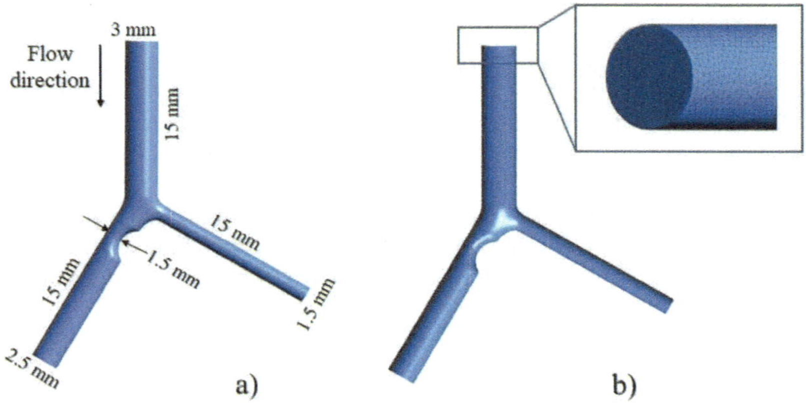

Fig. 8.1 (a) 3D geometry of the idealized RCA constructed in *SolidWorks®*; (b) 3D computational mesh obtained through *Meshing Ansys®* software.

Skewness equal to 0.58 with 182909 elements, which is considered accurate for numerical simulations.

8.2.4 Boundary Conditions

Boundary conditions must be imposed. At the inlet of the idealized geometry, a Womersley velocity profile was taken into account. This profile depends on the instant time of the cardiac cycle, the radial position at the inlet and the Womersley number:

$$W_o = R\sqrt{\frac{\rho\omega}{\mu}} \tag{8.14}$$

The Womersley number (W_o) is defined by the radius of the artery (R), the blood density (ρ), the viscosity of blood (μ) at infinite shear rate and the cardiac frequency (ω). For the present geometry, W_o is equal to 2.40 corresponding to an inlet diameter of the artery equal to 3 mm. At the outlet branches, pressure profiles were imposed. These profiles are dependent on the instant time of the cardiac cycle but radius-independent.

The boundary conditions, defined previously, for RCAs, were also implemented in UDFs in *Ansys®* software by some authors of the present paper (Pinho et al, 2019a,b). Fig. 9.2 shows the mean velocity profile imposed at the inlet of the idealized artery and the pressure profile at the outlet branches.

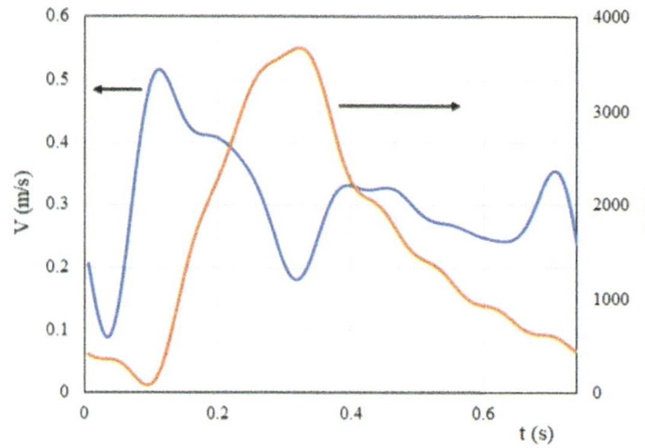

Fig. 8.2 Mean velocity profile imposed at the inlet of the idealized RCA (blue line) and pressure profile imposed at the outlet branches (orange line).

8.2.5 Numerical Method

Ansys® Fluent software was used to perform computational fluid dynamic (CFD) simulations of unsteady blood flow. Navier–Stokes equations were solved in a laminar regime, since Reynolds number in the systolic peak does not exceed the value of 1000. The velocity-pressure coupled equations were solved by the SIMPLE algorithm. The momentum equations with the implemented source terms were discretized by the second-order upwind scheme. The analysis was performed considering a total time of the cardiac cycle equal to 0.74 s, using 148 times steps, each one equal to 0.05 s; the number of iterations for each time step was equal to 20. The simulation process was completed according the convergence criteria of 1×10^{-4}.

8.3 Results and Discussion

Fig. 9.3 represents the velocity contours, in the systolic peak (maximum velocity of the cardiac cycle), along the middle plane of the 3D idealized coronary geometry, for three different rheological models: a Generalized Newtonian model, purely shear-thinning model, through Carreau model; and two viscoelastic non-Newtonian models as the Generalized Oldroyd-B and the multi-mode Giesekus.

For the three cases, there is an acceleration of blood flow in the stenotic region, where the maximum velocity is 1.40 m/s, and there are also recirculation regions just after the stenosis. However, Fig. 9.3 shows that the effect of the viscoelastic components of stress decreases the velocity of blood flow in the stenosis and increases the velocity in the recirculation regions. This effect is highlighted considering the multi-mode Giesekus model.

The Generalized Oldroyd-B model assumes that both mobility factor and extensibility coefficient are equal to 0 and only one mode. It is considered a quasi-linear model (Bird et al, 1987). The multi-mode Giesekus model takes into account four

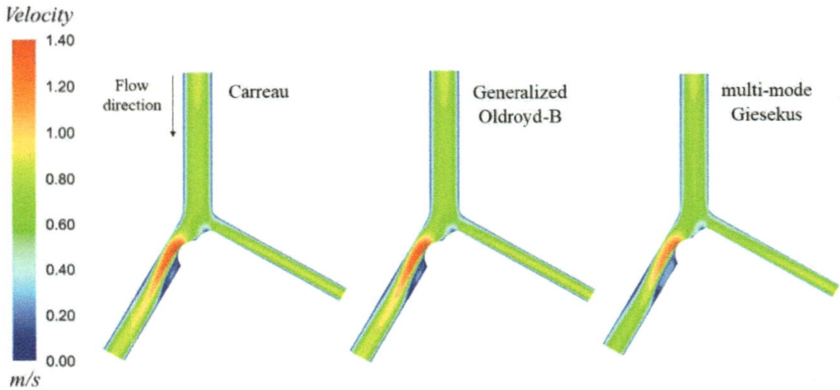

Fig. 8.3 Velocity contours, in the systolic peak, along the middle plane of the 3D idealized coronary geometry for the different rheological models.

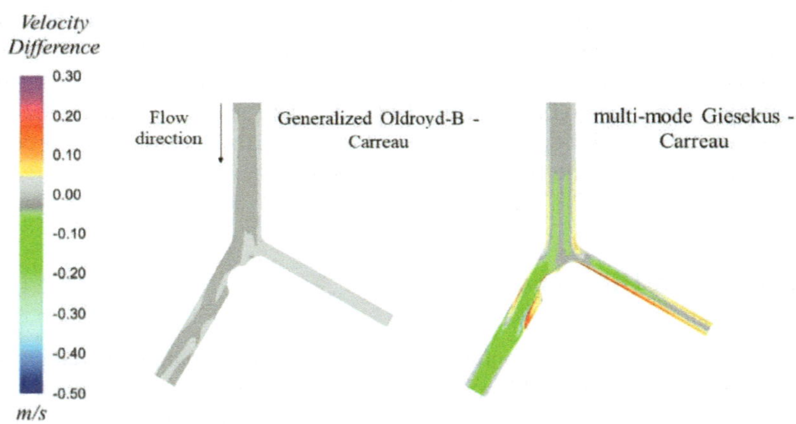

Fig. 8.4 Velocity difference between the Viscoelastic models (Generalized Oldroyd-B and multi-mode Giesekus) and the Generalized Newtonian model (Carreau), in the systolic peak.

modes with four different mobility factors (only the extensibility coefficient equal to 0). This model is a non-linear model (Larson, 1988). Therefore, Fig. 9.4 shows that velocity differences considering the Generalized Oldroyd-B model (quasi-linear viscoelastic model) and Carreau model are almost null (grey region), which means that viscoelastic effects using Generalized Oldroyd-B for blood are not so pronounced. These conclusions are in concordance with those of Bodnár et al (2011). Bodnár et al (2011) consider a different geometry and different boundary conditions (constant flow rate); however, they also concluded that for a higher flow rate, equal to 2 cm^3/s, velocities almost overlap considering these two models. In the present paper, the velocities are also almost coincident for a maximum flow rate, in the systolic peak, equal to 3.9 cm^3/s.

Differences in velocity between using the multi-mode Giesekus and Carreau model are highlighted in Fig.9.4, where the viscoelasticity effects are well evident. The green regions, velocity differences around -0.20 m/s, mean that resulted velocities from simulations taking into account Carreau model are 0.20 m/s higher than using multi-mode Giesekus. In the recirculation regions, the opposite happens. In Fig.9.4, a velocity difference of 0.20 m/s (red regions) can be observed, meaning that velocity field using multi-mode Giesekus is higher than using Carreau model, in the recirculation region. These differences are significant, mainly in regions with high velocity gradients as stenotic regions, in a scale with maximum value of 1.40 m/s. Such results are not surprising since multi-mode Giesekus model is a non-linear model and well-known as one of the best to characterize viscoelastic fluids (Bird et al, 1987).

8.4 Conclusion

The viscoelastic non-Newtonian models, the Generalized Oldroyd-B and multi-mode Giesekus, characterizing the complex rheology of blood for accurate hemodynamic simulations, were implemented in UDFs in *Ansys*® software. The velocity contours, in the middle plane of a 3D idealized right coronary artery, were plotted considering the purely shear-thinning model, Carreau model, and the two viscoelastic non-Newtonian models. Using the Generalized Oldroyd-B and Carreau model, differences are almost null, meaning that the viscoelastic effects in Generalized Oldroyd-B are not pronounced. This model is a quasi-linear model. These results are in concordance with those obtained in literature. Comparing results considering multi-mode Giesekus and Carreau model, differences are significant and equal to 0.20 m/s under a maximum velocity of 1.40 m/s (14.3%). Using the multi-mode Giesekus model, the viscoelastic effects are highlighted in addition to the shear-thinning, mainly in regions with high velocity gradients as the stenotic region. Whether the viscoelastic models are emphasized in idealized geometry bifurcations, where the velocity gradients of flow are high, the same viscoelastic models will certainly be accurate in real models of arteries.

Acknowledgements Authors gratefully acknowledge the funding by *Fundação para a Ciência e Tecnologia* (FCT), Portugal, through the funding of the "Associated Laboratory of Energy, Transports and Aeronautics (LAETA)", UID/EMS/50022/2019, the Institute of Science and Innovation in Mechanical and Industrial Engineering (LAETA-INEGI), the Engineering Faculty of University of Porto (FEUP), the Cardiovascular R&D Unit of the Medicine Faculty of University of Porto (FMUP) and the Cardiology Department of Gaia/Espinho Hospital Centre.

References

Ansys (2013) Ansys Academic 16.0. ANSYS Fluent Tutorial Guide

Auffray N, dell'Isola F, Eremeyev VA, Madeo A, Rosi G (2015) Analytical continuum mechanics à la Hamilton–Piola least action principle for second gradient continua and capillary fluids. Mathematics and Mechanics of Solids 20(4):375–417

Baskurt OK, Meiselman HJ (2003) Blood rheology and hemodynamics. Seminars in thrombosis and hemostasis 29(5):435–450

Bird R, Armstrong R, Hassager O (1987) Dynamics of Polymeric Liquids, Volume 1: Fluid Mechanics, 2nd edn. John Wiley and Sons Inc

Bodnár T, Sequeira A, Prosi M (2011) On the shear-thinning and viscoelastic effects of blood flow under various flow rates. Applied Mathematics and Computation 217(11):5055–5067

Campo-Deaño L, Dullens RPA, Aarts DGAL, Pinho FT, Oliveira MSN (2013) Viscoelasticity of blood and viscoelastic blood analogues for use in polydymethylsiloxane in vitro models of the circulatory system. Biomicrofluidics 7(3):34,102

Campo-Deaño L, Oliveira M, T Pinho F (2015) A Review of Computational Hemodynamics in Middle Cerebral Aneurysms and Rheological Models for Blood Flow. Applied Mechanics Reviews 67(3):30,801

De Santis G, Conti M, Trachet B, De Schryver T, De Beule M, Degroote J, Vierendeels J, Auricchio F, Segers P, Verdonck P, Verhegghe B (2013) Haemodynamic impact of stent-vessel (mal)apposition following carotid artery stenting: mind the gaps! Computer methods in biomechanics and biomedical engineering 16(6):648–659

Eremeyev VA, Altenbach H (2014) Equilibrium of a second-gradient fluid and an elastic solid with surface stresses. Meccanica 49(11):2635–2643

Johnston BM, Johnston PR, Corney S, Kilpatrick D (2004) Non-Newtonian blood flow in human right coronary arteries: steady state simulations. Journal of biomechanics 37(5):709–720

Larson RGBT (1988) Constitutive Equations for Polymer Melts and Solutions. In: Larson RGBT (ed) Butterworths Series in Chemical Engineering, Butterworth-Heinemann

Lee SW, Antiga L, Spence JD, Steinman DA (2008) Geometry of the carotid bifurcation predicts its exposure to disturbed flow. Stroke 39(8):2341–2347

Morbiducci U, Gallo D, Massai D, Ponzini R, Deriu MA, Antiga L, Redaelli A, Montevecchi FM (2011) On the importance of blood rheology for bulk flow in hemodynamic models of the carotid bifurcation. Journal of biomechanics 44(13):2427–2438

Pinho N, Castro CF, António CC, Bettencourt N, Sousa LC, Pinto SIS (2019a) Correlation between geometric parameters of the left coronary artery and hemodynamic descriptors of atherosclerosis: FSI and statistical study. Medical & biological engineering & computing 57(3):715–729

Pinho N, Sousa LC, Castro CF, António CC, Carvalho M, Ferreira W, Ladeiras-Lopes R, Ferreira ND, Braga P, Bettencourt N, Pinto SIS (2019b) The Impact of the Right Coronary Artery Geometric Parameters on Hemodynamic Performance. Cardiovascular engineering and technology 10(2):257–270

Rickert W, Vilchevskaya E, Müller W (2019) A note on Couette flow of micropolar fluids according to Eringen's theory. Mathematics and Mechanics of Complex Systems 7(1):25–50

Thanapong Chaichana, Zhonghua Sun JJ (2012) Computational Fluid Dynamics Analysis of the Effect of Plaques in the Left Coronary Artery. Computational and Mathematical Methods in Medicine 2012:9

Van Canneyt K, Morbiducci U, Eloot S, De Santis G, Segers P, Verdonck P (2013) A computational exploration of helical arterio-venous graft designs. Journal of biomechanics 46(2):345–353

Chapter 9
WSS Descriptors in a Patient RCA Taking into Account the Non-linear Viscoelasticity of Blood

Ernesto Romano, Luísa C. Sousa, Carlos C. António, Catarina F. Castro, and Sónia Isabel Silva Pinto

Abstract Hemodynamic analyses, in patient-specific right coronary arteries (RCA) considering non-linear viscoelastic property of blood, have been underexplored in literature. Therefore, authors intend to evaluate and compare the relative residence time hemodynamic descriptor, and the tendency for atherosusceptible regions, in arteries with absence of atherosclerosis, when a purely shear-thinning model for blood (Carreau Model) or a non-linear viscoelastic model (multi-mode Giesekus model) are used for hemodynamic simulations. *Ansys*® software was chosen since it is a user-friendly software widely used by other authors who are interested in this research field. Carreau model is incorporated in *Ansys*® package software; however, the multi-mode Giesekus model was implemented and validated by authors in a previous work. Using the non-linear viscoelastic property of blood, the critical atherosusceptible regions in a patient-specific artery are larger than using the purely shear-thinning model. In the present patient case, some regions with high velocity gradients, mainly in bifurcations, have an increase of 87.5%. Therefore, the multi-mode Giesekus model, taking into account the complex property of blood, should be used in order to obtain an accurate hemodynamic simulation and accurate atherosusceptible regions.

Keywords: Non-linear model · Viscoelastic property · Blood rheology · Hemodynamics · Patient-specific right coronary artery · Wall shear stress descriptors

---

E. Romano, L. C. Sousa, C. C. António, C. F. Castro, S. I. S. Pinto
Engineering Faculty, University of Porto, Porto, Portugal
Institute of Science and Innovation in Mechanical and Industrial Engineering (LAETA-INEGI), Porto, Portugal
e-mail: up201404532@fe.up.pt, lcsousa@fe.up.pt, cantonio@fe.up.pt, ccastro@fe.up.pt, spinto@fe.up.pt

B. E. Abali and I. Giorgio (eds.), *Developments and Novel Approaches in Biomechanics and Metamaterials*, Advanced Structured Materials 132,
https://doi.org/10.1007/978-3-030-50464-9_9

9.1 Introduction

Cardiovascular diseases, such as atherosclerosis development in arteries, are one of the most common causes of death in developed countries (Mozaffarian et al, 2015). It is well-known from clinical practice that specific sites are sensitive to develop atherosclerosis, the accumulation of lipoproteins inside the arteries. Thus, the stenosis blocks the normal circulation of blood flow (Ku, 1997). The Computed Tomography (CT) scans can give information about the geometry of the artery and location of the disease; however, do not explain the hemodynamic with detail. The study of blood flow in patient-specific cases has become highly relevant in clinical practice due to a need for understanding the mechanical stresses induced on the vessels. Hemodynamic simulations using CT image-based geometries of arteries, with all the conditions as close as possible to the patient, have been a powerful auxiliary tool for prevention, diagnosis and treatment of atherosclerosis disease.

Some authors have considered generalized models that can be useful to simulate hemodynamics in blood vessels. In studies of Auffray et al (2015), a description for second gradient continua is formulated for the case of capillary fluids, i.e., fluids for which the deformation energy depends on the second gradient of placement. Therefore, a Lagrangian action was introduced in both the material and spatial descriptions and the corresponding Euler–Langrange equations and boundary conditions were found. These conditions were formulated in terms of an objective deformation energy volume density. The second-gradient theory was also applied to Cahn–Hilliard fluids in studies of Seppecher (2000). The second gradient theory is necessary when describing a particular class of materials: when the energy depends on the second gradient of the displacement. They concluded that the use of extended thermodynamics is similar to describe a linear elastic material without invoking the Cauchy stress tensor. Moreover, Seppecher (1993) applied the Cahn–Hilliard model for another type of studies: multiphase fluid. He studied the influence of the wetting properties of the fluid upon the stability of a thin liquid film. Ivanova and Vilchevskaya (2016) formulated an additional constitutive equation in order to model structural transformations due to the consolidation or defragmentation of particles or anisotropic changes. Thus, the tensor of inertia of the elementary volume may change. Ivanova and Vilchevskaya (2016) suggest kinetic equations for the tensor of inertia of the elementary volume. They also discuss the specificity of the inelastic polar continuum description within the framework of the spatial distribution.

However, in the present work, authors intend to use, for hemodynamic simulations, the most accurate behavior of blood, well-known from literature (Thanapong Chaichana, Zhonghua Sun, 2012; De Santis et al, 2013; Lee et al, 2009; Morbiducci et al, 2011; Van Canneyt et al, 2013). Blood can be modelled by a viscoelastic non-Newtonian fluid, through a constitutive equation that combines viscosity and elasticity in the total extra stress tensor. Therefore, this total tensor can be splitted into the sum of the suspending fluid and the elastic contributions. The elastic contribution is dependent on the upper-convective derivative of the elastic stress tensor, among other variables. This model must be taken into account, especially, when considering the pulsatile nature of blood flow in arteries with stenosis or aneurysms, i.e., regions of high velocity gradients. However, the rheological behavior of blood, in the cardiovascular system, is frequently characterized by many authors as a Newtonian or a purely shear-thinning Non-Newtonian fluid (Thanapong Chaichana, Zhonghua Sun, 2012; De Santis et al, 2013; Lee et al, 2009; Morbiducci et al, 2011; Van Canneyt et al, 2013) which cannot represent with accuracy blood behavior.

The present work is focused on numerical studies of blood flow and wall shear stress (WSS) in patient-specific RCAs, arteries with low caliber and with regions of high velocity gradients, taking into account the viscoelastic property of blood. *Ansys*® software was used for hemodynamic simulations since it is a user-friendly software widely used by many authors who are interested in this research field. Non-linear viscoelastic models are not incorporated in *Ansys*® package. However, authors of the present paper implemented in user-defined functions (UDFs) and validated the viscoelastic model, in previous work, through 3D idealized geometries with stenosis (Romano et al, 2019).

The hemodynamic behavior under two different rheological models for blood was considered: a purely shear-thinning model incorporated in the software, Carreau model, and a non-linear viscoelastic model (shear-thinning plus elasticity) implemented and validated by Romano et al (2019), multi-mode Giesekus model. The goal is to compare, after hemodynamic simulations, the effects in the WSS of patient-specific RCAs when using such different models and analyze the importance of considering the elastic property of blood. Moreover, boundary conditions as close as possible to the patient-specific right coronary artery were imposed. Pulsatile flow, space and time-dependent velocity, was considered at the inlet (ostium); and the pressure profile, for each side-branch, dependent on the time of the cardiac cycle with radius-independent was taken into account at the outlet branches.

9.2 Materials and Methods

9.2.1 Study Case

From a population of patients referred to Coronary CT Angiography at Gaia/Espinho Hospital Centre, one male individual aged 41 and absence of coronary atherosclerotic disease was selected for this study (Fig. 9.1). This individual has presented a total absence of any calcification and absence of any non-calcified plaque or luminal irregularity. This subject gave informed consent and the present research was approved by the institutional ethical committee.

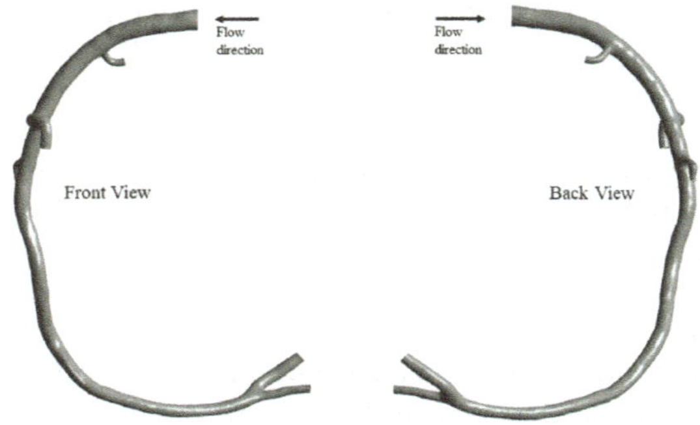

Fig. 9.1 Model of the patient-specific RCA in study (front and back view).

9.2.2 Geometry Reconstruction

The CT scans provided by the medical team were analyzed, through the commercial software *Mimics*®, by Pinho et al (2019b). The semi-automatic algorithm allows the reconstruction of the lumen through the manual selection of the Aorta domain and the RCA points of interest as ostium (inlet) and the multiple side-branches. After the manual selection of all the interested points, the algorithm adjusts the coronary path and the lumen area to produce the 3D mask of the RCA model (Fig. 9.2). Then, the 3D mask was processed in *3-Matic*® software to produce a smooth model of the lumen (Pinho et al, 2019b).

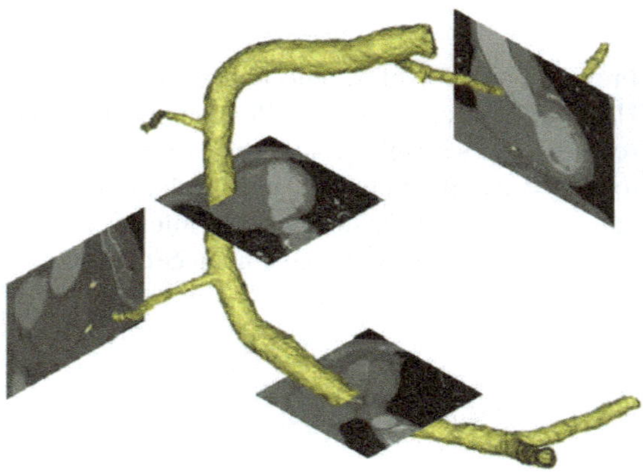

Fig. 9.2 Schematic representation of the geometry reconstruction based on CT scans (Siemens SOMATOM Force®, Erlanger, Germany) provided by the Cardiology Department of Gaia/Espinho Hospital Centre (Pinho et al, 2019b).

9.2.3 Mesh Generation

The patient-specific RCA model was discretized using *Meshing Ansys*® software with tetrahedral elements for the fluid domain. A *Patch Independent Method* was used in order to uniform the tetrahedral elements (see Fig. 9.3). Moreover, the mesh quality was evaluated through the statistical parameter *Maximum Skewness*. A value of the *Maximum Skewness* equal to 0 points to an equilateral cell, which is the best case scenario. A value of *Maximum Skewness* equal to 1 points to a completely degenerated cell. The accuracy of the mesh is achieved when the *Maximum Skewness* is lower than 0.95 (Ansys, 2013). The computational mesh of the present work has a *Maximum Skewness* equal to 0.59 with 399533 elements. Following the tutorial guide of *Ansys*® *Fluent* (Ansys, 2013), the mesh is considered accurate for numerical simulations.

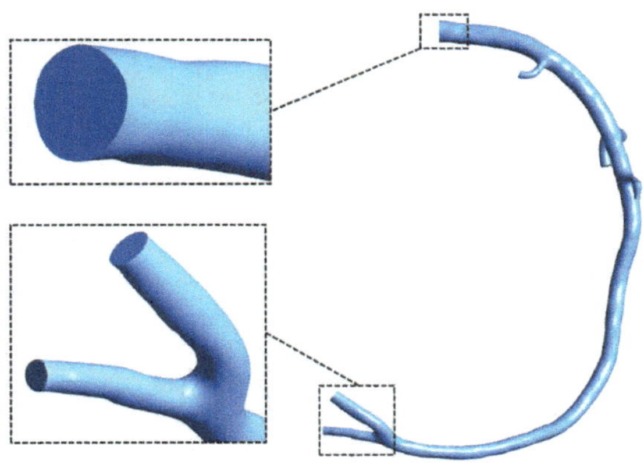

Fig. 9.3 Computational mesh of the fluid domain and detail of the uniform tetrahedral mesh at the inlet (ostium) and in a bifurcation.

9.2.4 Boundary Conditions

At the inlet boundary condition (ostium), a Womersley velocity profile was imposed. The profile depends, mainly, on the time of the cardiac cycle (t), the position of a given point (r) and the Womersley number (W_o) specific for a patient-specific artery:

$$u(r,t) = \frac{AR^2}{i\mu W_o^2}\left(1 - \frac{J_o\left(i^{3/2}W_o\frac{r}{R}\right)}{J_o\left(i^{3/2}W_o\right)}\right)e^{i\omega t} \tag{9.1}$$

where r is the radial distance from the center of the artery; J_o is the first order Bessel function; $A = \frac{1}{\rho}\frac{\partial P}{\partial r}$ is the pressure gradient, ρ is the blood density and the Womersley number (W_o) is defined by:

$$W_o = R\sqrt{\frac{\rho\omega}{\mu}} \tag{9.2}$$

depending on the radius of the artery (R), the cardiac frequency (ω) and the viscosity of blood (μ) at infinite shear rate. For this patient-specific case, W_o is equal to 2.72 corresponding to a diameter of the ostium (inlet) equal to 3.40 mm. The outlet pressure profiles, for each side-branch, depend on the time of the cardiac cycle and are also radius-independent.

The boundary conditions, referred previously, for patient-specific right coronary arteries, were already implemented in user-defined functions (UDFs) in *Ansys*® software by some authors of the present paper (Pinho et al, 2019a,b). Fig. 9.4 shows the pressure profile imposed at the outlet branches and the mean velocity profile at the inlet (ostium) for the patient-specific inlet radius and Womersley number.

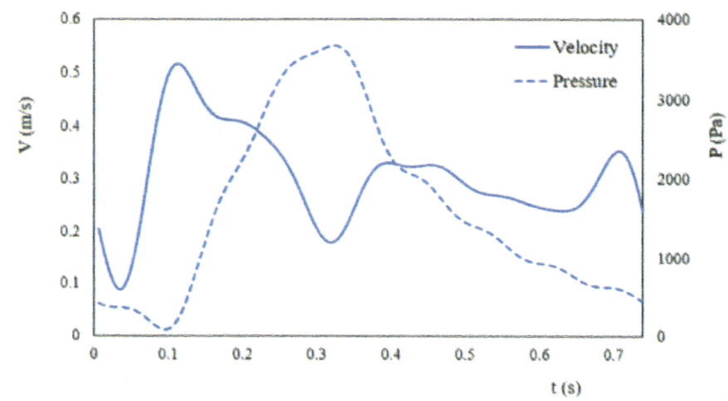

Fig. 9.4 Mean velocity profile imposed at the inlet (ostium) of the patient-specific RCA and pressure profile imposed at the outlet side-branches.

9.2.5 Blood Properties

Blood has been considered to be isotropic, incompressible, homogeneous, constant density (1060 kg/m^3) and non-Newtonian fluid. It is well-known from literature that blood is a complex fluid with the shear-thinning property, solvent part ($\boldsymbol{\tau}_s$), and the elastic property ($\boldsymbol{\tau}_e$) (Baskurt and Meiselman, 2003; Bodnár et al, 2011; Campo-Deaño et al, 2013, 2015).

Thus, in the governing equations where **u** is the velocity vector:

$$\begin{aligned} \nabla \cdot \mathbf{u} &= 0 \\ \rho\left(\frac{\partial \mathbf{u}}{\partial t} + \mathbf{u}\cdot\nabla\mathbf{u}\right) &= -\nabla p + \nabla\cdot\boldsymbol{\tau} \end{aligned} \tag{9.3}$$

the viscoelastic non-Newtonian behavior of blood is represented by a constitutive equation that combines viscosity and elasticity in the total extra stress tensor ($\boldsymbol{\tau}$). Therefore, this total tensor can be split into the sum of the suspending fluid ($\boldsymbol{\tau}_s$) and the elastic ($\boldsymbol{\tau}_e$) contributions:

$$\boldsymbol{\tau} = \boldsymbol{\tau}_s + \boldsymbol{\tau}_e \tag{9.4}$$

$\boldsymbol{\tau}_s$ is equal to:

$$\boldsymbol{\tau}_s = 2\mu_s \mathbf{D} \tag{9.5}$$

which depends on the viscosity of the solvent part (μ_s) and the strain rate tensor (**D**).

The elastic stress, $\boldsymbol{\tau}_e$, defined through Giesekus model, satisfies the following equation:

$$\boldsymbol{\tau}_{e_k} + \lambda_k \overset{\nabla}{\boldsymbol{\tau}}_{e_k} + \frac{\alpha_k \lambda_k}{\mu_{e_k}} (\boldsymbol{\tau}_{e_k} \cdot \boldsymbol{\tau}_{e_k}) = 2\mu_{e_k} \mathbf{D} \tag{9.6}$$

depending on the upper-convected derivative in the elastic contribution of the extra stress tensor ($\overset{\nabla}{\boldsymbol{\tau}}_{e_k}$), the relaxation time ($\lambda_k$), the viscosity related to the elastic part of the fluid (μ_{e_k}) and the mobility factor (α_k) resulting in a non-linear model (Larson, 1988). Moreover, Giesekus model, defining shear-thinning and elastic properties of blood, was used in multi-mode form. Each mode number is defined by *k*. Therefore, the total elastic stress ($\boldsymbol{\tau}_e$) is the sum of the several *k* modes ($\boldsymbol{\tau}_{e_k}$) in the total of *m* modes:

$$\boldsymbol{\tau}_e = \sum_{k=1}^{m} \boldsymbol{\tau}_{e_k} \tag{9.7}$$

Parameters of the multi-mode Giesekus model for human blood were determined in vitro by Campo-Deaño et al (2013) and can be shown in Table 9.1.

Table 9.1 Parameters of the multi-mode Giesekus model for human blood (Campo-Deaño et al, 2013)

Mode	μ_{e_k} $[Pa \cdot s]$	λ_k $[s]$	α_k
1	0.05	7	0.06
2	0.001	0.4	0.001
3	0.001	0.04	0.001
4	0.0016	0.006	0.001
Solvent	$\mu_s = 0.0012$ Pa s		

The hemodynamics taking into account the non-linear viscoelastic property of blood (shear-thinning plus elastic property), the multi-mode Giesekus model, will be compared with that obtained considering a purely shear-thinning model (without viscoelasticity)—Carreau model. In this model, $\boldsymbol{\tau} = \boldsymbol{\tau}_s$ and $\boldsymbol{\tau}_e$ is equal to 0. The viscosity (μ) can be defined as:

$$\mu(\dot{\gamma}) = \mu_\infty + (\mu_0 - \mu_\infty) \times [1 + (\lambda\dot{\gamma})^2]^{\frac{n-1}{2}} \tag{9.8}$$

$\dot{\gamma}$ being the shear rate. For blood at 37°C, the parameters are well defined by Johnston et al (2004) where the viscosity at infinite shear rate (μ_∞) is 0.00345 Pa s, the viscosity at zero shear rate (μ_0) is equal to 0.056 Pa s, the relaxation time (λ) is 3.313 s and the power index (n) is equal to 0.3568.

Carreau model is included in *Ansys®* package software and only the previous parameters for blood must be input. However, the so-called non-linear viscoelastic property of blood (shear-thinning plus elasticity) does not take part of the *Ansys®* package and must be implemented through the constitutive equation (9.6) and incorporated in the conservative equations. The source terms were modeled, by authors of the present work, in UDFs and added to the momentum equations in *Ansys ®* software (Romano et al, 2019). The source term for each stress component, and for each mode, is defined by:

$$S_{\tau_{ij_k}} = \frac{2\mu_{e_k} D_{ij}}{\lambda_k} - \frac{\alpha_k}{\mu_{e_k}}(\tau_{in_k} \cdot \tau_{nj_k}) + \tau_{nj_k}\frac{\partial u_i}{\partial x_n} + \tau_{in_k}\frac{\partial u_j}{\partial x_n} \tag{9.9}$$

9.2.6 Numerical Method

Ansys® Fluent software, was used to perform unsteady hemodynamic simulations. Navier–Stokes equations were solved in a laminar regime since Reynolds number in the systolic peak of the patient-specific case is not higher than 1000. The velocity-pressure coupled equations were calculated by SIMPLE algorithm and the momentum equations with the implemented source terms were discretized by the second-order upwind scheme. Three cardiac cycles, of 0.74 s each, were simulated (2.22 s in the total). The time step size considered was 0.005 s/time step, the time step number was 444 (148 time steps for each cardiac cycle) and the number of iterations for each time step was equal to 20. The simulation process was completed following the convergence criteria of 1×10^{-4}.

9.2.7 WSS Hemodynamic Descriptors

The WSS-based hemodynamic descriptors are parameters used to summarize the hemodynamic behavior along a cardiac cycle. They depend on the location in the artery s, the instant time of the cardiac cycle t, and the total time of the cardiac cycle T (Arzani and Shadden, 2016; Pinho et al, 2019a). The most used wall shear descriptors are the time average wall shear stress ($TAWSS$), the oscillatory shear index (OSI) and the relative residence time (RRT).

The Time Averaged Wall Shear Stress ($TAWSS$) evaluates the mean value of the WSS magnitude along the cardiac cycle:

$$TAWSS(s) = \frac{1}{T}\int_0^T |WSS(s,t)|\ dt \tag{9.10}$$

Low values of $TAWSS$, below 0.4 Pa, indicate a higher probability of plaque formation (Malek et al, 1999).

The Oscillatory Shear Index (OSI) is a dimensionless parameter that describes the disturbed flow near the wall. This hemodynamic descriptor ranges from 0, with no oscillation and unidirectional WSS, to 0.5 corresponding to a highly disturbed flow with 180° deflections (He and Ku, 1996; Soulis et al, 2011).

$$OSI(s) = 0.5 \left[1 - \frac{\left| \int_0^T \mathbf{WSS}(s,t)\ dt \right|}{\int_0^T |WSS(s,t)|\ dt} \right] \tag{9.11}$$

The Relative Residence Time descriptor (RRT) indicates the residence time of particles near the arterial wall. This hemodynamic descriptor is directly dependent on the OSI and inversely dependent on the $TAWSS$:

$$RRT(s) = \frac{1}{(1 - 2 \times OSI) \times TAWSS} \tag{9.12}$$

Thus, wall regions with low values of WSS and high variation of WSS direction (high values of OSI) are susceptible to atherosclerosis appearance. RRT values over the threshold of 8 Pa^{-1} indicate zones of risk. Once this parameter is a combination of $TAWSS$ and OSI, it is considered to be the strongest metric of assessing blood flow disruptions (Lee et al, 2009).

9.3 Results and Discussion

Fig. 9.5 shows the RRT spatial distribution, the strongest metric for assessing prone regions of atherosclerosis formation, in a patient-specific RCA. RRT is represented taking into account the purely shear-thinning model, Carreau model, and the non-linear viscoelastic model for blood (shear-thinning plus elastic property), multi-mode Giesekus model, in the hemodynamic simulations. Moreover, differences in the RRT spatial distribution between considering multi-mode Giesekus and Carreau model can be visualized on the right of Fig. 9.5.

Using the non-linear viscoelastic model or the purely shear-thinning model in the hemodynamic simulations, the critical regions (high RRT) are in similar specific locations. However, when the non-linear viscoelastic model, multi-mode Giesekus model, is considered, these critical regions are larger corresponding to higher values of RRT. Some regions, mainly bifurcations, RRT value varies from 1 Pa^{-1} (using Carreau model) to 8 Pa^{-1} (using multi-mode Giesekus model), representing an increase of 87.5%. Since multi-mode Giesekus model takes into account the shear-thinning and elastic property of blood, simultaneously, and Carreau model only considers the shear-thinning, significant differences observed (see Fig. 9.5 on

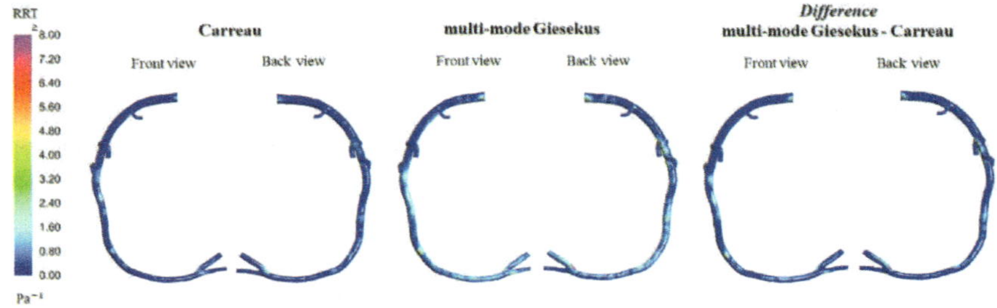

Fig. 9.5 RRT spatial distribution considering the purely shear-thinning model - Carreau model (left), the non-linear viscoelastic model - multi-mode Giesekus model (center), and differences between the models (right).

the right) are due to the elastic property of blood.

Therefore, considering the multi-mode Giesekus model in the hemodynamic simulations, regions of RRT higher than 8 Pa^{-1} were observed, meaning a high tendency for atherosclerosis appearance and the importance of using the non-linear viscoelastic model to obtain an accurate hemodynamic and accurate atherosusceptible regions in patient-specific RCAs.

9.4 Conclusion

The present work is focused on numerical studies of blood flow in patient-specific right coronary arteries, using *Ansys*® software. More precisely, the study presents the hemodynamic behavior in a patient-specific RCA, of an individual of 41 years old, under two different rheological models for blood: a purely shear-thinning model (Carreau model) and a non-linear viscoelastic model (multi-mode Giesekus). Considering these two models, RRT special distribution, the strongest metric for assessing prone regions of atherosclerosis formation, were obtained and compared through *Ansys*® software. Using the purely shear-thinning model and the non-linear viscoelastic model, the critical regions are in similar specific locations. However, when the non-linear viscoelastic model (multi-mode Giesekus model) is considered, these critical regions are larger. Mainly in bifurcations, the increase of RRT values is 87.5% in this patient case. This means a higher tendency for atherosclerosis appearance and the importance of the viscoelastic model's use for an accurate hemodynamic.

Acknowledgements Authors gratefully acknowledge the funding by *Fundação para a Ciência e Tecnologia* (FCT), Portugal, through the funding of the "Associated Laboratory of Energy, Transports and Aeronautics (LAETA)", UID/EMS/50022/2019, the Institute of Science and Innovation in Mechanical and Industrial Engineering (LAETA-INEGI), the Engineering Faculty of University of Porto (FEUP), the Cardiovascular R&D Unit of the Medicine Faculty of University of Porto (FMUP) and the Cardiology Department of Gaia/Espinho Hospital Centre.

References

Ansys (2013) Ansys Academic 16.0. ANSYS Fluent Tutorial Guide

Arzani A, Shadden SC (2016) Characterizations and Correlations of Wall Shear Stress in Aneurysmal Flow. Journal of biomechanical engineering 138(1)

Auffray N, dell'Isola F, Eremeyev VA, Madeo A, Rosi G (2015) Analytical continuum mechanics à la Hamilton–Piola least action principle for second gradient continua and capillary fluids. Mathematics and Mechanics of Solids 20(4):375–417

Baskurt OK, Meiselman HJ (2003) Blood rheology and hemodynamics. Seminars in thrombosis and hemostasis 29(5):435–450

Bodnár T, Sequeira A, Prosi M (2011) On the shear-thinning and viscoelastic effects of blood flow under various flow rates. Applied Mathematics and Computation 217(11):5055–5067

Campo-Deaño L, Dullens RPA, Aarts DGAL, Pinho FT, Oliveira MSN (2013) Viscoelasticity of blood and viscoelastic blood analogues for use in polydymethylsiloxane in vitro models of the circulatory system. Biomicrofluidics 7(3):34,102

Campo-Deaño L, Oliveira M, T Pinho F (2015) A Review of Computational Hemodynamics in Middle Cerebral Aneurysms and Rheological Models for Blood Flow. Applied Mechanics Reviews 67(3):30,801

De Santis G, Conti M, Trachet B, De Schryver T, De Beule M, Degroote J, Vierendeels J, Auricchio F, Segers P, Verdonck P, Verhegghe B (2013) Haemodynamic impact of stent-vessel (mal)apposition following carotid artery stenting: mind the gaps! Computer methods in biomechanics and biomedical engineering 16(6):648–659

He X, Ku DN (1996) Pulsatile flow in the human left coronary artery bifurcation: average conditions. Journal of biomechanical engineering 118(1):74–82

Ivanova EA, Vilchevskaya EN (2016) Micropolar continuum in spatial description. Continuum Mechanics and Thermodynamics 28(6):1759–1780

Johnston BM, Johnston PR, Corney S, Kilpatrick D (2004) Non-Newtonian blood flow in human right coronary arteries: steady state simulations. Journal of biomechanics 37(5):709–720

Ku DN (1997) Blood Flow In Arteries. Annual Review of Fluid Mechanics 29(1):399–434

Larson RGBT (1988) Constitutive Equations for Polymer Melts and Solutions. In: Larson RGBT (ed) Butterworths Series in Chemical Engineering, Butterworth-Heinemann

Lee SW, Antiga L, Steinman DA (2009) Correlations among indicators of disturbed flow at the normal carotid bifurcation. Journal of biomechanical engineering 131(6):61,013

Malek AM, Alper SL, Izumo S (1999) Hemodynamic shear stress and its role in atherosclerosis. JAMA 282(21):2035–2042

Morbiducci U, Gallo D, Massai D, Ponzini R, Deriu MA, Antiga L, Redaelli A, Montevecchi FM (2011) On the importance of blood rheology for bulk flow in hemodynamic models of the carotid bifurcation. Journal of biomechanics 44(13):2427–2438

Mozaffarian D, Benjamin EJ, Go AS, Arnett DK, Blaha MJ, Cushman M, de Ferranti S, Despres JP, Fullerton HJ, Howard VJ, Huffman MD, Judd SE, Kissela BM, Lackland DT, Lichtman JH, Lisabeth LD, Liu S, Mackey RH, Matchar DB, McGuire DK, Mohler ERr, Moy CS, Muntner P, Mussolino ME, Nasir K, Neumar RW, Nichol G, Palaniappan L, Pandey DK, Reeves MJ, Rodriguez CJ, Sorlie PD, Stein J, Towfighi A, Turan TN, Virani SS, Willey JZ, Woo D, Yeh RW, Turner MB (2015) Heart disease and stroke statistics–2015 update: a report from the American Heart Association. Circulation 131(4):e29–322

Pinho N, Castro CF, António CC, Bettencourt N, Sousa LC, Pinto SIS (2019a) Correlation between geometric parameters of the left coronary artery and hemodynamic descriptors of atherosclerosis: FSI and statistical study. Medical & biological engineering & computing 57(3):715–729

Pinho N, Sousa LC, Castro CF, António CC, Carvalho M, Ferreira W, Ladeiras-Lopes R, Ferreira ND, Braga P, Bettencourt N, Pinto SIS (2019b) The Impact of the Right Coronary Artery Geometric Parameters on Hemodynamic Performance. Cardiovascular engineering and technology 10(2):257–270

Romano E, Sousa LC, António CC, Castro CF, Pinto SIS (2019) Non-linear or quasi-linear viscoelastic property of blood for hemodynamic simulations. Advanced Structured Materials: Proceedings of ICoNSoM 2019 (submitted)

Seppecher P (1993) Equilibrium of a Cahn-Hilliard fluid on a wall: influence of the wetting properties of the fluid upon the stability of a thin liquid film. European journal of mechanics series B fluids 12:69–69

Seppecher P (2000) Second-gradient theory: application to Cahn-Hilliard fluids. In: Continuum thermomechanics, Springer, pp 379–388

Soulis J, Fytanidis D, Papaioannou V, Giannoglou G (2011) Oscillating shear index, wall shear stress and low density lipoprotein accumulation in human RCAs. Brunel University

Thanapong Chaichana, Zhonghua Sun JJ (2012) Computational Fluid Dynamics Analysis of the Effect of Plaques in the Left Coronary Artery. Computational and Mathematical Methods in Medicine 2012:9

Van Canneyt K, Morbiducci U, Eloot S, De Santis G, Segers P, Verdonck P (2013) A computational exploration of helical arterio-venous graft designs. Journal of biomechanics 46(2):345–353

Chapter 10
A Method for the Quantification of Architectural Anisotropy in Cancellous Bone Samples Using CT Images

Nicolas Rogalski, Christophe Cluzel, and Sébastien Laporte

Abstract In osteoporotic patients, the sites most prone to fracture are characterized by a predominance of cancellous bone. It has been shown that trabecular architecture plays a crucial role in the mechanical performance of this type of bone. It therefore appears necessary to be able to describe this architecture within an elementary volume. This requires a quantification of the anisotropy of the bone, i.e. a description of the preferred orientations of its architectural elements. Classical techniques are based on the use of a symmetrical second-order tensor, implying orthotropic symmetry within the bone, which is not always the case. In order to properly characterize the architectural anisotropy of cancellous bone samples, this paper presents a new approach for computing the preferred orientations of CT-scanned samples. This method is based on the skeletonization of surfaces reconstructed from binarized CT images, on the use of a projector on the directing vectors of the skeletal rods and finally on the 3D surface analysis of the distribution of the orientations and lengths of the rods. The method introduced has made it possible to obtain the preferred directions of 52 samples of bovine cancellous bone and to show that, in general, these directions are not orthogonal, refuting the oversimplification of orthotropic models.

Keywords: Cancellous bone · Anisotropy · CT images

N. Rogalski, S. Laporte
Arts et Metiers Institute of Technology, Université Sorbonne Paris Nord, IBHGC - Institut de Biomécanique Humaine Georges Charpak, F-75013 Paris, France
e-mail: nicolas.rogalski@ensam.eu, sebastien.laporte@ensam.eu

C. Cluzel
Laboratoire de Mécanique et Technologie, Ecole Normale Supérieure Paris Saclay, Cachan, France
Département Science et Genie des Materiaux, Institut Universitaire de Technologie d'Evry Vald'Essonne, Every, France
e-mail: christophe.cluzel@ens-paris-saclay.fr

B. E. Abali and I. Giorgio (eds.), *Developments and Novel Approaches in Biomechanics and Metamaterials*, Advanced Structured Materials 132,
https://doi.org/10.1007/978-3-030-50464-9_10

10.1 Introduction

In osteoporosis, the sites most prone to fracture, whether hip, femur or vertebrae, are bone regions characterized by a predominance of cancellous bone (Liu et al, 2006). Cancellous bone plays a crucial role in charge transmission and energy absorption (Silva et al, 1997). However, despite the relationship between an increased risk of fracture and a decrease in bone mineral density (Cummings et al, 1993), more than half of postmenopausal women who have had a hip fracture have a density above the World Health Organization osteoporosis threshold (Wainwright et al, 2005). On the other hand, a study on microarchitectural differences between white women and Chinese American women showed that despite a lower bone mineral density, Chinese American women have fewer fractures than white women (Liu et al, 2011). The study of trabecular microstructure therefore contributes to the understanding of lesion mechanisms and the prediction of fracture risk.

To perform the mechanical study of trabecular microstructure, finite element calculations are currently used (Pottecher et al, 2016). However, these models do not integrate the microarchitectural specificities of trabecular bone. Another approach is to quantify the structural and mechanical anisotropy of bone (Gomez-Benito et al, 2005). Current methods allow to determine a symmetric, positive definite, second-rank fabric tensor, from which can be extracted estimates of principal component directions and magnitudes (Moreno et al, 2014) (the Mean Interception Length (MIL) (Whitehouse, 1974), the Volume Orientation (VO) (Odgaard et al, 1990), the Star Length Distribution (SLD) (Smit et al, 1998)). Although a second-rank tensor provides accurate information about a continuum material, it defines three orthogonal axes whereas the alignment directions of the elements of the trabecular network are not necessarily orthogonal (Ketcham and Ryan, 2004). It can also mask the secondary reinforcing directions, leading to an oversimplification in the case of porous materials such as trabecular bone.

The objective of this paper is therefore to introduce a method allowing the characterization of architectural anisotropy of cancellous bone samples from micro-scanner images in order to obtain the privileged orientations, but also to address the gaps among the various parameters described in the literature, used to analyze the microarchitecture of cancellous bone (Bouxsein et al, 2010); none provides information, at the scale of a representative elementary volume, on the directions of trabecular networks. The proposed tool is based on an analysis of three-dimensional geometric anisotropy in terms of trabecular orientation.

10.2 Material and Methods

10.2.1 Trabecular Bone Image Processing

The use of a CT-scan (Phoenix v | Tome | x L240 / NF180: 70 kV, 350 μA, acquisition time of 500 ms per image, resolution of 80 μm), allowed the acquisition of 2D images of cancellous bone. These images were binarized using a multi-level Otsu method (Otsu, 1979) (see Fig. 10.1). Using a segmentation algorithm, the 3D surfaces of the bone samples were obtained. Finally, a thinning algorithm (Lee et al, 1994) was used to skeletonize the surface.

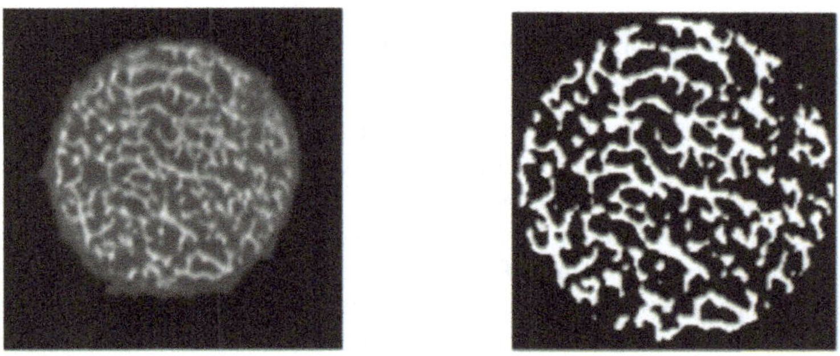

Fig. 10.1 2D Image from the CT-scan (left) and binarized image (right).

The curved lines of the skeleton were approximated by their chord to obtain a network of straight lines (see Fig. 10.2).

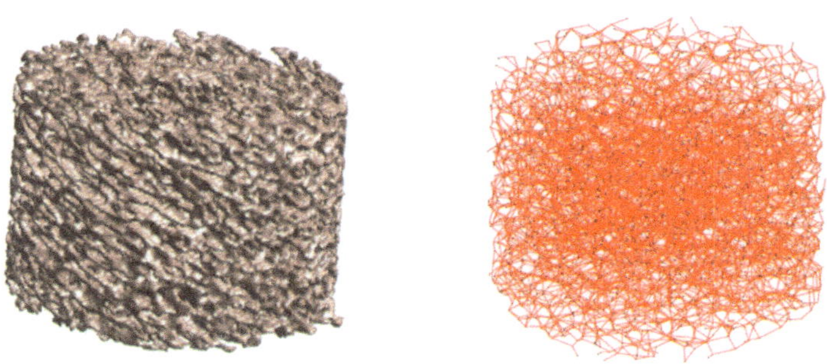

Fig. 10.2 Reconstructed surface (left) and sample after skeletonization (right).

10.2.2 The Projector

To analyze the architectural anisotropy of the samples in order to obtain the preferred alignment directions of the rods, as well as to obtain the distribution of their average lengths, a projector was introduced. The purpose of the projector, defined by equation (15.1), is to account for the relative contributions of rods of different orientations using their respective directing vectors. It takes the form:

$$k(\alpha_T, \alpha_0, p) = A\left(1 - e^{-\left(\frac{\pi - 2\alpha_T}{\pi - 2\alpha_0}\right)^p}\right) \tag{10.1}$$

- α_T: angle between the direction of interest and the mechanical axis of the trabeculae (radians).
- α_0: angular width of the projector (radians).
- p: power of the projector.

The principle is that of a filter: a direction of interest is first chosen. Then, if the angle between the axis (the directing vector) of a trabecula and the direction of interest is between 0 and α_0, the axis is projected on the direction of interest. This allows to group the similar contributions of orientation together in order to highlight the privileged directions. The p-factor provides better filtration quality. When p increases, the projector values for angles greater than α_0 tend towards 0 (see Fig. 10.3). It is therefore a question of finding a compromise between precision, relevance in the choice of secondary directions and calculation time.

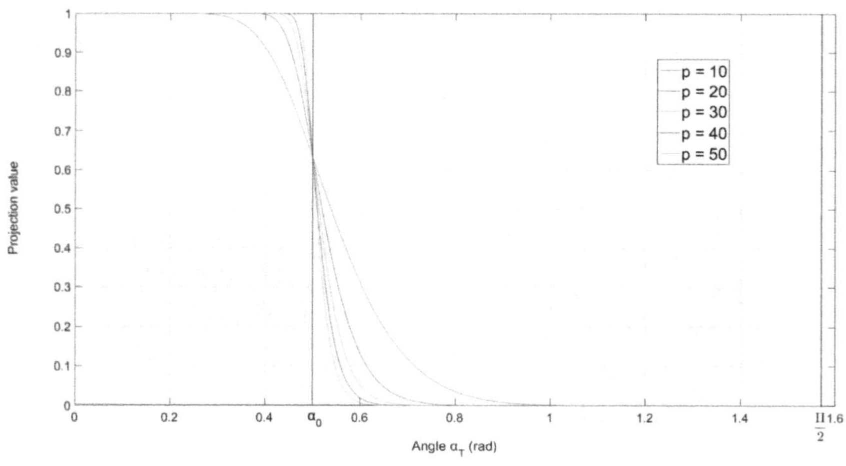

Fig. 10.3 Projector values as a function of angle α_T for different values of p.

From the projector, the 3D distribution of the trabeculae in the direction n is given by equation (15.2)

$$d_n(\alpha_0, p) = \frac{1}{N_{Trab}} \sum_{T=1}^{N_{Trab}} k(\alpha_T, \alpha_0, p) \tag{10.2}$$

The mean length of the trabeculae in the direction n is given by equation (15.3)

$$L_n(\alpha_0, p) = \frac{\sum_{T=1}^{N_{Trab}} L_T\, k(\alpha_T, \alpha_0, p)}{\sum_{T=1}^{N_{Trab}} k(\alpha_T, \alpha_0, p)} \tag{10.3}$$

Using the principle of 3D histogram on a continuous surface (cubed sphere Bruno, 2020) and the projector, it was possible to obtain 3D representations of the distribution of the orientations of the directing vectors of the rods as well as their lengths. The preferred orientations and rod lengths were then computed by analyzing the local extrema of the surfaces obtained, the low extrema being neglected. Finally, the samples were classified according to their number of preferred orientations.

10.2.3 Samples of Cancellous Bone

The image processing method and then the projector were used on cancellous bone samples from cattle aged about 4 years. 52 cylindrical samples (height: 7.5 mm, diameter: 10.5 mm) from the femoral head and greater trochanter of 6 different femurs were collected in collaboration with the BISRU laboratory at the University of Cape Town, South Africa (Prot, 2015) and processed.

10.3 Results

10.3.1 Influence of Projector Parameters

It is possible to play with the projector parameters (angular width and power) in order to monitor the effects of the filtration process and thus modify the shape of the 3D distribution of orientations. Decreasing the width allows to take into account privileged directions closer to each other and vice-versa. The parameter p is used to control the quality of filtration : it is possible to switch from binary filtration with a high power (the directing vector of the rod is included in the angular opening around the direction of interest, or not) to a softer filtration, for a lower power (see Fig. 10.3, 10.4).

The principle is the same for the distribution of lengths (see Fig. 10.5).

Fig. 10.4 Influence of the projector parameters on the 3D distribution of the rod orientations. (a): α_0=10°, p=20 (b): α_0=2°, p=20 (c): α_0=2°, p=40.

10.3.2 Use of the Projector on Bone Samples

By using the projector on bone samples treated with the method described in 10.2.1 and computing the local extrema of the resulting surface, the preferred orientations, the mean rod lengths, the rod length standard deviations and the mean angles between the directing vectors of the preferred orientations were obtained. The samples were sorted in three groups according to their number of preferred orientations (Table 15.1, 15.2).

Table 10.1 Mean trabecular lengths, length standard deviations for the different groups of cancellous bone samples.

Number of preferred orientations	Mean trabecular length (μm)	Length standard deviation (μm)
4	484.72	13.89
3	476.77	11.29
2	476.54	13.98

Five samples were not referenced because the number of orientations was greater than 4. In this case, the relative importance of the orientations within the network decreases as the bone approaches the isotropic case, which makes their determination more difficult. Fig. 10.6 illustrates the different possible configurations within the samples, in terms of preferred orientations.

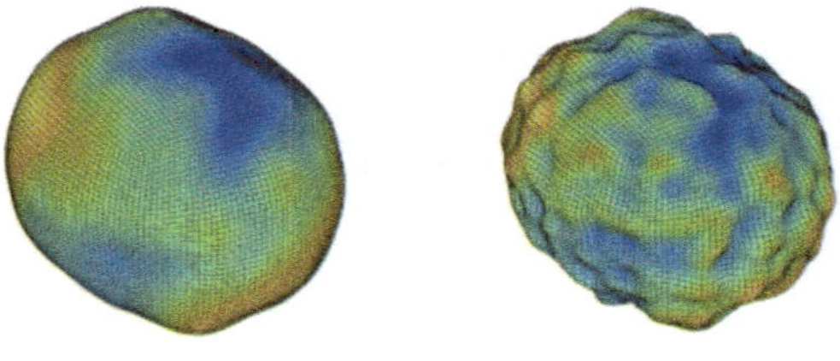

Fig. 10.5 Influence of the projector parameters on the 3D distribution of rod lengths. (a): α_0=10°, p=20, (b): α_0=5°, p=40.

Table 10.2 Mean angles between the directing vectors of preferred orientations for the different groups of cancellous bone samples. Angles of directing vectors (deg).

Number of preferred orientations	$(\widehat{\mathbf{v}_1\mathbf{v}_2})$	$(\widehat{\mathbf{v}_1\mathbf{v}_3})$	$(\widehat{\mathbf{v}_2\mathbf{v}_3})$	$(\widehat{\mathbf{v}_1\mathbf{v}_4})$	$(\widehat{\mathbf{v}_2\mathbf{v}_4})$	$(\widehat{\mathbf{v}_3\mathbf{v}_4})$
4	60.03	79.02	64.96	66.26	75.89	53.38
3	72.95	72.76	76.6			
2	74.42					

10.4 Discussion

In this paper, an approach allowing the quantification of architectural anisotropy as well as obtaining information on the average length of rods in a sample of cancellous bone was presented. The method relies on the use of a set of images acquired by a CT-scan, processed to reconstruct a surface and a tool based on the principle of projection of the directing vectors of the rods after skeletonization of the surface.

The projector offers the possibility of computing non-orthogonal privileged directions and secondary directions, contrary to conventional techniques for quantifying architectural anisotropy, such as the MIL, the VO or the SLD. It has been shown that the measured angles between the preferred directions are on average quite far apart from 90 degrees (Table 15.2), which proves that the hypothesis of orthotropy within the trabecular network is over-simplifying in most cases. Highlighting the actual preferred directions is essential because the mechanics of cancellous bone is driven by its microstructure.

The whole approach presents some drawbacks though. In particular, the method presents a sensitivity due to the imaging techniques used. The resolution of the scan must be sufficient to preserve a maximum of information on the architecture of the

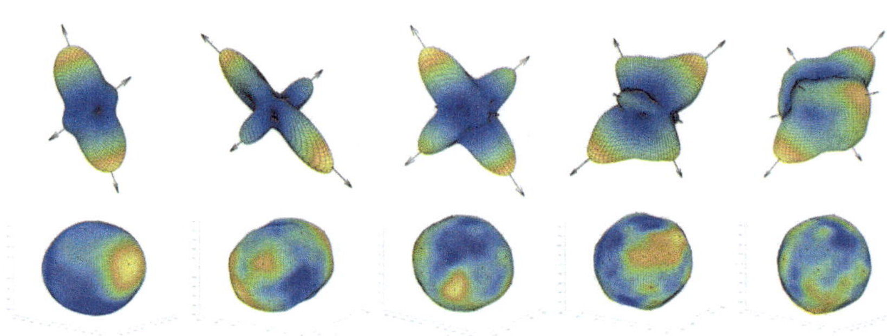

Fig. 10.6 3D distribution of orientations and trabecular length (in µm) for bone samples (α_0=10°, p=20). From left to right: 2 non-orthogonal orientations, 2 orthogonal orientations, 3 orthogonal orientations, 3 non-orthogonal orientations, 4 orientations.

bone, but the possibility of clinical use must also be taken into account, which limits the minimum usable resolution. Moreover, the resolution also affects the choice of the projector parameters. Indeed, the chosen voxel size allows to obtain a minimum angular width value below which one cannot work, this value corresponding to the positioning inaccuracy of the bone sample in the scanner (see Fig. 10.7).

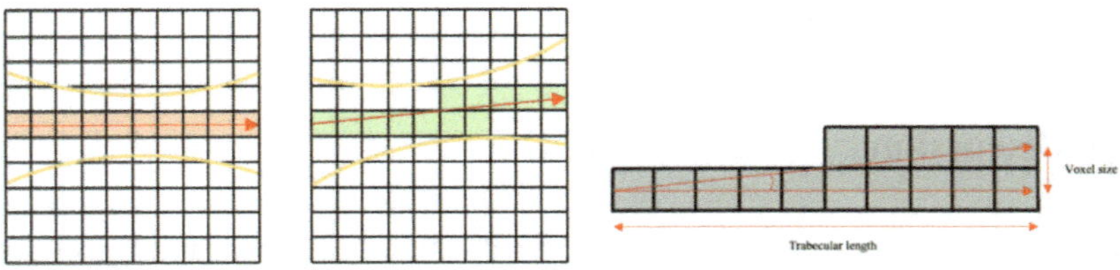

Fig. 10.7 Difference in rod orientation obtained after skeletonization by considering the positioning uncertainty in the CT-scan.

Considering an average rod length of 480 μm (see Table 15.1) as well as a voxel size of 80 μm, the angle corresponding to this uncertainty equals approximately 9.5 degrees. The voxel size therefore limits the angular width of the projector.

In addition, the thinning algorithm used does not differentiate the plate and rods elements of the samples. The skeleton shows only rods. A more complex algorithm can be used to preserve this distinction in the skeletonization process, as it was shown that plates play a crucial role in the mechanical behavior of cancellous bone in several regions of the body (Wang et al, 2013). Another point is that the rods are approximated by their chord, their curves are neglected. It is advisable to take this into account later on, in order to obtain a more accurate geometric description and to be able to envisage damage phenomena linked to microarchitecture (buckling for example). Finally, the application of this tool will be more difficult to implement than the above-mentioned techniques because it will rely on the use of higher order tensors, in order to be able to use all the geometrical information obtained from the projector.

Once the improvements listed have been incorporated, it will then be possible to compare the method presented to other methods that do not rely on a skeletonization process, such as the use of facet normals of a STL file (Cluzel and Allena, 2018). It would also be interesting to carry out a mapping of the anisotropic field of coordinates on the femurs where the samples were taken, similarly to the work done on cortical bone for orthotropic directions (Allena and Cluzel, 2018). This would allow to show the alignment of the privileged orientations with the mechanical loading axes, which would constitute a validation of the interest of the approach based on physiology. A method allowing the detection of extrema based on the norm of orientation vectors could be considered in order to automatically neglect low intensity directions within the network.

References

Allena R, Cluzel C (2018) Heterogeneous directions of orthotropy in three-dimensional structures: finite element description based on diffusion equations. Mathematics and Mechanics of Complex Systems 6(4):339–351

Bouxsein ML, Boyd SK, Christiansen BA, Guldberg RE, Jepsen KJ, Müller R (2010) Guidelines for assessment of bone microstructure in rodents using micro-computed tomography. Journal of Bone and Mineral Research: The Official Journal of the American Society for Bone and Mineral Research 25(7):1468–1486

Bruno L (2020) Cubed Sphere

Cluzel C, Allena R (2018) A general method for the determination of the local orthotropic directions of heterogeneous materials: application to bone structures using μCT images. Mathematics and mechanics of complex systems 6(4):353–367

Cummings SR, Browner W, Cummings SR, Black DM, Nevitt MC, Browner W, Genant HK, Cauley J, Ensrud K, Scott J, Vogt TM (1993) Bone density at various sites for prediction of hip fractures. The Lancet 341(8837):72–75

Gomez-Benito MJ, Garcıa-Aznar JM, Doblare M (2005) Finite Element Prediction of Proximal Femoral Fracture Patterns Under Different Loads. Journal of Biomechanical Engineering 127(1):9–14

Ketcham RA, Ryan TM (2004) Quantification and visualization of anisotropy in trabecular bone. Journal of Microscopy 213(2):158–171

Lee TC, Kashyap RL, Chu CN (1994) Building Skeleton Models via 3-D Medial Surface Axis Thinning Algorithms. CVGIP: Graphical Models and Image Processing 56(6):462–478

Liu XS, Sajda P, Saha PK, Wehrli FW, Guo XE (2006) Quantification of the Roles of Trabecular Microarchitecture and Trabecular Type in Determining the Elastic Modulus of Human Trabecular Bone. Journal of bone and mineral research : the official journal of the American Society for Bone and Mineral Research 21(10):1608–1617

Liu XS, Walker MD, McMahon DJ, Udesky J, Liu G, Bilezikian JP, Guo XE (2011) Better skeletal microstructure confers greater mechanical advantages in Chinese-American women versus white women. Journal of Bone and Mineral Research: The Official Journal of the American Society for Bone and Mineral Research 26(8):1783–1792

Moreno R, Borga M, Smedby O (2014) Techniques for Computing Fabric Tensors: A Review. In: Westin CF, Vilanova A, Burgeth B (eds) Visualization and Processing of Tensors and Higher Order Descriptors for Multi-Valued Data, Springer, Berlin, Heidelberg, Mathematics and Visualization, pp 271–292

Odgaard A, Jensen EB, Gundersen HJG (1990) Estimation of structural anisotropy based on volume orientation. A new concept. Journal of Microscopy 157(2):149–162

Otsu N (1979) A Threshold Selection Method from Gray-Level Histograms. IEEE Transactions on Systems, Man, and Cybernetics 9(1):62–66

Pottecher P, Engelke K, Duchemin L, Museyko O, Moser T, Mitton D, Vicaut E, Adams J, Skalli W, Laredo JD, Bousson V (2016) Prediction of Hip Failure Load: In Vitro Study of 80 Femurs Using Three Imaging Methods and Finite Element Models-The European Fracture Study (EFFECT). Radiology 280(3):837–847

Prot M (2015) Comportement mécanique de l'os spongieux à différentes vitesses de déformation. : relations entre architecture et réponse mécanique. thesis, Paris, ENSAM

Silva MJ, Keaveny TM, Hayes WC (1997) Load sharing between the shell and centrum in the lumbar vertebral body. Spine 22(2):140–150

Smit, Schneider, Odgaard (1998) Star length distribution: a volume-based concept for the characterization of structural anisotropy. Journal of Microscopy 191(3):249–257

Wainwright SA, Marshall LM, Ensrud KE, Cauley JA, Black DM, Hillier TA, Hochberg MC, Vogt MT, Orwoll ES, Study of Osteoporotic Fractures Research Group (2005) Hip fracture in women without osteoporosis. The Journal of Clinical Endocrinology and Metabolism 90(5):2787–2793

Wang J, Zhou B, Parkinson I, Thomas CDL, Clement JG, Fazzalari N, Guo XE (2013) Trabecular Plate Loss and Deteriorating Elastic Modulus of Femoral Trabecular Bone in Intertrochanteric Hip Fractures. Bone Research 1(1):346–354

Whitehouse WJ (1974) The quantitative morphology of anisotropic trabecular bone. Journal of Microscopy 101(Pt 2):153–168

Chapter 11
A Model of Integrin and VEGF Receptors Recruitment on Endothelial Cells

Mattia Serpelloni, Matteo Arricca, Valentina Damioli, Cosetta Ravelli, Elisabetta Grillo, Stefania Mitola, and Alberto Salvadori

Abstract Angiogenesis is a multistep process in which endothelial cells (ECs) are affected by several extracellular stimuli, including growth factors, extracellular matrix (ECM), and parenchymal and stromal cells. In this process, growth factor receptors as well as adhesion receptors convey the extracellular signaling in a coordinate intracellular pathway. The Vascular Endothelial Growth Factor (VEGF), by binding the Vascular endothelial growth factor receptor 2 (VEGFR-2), promotes EC proliferation, migration, and their reorganization in active vessels. Once engaged, the activation of VEGFR-2 is modulated by its interaction with β_3 integrin. Although the ability of VEGFR-2 to participate in a complex with β_3 integrin is well known, the close correlation between their activation and the multiphysical phenomena regulating EC dynamics remains still very restricted. Here we computationally model the VEGFR-2 and β_3 integrin membrane dynamics by a multi-physics model, to identify how ligands stimulation induces the polarization of receptors in cell protrusions and in the basal aspect of ECs plated on a ligand-enriched ECM. The research for new anti-angiogenic solutions through the controlled activation of the ECs could arise from the determination of the laws that govern the polarization of the receptors.

M. Serpelloni, M. Arricca, A. Salvadori
Università degli Studi di Brescia, Brescia, 25123, Italy
Department of Mechanical and Industrial Engineering (DIMI)
e-mail: m.serpelloni002@unibs.it, m.arricca@unibs.it, alberto.salvadori@unibs.it

V. Damioli
SISSA - International School for Advanced Studies, Trieste, 34136 Italy
mathLab
e-mail: vdamioli@sissa.it

C. Ravelli, E. Grillo, S. Mitola
Università degli Studi di Brescia, Brescia, 25123, Italy
Department of Molecular and Translational Medicine (DMMT)
Laboratory for Preventive and Personalized Medicine (MPP Lab)
e-mail: cosetta.ravelli@unibs.it, elisabetta.grillo@unibs.it, stefania.mitola@unibs.it

B. E. Abali and I. Giorgio (eds.), *Developments and Novel Approaches in Biomechanics and Metamaterials*, Advanced Structured Materials 132,
https://doi.org/10.1007/978-3-030-50464-9_11

Keywords: Receptor dynamics · Continuum thermodynamics · Finite element method

11.1 Introduction

Tumor neovascularization and angiogenesis-dependent diseases are characterized by the uncontrolled release of angiogenic growth factors, leading to EC activation. Although growth factors are considered soluble molecules, many of them bind extracellular matrix components leading to the formation of immobilized ECM-bound complexes. Among these, the Vascular endothelial growth factor (VEGF-A), a major pro-angiogenic molecule, is produced by tumor or inflammatory cells, and accumulates in extracellular matrix, and interacts with its tyrosine kinase receptor VEGFR-2 expressed on EC surface. The ECM-anchored VEGF induces the recruitment of VEGFR-2 at the basal portion of the basal membrane, receptor dimerization and the activation of an intracellular signaling through a series of phosphorylations.

The EC response to this ligand differs upon the physical state of the ECs. Upon VEGF stimulation, the phosphorylation grade of VEGFR-2 is slighter in confluent cells compared to that observed in sparse cells (Neufeld et al, 1999). Also, adhesion to different ECMs modulates the VEGFR-2 response (Ravelli et al, 2015; Soldi et al, 1999). Such distinctive feature may be attributed to the VEGFR-2 association with different transmembrane proteins, which form distinct multi-molecular complexes that interact with cytosolic transducers. VEGFR-2 forms a complex with Ve-Cadherin and -catenin into cell-cell contacts (Carmeliet et al, 1999), with the vascular endothelial-phosphotyrosine phosphatase (VE-PTP) into cell-cell junctions (Esser et al, 1998), with neuropilin 1 (Peach et al, 2018) and $\alpha_v\beta_3$ in lipid raft domains (Ravelli et al, 2015). $\alpha_v\beta_3$ exists in an inactive form with a bent-clasped conformation (low-affinity integrin), in an activable form with an unbent-clasped conformation or in an active form with an unbent and unclasped conformation (high-affinity configuration) (Eliceiri, 2001; Hynes, 2002; Valdembri and Serini, 2012). The conformation and the affinity of $\alpha_v\beta_3$ for its ligands fibrin(ogen), fibronectin, thrombospondin, von Willebrand factor, and vitronectin is subjected to short-term modulation by phorbol esters, Mn2+, ADP, vascular endothelial growth factor, basic fibroblast growth factor, and elevations in intracellular cyclic AMP (cAMP) (Calderwood, 2004). $\alpha_v\beta_3$ integrin mediates cell-matrix interactions, and, similarly to VEGFR-2, transmits "outside-in" signals to the cell, which trigger a large array of intracellular signaling events. $\alpha_v\beta_3$ integrin is one of the most important survival system for nascent vessels and participates in the full activation of VEGFR-2 triggered by VEGF-A or gremlin, which are important angiogenic inducers in tumor, inflammation and tissue regeneration. Although the role of productive crosstalk between VEGFR-2 and $\alpha_v\beta_3$ in the angiogenic response is well characterized in terms of intracellular signaling, ECs migration and proliferation, the effects of complex formation on the membrane dynamic of both receptors is still missing.

Starting from imaging experimental data and a rigorous thermodynamic approach (Salvadori et al, 2018b), we modeled in recent publications (Damioli et al, 2017; Salvadori et al, 2018a) the relocation of VEGFR-2 on ECs membrane during the angiogenic process. The developed chemo-transport-mechanics model captures the VEGFR-2 recruitment at the basal portion of EC in active blood vessels and highlights three different phases in receptors relocation, driven by three main regulatory factors: extracellular ligand/receptor chemical interaction, cell mechanical deformation, and receptor diffusion.

It is known that VEGF-A (Soldi et al, 1999) or Gremlin (Ravelli et al, 2013) trigger the polarization of VEGFR-2. It has been also proved that integrin participates in VEGFR-2 full activation, sparking the propagation of intracellular signaling cascades that affect the cell mechanical response. This correlation shows to be particularly efficient in prolonging and strengthening the intracellular signal released by VEGFR-2-Ligand complex.

In the present work, we aim at theoretically modeling and at numerically simulating the interplay between VEGFR-2, $\alpha_v\beta_3$ in its different conformations, VEGF-A or Gremlin, and ECM components (e.g. fibrinogen). The receptor dynamics is combined with the cell mechanical deformation and with the chemical interactions in the framework of the thermodynamics of continua (Gurtin et al, 2010).

The multi-physics model, described in Section 11.2, stems from the balance equations for mass, energy and entropy, includes thermodynamic restrictions and constitutive choices together with standard chemical kinetics. The equations that govern the problem of relocation of VEGFR-2 and Integrin motion on the membrane driven by their specific ligands will ultimately be expressed in a strong and dimensionless formalism, in terms of four space-time unknown scalar concentration fields. In order to enable a numerical approximation of the partial differential equations above, a weak form will be devised in Section 11.3. Such a weak form naturally leads to a semi-discrete problem, by approximating the space-time unknown scalar concentration fields into suitable finite dimensional spaces, thus rephrasing the weak form into a system of ordinary differential equations, whose solution is an approximation of the exact solution for each time. A Backward Euler scheme for the finite element approximation of the chemo-transport model was implemented. To avoid poor numerical accuracy, numerical simulations require a very small time discretization step, thus leading to a very high computational cost in properly capturing the spreading process. The time-dependent partial differential equations have been therefore rephrased in order to apply higher order time integration schemes. The non linear, discretized equations have been implemented in a high performance computing environment, exploiting the deal.ii open software library, in Section 11.4. Final remarks conclude the paper.

11.2 Modeling VEGFR-2 and Integrin Motion Driven by Their Specific Ligands

The equations that model the motion of integrins and VEGFR-2 on the lipid membrane are detailed in this section. The relocation of these proteins is assumed to take place merely along the lipid bilayer, i.e. internalization processes as well as supply of proteins from the cytosol are neglected. Governing equations emanate from a chemo-transport-mechanical model in terms of balance equations coupled to thermodynamic restrictions. These two items will be separately dealt with in the next two subsections.

11.2.1 Chemical Reactions

Proteins relocation is guided by their corresponding ligands, i.e. *VEGF-A* or *Gremlin* for VEGFR-2 and *fibronectin* for integrins. The interactions between ligands and receptors are modeled as three chemical reactions, accompanied by the mass balance equations. They are defined on the cell membrane, which will be henceforth denoted with Ω.

The first chemical reaction

$$I + L_I \underset{k_b^1}{\overset{k_f^1}{\leftrightarrows}} \underbrace{I_h L_I}_{C_1} \tag{11.1}$$

accounts for the interaction between the diffusing, not engaged integrin receptors (I) with fibronectin (L_I), which leads to the formation of complex C_1. The subscript h emphasizes that integrin molecules bound to fibronectin within the complex C_1 are in a hight affinity state, i.e. they are trans-membrane proteins that manifest modest relocation propensity on the lipid bilayer. Reaction (11.1) induces a cascade of intracellular signals, which lead to the formation of macromolecular clusters (so-called focal adhesions) through which mechanical forces and regulatory signals are transmitted between the ECM and ECs.

The second chemical reaction

$$R + L_R \underset{k_b^2}{\overset{k_f^2}{\leftrightarrows}} C_2 \tag{11.2}$$

describes the interaction between VEGFR-2 (R) and VEGF-A or Gremlin (L_R), which provides complex C_2 and induces a cascade of intracellular signals that ultimately lead to the activation of angiogenesis. Reaction (11.2) alone was studied thoroughly in the two publications (Damioli et al, 2017; Salvadori et al, 2018a).

The last chemical reaction that this work concerns with, i.e.

$$I + C_2 \underset{k_b^3}{\overset{k_f^3}{\leftrightarrows}} C_3 \tag{11.3}$$

depicts the interplay between the complex C_2, supplied by reaction (11.2), and the integrin receptors I. Reaction 11.3 ultimately provides another complex, denoted with C_3, which drives the long-term VEGFR-2 phosphorylation required to trigger the first intra-cellular signal.

11.2.2 Mass Balance Equations

In order to include the reaction rate of reactions (11.1)–(11.3) depicted in section 11.2.1 into appropriate mass balance equations, the three chemical reactions will be here stated in the following abstract form

$$\mathrm{A} + \mathrm{B} \underset{k_b}{\overset{k_f}{\leftrightarrows}} \mathrm{C} \, . \tag{11.4}$$

The integral form of the mass balance equation for the generic species I is defined as follows:

$$\frac{\mathrm{d}}{\mathrm{d}t} \int_{\Omega(t)} c_I(\mathbf{x}, t) \, \mathrm{d}S = - \oint_{\Gamma(t)} \mathbf{h}_I \cdot \mathbf{t}_\perp \, \mathrm{d}l + \int_{\Omega(t)} s_I(\mathbf{x}, t) \, \mathrm{d}S \, , \tag{11.5}$$

on a subpart $\Omega(t)$ of the membrane, as shown in Fig. 11.1. In eq. (11.5), vector $\mathbf{t}_\perp$ is normal to the curve $\Gamma(t)$ at a generic point $\mathbf{x}$ on the cell membrane. As shown in Fig. 11.1, the so-called (mobile) trihedron of Frenet is completed by the vectors $\mathbf{n}$ and $\mathbf{t}_{\shortparallel}$, i.e. the cell membrane unit normal and the vector tangent to the curve $\Gamma(t)$, respectively. Furthermore, c_I is the molarity of species I (i.e. the number of molecules per unit area); $\mathbf{h}_I$ is the mass flux in terms of molecules, i.e. the number of molecules of species I measured per unit length per unit time, and is a tangent vector field on the membrane; s_I is the rate in number of molecules per unit volume per unit time at which species I is generated by sources, and t is the time.

Exploiting Stokes's theorem, the line integral in equation (11.5) can be written as

$$\begin{aligned} \oint_\Gamma \mathbf{h}_I \cdot \mathbf{t}_\perp \, \mathrm{d}l = \oint_\Gamma \mathbf{h}_I \cdot (\mathbf{t}_{\shortparallel} \times \mathbf{n}) \, \mathrm{d}l = \oint_\Gamma (\mathbf{n} \times \mathbf{h}_I) \cdot \mathbf{t}_{\shortparallel} \, \mathrm{d}l = \\ \oint_\Gamma (\mathbf{n} \times \mathbf{h}_I) \cdot \mathbf{dl} = \int_\Omega \mathrm{curl} \, [\, \mathbf{n} \times \mathbf{h}_I \,] \cdot \mathbf{n} \, \mathrm{d}S \, . \end{aligned} \tag{11.6}$$

By denoting as usual with

$$\mathrm{div}_\Omega \, [\, \mathbf{h}_I \,] = \mathrm{curl} \, [\, \mathbf{n} \times \mathbf{h}_I \,] \cdot \mathbf{n} \, , \tag{11.7}$$

the mass balance equation can be finally recasted in the form

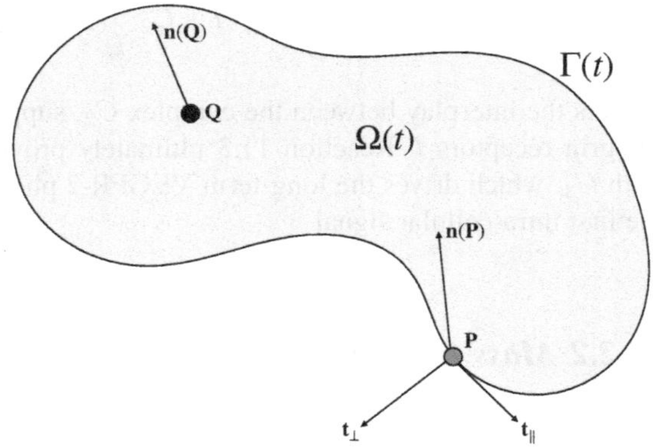

Fig. 11.1 Membrane surface subpart Ω at time t and its closed boundary Γ. The unit vector normal $\mathbf{n}$ and the normal and tangent vectors to the curve $\Gamma(t)$, i.e. $\mathbf{t}_\perp$ and $\mathbf{t}_{\shortparallel}$, are depicted as well.

$$\frac{\mathrm{d}}{\mathrm{d}t}\int_{\Omega(t)} c_I(\mathbf{x},t)\,\mathrm{d}S + \int_{\Omega(t)} \mathrm{div}_\Omega\left[\,\mathbf{h}_I\,\right]\,\mathrm{d}S = \int_{\Omega(t)} s_I(\mathbf{x},t)\,\mathrm{d}S\;. \tag{11.8}$$

It is straightforward to show that Reynold's theorem on $\Omega(t)$ reads as follows:

$$\frac{\mathrm{d}}{\mathrm{d}t}\int_{\Omega(t)} c_I\,\mathrm{d}S = \int_{\Omega(t)} \frac{\mathrm{d}\,c_I}{\mathrm{d}\,t} + c_I\,\mathrm{div}_\Omega\left[\,\mathbf{v}_b\,\right]\,\mathrm{d}S\;, \tag{11.9}$$

where $\mathbf{v}_b(\mathbf{x},t)$ is the velocity of advection at point $\mathbf{x}$ and time t. Replacing eq. (11.9) into eq. (11.8) leads to localize the mass balance equation (11.5) for species I at point $\mathbf{x}$ and time t in the form

$$\frac{\mathrm{d}\,c_I}{\mathrm{d}\,t} + c_I\,\mathrm{div}_\Omega\left[\,\mathbf{v}_b\,\right] + \mathrm{div}_\Omega\left[\,\mathbf{h}_I\,\right] = s_I(\mathbf{x},t)\;\;, \tag{11.10}$$

which can be pulled back to a reference configuration using standard arguments of continuum mechanics (Gurtin et al, 2010) as follows:

$$\frac{\partial c_{I_R}}{\partial t} + \mathrm{Div}_\Omega\left[\,\mathbf{h}_{I_R}\,\right] = s_{I_R}(\mathbf{X},t)\;\;. \tag{11.11}$$

The apex $_R$ denotes quantities in the reference configuration. In this work, however, we will follow a different path of reasoning, mutuated from Damioli et al (2017); Salvadori et al (2018a).

11.2.3 Surrogated Mechanics

The geometrical evolution of the cell during its spreading on the substrate couples with the relocation of receptors along the lipid membrane. Mathematically, the coupling occurs through the velocity of advection in eq. (11.10) and through the mass supply $s_{I_R}(\mathbf{X}, t)$ in eq. (11.11). Specifically, the latter accounts for the spreading by setting the amount of the cell geometry that, being in contact with the substrate, interacts with ligands, ensuring that the chemical reaction (11.4) takes place.

Assigning a given value for $s_{I_R}(\mathbf{X}, t)$ is equivalent to surrogate the mechanics on a spherical geometry, a priori imposing a supply of ligands L_R and L_I (see equations (11.1)–(11.2)) that is calibrated from experimental investigations of the mechanics of spreading. We chose this form[1]

$$\begin{aligned} s_{L_R}(x,t) &= \frac{\overline{c_{L_R}}}{\bar{t}} \mathcal{H}\left[t - \frac{x}{v}\right] \mathcal{H}\left[\bar{t} - t + \frac{x}{v}\right] \\ s_{L_I}(x,t) &= \frac{\overline{c_{L_I}}}{\bar{t}} \mathcal{H}\left[t - \frac{x}{v}\right] \mathcal{H}\left[\bar{t} - t + \frac{x}{v}\right] \end{aligned} \tag{11.12}$$

where:

- $\overline{c_{L_R}}$ and $\overline{c_{L_I}}$ are the concentrations of substrate immobilized ligands;
- $\mathcal{H}[-]$ is the Heaviside step function;
- t_f is the time experimentally required to a complete mechanical deformation;
- $v = \dfrac{\pi r}{2 t_f}$ is the velocity of the mechanical deformation (assumed constant up to t_f);
- r is the cell radius;
- $\bar{t} \ll t_f$ is a parameter that identifies a finite time required for binding;
- x is the curvilinear abscissa on the meridian plane of the sphere;
- t is the generic time.

By this approach, already elaborated in our previous works (Damioli et al, 2017; Salvadori et al, 2018a), the relocation of receptors can be solved without numerical simulations of the real spreading process. Whereby questionable quantitative response is expected, the physics of the relocation of receptors is captured with fidelity.

Application of the surrogated mechanics approach to equations (11.4) and (11.11) finally leads to the following set of paradigmatic mass balance equations

$$\begin{aligned} \frac{\partial c_A}{\partial t} + \operatorname{div}_\Omega [\, \mathbf{h}_A \,] + w^{(11.4)} &= s_A \,, \\ \frac{\partial c_B}{\partial t} + \operatorname{div}_\Omega [\, \mathbf{h}_B \,] + w^{(11.4)} &= s_B \,, \\ \frac{\partial c_C}{\partial t} + \operatorname{div}_\Omega [\, \mathbf{h}_C \,] - w^{(11.4)} &= s_C \,. \end{aligned} \tag{11.13}$$

As a result of the assumptions that the ligands on the substrate are not free to move, and neither are the complex molecules after activation, it follows that equation (11.13) particularize to the integrin - VEGFR reactions (11.1)-(11.3) in the following form:

---

[1] Following this path of reasoning, the identification of a reference configuration is unnecessary. Therefore, the suffix $_R$ will be removed from now on. The notation $\operatorname{div}_\Omega [\,]$ is equivalent to $\operatorname{Div}_\Omega [\,]$.

$$
\begin{aligned}
&\frac{\partial c_I}{\partial t} + \mathrm{div}_\Omega \left[\mathbf{h}_I \right] + w^{(1)} + w^{(3)} = 0 \\
&\frac{\partial c_{L_I}}{\partial t} + w^{(1)} = s_{L_I} \\
&\frac{\partial c_{C_1}}{\partial t} - w^{(1)} = 0 \\
&\frac{\partial c_R}{\partial t} + \mathrm{div}_\Omega \left[\mathbf{h}_R \right] + w^{(2)} = 0 \\
&\frac{\partial c_{L_R}}{\partial t} + w^{(2)} = s_{L_R} \\
&\frac{\partial c_{C_2}}{\partial t} - w^{(2)} + w^{(3)} = 0 \\
&\frac{\partial c_{C_3}}{\partial t} - w^{(3)} = 0
\end{aligned}
\tag{11.14}
$$

11.2.4 Constitutive Assumptions and Chemical Kinetics

Governing equations emanate from the mass balance equations (**??**) after a rigorous thermodynamic derivation of the following constitutive equations of Fick's type for the receptors fluxes

$$
\begin{aligned}
\mathbf{h}_I &= - \mathcal{D}_I \nabla_\Omega \left[c_I \right] \\
\mathbf{h}_R &= - \mathcal{D}_R \nabla_\Omega [c_R] \,.
\end{aligned}
\tag{11.15}
$$

Fick's constitutive equations satisfy thermodynamic restrictions, as well known; their derivation will not be detailed here, see e.g., Gurtin et al (2010). Note, though, that the gradient operator in equations (11.15) is defined on the surface, i.e.

$$
\nabla_\Omega \left[c \right] = \nabla \left[c \right] - \left(\mathbf{n} \cdot \nabla \left[c \right] \right) \mathbf{n}
\tag{11.16}
$$

The chemical kinetics of the paradigmatic reaction (11.4) is modeled as for ideal systems (De Groot and Mazur, 1984) via the law of mass action:

$$
w^{(11.4)} = k_f \, \frac{\vartheta_B}{(1-\vartheta_B)} \, \frac{\vartheta_A}{(1-\vartheta_A)} - k_b \, \frac{\vartheta_C}{(1-\vartheta_C)}
\tag{11.17}
$$

In eq. (11.17), ϑ_A denotes the non-dimensional ratio between the concentration of species A and its amount c_A^{max} at saturation,

$$
\vartheta_A = c_A / c_A^{max} \,.
$$

Similarly ϑ_B and ϑ_C for species B and C. At chemical equilibrium, as $w^{(11.4)} = 0$, the concentrations obey the relation

$$
\frac{k_f}{k_b} = \frac{\vartheta_C^{\mathrm{eq}}}{(1-\vartheta_C^{\mathrm{eq}})} \, \frac{(1-\vartheta_R^{\mathrm{eq}})}{\vartheta_R^{\mathrm{eq}}} \, \frac{(1-\vartheta_L^{\mathrm{eq}})}{\vartheta_L^{\mathrm{eq}}} = K_{\mathrm{eq}}^{(11.4)} \,,
\tag{11.18}
$$

which defines the constant of equilibrium $K_{\rm eq}^{(11.4)}$ of reaction (11.4). The description of the chemical reaction (11.4) in terms of activities can be found in Salvadori et al (2018a,b).

Far from the saturation limit, $(1 - \vartheta_A) \sim 1$ (the same for species B and C). The mass action law (11.17) thus simplifies as

$$w^{(11.4)} = \tilde{k}_f \, c_B \, c_A - \tilde{k}_b \, c_C \tag{11.19}$$

having denoted with

$$\tilde{k}_f = k_f (c_B^{max} c_A^{max})^{-1} \, , \qquad \tilde{k}_b = k_b (c_C^{max})^{-1} \, .$$

Either eq. (11.17) or eq. (11.19) provide a set of three balance laws in terms of concentrations c_A, c_B, and c_C once inserted into eq. (11.13). Provided that the mass fluxes are constitutively related to concentrations, as for Fick's laws (11.15), eq. (11.13) turn out to be three governing equations.

Experimental evidences (Damioli et al, 2017) show that the equilibrium constant, paradigmatically defined by eq. (11.18), is high for all the three chemical reactions (11.1)–(11.3). This favors the formation of products and the depletions of receptors and ligands. Moreover, the diffusion of receptors on the cell membrane appears to be much slower than the interaction kinetics, i.e. the time required to reach chemical equilibrium is orders of magnitude smaller than the time-scale of other processes. For these reasons, the concentrations of species may be governed by thermodynamic equilibrium at all times.

Making reference again to the paradigmatic equation (11.19), the concentration of complex c_C relates to the others in the case of *infinitely fast kinetics*. Far from saturation, equating the reaction rate to zero, i.e. $w^{(11.4)} = 0$, implies

$$c_C = \frac{c_B \, c_A}{\alpha} \, , \tag{11.20}$$

having denoted with α the following constant:

$$\alpha = \frac{\tilde{k}_b}{\tilde{k}_f} = \frac{c_A^{max} \, c_B^{max}}{c_C^{max}} \, \frac{1}{K_{\rm eq}^{(11.4)}} \, . \tag{11.21}$$

11.2.5 Governing Equations

In conclusion, therefore, the paradigmatic problem (11.13) in the surrogated mechanics approach can be written in terms of concentrations after imposing that species B and C are immobile, that the Fick law $(11.15)_1$ relates the mass flux $\mathbf{h}_A$ to the concentration c_A, and either imposing the mass action law in the form (11.17) ((11.19) when far from saturation) or assuming infinitely fast kinetics in the form (11.20). In this last case, the two concentrations c_A and c_B describe the paradigmatic problem

(11.13), as follows:

$$\begin{aligned} \frac{\partial c_A}{\partial t} \left(1+\frac{c_B}{\alpha}\right) + \frac{c_A}{\alpha}\frac{\partial c_B}{\partial t} + \mathrm{div}_\Omega\left[-\mathcal{D}_A \nabla_\Omega\left[c_A\right]\right] &= 0 \\ \frac{\partial c_B}{\partial t} \left(1+\frac{c_A}{\alpha}\right) + \frac{c_B}{\alpha}\frac{\partial c_A}{\partial t} &= s_B \, . \end{aligned} \tag{11.22}$$

Parameter α depicts the influence of chemistry at infinitely fast kinetics; parameter $\mathcal{D}_A$ is associated with the diffusion process on the surface; function s_B accounts for mechanics, in a surrogated way. All these multi-physics processes impact onto the discretization parameters for the numerical simulation.

Equations (11.22) can be properly rephrased to model to the integrin - VEGFR reactions (11.1)–(11.3). Four unknown fields, $c_I(\mathbf{x},t)$, $c_{L_I}(\mathbf{x},t)$, $c_R(\mathbf{x},t)$, $c_{L_R}(\mathbf{x},t)$ describe the evolution of the system through the following non-linear partial differential equations:

$$\begin{aligned} &\frac{\partial c_{L_R}}{\partial t}\left(1+\frac{c_R}{\alpha_2}+\frac{c_R}{\alpha_2}\frac{c_I}{\alpha_3}\right) + \frac{c_R c_{L_R}}{\alpha_2\alpha_3}\frac{\partial c_I}{\partial t} + \frac{\partial c_R}{\partial t}\left(\frac{c_{L_R}}{\alpha_2}+\frac{c_{L_R}}{\alpha_2}\frac{c_I}{\alpha_3}\right) - s_{L_R} = 0 \\ &\frac{\partial c_R}{\partial t} + \mathrm{div}_\Omega\left[-\mathcal{D}_R\nabla_\Omega\left[c_R\right]\right] - \frac{\partial c_{L_R}}{\partial t} + s_{L_R} = 0 \\ &\frac{\partial c_I}{\partial t}\left(1+\frac{c_R c_{L_R}}{\alpha_2\alpha_3}\right) + \mathrm{div}_\Omega\left[-\mathcal{D}_I\nabla_\Omega\left[c_I\right]\right] - \frac{\partial c_{L_I}}{\partial t} + \\ &\qquad \frac{c_I}{\alpha_3}\frac{c_{L_R}}{\alpha_2}\frac{\partial c_R}{\partial t} + \frac{c_I}{\alpha_3}\frac{c_R}{\alpha_2}\frac{\partial c_{L_R}}{\partial t} + s_{L_I} = 0 \\ &\frac{\partial c_{L_I}}{\partial t}\left(1+\frac{c_I}{\alpha_1}\right) + \frac{c_{L_I}}{\alpha_1}\frac{\partial c_I}{\partial t} - s_{L_I} = 0 \end{aligned} \tag{11.23}$$

Initial conditions are dictated by experimental evidences and hold

$$c_I(\mathbf{x},0) = 8, \quad c_{L_I}(\mathbf{x},0) = 0, \quad c_R(\mathbf{x},0) = 4.8, \quad c_{L_R}(\mathbf{x},0) = 0 \tag{11.24}$$

11.2.5.1 Dimensionless Equations

The interplay among the multi-physics processes in the governing equations (11.23) can be better pointed out in a dimensionless formulation. To this aim, denote with L_d a reference length, with t_d a problem timescale, with c_{bulk} a reference concentration and finally define the dimensionless (starred) amounts

$$\mathbf{x} = L_d\,\mathbf{x}^\star \, , \quad t = t_d\, t^\star \, , \quad c_A = c_{bulk}\, c_A^\star \, , \quad \alpha_i = c_{bulk}\alpha_i^\star \tag{11.25}$$

with $i = 1, 2, 3$. Exploiting the simple identities

$$\frac{\partial c_A}{\partial t} = \frac{c_{bulk}}{t_d} \frac{\partial c^\star_A}{\partial t^\star} , \qquad s_A = \frac{c_{bulk}}{t_d} s^\star_A ,$$
$$\nabla_\Omega [c_a] = \frac{c_{bulk}}{L_d} \nabla^\star_\Omega [c^\star_A] , \qquad \mathrm{div}_\Omega [f] = \frac{1}{L_d} \mathrm{div}^\star_\Omega [f] \tag{11.26}$$

and defining the Peclet number (Quarteroni and Valli, 2008) for a generic species A as

$$\mathcal{P}^A_e = \frac{L^2_d}{\mathcal{D}_A t_d} , \tag{11.27}$$

the following dimensionless governing equations of the problem come out

$$\begin{aligned}
&\frac{\partial c^\star_{L_R}}{\partial t^\star} \left(1 + \frac{c^\star_R}{\alpha^\star_2} + \frac{c^\star_R}{\alpha^\star_2} \frac{c^\star_I}{\alpha^\star_3} \right) + \frac{c^\star_R c^\star_{L_R}}{\alpha^\star_2 \alpha^\star_3} \frac{\partial c^\star_I}{\partial t^\star} + \frac{\partial c^\star_R}{\partial t^\star} \left(\frac{c^\star_{L_R}}{\alpha^\star_2} + \frac{c^\star_{L_R}}{\alpha^\star_2} \frac{c^\star_I}{\alpha^\star_3} \right) - s^\star_{L_R} = 0 , \\
&\frac{\partial c^\star_R}{\partial t^\star} - \mathrm{div}^\star_\Omega \left[\frac{1}{\mathcal{P}^R_e} \nabla^\star_\Omega [c^\star_R] \right] - \frac{\partial c^\star_{L_R}}{\partial t^\star} + s^\star_{L_R} = 0 , \\
&\frac{\partial c^\star_I}{\partial t^\star} \left(1 + \frac{c^\star_R c^\star_{L_R}}{\alpha^\star_2 \alpha^\star_3} \right) - \mathrm{div}^\star_\Omega \left[\frac{1}{\mathcal{P}^I_e} \nabla^\star_\Omega [c^\star_I] \right] - \\
&\qquad \frac{\partial c^\star_{L_I}}{\partial t^\star} + \frac{c^\star_I}{\alpha^\star_3} \frac{c^\star_{L_R}}{\alpha^\star_2} \frac{\partial c^\star_R}{\partial t^\star} + \frac{c^\star_I}{\alpha^\star_3} \frac{c^\star_R}{\alpha^\star_2} \frac{\partial c^\star_{L_R}}{\partial t^\star} + s^\star_{L_I} = 0 , \\
&\frac{\partial c^\star_{L_I}}{\partial t^\star} \left(1 + \frac{c^\star_I}{\alpha^\star_1} \right) + \frac{c^\star_{L_I}}{\alpha^\star_1} \frac{\partial c^\star_I}{\partial t^\star} - s^\star_{L_I} = 0 .
\end{aligned} \tag{11.28}$$

These initial values PDEs will be solved for the unknown fields $c^\star_{L_R}$, $c^\star_R$, $c^\star_I$, and $c^\star_{L_I}$ once initial and boundary conditions will be given. Whereby the transport process is ruled by the Peclet's numbers $\mathcal{P}^R_e$ and $\mathcal{P}^I_e$, several other dimensionless amounts control chemistry (the three parameters $\alpha^\star_i$) and spreading ($s^\star_{L_R}$ and $s^\star_{L_I}$). The ratios between those numbers define the limiting factors during the evolution in time of the whole system. It will be clarified in discussing the numerical approximation of problem (11.28), that stems from its weak form.

11.3 Weak Formulation and Finite Elements Discretization

This section aims at building up the weak form of the governing equations to enable a numerical approximation of the partial differential equations problem (11.28) equipped with appropriate initial and boundary conditions. Formally speaking, the weak formulation is obtained after multiplication of the strong form of the governing equations by a suitable set of time independent test functions (expressed here with a superposed caret), and performing an integration upon the domain, exploiting Green's formula with the aim of reducing the order of differentiation. Such a weak form, in terms of the unknown fields $c^\star_{L_R}$, $c^\star_R$, $c^\star_I$, and $c^\star_{L_I}$, reads as follows:

$$
\begin{aligned}
&\int_{\Omega^\star} \hat{c}^\star_{L_R} \left[\frac{\partial c^\star_{L_R}}{\partial t^\star} \left(1 + \frac{c^\star_R}{\alpha^\star_2} + \frac{c^\star_R}{\alpha^\star_2} \frac{c^\star_I}{\alpha^\star_3} \right) + \frac{c^\star_R c^\star_{L_R}}{\alpha^\star_2 \alpha^\star_3} \frac{\partial c^\star_I}{\partial t^\star} + \frac{\partial c^\star_R}{\partial t^\star} \left(\frac{c^\star_{L_R}}{\alpha^\star_2} + \frac{c^\star_{L_R}}{\alpha^\star_2} \frac{c^\star_I}{\alpha^\star_3} \right) \right] + \\
&\quad \hat{c}^\star_R \left[\frac{\partial c^\star_R}{\partial t^\star} - \frac{\partial c^\star_{L_R}}{\partial t^\star} \right] + \frac{1}{\mathcal{P}^R_e} \nabla^\star_\Omega [\, c^\star_R \,] \cdot \nabla^\star_\Omega [\, \hat{c}^\star_R \,] + \frac{1}{\mathcal{P}^I_e} \nabla^\star_\Omega [\, c^\star_I \,] \cdot \nabla^\star_\Omega [\, \hat{c}^\star_I \,] + \\
&\quad \hat{c}^\star_I \left[\frac{\partial c^\star_I}{\partial t^\star} \left(1 + \frac{c^\star_R c^\star_{L_R}}{\alpha^\star_2 \alpha^\star_3} \right) - \frac{\partial c^\star_{L_I}}{\partial t^\star} + \frac{c^\star_I}{\alpha^\star_3} \frac{c^\star_{L_R}}{\alpha^\star_2} \frac{\partial c^\star_R}{\partial t^\star} + \frac{c^\star_I}{\alpha^\star_3} \frac{c^\star_R}{\alpha^\star_2} \frac{\partial c^\star_{L_R}}{\partial t^\star} \right] + \\
&\quad \hat{c}^\star_{L_I} \left[\frac{\partial c^\star_{L_I}}{\partial t^\star} \left(1 + \frac{c^\star_I}{\alpha^\star_1} \right) + \frac{c^\star_{L_I}}{\alpha^\star_1} \frac{\partial c^\star_I}{\partial t^\star} \right] = \\
&\quad \int_{\Omega^\star} \hat{c}^\star_{L_R} \, s^\star_{L_R} - \hat{c}^\star_R \, s^\star_{L_R} - \hat{c}^\star_I \, s^\star_{L_I} + \hat{c}^\star_{L_I} \, s^\star_{L_I} \, \mathrm{d}\Omega^\star
\end{aligned}
\tag{11.29}
$$

Note that there is no contribution defined on the boundary, because the cell membrane Ω is a closed surface. The weak form above can be written in a more abstract setting: we seek for unknown fields $c^\star_A(\mathbf{x}^\star, t^\star)$ —with A denoting L_R, R, I, or L_I— in the functional space $\mathcal{V}^{[0,t^\star_f]}$ that satisfy initial conditions $c^\star_A(\mathbf{x}^\star, 0) = c_0{}^\star_A(\mathbf{x}^\star)$ such that

$$
a\left(\left\{ \frac{\partial}{\partial t^\star} c^\star_A(\mathbf{x}^\star, t^\star), \; c^\star_A(\mathbf{x}^\star, t^\star) \right\}, \hat{c}^\star_A(\mathbf{x}^\star) \right) = \left(f_A(\mathbf{x}^\star, t^\star), \hat{c}^\star_A(\mathbf{x}^\star) \right) \tag{11.30}
$$

for all $\hat{c}^\star_A(\mathbf{x}^\star)$ that belong to a suitable functional space $\mathcal{V}$. In the abstract form (11.30), $a(\cdot, \cdot)$ is a non linear functional and (f, c) denotes the standard scalar product between f and c. The identification of the functional spaces $\mathcal{V}^{[0,t^\star_f]}$, $\mathcal{V}$ falls beyond the scope of the present paper.

The weak form (11.30) naturally leads to a semi-discrete problem, by approximating the space $\mathcal{V}$ by a finite dimensional space $\mathcal{V}_h$. To this aim, unknown fields $c^\star_A(\mathbf{x}^\star, t^\star)$ —with A denoting L_R, R, I, or L_I— will be approximated as a product of separated variables, by means of a basis $\{\varphi^A_i(\mathbf{x}^\star)\}$ of spatial shape functions and nodal unknowns that depend solely on time

$$
c_h{}^\star_A(\mathbf{x}^\star, t^\star) = \varphi^A_i(\mathbf{x}^\star) \; c^{\star A}_i(t^\star) \, . \tag{11.31}
$$

The Einstein summation convention is taken for repeated indexes. The semi-discrete approximate problem reads as follows: given $c_{0h}{}^\star_A(\mathbf{x}^\star)$ a suitable approximation of the initial datum $c_0{}^\star_A(\mathbf{x}^\star)$, for each $t^\star \in [0, t^\star_f]$ find $c_h{}^\star_A$ such that

$$
a\left(\varphi^A_j(\mathbf{x}^\star) \; \left\{ \frac{\partial}{\partial t^\star} c^{\star A}_j(t^\star), \; c^{\star A}_j(t^\star) \right\}, \varphi^A_i(\mathbf{x}^\star) \right) = \left(f_A(\mathbf{x}^\star, t^\star), \varphi^A_i(\mathbf{x}^\star) \right) . \tag{11.32}
$$

The weak form (11.29) is thus rephrased into the system of ordinary differential equations (11.32), whose solution is an approximation of the exact solution for each $t^\star$.

In order to obtain a full discretization of the weak form (11.29), we consider a uniform mesh for the time variable $t^\star$ and define $t^\star_n = n \, \Delta t^\star$ with $n = 0, 1, \ldots,$

and $\Delta t^\star > 0$ being the time step. The time derivative will be replaced by suitable difference quotients

$$\frac{\partial c_i^{\star A}}{\partial t^\star} \simeq \frac{c_i^{\star A}(t_n^\star + \Delta t^\star) - c_i^{\star A}(t_n^\star)}{\Delta t^\star} \,, \tag{11.33}$$

thus constructing a sequence $c_{h\,A}^{n\star}(\mathbf{x}^\star)$ that approximates the exact solution $c_A^\star(\mathbf{x}^\star, t_n^\star)$ (Quarteroni and Valli, 2008). Making recourse to the Backward Euler method leads to the non linear problem depicted in appendix 11.7.

The Backward Euler scheme for the finite element approximation of the chemo-transport model was implemented exploiting the high performance computing open source library deal.ii (*https://www.dealii.org/*). Numerical simulations shown that it is mandatory to use a very small time discretization step $\Delta t^\star$ when a first order accuracy numerical integration scheme is used. The limiting time scale is imposed by the mechanical deformation of the cell (identified within the equations by the terms of fictitious sources $s_{L_R}^\star$ and $s_{L_I}^\star$). Therefore, properly capturing the spreading process leads to a very high computational cost.

The time-dependent partial differential equations (11.28) can be conveniently rephrased in order to apply higher order time integration schemes. A variable change will be adopted, namely

$$c_A^\star = c_{L_R}^\star + c_{C_2}^\star + c_{C_3}^\star \,, \quad c_B^\star = c_R^\star - c_{L_R}^\star \,, \quad c_D^\star = c_I^\star + c_{C_3}^\star - c_{L_I}^\star \,, \tag{11.34}$$

having denoted with

$$c_{C_1}^\star = \frac{c_I^\star c_{L_I}^\star}{\alpha_1^\star} \,, \quad c_{C_2}^\star = \frac{c_R^\star c_{L_R}^\star}{\alpha_2^\star} \,, \quad c_{C_3}^\star = \frac{c_I^\star c_R^\star c_{L_R}^\star}{\alpha_2^\star \alpha_3^\star} \,. \tag{11.35}$$

Simple algebra leads to the following set of time-dependent PDEs,

$$\frac{\partial c_A^\star}{\partial t^\star} - s_{L_R}^\star = 0 \tag{11.36a}$$

$$\frac{\partial c_B^\star}{\partial t^\star} - \frac{1}{\mathcal{P}_e^R} \mathrm{div}_\Omega^\star \left[\nabla_\Omega^\star \left[c_R^\star \right] \right] + s_{L_R}^\star = 0 \tag{11.36b}$$

$$\frac{\partial c_D^\star}{\partial t^\star} - \frac{1}{\mathcal{P}_e^I} \mathrm{div}_\Omega^\star \left[\nabla_\Omega^\star \left[c_I^\star \right] \right] + s_{L_I}^\star = 0 \tag{11.36c}$$

$$\frac{\partial c_E^\star}{\partial t^\star} - s_{L_I}^\star = 0 \tag{11.36d}$$

subject to the constraints

$$\begin{aligned} &c_R^\star c_R^\star \left(\alpha_3^\star + c_I^\star \right) + c_R^\star \left(\alpha_2^\star \alpha_3^\star - \alpha_3^\star c_B^\star - c_I^\star c_B^\star \right) - \alpha_2^\star \alpha_3^\star \left(c_B^\star + c_A^\star \right) = 0, \\ &\quad c_I^\star c_I^\star \left(\alpha_2^\star \alpha_3^\star + c_R^\star c_R^\star - c_R^\star c_B^\star \right) + c_I^\star \big(- \alpha_2^\star \alpha_3^\star c_D^\star + \alpha_1^\star \alpha_2^\star \alpha_3^\star + \\ &\qquad \alpha_1^\star c_R^\star c_R^\star - \alpha_1^\star c_R^\star c_B^\star \big) - \alpha_1^\star \alpha_2^\star \alpha_3^\star \left(c_E^\star + c_D^\star \right) = 0, \end{aligned} \tag{11.37}$$

and initial conditions

$$c^\star_A(0) = 0 \,, \qquad c^\star_B(0) = \frac{4.8}{c_{bulk}} \,, \qquad c^\star_D(0) = \frac{8}{c_{bulk}} \,, \qquad c^\star_E(0) = 0 \,. \quad (11.38)$$

Analytical integration of (11.36a) and (11.36d) together with relevant initial conditions (11.38) leads to

$$c^\star_A(\mathbf{x}^\star, t^\star) = S^\star_{L_R}(\mathbf{x}^\star, t^\star) \,, \qquad c^\star_E(\mathbf{x}^\star, t^\star) = S^\star_{L_I}(\mathbf{x}^\star, t^\star) \,, \qquad (11.39)$$

where $S^\star_{L_R}$ and $S^\star_{L_I}$ are the integral over time of the source terms. They correspond to the value of ligands that at every instant $t^\star$ are in contact with the cell membrane at point $\mathbf{x}^\star$ due to the (surrogated) mechanics of spreading and are also available for the reaction with relevant counterparts. As such, they shall be assumed as experimental data, whereby the source terms $s^\star_{L_R}$ and $s^\star_{L_I}$ are much harder to identify.

Since $c^\star_A$ and $c^\star_E$ have been analytically solved, the weak form of problem (11.36) can be recast in a standard (Quarteroni and Valli, 2008) abstract setting: we seek for unknown fields $c^\star_Z(\mathbf{x}^\star, t^\star)$ —with Z denoting B or D— in the functional space $\mathcal{V}^{[0,t^\star_f]}$ that satisfy initial conditions (11.38) and subject to the constraints (11.37) such that

$$\frac{\partial}{\partial t^\star} b\left(c^\star_Z(\mathbf{x}^\star, t^\star), \hat{c}^\star_Z(\mathbf{x}^\star)\right) + a\left(c^\star_Z(\mathbf{x}^\star, t^\star), \hat{c}^\star_Z(\mathbf{x}^\star)\right) = \left(f_Z(\mathbf{x}^\star, t^\star), \hat{c}^\star_Z(\mathbf{x}^\star)\right) \qquad (11.40)$$

for all $\hat{c}^\star_Z(\mathbf{x}^\star)$ that belong to a suitable functional space $\mathcal{V}$. In the abstract form (11.40), $a(\cdot,\cdot)$ and $b(\cdot,\cdot)$ are the usual bilinear forms of the Laplace operator written on a non-Riemann manifold $\Omega^\star$. The weak form (11.40) can be conveniently integrated in time, providing the following approximation scheme

$$\begin{aligned} & b\left(c^\star_Z(\mathbf{x}^\star, t^\star_n), \hat{c}^\star_Z(\mathbf{x}^\star)\right) - b\left(c^\star_Z(\mathbf{x}^\star, t^\star_{n-1}), \hat{c}^\star_Z(\mathbf{x}^\star)\right) + \\ & \int_{t^\star_{n-1}}^{t^\star_n} a\left(c^\star_Z(\mathbf{x}^\star, \tau^\star), \hat{c}^\star_Z(\mathbf{x}^\star)\right) \mathrm{d}\tau^\star = \left(\int_{t^\star_{n-1}}^{t^\star_n} f_Z(\mathbf{x}^\star, \tau^\star)\mathrm{d}\tau^\star, \hat{c}^\star_Z(\mathbf{x}^\star)\right) \end{aligned} \qquad (11.41)$$

for all $\hat{c}^\star_Z(\mathbf{x}^\star)$, whereby $c^\star_Z(\mathbf{x}^\star, \tau^\star)$ must satisfy initial conditions (11.38) and is subject to the constraints (11.37). Note that the integral

$$\int_{t^\star_{n-1}}^{t^\star_n} f_Z(\mathbf{x}^\star, \tau^\star)\mathrm{d}\tau^\star$$

is given in closed form and involves the functions $S^\star_{L_R}(\mathbf{x}^\star, t^\star)$ and $S^\star_{L_I}(\mathbf{x}^\star, t^\star)$ evaluated in $t^\star_{n-1}$ and $t^\star_n$. In view of this feature, the contribution of cell spreading in terms of surrogated mechanics is captured with high accuracy, whereas the accuracy in the approximation of the transport term

$$\int_{t^\star_{n-1}}^{t^\star_n} a\left(c^\star_Z(\mathbf{x}^\star, \tau^\star), \hat{c}^\star_Z(\mathbf{x}^\star)\right) \mathrm{d}\tau^\star$$

appears not to require too small time steps in view of the longer timescale of the diffusivity compared to the other two physics involved (see also the similar conclusions reached in Damioli et al (2017)). The extended version of the weak form (11.41) can be found in Appendix 11.7.

11.4 Simulations

The relocation of VEGFR-2 (i.e. the main pro-angiogenic receptor expressed on the endothelial cells) and integrin (i.e. the cardinal regulator of mechanical signal transduction between endothelial cells and the extracellular matrix (ECM)) along the cell membrane of an endothelial cell during its adhesion onto a ligands-enriched substrate was simulated implementing a fully coupled Newton–Raphson solver for the discretized weak form (11.41) exploiting the high performance computing library deal.ii (see *https://www.dealii.org*).

Parameters and Data

The process was investigated over a time span of $t = 3900$ s, an adequate extent for the deployment of the experimentally observed mechanisms. The parameter t_f, which represents the time required for a complete adhesion, was calibrated as $t_f = 600$ s from experimental evidences. Since spreading has been mechanically surrogated in the present note, half of the sphere that geometrically recovers the shape of the membrane is taken as covered by ligands at time t_f. The finite time required for binding two reactants has been set to $\bar{t} = 1$ s .

The size $l = 20$ μm of the radius of the cell-sphere has been deduced from an average of 50 measurements on different endothelial cells. Utilizing two different experimental techniques, Fluorescence Recovery After Photobleaching (FRAP) and Surface Plasmon Resonance (SPR), it was possible to measure the diffusivity of the VEGFR-2 $\mathcal{D}_R = 0.21$ μm^2/s and the kinetic parameter $K_{eq}^{(2)} = 354059$ for Gremlin/VEGFR-2.

The amount of VEGFR-2 on the cell membrane per μm^2 at the beginning of the process, $c_R(\mathbf{x}, 0) = 4.8$ molecules/μm^2, come out dividing the number of high-affinity binding sites for cell surface area. Consistent with in vitro observations, it was taken $c_{L_R}^{max} = 16000$ molecules/μm^2 assuming that $c_R^{max} = c_{C_2}^{max} = c_{C_3}^{max}$ and $c_I^{max} = c_{C_1}^{max}$. Furthermore, the following parameters have been deduced from preliminary experimental results:

- $\mathcal{D}_I$, the diffusivity of integrin: here assumed equal to 0.23 μm^2/s;
- $K_{eq}^{(1)}$ and $K_{eq}^{(3)}$, the equilibrium constants for the chemical reactions 11.1 and 11.3: here taken equal to $K_{eq}^{(2)}$;

- $c_I(\mathbf{x}, 0)$, the initial concentration of free integrin that has been taken to be equal to 8 molecules/μm^2 (a value that is reasonably larger than the initial concentration of VEGFR-2 but without experimental confirmation);
- $c_{L_I}^{max}$ and c_I^{max}, saturation limits for the fibronectin and integrin: here assumed equal to $c_{L_R}^{max}$, the saturation limit for Gremlin.

Numerical Approximation of s_{L_R} and s_{L_I}

To avoid discontinuities, a "Gaussian" approximation on the relationships (11.12) has been implemented:

$$\begin{aligned} s_{L_R} &\simeq a \exp\left[-((t-b)^2)/(c^2)\right] , \\ s_{L_I} &\simeq d \exp\left[-((t-e)^2)/(f^2)\right] , \end{aligned} \tag{11.42}$$

in which the coefficients $a,\ b,\ c,\ d,\ e,\ f$ characterize the shape of s_{L_R} and s_{L_I}. The latter can be integrated in time:

$$\begin{aligned} \int_0^t s_{L_R}(\mathbf{x},\tau)d\tau &= S_{L_R}(\mathbf{x},t) - S_{L_R}(\mathbf{x},0) = \frac{1}{2}ac\sqrt{\pi}(Erf[b/c] - Erf[(b-t)/c]) \\ \int_0^t s_{L_I}(\mathbf{x},\tau)d\tau &= S_{L_I}(\mathbf{x},t) - S_{L_I}(\mathbf{x},0) = \frac{1}{2}df\sqrt{\pi}(Erf[e/f] - Erf[(e-t)/f]) \ . \end{aligned} \tag{11.43}$$

Expressions (11.43) are consistent with the requirements $S_{L_R}(\mathbf{x}, 0) = 0$ and $S_{L_I}(\mathbf{x}, 0) = 0$. The function (11.43) provide a supply of ligands, at the end of the cell-substrate contact dynamic process, equal to 16 molecules/μm^2. Such a total density of ligands is almost as twice as compared to the initial concentration of integrins (here fixed at 8 molecules/μm^2) and more than three times compared to VEGFR-2 (here fixed at 4.8 molecules/μm^2).

Discretization of Geometry

The tessellation of the cell membrane consists of 37650 quadrilaterals, uniformly distributed over the spherical surface. The discretization, depicted in Fig. 11.2, remained unchanged throughout the analysis, with no remeshing.

11.4.1 Outcomes

In addition to the values of $c_R^\star$ and $c_I^\star$, which are derived directly from the linear system solution together with $c_A^\star, c_B^\star, c_D^\star, c_E^\star$, the values of $c_{L_R}^\star, c_{L_I}^\star, c_{C_1}^\star, c_{C_2}^\star, c_{C_3}^\star$ will be evaluated and converted back into dimensional quantities, at each Gauss point, during the post-processing, by means of the identities (11.34) and (11.35).

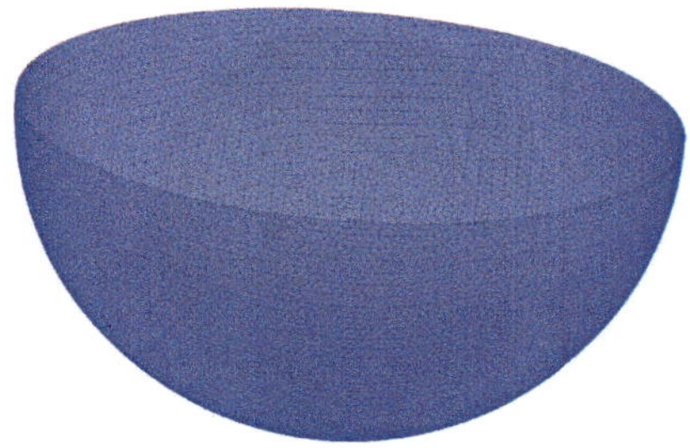

Fig. 11.2 Half-sphere tessellation via hexagons.

Evolution in Space and in Time of Free Receptors and Complexes

The evolution in space and time of free receptors and complexes is regulated by three mechanism:

- *chemistry*—chemical interaction among ligands-receptors (chemical reactions (11.1), (11.2)) and receptor-complex (chemical reaction (11.3)),
- *mechanics*—mechanical spreading of the cell that puts ligand into contact with receptors, and
- *diffusion*—Brownian motion of unbound receptors.

Chemical reaction (11.3) connects the reactions (11.1) and (11.2). Accordingly, the evolution in space and time of free receptors is connected to the history of complexes.

Depletion of VEGFR-2

A major depletion of free VEGFR-2 on the entire membrane is observed in numerical simulations (Fig. 11.3): the concentration of free VEGFR-2 decreases from the initial amount (4.8 molecules/μm^2) to a value one order of magnitude smaller on the apical side (after 3900 s). Free VEGFR-2 receptors on the basal membrane are essentially absent.

Concerning the chemical interactions involving VEGFR-2, since there is a much higher quantity of integrin and Gremlin than VEGFR-2, it emerges that:

1. the equilibrium constants of the reactions (11.2) and (11.3), are extremely high (and comparable in magnitude). As such, they favor the production of complexes;
2. when a receptor c_R gets in contact with the corresponding ligand c_{L_R}, c_{C_2} is immediately generated;
3. the newly produced complex instantly interacts with the available integrin, giving c_{C_3}.

This chemical-loop concerning c_R, c_{L_R}, c_I, c_{C_2} and c_{C_3}, occurs whenever two reactants of a specific chemical reaction meet on the cell membrane.

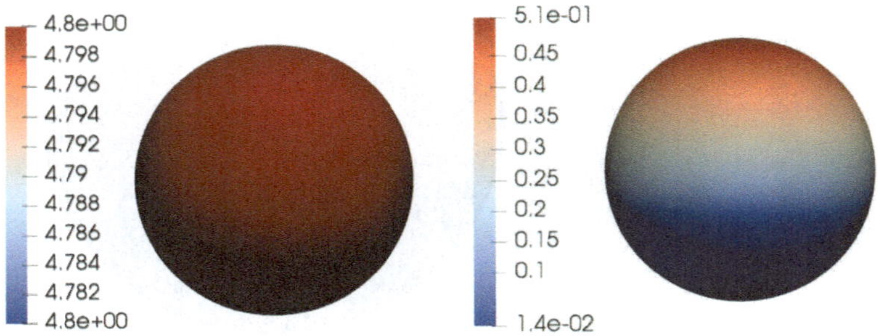

(a) t=0 s **(b)** t=3900 s

Fig. 11.3 Comparison between the concentration distribution of VEGFR2 at the initial and final time of the analyses.

The physics that preside the motion of the reagents change according to the time of observation of the events. Because the time scale of the mechanics is much faster than that of diffusion, one discriminates a first *chemo-mechanically-dominated* period, which begins at the instant $t = 0$ s (pure chemo-dominant) and terminates at the end of the mechanical spreading of the cell ($t_f = 600$ s). During this time, the basal part of the cell completely adheres to the substrate, and the VEGFR-2 receptors get trapped (Salvadori et al, 2018b) by the ligands.

A second phase, *chemo-diffusive-dominated*, exists from $t = t_f$ until the end of the analysis, at $t = 3900$ s. During this period, free receptors diffuse across the cell membrane, from the apical to the basal side of the cell. As soon as that they meet ligands, they are immediately captured on the substrate.

In order to visualize the overall information mentioned above, a few frames of the evolution of VEGFR-2 over time are shown in Fig. 11.4.

Evolution of C_2 and C_3

From the frames in Fig. 11.5 it is seen that two "rings" of complexes C_2 exist on the cell membrane in contact with ligands. Moving from the pole at the basal side of the cell towards the equatorial area of the sphere, three homogeneous bands are clearly visible. They show different intensities: medium (about 3.2 molecules/μm^2), high (about 4.2 molecules/μm^2) and low (about 2.2 molecules/μm^2), respectively. These different areas of complexes accumulation generate progressively. The medium zone is essentially completed at the end of the chemo-mechanical-dominated phase ($t_f = 600$ s), whereupon the two rings grow throughout the chemo-diffusive-dominated phase.

Once again, we highlight that it is not possible to study the outcome of a single reaction, because all three are intimately linked. If one considers the distribution of the complex C_3, in Fig. 11.6, an accumulation at the edge that delimitates the area of contact between cell and substrate becomes evident. It is easy to figure out that

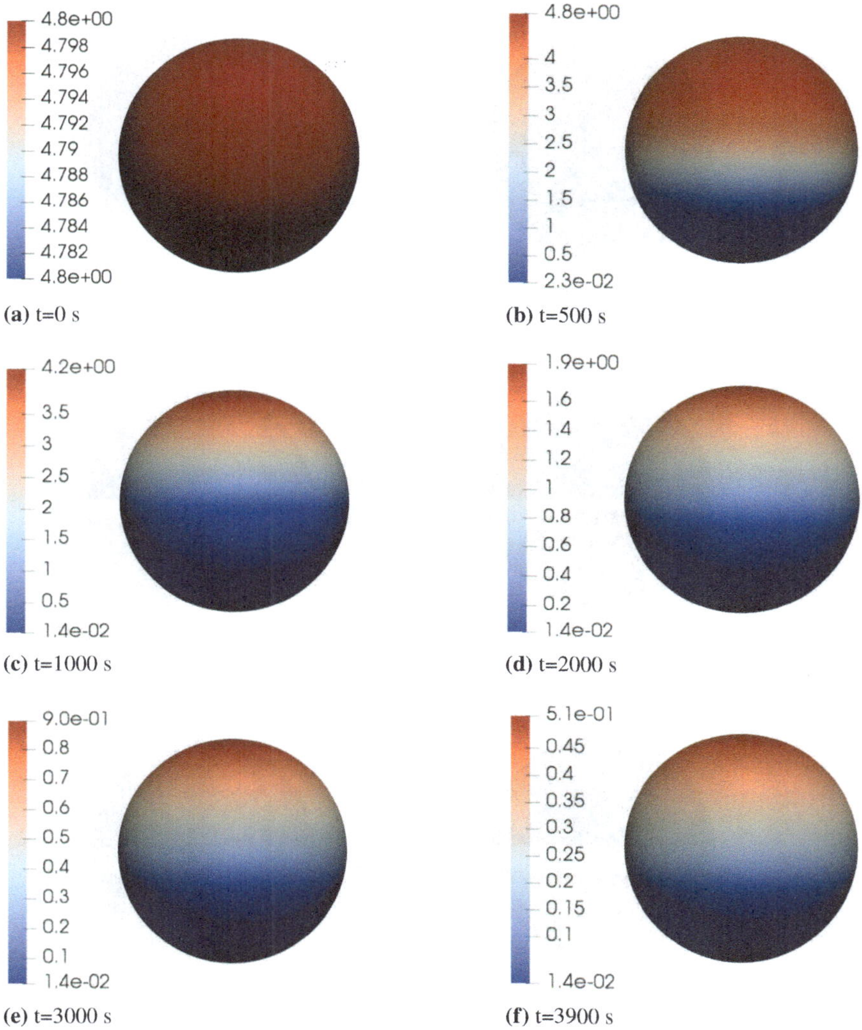

Fig. 11.4 Distribution of the concentration of the VEGFR-2 at times $t = 0, 500, 1000, 2000, 3000, 3900$ s.

the low concentration ring for C_2 in Fig. 11.5 is, at the same time, a zone for high accumulation of complex C_3.

In fact, C_2 would accumulate at the edge that delimitates the area of contact between cell and substrate. This phenomenon has been illustrated in depth in Damioli et al (2017) and motivated by the diffusion of VEGFR-2 from the apical side toward the basal membrane coupled to the chemical reaction (11.2), which highly favors the production of C_2. As soon as the latter is formed, reaction (11.3) is triggered

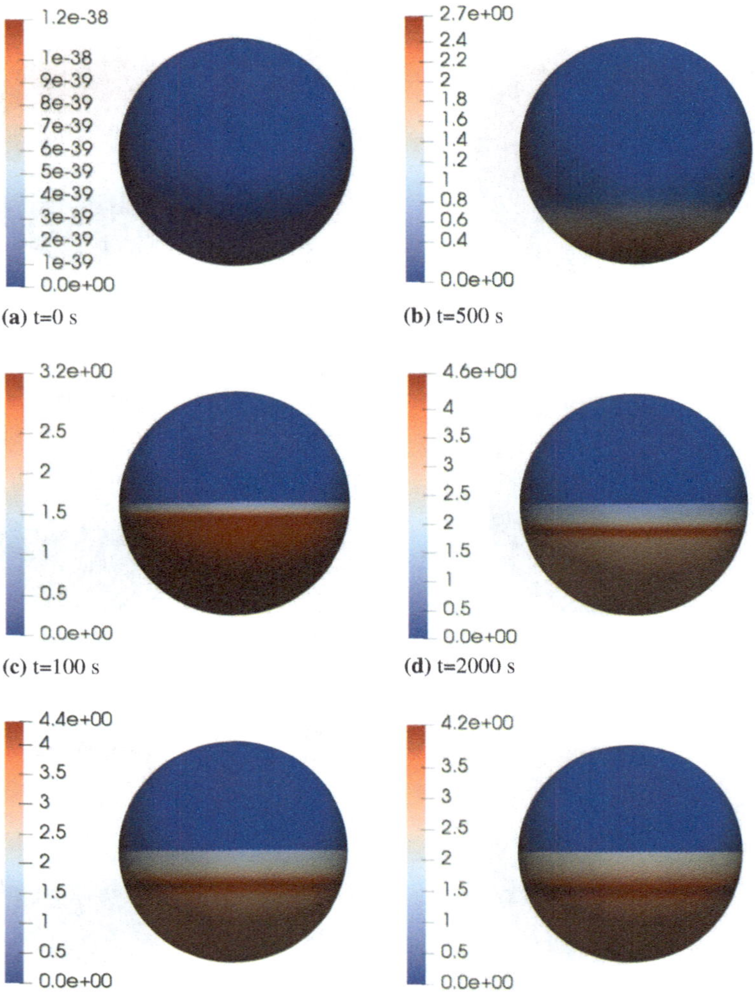

(a) t=0 s **(b)** t=500 s

(c) t=100 s **(d)** t=2000 s

(e) t=3000 s **(f)** t=3900 s

Fig. 11.5 Distribution of the concentration of the complex c_{C_2} at time $t = 0,\ 500,\ 1000,\ 2000,\ 3000,\ 3900$ s.

and it proceeds until integrin receptors are available. The latter, however, are rapidly consumed by reactions (11.1) and (11.3) and once they are fully depleted, C_2 cannot be converted to C_3 anymore and it stores itself in its own accumulation ring.

Note that in view of the diffusion of VEGFR-2, the C_2 accumulation ring moves in time towards the basal pole of the cell.

In conclusion, the complexes c_{C_2} and c_{C_3}, evolve driven by a large flow of receptors from the apical part of the cell towards the basal part. These free receptors, meeting the corresponding reactants in the cell-substrate contact zone, generate complexes and trap. In this regard, it is useful to analyze the trend of the integrin receptor and the corresponding complex c_{C_1} generated in the chemical reaction (11.1).

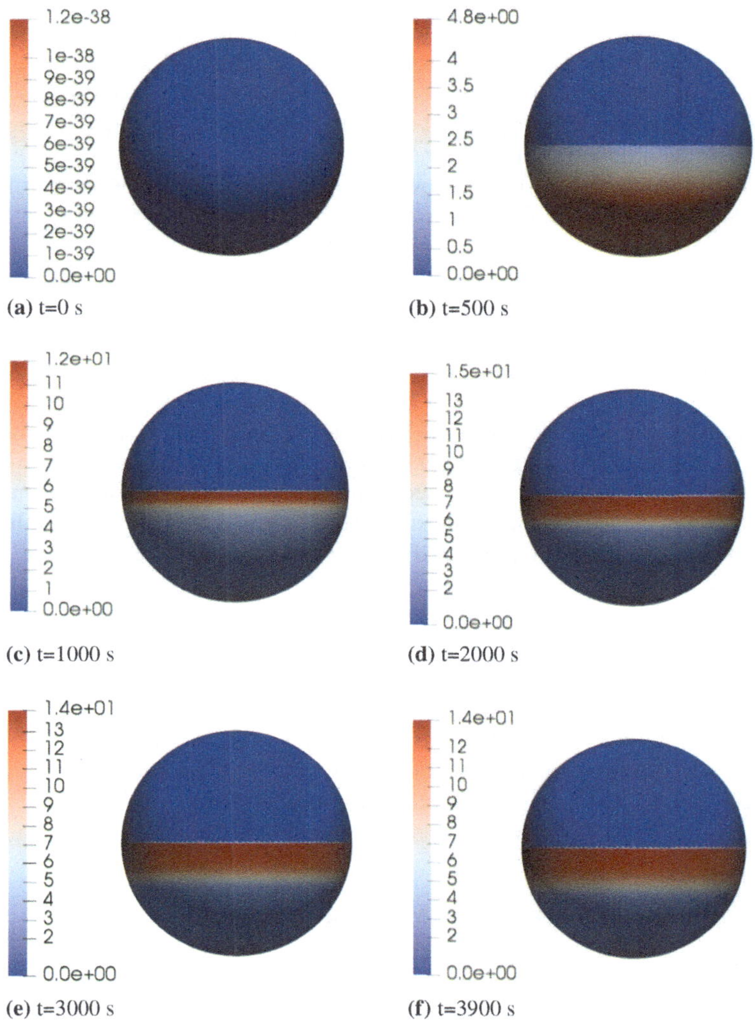

Fig. 11.6 Distribution of the concentration of the complex c_{C_3} at time $t = 0, 500, 1000, 2000, 3000, 3900$ s.

Evolution of C_I

The free integrin receptor I and VEGFR-2 undergo a similar evolution[2]. Free integrin concentration ranges from 8 molecules/μm^2 (value at time $t = 0$ s) to a value of the order of magnitude of $10^{-1} - 10^{-2}$ molecules/μm^2 at the end of the analysis, at time $t = 3900$ s. The reasons for the global consumption of integrin, visible in Fig. 11.7, are to be found in the receptors-ligands chemical interactions (11.1)–(11.3).

Figures 11.6 and 11.8 show an accumulation of complexes C_1 and C_3 at the edge between cell and substrate. The magnitude of the concentrations is similar, because similar are the equilibrium constants of chemical reactions (11.1) and (11.3) and the saturation limits of integrins and fibronectin.

Unlike the complex C_2, which shows three homogeneous areas with different magnitude, the concentration of C_1 and C_3 locates in two (high intensity and low intensity) zones.

Diffusion guides the free integrin and the VEGFR-2 from the apical part towards the basal one, leading to the accumulation of complexes in a relatively narrow zone of high availability of ligands and C_2. When free integrins cross the border that identifies the cell-substrate contact zone, they are immediately captured by L_I and C_2, which have the same affinity with I. The difference in concentrations between c_{C_1} and c_{C_3} ($c_{C_1} \simeq 14$ molecules/μm^2 is greater than $c_{C_3} \simeq 13.5$ molecules/μm^2) is due to the different physics that supply ligands and C_2. The (larger) availability of fibronectin is provided by the chemo-mechanics, whereas the C_2 at hand is generated from 11.2 and in the high intensity narrow zone is mostly due to the chemo-diffusive phase that cause the migration of VEGFR-2 from the apical to the basal side of the cell.

In the low intensity zone c_{C_1} and c_{C_3} reach the values of about 7 molecules/μm^2 and 2.6 molecules/μm^2, respectively. The gap is due again to the larger availability of fibronectin for the reaction (11.1) compared to the C_2 at hand for the reaction (11.3).

It should be emphasized that the flux of integrins is higher than that of VEGFR-2 because $\mathcal{D}_I$ is bigger than $\mathcal{D}_R$ and integrins are consumed concurrently by two chemical reactions. The latter justifies a large concentration gradient of I.

11.5 Remarks on Surrogated Mechanics

During processes such as tissue regeneration and wound healing, cell adhesion, spreading, and migration are controlled by receptor-mediated interactions with the extracellular matrix. The synergy among soluble growth factors, cytokines, and the extracellular matrix acts to regulate changes in cell shape, growth, proliferation, and motility.

[2] As for the VEGFR-2, a first chemo-mechanical-dominated period and a following chemo-diffusive-dominated phase hold also for the integrin.

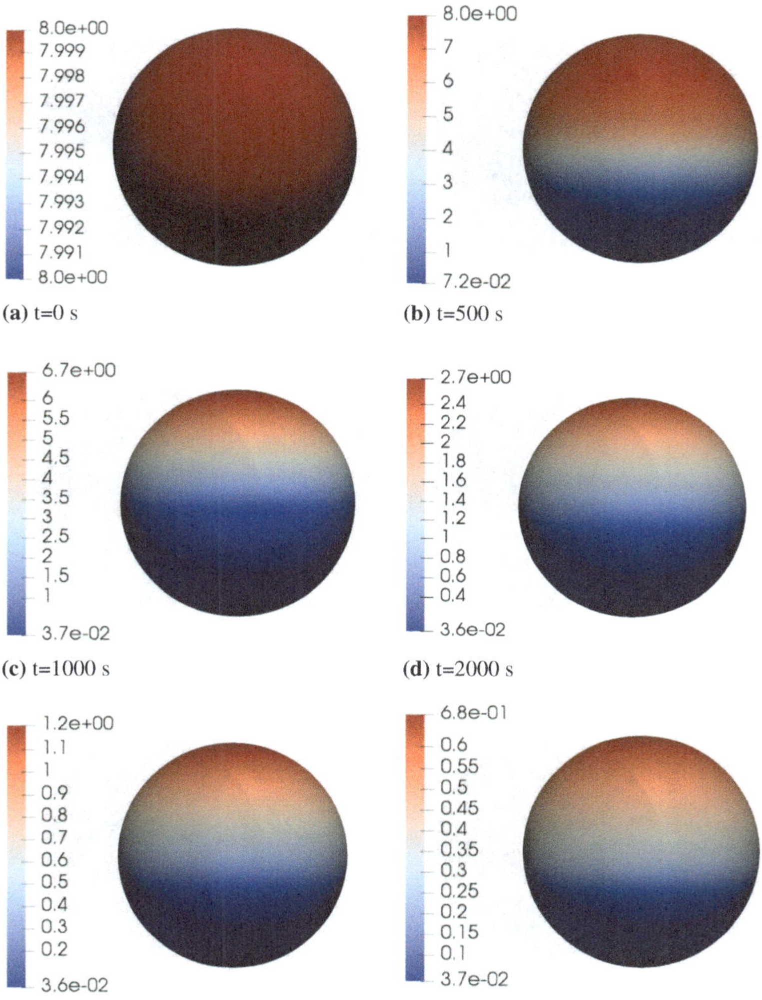

(a) t=0 s **(b)** t=500 s

(c) t=1000 s **(d)** t=2000 s

(e) t=3000 s **(f)** t=3900 s

Fig. 11.7 Distribution of the concentration of integrin at time $t = 0, 500, 1000, 2000, 3000, 3900$ s.

In order to identify how ligands stimulation induces the polarization of receptors in cell protrusions and in the basal aspect of ECs plated on a ligand-enriched ECM, a chemo-transport model has been devised in equations (11.10)–(11.11), in a finite strain formulation.

The interaction between a cell and its substrate has a specific, important ability to control cell morphology via traction exerted through cellular receptors onto the ECM (Reinhart-King et al, 2005). These interactions during spreading have not been accounted for explicitly in the present paper, rather we surrogated the mechanics

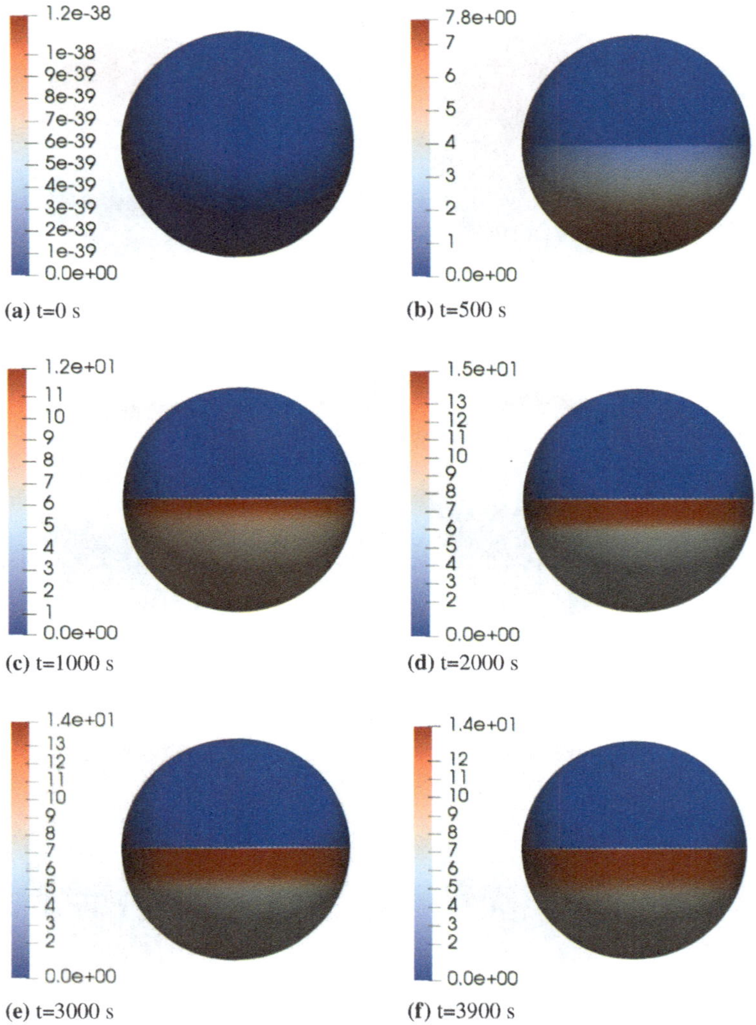

Fig. 11.8 Distribution of the concentration of the complex c_{C_1} at time $t = 0, 500, 1000, 2000, 3000, 3900$ s.

for the sake of focusing on fundamental membrane dynamics of the relocation of VEGFR-2 and beta3 integrin and their time-scales.

Mechanical aspects of cell deformation have been treated more extensively by many authors. The mechanical response of the cell during its spreading has been attributed merely to the lipid membrane (by means of the underlying cortical actin network) in several contributions. In the so-called *active gel theory* (Joanny et al, 2013; Kruse et al, 2005; Marchetti et al, 2013; Prost et al, 2015), the lipid membranes were regarded as two-dimensional liquid crystals whose structure is conferred by the oriented lipid molecules, an approach apparently originated in Helfrich (1973). General cell-substrate contact conditions have been developed for lipid membranes interacting with curved substrates along their edges in Agrawal and Steigmann (2009), through a variational energy principle. The same authors extended these ideas to the electromechanics of lipid bilayers in Steigmann and Agrawal (2016), accounting for flexo-electricity so to include deformations in the presence of applied electric fields, as well as to include the effects of a continuous distribution of transmembrane proteins (Agrawal and Steigmann, 2011).

Mechano-sensitive cellular contractility models that account for the stress fibers (SFs) reorganization in the bulk of the cell have also been proposed in a number of phenomenological models (to cite a few see the large repertoire of Deshpande et al, 2011, 2008, 2006, 2007; McMeeking and Deshpande, 2017; Ronan et al, 2012, 2014; Obbink-Huizer et al, 2014; Vernerey and Farsad, 2011). SFs are bundles of actin, capable to generate contractile force and linked to integrins by means of focal adhesions (FAs), which are able to convey a chemo-mechanical cue between receptors and the SFs themselves. The model proposed in Vigliotti et al (2016) quantifies the traction forces activated by the SFs and transmitted via FAs onto the integrins. To this aim, the authors developed a thermodynamically consistent framework to describe the stress, strain, and strain rate dependence of SFs formation and remodeling. A steady-state implementation of this thermodynamically motivated framework, coupled with an active thermodynamically consistent focal adhesion assembly, predicts cell cytoskeletal distributions on micro-patterned substrates (McEvoy et al, 2017). All of these cell mechanics formulations differ from active gels theory since they use a phenomenological Hill-type stress-strain rate relationship to represent active contractility of SFs.

11.6 Conclusion

Recent technological advances made available a large number of experimental data in biology. On the one hand, this is a motivation for great excitement. On the other, it pushes interpretative abilities to the limit. For this sake, the ability of multi-physics models to predict the time-space evolution of complex processes and to unravel their intimate nature is more and more becoming of pivotal importance in science.

The present work provides a contribution in understanding how VEGFR-2 and integrins regulate tumor angiogenesis, by means of a chemo-transport-mechanical model, set in the thermodynamics of continua. This model couples three chemical reactions to continuity equations for mass, energy, and entropy. Thermodynamic restrictions (Clausius–Duhem inequality) set limits for the Helmholtz free-energy, and ultimately, on constitutive relations.

Two simplifying assumptions are taken in this model: *surrogated mechanics* and *infinitely fast kinetics*. By the first one, the cell-substrate contact dynamics is surrogated by means of a fictitious term in the mass balance equations of the ligands.

The second makes chemical reactions always at equilibrium; such a scenario stems from experimental evidences, which show that the time required to reach the chemical equilibrium is orders of magnitude lower than the timescale of all other physical processes.

The formal derivation of the governing equations has been achieved both in strong and weak forms. The latter has been discretized and implemented in the deal.ii framework, an open-source C++ software library that supports the creation of high performance computing finite element codes.

The different time scales that characterize the chemical, mechanical and diffusive phenomena lead to a clear identification of two predominant phases of the trapping of free receptors by the corresponding ligands. The time-space evolution of the receptor dynamics has been discussed at large in the simulations section.

The achieved comprehension, in spite of the two major simplifying assumptions taken, encourages us to devote efforts towards more sophisticated formulations, so to further lessen the numerical approximations introduced herein. This improvement will enhance the predictive capabilities of the current model by, for instance, properly accounting for the mechanical deformation of the cell. The chemo-diffusive model presented in the current work will be improved enriching the chemical potential of bound integrins with the potential energy of the load applied to integrins. The emerging model would be able to capture the cell's active and passive mechanical behavior and couple it with the receptor dynamics of integrins and VEGFR-2 proposed here, providing an in silico model of how the mechanical processes of an endothelial cell can influence the activity of the main angiogenic regulator (VEGFR-2).

Acknowledgements Authors are gratefully indebted with the company *Ferriera Valsabbia* for the support to their research through a devoted fellowship. MS acknowledges the support of Guido Berlucchi Foundation Young Researchers Mobility Programme.

11.7 Appendix

A: Backward Euler Formulation

First Equation

$$
\begin{aligned}
&\int_{\Omega^\star} \varphi_i^{\star L_R}(\mathbf{x}^\star)\varphi_l^{\star L_R}(\mathbf{x}^\star)d\Omega^\star \left(\frac{c_l^{\star L_R}(t^\star+\Delta t^\star)-c_l^{\star L_R}(t^\star)}{\Delta t^\star}\right)+\\
&\int_{\Omega^\star} \varphi_i^{\star L_R}(\mathbf{x}^\star)\varphi_l^{\star L_R}(\mathbf{x}^\star)\varphi_k^{\star R}(\mathbf{x}^\star)d\Omega^\star \left(\frac{c_l^{\star L_R}(t^\star+\Delta t^\star)-c_l^{\star L_R}(t^\star)}{\Delta t^\star}\right)\frac{c_k^{\star R}(t^\star+\Delta t^\star)}{\alpha_2^\star}+\\
&\int_{\Omega^\star} \varphi_i^{\star L_R}(\mathbf{x}^\star)\varphi_k^{\star R}(\mathbf{x}^\star)\varphi_l^{\star L_R}(\mathbf{x}^\star)d\Omega^\star \left(\frac{c_k^{\star R}(t^\star+\Delta t^\star)-c_k^{\star R}(t^\star)}{\Delta t^\star}\right)\frac{c_l^{\star L_R}(t^\star+\Delta t^\star)}{\alpha_2^\star}+\\
&\int_{\Omega^\star} \varphi_i^{\star L_R}(\mathbf{x}^\star)\varphi_l^{\star L_R}(\mathbf{x}^\star)\varphi_k^{\star R}(\mathbf{x}^\star)\varphi_p^{\star I}(\mathbf{x}^\star)d\Omega^\star \frac{c_l^{\star L_R}(t^\star+\Delta t^\star)}{\Delta t^\star}\frac{c_k^{\star R}(t^\star+\Delta t^\star)c_p^{\star I}(t^\star+\Delta t^\star)}{\alpha_2^\star\alpha_3^\star}-\\
&\int_{\Omega^\star} \varphi_i^{\star L_R}(\mathbf{x}^\star)\varphi_l^{\star L_R}(\mathbf{x}^\star)\varphi_k^{\star R}(\mathbf{x}^\star)\varphi_p^{\star I}(\mathbf{x}^\star)d\Omega^\star \frac{c_l^{\star L_R}(t^\star)}{\Delta t^\star}\frac{c_k^{\star R}(t^\star+\Delta t^\star)c_p^{\star I}(t^\star+\Delta t^\star)}{\alpha_2^\star\alpha_3^\star}+\\
&\int_{\Omega^\star} \varphi_i^{\star L_R}(\mathbf{x}^\star)\varphi_p^{\star I}(\mathbf{x}^\star)\varphi_k^{\star R}(\mathbf{x}^\star)\varphi_l^{\star L_R}(\mathbf{x}^\star)d\Omega^\star \frac{c_p^{\star I}(t^\star+\Delta t^\star)}{\Delta t^\star}\frac{c_k^{\star R}(t^\star+\Delta t^\star)c_l^{\star L_R}(t^\star+\Delta t^\star)}{\alpha_2^\star\alpha_3^\star}-\\
&\int_{\Omega^\star} \varphi_i^{\star L_R}(\mathbf{x}^\star)\varphi_p^{\star I}(\mathbf{x}^\star)\varphi_k^{\star R}(\mathbf{x}^\star)\varphi_l^{\star L_R}(\mathbf{x}^\star)d\Omega^\star \frac{c_p^{\star I}(t^\star)}{\Delta t^\star}\frac{c_k^{\star R}(t^\star+\Delta t^\star)c_l^{\star L_R}(t^\star+\Delta t^\star)}{\alpha_2^\star\alpha_3^\star}+\\
&\int_{\Omega^\star} \varphi_i^{\star L_R}(\mathbf{x}^\star)\varphi_k^{\star R}(\mathbf{x}^\star)\varphi_l^{\star L_R}(\mathbf{x}^\star)\varphi_p^{\star I}(\mathbf{x}^\star)d\Omega^\star \frac{c_k^{\star R}(t^\star+\Delta t^\star)}{\Delta t^\star}\frac{c_l^{\star L_R}(t^\star+\Delta t^\star)c_p^{\star I}(t^\star+\Delta t^\star)}{\alpha_2^\star\alpha_3^\star}-\\
&\int_{\Omega^\star} \varphi_i^{\star L_R}(\mathbf{x}^\star)\varphi_k^{\star R}(\mathbf{x}^\star)\varphi_l^{\star L_R}(\mathbf{x}^\star)\varphi_p^{\star I}(\mathbf{x}^\star)d\Omega^\star \frac{c_k^{\star R}(t^\star)}{\Delta t^\star}\frac{c_l^{\star L_R}(t^\star+\Delta t^\star)c_p^{\star I}(t^\star+\Delta t^\star)}{\alpha_2^\star\alpha_3^\star}-\\
&\int_{\Omega^\star} \varphi_i^{\star L_R}(\mathbf{x}^\star)s_{L_R}^\star d\Omega^\star = 0
\end{aligned}
$$

$$
c_{L_R}^\star(\mathbf{x}^\star,0)=0
$$

Second Equation

$$\int_{\Omega^\star} \varphi_i^{\star R}(\mathbf{x}^\star)\varphi_k^{\star R}(\mathbf{x}^\star)d\Omega^\star \left(\frac{c_k^{\star R}(t^\star+\Delta t^\star)-c_k^{\star R}(t^\star)}{\Delta t^\star}\right) -$$

$$\int_{\Omega^\star} \varphi_i^{\star R}(\mathbf{x}^\star)\varphi_l^{\star L_R}(\mathbf{x}^\star)d\Omega^\star \left(\frac{c_l^{\star L_R}(t^\star+\Delta t^\star)-c_l^{\star L_R}(t^\star)}{\Delta t^\star}\right) +$$

$$\int_{\Omega^\star} \nabla_\Omega^\star\left[\varphi_k^{\star R}(\mathbf{x}^\star)\right]\cdot\nabla_\Omega^\star\left[\varphi_i^{\star R}(\mathbf{x}^\star)\right]d\Omega^\star \frac{c_k^{\star R}(t^\star+\Delta t^\star)}{\mathcal{P}_e^R} +$$

$$\int_{\Omega^\star} \varphi_i^{\star R}(\mathbf{x}^\star)\varphi_l^{\star L_R}(\mathbf{x}^\star)s_{L_R}^\star d\Omega^\star = 0$$

$$c_R^\star(\mathbf{x}^\star,0) = \frac{4.8}{c_{bulk}}$$

Third Equation

$$\int_{\Omega^\star} \varphi_i^{\star I}(\mathbf{x}^\star)\varphi_p^{\star I}(\mathbf{x}^\star)d\Omega^\star \left(\frac{c_p^{\star I}(t^\star+\Delta t^\star)-c_p^{\star I}(t^\star)}{\Delta t^\star}\right) -$$

$$\int_{\Omega^\star} \varphi_i^{\star I}(\mathbf{x}^\star)\varphi_m^{\star L_I}(\mathbf{x}^\star)d\Omega^\star \left(\frac{c_m^{\star L_I}(t^\star+\Delta t^\star)-c_m^{\star L_I}(t^\star)}{\Delta t^\star}\right) +$$

$$\int_{\Omega^\star} \varphi_i^{\star I}(\mathbf{x}^\star)\varphi_p^{\star I}(\mathbf{x}^\star)\varphi_k^{\star R}(\mathbf{x}^\star)\varphi_l^{\star L_R}(\mathbf{x}^\star)d\Omega^\star \frac{c_p^{\star I}(t^\star+\Delta t^\star)}{\Delta t^\star}\frac{c_k^{\star R}(t^\star+\Delta t^\star)c_l^{\star L_R}(t^\star+\Delta t^\star}{\alpha_2^\star\alpha_3^\star}$$

$$\int_{\Omega^\star} \varphi_i^{\star I}(\mathbf{x}^\star)\varphi_p^{\star I}(\mathbf{x}^\star)\varphi_k^{\star R}(\mathbf{x}^\star)\varphi_l^{\star L_R}(\mathbf{x}^\star)d\Omega^\star \frac{c_p^{\star I}(t^\star)}{\Delta t^\star}\frac{c_k^{\star R}(t^\star+\Delta t^\star)c_l^{\star L_R}(t^\star+\Delta t^\star)}{\alpha_2^\star\alpha_3^\star} +$$

$$\int_{\Omega^\star} \varphi_i^{\star I}(\mathbf{x}^\star)\varphi_k^{\star R}(\mathbf{x}^\star)\varphi_p^{\star I}(\mathbf{x}^\star)\varphi_l^{\star L_R}(\mathbf{x}^\star)d\Omega^\star \frac{c_k^{\star R}(t^\star+\Delta t^\star)}{\Delta t^\star}\frac{c_p^{\star I}(t^\star+\Delta t^\star)c_l^{\star L_R}(t^\star+\Delta t^\star}{\alpha_2^\star\alpha_3^\star}$$

$$\int_{\Omega^\star} \varphi_i^{\star I}(\mathbf{x}^\star)\varphi_k^{\star R}(\mathbf{x}^\star)\varphi_p^{\star I}(\mathbf{x}^\star)\varphi_l^{\star L_R}(\mathbf{x}^\star)d\Omega^\star \frac{c_k^{\star R}(t^\star)}{\Delta t^\star}\frac{c_p^{\star I}(t^\star+\Delta t^\star)c_l^{\star L_R}(t^\star+\Delta t^\star)}{\alpha_2^\star\alpha_3^\star} +$$

$$\int_{\Omega^\star} \varphi_i^{\star I}(\mathbf{x}^\star)\varphi_l^{\star L_R}(\mathbf{x}^\star)\varphi_p^{\star I}(\mathbf{x}^\star)\varphi_k^{\star R}(\mathbf{x}^\star)d\Omega^\star \frac{c_l^{\star L_R}(t^\star+\Delta t^\star)}{\Delta t^\star}\frac{c_p^{\star I}(t^\star+\Delta t^\star)c_k^{\star R}(t^\star+\Delta t^\star)}{\alpha_2^\star\alpha_3^\star}$$

$$\cdots \int_{\Omega^\star} \varphi_i^{\star I}(\mathbf{x}^\star)\varphi_l^{\star L_R}(\mathbf{x}^\star)\varphi_p^{\star I}(\mathbf{x}^\star)\varphi_k^{\star R}(\mathbf{x}^\star)d\Omega^\star \frac{c_l^{\star L_R}(t^\star)}{\Delta t^\star}\frac{c_p^{\star I}(t^\star+\Delta t^\star)c_k^{\star R}(t^\star+\Delta t^\star)}{\alpha_2^\star\alpha_3^\star}+$$

$$\int_{\Omega^\star} \nabla_\Omega^\star\left[\varphi_p^{\star I}(\mathbf{x}^\star)\right]\cdot\nabla_\Omega^\star\left[\varphi_i^{\star I}(\mathbf{x}^\star)\right]d\Omega^\star\frac{c_p^{\star I}(t^\star+\Delta t^\star)}{\mathcal{P}_e^I}+\int_{\Omega^\star}\varphi_i^{\star I}(\mathbf{x}^\star)\varphi_p^{\star I}(\mathbf{x}^\star)s_{L_I}^\star d\Omega^\star=0$$

$$c_I^\star(\mathbf{x}^\star,0)=\frac{8}{c_{bulk}}$$

Fourth Equation

$$\int_{\Omega^\star}\varphi_i^{\star L_I}(\mathbf{x}^\star)\varphi_m^{\star L_I}(\mathbf{x}^\star)d\Omega^\star\left(\frac{c_m^{\star L_I}(t^\star+\Delta t^\star)-c_m^{\star L_I}(t^\star)}{\Delta t^\star}\right)+$$

$$\int_{\Omega^\star}\varphi_i^{\star L_I}(\mathbf{x}^\star)\varphi_m^{\star L_I}(\mathbf{x}^\star)\varphi_p^{\star I}(\mathbf{x}^\star)d\Omega^\star\left(\frac{c_m^{\star L_I}(t^\star+\Delta t^\star)-c_m^{\star L_I}(t^\star)}{\Delta t^\star}\right)\frac{c_p^{\star I}(t^\star+\Delta t^\star)}{\alpha_1^\star}+$$

$$\int_{\Omega^\star}\varphi_i^{\star L_I}(\mathbf{x}^\star)\varphi_p^{\star I}(\mathbf{x}^\star)\varphi_m^{\star L_I}(\mathbf{x}^\star)d\Omega^\star\left(\frac{c_p^{\star I}(t^\star+\Delta t^\star)-c_p^{\star I}(t^\star)}{\Delta t^\star}\right)\frac{c_m^{\star L_I}(t^\star+\Delta t^\star)}{\alpha_1^\star}-$$

$$\int_{\Omega^\star}\varphi_i^{\star L_I}(\mathbf{x}^\star)s_{L_I}^\star d\Omega^\star=0$$

$$c_{L_I}^\star(\mathbf{x}^\star,0)=0$$

B: Weak Form

First Equation

$$\int_{\Omega^\star}\varphi_i^{\star A}(\mathbf{x}^\star)\varphi_j^{\star A}(\mathbf{x}^\star)c_j^{\star A}(t^\star)d\Omega^\star-\int_{\Omega^\star}\varphi_i^{\star A}(\mathbf{x}^\star)S_{L_R}^\star(\mathbf{x}^\star,t^\star)d\Omega^\star=0$$

$$c_A^\star(\mathbf{x}^\star,0)=0$$

Second Equation

$$\int_{\Omega^\star} \varphi_i^{\star B}(\mathbf{x}^\star)\varphi_l^{\star B}(\mathbf{x}^\star)c_l^{\star B}(t_n^\star)d\Omega^\star - \int_{\Omega^\star} \varphi_i^{\star B}(\mathbf{x}^\star)\varphi_l^{\star B}(\mathbf{x}^\star)c_l^{\star B}(t_{n-1}^\star)d\Omega^\star +$$
$$\frac{1}{\mathcal{P}_e^R}\int\limits_{\Omega^\star} \nabla_\Omega^\star\left[\varphi_i^{\star B}(\mathbf{x}^\star)\right]\cdot\nabla_\Omega^\star\left[\varphi_k^{\star R}(\mathbf{x}^\star)\right]d\Omega^\star \int\limits_{t_{n-1}^\star}^{t_n^\star} c_k^{\star R}(\tau^\star)d\tau^\star +$$
$$\int_{\Omega^\star} \varphi_i^{\star B}(\mathbf{x}^\star)S_{L_R}^\star(\mathbf{x}^\star, t_n^\star)d\Omega^\star - \int_{\Omega^\star} \varphi_i^{\star B} S_{L_R}^\star(\mathbf{x}^\star, t_{n-1}^\star)d\Omega^\star = 0$$
$$c_B^\star(\mathbf{x}^\star, 0) = \frac{4.8}{c_{bulk}}$$

Third Equation

$$\int_{\Omega^\star} \varphi_i^{\star D}(\mathbf{x}^\star)\varphi_m^{\star D}(\mathbf{x}^\star)c_m^{\star D}(t_n^\star)d\Omega^\star - \int_{\Omega^\star} \varphi_i^{\star D}(\mathbf{x}^\star)\varphi_m^{\star D}(\mathbf{x}^\star)c_m^{\star D}(t_{n-1}^\star)d\Omega^\star +$$
$$\frac{1}{\mathcal{P}_e^I}\int\limits_{\Omega^\star} \nabla_\Omega^\star\left[\varphi_i^{\star D}(\mathbf{x}^\star)\right]\cdot\nabla_\Omega^\star\left[\varphi_p^{\star I}(\mathbf{x}^\star)\right]d\Omega^\star \int\limits_{t_{n-1}^\star}^{t_n^\star} c_p^{\star I}(\tau^\star)d\tau^\star +$$
$$\int_{\Omega^\star} \varphi_i^{\star D}(\mathbf{x}^\star)S_{L_I}^\star(\mathbf{x}^\star, t_n^\star)d\Omega^\star - \int_{\Omega^\star} \varphi_i^{d} S_{L_I}^\star(\mathbf{x}^\star, t_{n-1}^\star)d\Omega^\star = 0$$
$$c_D^\star(\mathbf{x}^\star, 0) = \frac{8}{c_{bulk}}$$

Fourth Equation

$$\int_{\Omega^\star} \varphi_i^{\star E}(\mathbf{x}^\star)\varphi_s^{\star E}(\mathbf{x}^\star)c_s^{\star E}(t^\star)d\Omega^\star - \int_{\Omega^\star} \varphi_i^{\star E}(\mathbf{x}^\star)S_{L_I}^\star(\mathbf{x}^\star, t^\star)d\Omega^\star = 0$$
$$c_E^\star(\mathbf{x}^\star, 0) = 0$$

Fifth Equation

$$\alpha_3^\star \int_{\Omega^\star} \varphi_i^{\star R}(\mathbf{x}^\star)\varphi_k^{\star R}(\mathbf{x}^\star)\varphi_k^{\star R}(\mathbf{x}^\star)c_k^{\star R}(t^\star)c_k^{\star R}(t^\star)d\Omega^\star +$$
$$\int_{\Omega^\star} \varphi_i^{\star R}(\mathbf{x}^\star)\varphi_k^{\star R}(\mathbf{x}^\star)\varphi_k^{\star R}(\mathbf{x}^\star)\varphi_p^{\star I}(\mathbf{x}^\star)c_k^{\star R}(t^\star)c_k^{\star R}(t^\star)c_p^{\star I}(t^\star)d\Omega^\star +$$
$$\alpha_2^\star\alpha_3^\star \int_{\Omega^\star} \varphi_i^{\star R}(\mathbf{x}^\star)\varphi_k^{\star R}(\mathbf{x}^\star)c_k^{\star R}(t^\star)d\Omega^\star - \ldots$$

$$\ldots\alpha_3^\star \int_{\Omega^\star} \varphi_i^{\star R}(\mathbf{x}^\star)\varphi_k^{\star R}(\mathbf{x}^\star)\varphi_l^{\star B}(\mathbf{x}^\star)c_k^{\star R}(t^\star)c_l^{\star B}(t^\star)d\Omega^\star -$$

$$\int_{\Omega^\star} \varphi_i^{\star R}(\mathbf{x}^\star)\varphi_k^{\star R}(\mathbf{x}^\star)\varphi_p^{\star I}(\mathbf{x}^\star)\varphi_l^{\star B}(\mathbf{x}^\star)c_k^{\star R}(t^\star)c_p^{\star I}(t^\star)c_l^{\star B}(t^\star)d\Omega^\star -$$

$$\alpha_2^\star\alpha_3^\star \int_{\Omega^\star} \varphi_i^{\star R}(\mathbf{x}^\star)\varphi_l^{\star B}(\mathbf{x}^\star)c_l^{\star B}(t^\star)d\Omega^\star -$$

$$\alpha_2^\star\alpha_3^\star \int_{\Omega^\star} \varphi_i^{\star R}(\mathbf{x}^\star)\varphi_j^{\star A}(\mathbf{x}^\star)c_j^{\star A}(t^\star)d\Omega^\star = 0$$

$$c_R^\star(\mathbf{x}^\star, 0) = \frac{4.8}{c_{bulk}}$$

Sixth Equation

$$\alpha_2^\star\alpha_3^\star \int_{\Omega^\star} \varphi_i^{\star I}(\mathbf{x}^\star)\varphi_p^{\star I}(\mathbf{x}^\star)\varphi_p^{\star I}(\mathbf{x}^\star)c_p^{\star I}(t^\star)c_p^{\star I}(t^\star)d\Omega^\star +$$

$$\int_{\Omega^\star} \varphi_i^{\star I}(\mathbf{x}^\star)\varphi_p^{\star I}(\mathbf{x}^\star)\varphi_p^{\star I}(\mathbf{x}^\star)\varphi_k^{\star R}(\mathbf{x}^\star)\varphi_k^{\star R}(\mathbf{x}^\star)c_p^{\star I}(t^\star)c_p^{\star I}(t^\star)c_k^{\star R}(t^\star)c_k^{\star R}(t^\star)d\Omega^\star -$$

$$\int_{\Omega^\star} \varphi_i^{\star I}(\mathbf{x}^\star)\varphi_p^{\star I}(\mathbf{x}^\star)\varphi_p^{\star I}(\mathbf{x}^\star)\varphi_k^{\star R}(\mathbf{x}^\star)\varphi_l^{\star B}(\mathbf{x}^\star)c_p^{\star I}(t^\star)c_p^{\star I}(t^\star)c_k^{\star R}(t^\star)c_l^{\star B}(t^\star)d\Omega^\star -$$

$$\alpha_2^\star\alpha_3^\star \int_{\Omega^\star} \varphi_i^{\star I}(\mathbf{x}^\star)\varphi_p^{\star I}(\mathbf{x}^\star)\varphi_m^{\star D}(\mathbf{x}^\star)c_p^{\star I}(t^\star)c_m^{\star D}(t^\star)d\Omega^\star +$$

$$\alpha_1^\star\alpha_2^\star\alpha_3^\star \int_{\Omega^\star} \varphi_i^{\star I}(\mathbf{x}^\star)\varphi_p^{\star I}(\mathbf{x}^\star)c_p^{\star I}(t^\star)d\Omega^\star +$$

$$\alpha_1^\star \int_{\Omega^\star} \varphi_i^{\star I}(\mathbf{x}^\star)\varphi_p^{\star I}(\mathbf{x}^\star)\varphi_k^{\star R}(\mathbf{x}^\star)\varphi_k^{\star R}(\mathbf{x}^\star)c_p^{\star I}(t^\star)c_k^{\star R}(t^\star)c_k^{\star R}(t^\star)d\Omega^\star -$$

$$\alpha_1^\star \int_{\Omega^\star} \varphi_i^{\star I}(\mathbf{x}^\star)\varphi_p^{\star I}(\mathbf{x}^\star)\varphi_k^{\star R}(\mathbf{x}^\star)\varphi_l^{\star B}(\mathbf{x}^\star)c_p^{\star I}(t^\star)c_k^{\star R}(t^\star)c_l^{\star B}(t^\star)d\Omega^\star -$$

$$\alpha_1^\star\alpha_2^\star\alpha_3^\star \int_{\Omega^\star} \varphi_i^{\star I}(\mathbf{x}^\star)\varphi_s^{\star E}(\mathbf{x}^\star)c_s^{\star E}(t^\star)d\Omega^\star -$$

$$\alpha_1^\star\alpha_2^\star\alpha_3^\star \int_{\Omega^\star} \varphi_i^{\star I}(\mathbf{x}^\star)\varphi_m^{\star D}(\mathbf{x}^\star)c_m^{\star D}(t^\star)d\Omega^\star = 0$$

$$c_I^\star(\mathbf{x}^\star, 0) = \frac{8}{c_{bulk}}$$

in which the following Newton–Cotes quadrature formula will be given

$$\int_{\Omega^\star} \frac{1}{\mathcal{P}_e^R} \nabla^\star_\Omega \left[\varphi_i^{\star B}(\mathbf{x}^\star) \right] \cdot \nabla^\star_\Omega \left[\varphi_k^{\star R}(\mathbf{x}^\star) \right] d\Omega^\star \int_{t^\star_{n-1}}^{t^\star_n} c_k^{\star R}(\tau^\star) d\tau^\star$$

$$\simeq \int_{\Omega^\star} \frac{1}{\mathcal{P}_e^R} \nabla^\star_\Omega \left[\varphi_i^{\star B}(\mathbf{x}^\star) \right] \cdot \nabla^\star_\Omega \left[\varphi_k^{\star R}(\mathbf{x}^\star) \right] d\Omega^\star \frac{\Delta t^\star}{2} \left(c_k^{\star R}(t^\star_n) + c_k^{\star R}(t^\star_{n-1}) \right)$$

$$\int_{\Omega^\star} \frac{1}{\mathcal{P}_e^I} \nabla^\star_\Omega \left[\varphi_i^{\star D}(\mathbf{x}^\star) \right] \cdot \nabla^\star_\Omega \left[\varphi_p^{\star I}(\mathbf{x}^\star) \right] d\Omega^\star \int_{t^\star_{n-1}}^{t^\star_n} c_p^{\star I}(\tau^\star) d\tau^\star$$

$$\simeq \int_{\Omega^\star} \frac{1}{\mathcal{P}_e^I} \nabla^\star_\Omega \left[\varphi_i^{\star D}(\mathbf{x}^\star) \right] \cdot \nabla^\star_\Omega \left[\varphi_p^{\star I}(\mathbf{x}^\star) \right] d\Omega^\star \frac{\Delta t^\star}{2} \left(c_p^{\star I}(t^\star_n) + c_p^{\star I}(t^\star_{n-1}) \right)$$

Finally, it is possible to deduce the following six formulations:

First equation

$$\int_{\Omega^\star} \varphi_i^{\star A}(\mathbf{x}^\star) \varphi_j^{\star A}(\mathbf{x}^\star) d\Omega^\star \left[c_j^{\star A}(t^\star) \right] - \int_{\Omega^\star} \varphi_i^{\star A}(\mathbf{x}^\star) S^\star_{L_R}(\mathbf{x}^\star, t^\star) d\Omega^\star = 0$$
$$c_A^\star(\mathbf{x}^\star, 0) = 0$$

Second equation

$$\int_{\Omega^\star} \varphi_i^{\star B}(\mathbf{x}^\star) \varphi_l^{\star B}(\mathbf{x}^\star) d\Omega^\star \left[c_l^{\star B}(t^\star_n) \right] - \int_{\Omega^\star} \varphi_i^{\star B}(\mathbf{x}^\star) \varphi_l^{\star B}(\mathbf{x}^\star) d\Omega^\star \left[c_l^{\star B}(t^\star_{n-1}) \right] +$$
$$\frac{1}{\mathcal{P}_e^R} \int_{\Omega^\star} \nabla^\star_\Omega \left[\varphi_i^{\star B}(\mathbf{x}^\star) \right] \cdot \nabla^\star_\Omega \left[\varphi_k^{\star R}(\mathbf{x}^\star) \right] d\Omega^\star \left[\frac{\Delta t^\star}{2} \left(c_k^{\star R}(t^\star_n) + c_k^{\star R}(t^\star_{n-1}) \right) \right] +$$
$$\int_{\Omega^\star} \varphi_i^{\star B}(\mathbf{x}^\star) S^\star_{L_R}(\mathbf{x}^\star, t^\star_n) d\Omega^\star - \int_{\Omega^\star} \varphi_i^{\star B} S^\star_{L_R}(\mathbf{x}^\star, t^\star_{n-1}) d\Omega^\star = 0$$
$$c_B^\star(\mathbf{x}^\star, 0) = \frac{4.8}{c_{bulk}}$$

Third equation

$$\int_{\Omega^\star} \varphi_i^{\star D}(\mathbf{x}^\star)\varphi_m^{\star D}(\mathbf{x}^\star)d\Omega^\star \left[c_m^{\star D}(t_n^\star)\right] - \int_{\Omega^\star} \varphi_i^{\star D}(\mathbf{x}^\star)\varphi_m^{\star D}(\mathbf{x}^\star)d\Omega^\star \left[c_m^{\star D}(t_{n-1}^\star)\right] +$$

$$\frac{1}{\mathcal{P}_e^I}\int_{\Omega^\star} \nabla_\Omega^\star\left[\varphi_i^{\star D}\right]\cdot\nabla_\Omega^\star\left[\varphi_p^{\star I}(\mathbf{x}^\star)\right]d\Omega^\star\left[\frac{\Delta t^\star}{2}\left(c_p^{\star I}(t_n^\star)+c_p^{\star I}(t_{n-1}^\star)\right)\right] +$$

$$+\int_{\Omega^\star} \varphi_i^{\star D}(\mathbf{x}^\star)S_{L_I}^\star(\mathbf{x}^\star,t_n^\star)d\Omega^\star - \int_{\Omega^\star} \varphi_i^{d}(\mathbf{x}^\star)S_{L_I}^\star(\mathbf{x}^\star,t_{n-1}^\star)d\Omega^\star = 0$$

$$c_D^\star(\mathbf{x}^\star,0) = \frac{8}{c_{bulk}}$$

Fourth equation

$$\int_{\Omega^\star} \varphi_i^{\star E}(\mathbf{x}^\star)\varphi_s^{\star E}(\mathbf{x}^\star)d\Omega^\star\left[c_s^{\star E}(t^\star)\right] - \int_{\Omega^\star} \varphi_i^{\star E}(\mathbf{x}^\star)S_{L_I}^\star(\mathbf{x}^\star,t^\star)d\Omega^\star = 0$$

$$c_E^\star(\mathbf{x}^\star,0) = 0$$

Fifth equation

$$\alpha_3^\star\int_{\Omega^\star} \varphi_i^{\star R}(\mathbf{x}^\star)\varphi_k^{\star R}(\mathbf{x}^\star)\varphi_k^{\star R}(\mathbf{x}^\star)d\Omega^\star\left[c_k^{\star R}(t^\star)c_k^{\star R}(t^\star)\right] +$$

$$\int_{\Omega^\star} \varphi_i^{\star R}(\mathbf{x}^\star)\varphi_k^{\star R}(\mathbf{x}^\star)\varphi_k^{\star R}(\mathbf{x}^\star)\varphi_p^{\star I}(\mathbf{x}^\star)d\Omega^\star\left[c_k^{\star R}(t^\star)c_k^{\star R}(t^\star)c_p^{\star I}(t^\star)\right] +$$

$$\alpha_2^\star\alpha_3^\star\int_{\Omega^\star} \varphi_i^{\star R}(\mathbf{x}^\star)\varphi_k^{\star R}(\mathbf{x}^\star)d\Omega^\star\left[c_k^{\star R}(t^\star)\right] -$$

$$\alpha_3^\star\int_{\Omega^\star} \varphi_i^{\star R}(\mathbf{x}^\star)\varphi_k^{\star R}(\mathbf{x}^\star)\varphi_l^{\star B}(\mathbf{x}^\star)d\Omega^\star\left[c_k^{\star R}(t^\star)c_l^{\star B}(t^\star)\right] -$$

$$\int_{\Omega^\star} \varphi_i^{\star R}\varphi_k^{\star R}(\mathbf{x}^\star)\varphi_p^{\star I}(\mathbf{x}^\star)\varphi_l^{\star B}(\mathbf{x}^\star)d\Omega^\star\left[c_k^{\star R}(t^\star)c_p^{\star I}(t^\star)c_l^{\star B}(t^\star)\right] -$$

$$\alpha_2^\star\alpha_3^\star\int_{\Omega^\star} \varphi_i^{\star R}(\mathbf{x}^\star)\varphi_l^{\star B}(\mathbf{x}^\star)d\Omega^\star\left[c_l^{\star B}(t^\star)\right] -$$

$$\alpha_2^\star\alpha_3^\star\int_{\Omega^\star} \varphi_i^{\star R}(\mathbf{x}^\star)\varphi_j^{\star A}(\mathbf{x}^\star)d\Omega^\star\left[c_j^{\star A}(t^\star)\right] = 0$$

$$c_R^\star(\mathbf{x}^\star,0) = \frac{4.8}{c_{bulk}}$$

Sixth equation

$$\alpha_2^\star \alpha_3^\star \int_{\Omega^\star} \varphi_i^{\star I}(\mathbf{x}^\star)\varphi_p^{\star I}(\mathbf{x}^\star)\varphi_p^{\star I}(\mathbf{x}^\star) d\Omega^\star \left[c_p^{\star I}(t^\star) c_p^{\star I}(t^\star)\right] +$$

$$\int_{\Omega^\star} \varphi_i^{\star I}(\mathbf{x}^\star)\varphi_p^{\star I}(\mathbf{x}^\star)\varphi_p^{\star I}(\mathbf{x}^\star)\varphi_k^{\star R}(\mathbf{x}^\star)\varphi_k^{\star R}(\mathbf{x}^\star) d\Omega^\star \left[c_p^{\star I}(t^\star) c_p^{\star I}(t^\star) c_k^{\star R}(t^\star) c_k^{\star R}(t^\star)\right] -$$

$$\int_{\Omega^\star} \varphi_i^{\star I}(\mathbf{x}^\star)\varphi_p^{\star I}(\mathbf{x}^\star)\varphi_p^{\star I}(\mathbf{x}^\star)\varphi_k^{\star R}(\mathbf{x}^\star)\varphi_l^{\star B}(\mathbf{x}^\star) d\Omega^\star \left[c_p^{\star I}(t^\star) c_p^{\star I}(t^\star) c_k^{\star R}(t^\star) c_l^{\star B}(t^\star)\right] -$$

$$\alpha_2^\star \alpha_3^\star \int_{\Omega^\star} \varphi_i^{\star I}(\mathbf{x}^\star)\varphi_p^{\star I}(\mathbf{x}^\star)\varphi_m^{\star D}(\mathbf{x}^\star) d\Omega^\star \left[c_p^{\star I}(t^\star) c_m^{\star D}(t^\star)\right] +$$

$$\alpha_1^\star \alpha_2^\star \alpha_3^\star \int_{\Omega^\star} \varphi_i^{\star I}(\mathbf{x}^\star)\varphi_p^{\star I}(\mathbf{x}^\star) d\Omega^\star \left[c_p^{\star I}(t^\star)\right] +$$

$$\alpha_1^\star \int_{\Omega^\star} \varphi_i^{\star I}(\mathbf{x}^\star)\varphi_p^{\star I}(\mathbf{x}^\star)\varphi_k^{\star R}(\mathbf{x}^\star)\varphi_k^{\star R}(\mathbf{x}^\star) d\Omega^\star \left[c_p^{\star I}(t^\star) c_k^{\star R}(t^\star) c_k^{\star R}(t^\star)\right] -$$

$$\alpha_1^\star \int_{\Omega^\star} \varphi_i^{\star I}(\mathbf{x}^\star)\varphi_p^{\star I}(\mathbf{x}^\star)\varphi_k^{\star R}(\mathbf{x}^\star)\varphi_l^{\star B}(\mathbf{x}^\star) d\Omega^\star \left[c_p^{\star I}(t^\star) c_k^{\star R}(t^\star) c_l^{\star B}(t^\star)\right] -$$

$$\alpha_1^\star \alpha_2^\star \alpha_3^\star \int_{\Omega^\star} \varphi_i^{\star I}(\mathbf{x}^\star)\varphi_s^{\star E}(\mathbf{x}^\star) d\Omega^\star \left[c_s^{\star E}(t^\star)\right] -$$

$$\alpha_1^\star \alpha_2^\star \alpha_3^\star \int_{\Omega^\star} \varphi_i^{\star I}(\mathbf{x}^\star)\varphi_m^{\star D}(\mathbf{x}^\star) d\Omega^\star \left[c_m^{\star D}(t^\star)\right] = 0$$

$$c_I^\star(\mathbf{x}^\star, 0) = \frac{8}{c_{bulk}}$$

References

Agrawal A, Steigmann DJ (2009) Boundary-value problems in the theory of lipid membranes. Continuum Mechanics and Thermodynamics 21(1):57–82

Agrawal A, Steigmann DJ (2011) A model for surface diffusion of trans-membrane proteins on lipid bilayers. Zeitschrift für angewandte Mathematik und Physik 62(3):549–563

Calderwood DA (2004) Integrin activation. Journal of cell science 117(5):657–666

Carmeliet P, Ng YS, Nuyens D, et al (1999) Impaired myocardial angiogenesis and ischemic cardiomyopathy in mice lacking the vascular endothelial growth factor isoforms VEGF 164 and VEGF 188. Nature medicine 5(5):495–502

Damioli V, Salvadori A, Beretta GP, Ravelli C, Mitola S (2017) Multi-physics interactions drive VEGFR2 relocation on endothelial cells. Scientific reports 7(1):1–11

De Groot SR, Mazur P (1984) Non-equilibrium thermodynamics. Dover

Deshpande R, Cheng YT, Verbrugge MW, Timmons A (2011) Diffusion induced stresses and strain energy in a phase-transforming spherical electrode particle. Journal of the Electrochemical Society 158(6):A718–A724

Deshpande VS, McMeeking RM, Evans AG (2006) A bio-chemo-mechanical model for cell contractility. Proceedings of the National Academy of Sciences 103(38):14,015–14,020

Deshpande VS, McMeeking RM, Evans AG (2007) A model for the contractility of the cytoskeleton including the effects of stress-fibre formation and dissociation. Proceedings of the Royal Society A: Mathematical, Physical and Engineering Sciences 463(2079):787–815

Deshpande VS, Mrksich M, McMeeking RM, Evans AG (2008) A bio-mechanical model for coupling cell contractility with focal adhesion formation. Journal of the Mechanics and Physics of Solids 56(4):1484–1510

Eliceiri BP (2001) Integrin and growth factor receptor crosstalk. Circulation research 89(12):1104–1110

Esser S, Lampugnani MG, Corada M, Dejana E, Risau W (1998) Vascular endothelial growth factor induces VE-cadherin tyrosine phosphorylation in endothelial cells. Journal of cell science 111(13):1853–1865

Gurtin ME, Fried E, Anand L (2010) The mechanics and thermodynamics of continua. Cambridge University Press

Helfrich W (1973) Elastic properties of lipid bilayers: theory and possible experiments. Zeitschrift für Naturforschung C 28(11-12):693–703

Hynes RO (2002) Integrins: bidirectional, allosteric signaling machines. cell 110(6):673–687

Joanny JF, Kruse K, Prost J, Ramaswamy S (2013) The actin cortex as an active wetting layer. The European Physical Journal E 36(5):1–6

Kruse K, Joanny JF, Jülicher F, Prost J, Sekimoto K (2005) Generic theory of active polar gels: a paradigm for cytoskeletal dynamics. The European Physical Journal E 16(1):5–16

Marchetti MC, Joanny JF, Ramaswamy S, Liverpool TB, Prost J, Rao M, Simha RA (2013) Hydrodynamics of soft active matter. Reviews of Modern Physics 85(3):1143–1189

McEvoy E, Deshpande VS, McGarry P (2017) Free energy analysis of cell spreading. Journal of the mechanical behavior of biomedical materials 74:283–295

McMeeking RM, Deshpande VS (2017) A bio-chemo-mechanical model for cell contractility, adhesion, signaling, and stress-fiber remodeling. In: Holzapfel G, Ogden R (eds) Biomechanics: Trends in Modeling and Simulation, vol 20, Springer, pp 53–81

Neufeld G, Cohen T, Gengrinovitch S, Poltorak Z (1999) Vascular endothelial growth factor (VEGF) and its receptors. The FASEB journal 13(1):9–22

Obbink-Huizer C, Oomens CW, Loerakker S, Foolen J, Bouten CV, Baaijens FP (2014) Computational model predicts cell orientation in response to a range of mechanical stimuli. Biomechanics and modeling in mechanobiology 13(1):227–236

Peach CJ, Kilpatrick LE, Friedman-Ohana R, Zimmerman K, Robers MB, Wood KV, Woolard J, Hill SJ (2018) Real-time ligand binding of fluorescent VEGF-A isoforms that discriminate between VEGFR2 and NRP1 in living cells. Cell chemical biology 25(10):1208–1218

Prost J, Jülicher F, Joanny JF (2015) Active gel physics. Nature Physics 11(2):111–117

Quarteroni A, Valli A (2008) Numerical approximation of partial differential equations, vol 23. Springer Science & Business Media

Ravelli C, Mitola S, Corsini M, Presta M (2013) Involvement of $\alpha_v\beta_3$ integrin in gremlin-induced angiogenesis. Angiogenesis 16(1):235–243

Ravelli C, Grillo E, Corsini M, Coltrini D, Presta M, Mitola S (2015) β_3 integrin promotes long-lasting activation and polarization of vascular endothelial growth factor receptor 2 by immobilized ligand. Arteriosclerosis, thrombosis, and vascular biology 35(10):2161–2171

Reinhart-King CA, Dembo M, Hammer DA (2005) The dynamics and mechanics of endothelial cell spreading. Biophysical journal 89(1):676–689

Ronan W, Deshpande VS, McMeeking RM, McGarry JP (2012) Numerical investigation of the active role of the actin cytoskeleton in the compression resistance of cells. Journal of the Mechanical Behavior of Biomedical Materials 14:143–157

Ronan W, Deshpande VS, McMeeking RM, McGarry JP (2014) Cellular contractility and substrate elasticity: a numerical investigation of the actin cytoskeleton and cell adhesion. Biomechanics and modeling in mechanobiology 13(2):417–435

Salvadori A, Damioli V, Ravelli C, Mitola S (2018a) Modeling and simulation of VEGF receptors recruitment in angiogenesis. Mathematical Problems in Engineering 2018:1–10

Salvadori A, McMeeking R, Grazioli D, Magri M (2018b) A coupled model of transport-reaction-mechanics with trapping. Part I–Small strain analysis. Journal of the Mechanics and Physics of Solids 114:1–30

Soldi R, Mitola S, Strasly M, Defilippi P, Tarone G, Bussolino F (1999) Role of $\alpha_v\beta_3$ integrin in the activation of vascular endothelial growth factor receptor-2. EMBO J 18(4):882–892

Steigmann D, Agrawal A (2016) Electromechanics of polarized lipid bilayers. Mathematics and Mechanics of Complex Systems 4(1):31–54

Valdembri D, Serini G (2012) Regulation of adhesion site dynamics by integrin traffic. Current opinion in cell biology 24(5):582–591

Vernerey FJ, Farsad M (2011) A constrained mixture approach to mechano-sensing and force generation in contractile cells. Journal of the mechanical behavior of biomedical materials 4(8):1683–1699

Vigliotti A, Ronan W, Baaijens FPT, Deshpande VS (2016) A thermodynamically motivated model for stress-fiber reorganization. Biomechanics and modeling in mechanobiology 15(4):761–789

Chapter 12
Designing Optimal Scaffold Topographies to Promote Nucleus-Guided Mechanosensitive Cell Migration Using *in Silico* Models

Maxime Vassaux, Laurent Pieuchot, Karine Anselme, Maxence Bigerelle, and Jean-Louis Milan

Abstract Computational models have become an essential part of exploratory protocols in cell biology, as a complement to *in vivo* or *in vitro* experiments. These virtual models have the twofold advantage of enabling access to new types of data and validate complex theories. The design of mechanically functionalized biomaterials or scaffolds, to promote cell proliferation and invasion in the absence or in the complement of synthetic chemical coatings, can certainly benefit from these hybrid testing approaches. The underlying fundamental process of cell migration and in particular its dependence on the cell mechanical/geometrical environment remains crudely understood. Currently at least two theories explain the migration patterns observed by cells on curved topographies, involving either polymerization dynamics of actin or assembly dynamics of focal adhesions. We recently proposed a third mechanism relying on nucleus mechanosensitivity, which has been tested extensively experimentally and computationally. We now explore the hypothesis that nucleosensitivity could be a mechanism for cells to optimally find microenvironments suited for mitosis, providing mechanical stability and relaxation. By means of a computational mechanical model with intracellular structure detail, we investigate how the complex interplay between this new migration mechanism and the microenvironment topography can lead to more relaxed cells and organelles. To go

M. Vassaux, J.-L. Milan
Aix Marseille Univ, CNRS, ISM, Marseille, France
e-mail: m.vassaux@ucl.ac.uk, jean-louis.milan@univ-amu.fr

L. Pieuchot, K. Anselme
Université de Haute-Alsace, CNRS, IS2M, UMR 7361, Mulhouse, France
Université de Strasbourg, Strasbourg, France
e-mail: laurent.pieuchot@uha.fr, karine.anselme@uha.fr

M. Bigerelle
Université de Valenciennes et du Hainaut Cambrésis, Laboratoire d'Automatique, de Mécanique et d'Informatique industrielle et Humaine (LAMIH), UMR-CNRS 8201, Le Mont Houy, Valenciennes, France
e-mail: Maxence.Bigerelle@uphf.fr

B. E. Abali and I. Giorgio (eds.), *Developments and Novel Approaches in Biomechanics and Metamaterials*, Advanced Structured Materials 132,
https://doi.org/10.1007/978-3-030-50464-9_12

further, we simulated in this study cell migration via a novel protocol in silico which generates dynamical ripple wave on a deformable substrate and changes topography over time. This kind of in silico protocols based on a new understanding of cell migration and nucleosensitivity could, therefore, inform the design of optimized scaffold topographies for cell invasion and proliferation.

Keywords: Biomechanics · Continuum mechanics · Finite Element Method (FEM)

12.1 Introduction

12.1.1 Intertwined Computational-Experimental Protocols

Computational physical models are large sets of equations that describe a controlled, reduced version of an experiment. Unlike analytical models, computational ones can integrate a more significant part of the complexity of living systems, as computational resources and methods allow to solve numerous and complex equations on large and heterogeneous systems. Nevertheless, *in silico* models remain far from the full complexity of in vivo experiments, and their validity relying on various assumptions can always be questioned. In vitro models are a first step toward breaking down the physics of living systems, disentangling that complexity, but *in silico* ones constitute a step further. Hypothesized multiple physics and the multiple scales involved in the mechanisms regulating the behavior of living systems can easily be integrated and tested as desired within *in silico* models. They have become an essential tool in theoretical biophysics, complementary to *in vivo* and *in vitro* models (Mogilner, 2009; Rodriguez et al, 2013; Rens and Merks, 2017). Indeed, *in silico* models enable integrating more complexity in a controlled way. For example, in discerning the origin of observed biological behavior, *in silico* models enable to sort active regulatory mechanism from passive physics (Nickaeen et al, 2019; Winkler et al, 2019). *In silico* models also provide complementary data, hardly accessible with in vitro and even less in vivo models: (i) at different scales, very small ones, for example, using methods solving the mechanics of clouds of electrons (Zink et al, 2013) (ii) and of different type, quantities that cannot be measured directly, called internal state variables in thermodynamics, among which forces and stresses can be found for example in Milan et al (2016).

12.1.2 Biomaterials Design for Tissue Engineering

The design of biocompatible materials for tissue engineering requires understanding how cells and materials interact to promote cell proliferation and invasion. Cell biology and more specifically migration are conditioned by the scaffold (or substrate) physics, which constitute a set of cues relying on chemistry (Zigmond and Hirsch, 1973; Dillon et al, 1995), electromagnetics (Adey, 1983) and mechanics (Isenberg et al, 2009). These physics play a role on multiple scales predominantly ranging from the characteristic scale of electrons, up to at least the characteristic scale of the cell. However, it is not possible yet to exclude larger scales, as macroscopic thermodynamical effects could certainly influence cell mechanobiology (Isenberg et al, 2009). Unravelling the interplay between these different cues at different scales and cell migration would certainly enable to engineer optimal biomaterials.

12.1.3 The Example of MAPS: Could Topography Cause Invasion?

Recently developed biocompatible scaffolds made of a microporous annealed particle (MAP) gel display promising levels of cell invasion and proliferation (Griffin et al, 2015; Darling et al, 2018). MAP gels are an example of biomaterial making use of the newly understood cell signaling cues. The mechanical properties of the gel can be tuned to steer the differentiation of cells depending on the type of tissue to repair Annealing particles of controlled sizes also enable to tune stability and therefore biodegradability of the scaffold. Nevertheless, the higher orders of cell invasion and proliferation are hardly explained by these features of the gels. More generally, microporous scaffolds tend to induce similar enhanced cell migration rates. In the meantime, recent pieces of research have highlighted *in vitro* and confirmed *in silico* how the geometry and the topography of the substrate can direct migration (Clark et al, 1991; Doyle et al, 2009; Czeisler et al, 2016). Specifically, the mechanical instability conveyed by convex topographies (Vassaux and Milan, 2017; Pieuchot et al, 2018) is shown to promote cell motility (Vassaux et al, 2019). Could a link more substantial than a correlation between the observed enhanced invasion and the topographical cue be established here? Potentially, causation involving underlying cell mechanics?

12.2 Understanding Cell Migration in Interaction with Extracellular Topography

12.2.1 Current Theories and in Silico Models Used to Explore Them

Cells have evolved multiple mechanisms to migrate in interaction with their environment. For instance, we reported recently a new cellular ability, which we termed “curvotaxis” that enables the cells to respond to cell-scale curvature variations, a ubiquitous trait of cellular biotopes (Pieuchot et al, 2018). Ascertaining enhanced invasion as a consequence of the specific geometry of microporous scaffolds lacks a mechanistic explanation of how cell migration is systematically influenced by the curvature of the underneath substrate. Topography and curvature influence cell physics provoking local confinement at cell-scale or below, from which originates cell polarization. At least three theories have been developed in the last two years and supported by means of computational models (Winkler et al, 2019; Vassaux et al, 2019; Schakenraad et al, 2019):

a) **Topography as confinement of actin polymerization during cell migration**. Winkler et al (2019) have shown that confinement breaks down the symmetry of actin polymerization in the cytoskeleton, and therefore favors a particular direction of the extension of the lamellipodium. This mechanism is endorsed by simulating physiological migration patterns using a continuum phase-field model of a single adherent cell and its internal actin organization. Actin ordering and polymerization seen as key factors of cell motility is not a recent discovery (Mogilner, 2009), however, as a source of persistence in confined environments definitely is. Even pieces of evidence in epithelia show a correlation between actin organization and topography, actin flowing away from parts of the cytoskeleton exposed to convex curvatures (Chen et al, 2019).
b) **The nucleus pushed away from convex topography indicates the direction for cell migration to more relax region**. We have rather focused on the role of the nucleus in Vassaux et al (2019). Confinement is shown to polarize the nucleus position inside the cell and we hypothesized that nucleus internal motility is a precursor of migration guidance. We had recently proved a correlation between the direction of nucleus motility and the direction of cell migration on sinusoidal surfaces (Fig. 12.1) (Pieuchot et al, 2018). We simulated cell adhesion on sinusoidal surfaces using our particle-based model of a single adherent cell with an explicit description of the nucleus (Vassaux and Milan, 2017; Vassaux, 2018). Simulations indicated a decentering of the nucleus toward the valleys of the sinus, which are concave regions of lower pressure (Fig. 12.2). We integrated secondly the mechanism of cell migration in the direction of intracellular nucleus displacement. Cell model reached concave regions whatever is its initial deposit location (Fig. 12.3). Simulations of persistent migration away from convex topographies and stabilization on symmetric concave niches supported our theory. This nucleus mechanosensitive mechanism could explain the intensive cell invasion and proliferation observed in MAP scaffolds, in which cells are solely exposed to convex surfaces.
c) **Cell migration between obstacles as Brownian particle movement involving repelling force**. Schakenraad et al (2019) have actually led an investigation of cell migration one scale above, to which the cell is modelled as an active Brownian particle, assuming that the influence of the environment's topography is cell-type independent. Cells consist of rigid disks with a finite area, imposed with a self-propelling velocity. The magnitude of the imposed velocity is constant, and its direction is defined as fluctuating randomly with some persistence. Cell migration is simulated on a substrate paved with obstacles which simply exert a repelling force modifying the overall cell motion, but not the imposed velocity. Schakenraad et al (2019) have shown that the observed guidance of

migration on such substrates could be caused entirely by the spatial modulation of obstacles, cells crudely migrating toward less confined spaces, independently organelles mechanosensitivity.

These three theories we presented hereabove are all valid, have redundancies and grey areas, therefore not mutually exclusive, but need to be sorted. These theories illustrate separate mechanisms of guided migration in anisotropic environments, at the scale of the organelle or of the whole cell. The actin cortex contractility is probably a redundant element of the two first theories, as it builds up the pressure gradient in the cytoskeleton polarizing either actin polymerization or nucleus position. In turn, cell migration may be, respectively, either a passive mechanism or a nucleus centering regulation mechanism (Almonacid et al, 2015).

12.2.2 How Conceptually Mechanical in Silico Cell Models Can Be Used at the Interface Between Materials Science and Cell Biophysics for Scaffold Design?

Harnessing cells polarization and migration mechanisms, the design of scaffolds could be improved. The scaffold topography and the induced confinement could be tuned to promote optimal motility depending on the cell lineage and its characteristic mechanical properties. *In silico* mechanical cell models constitute a tool of choice for scaffold design. Such models integrate simultaneously the topographical and mechanical complexity of the cell microenvironment that is the scaffold and the mechanosensititvity of the cell migration process. In turn, sensitivity analysis of migration rate and persistence to the scaffold design parameters is rendered easily tractable.

12.3 Going Deeper in the Understanding of Cell Migration with an *in Silico* Cell Model

12.3.1 Description of the Mechanical in Silico Cell Model

The *in silico* cell model we are developing integrates substrate and cell dynamics describing the mechanical structure as assembly of rigid particles (Vassaux and Milan, 2017). The model explicitly integrates actin, microtubules, intermediate filaments networks, contractile stress fibers, a contractile actomyosin cortex mingled in the cytoplasmic membrane, a viscous cytosol, and a viscoplastic nucleus (Fig 12.4.a). Each internal cell structure is modelled as an assembly of particles interacting via contact or springs. The parameterization of the model's interaction potentials has been largely verified and validated against indentation tests on mesenchymal stem cells (Vassaux and Milan, 2017). Complete details on the mathematical foundation

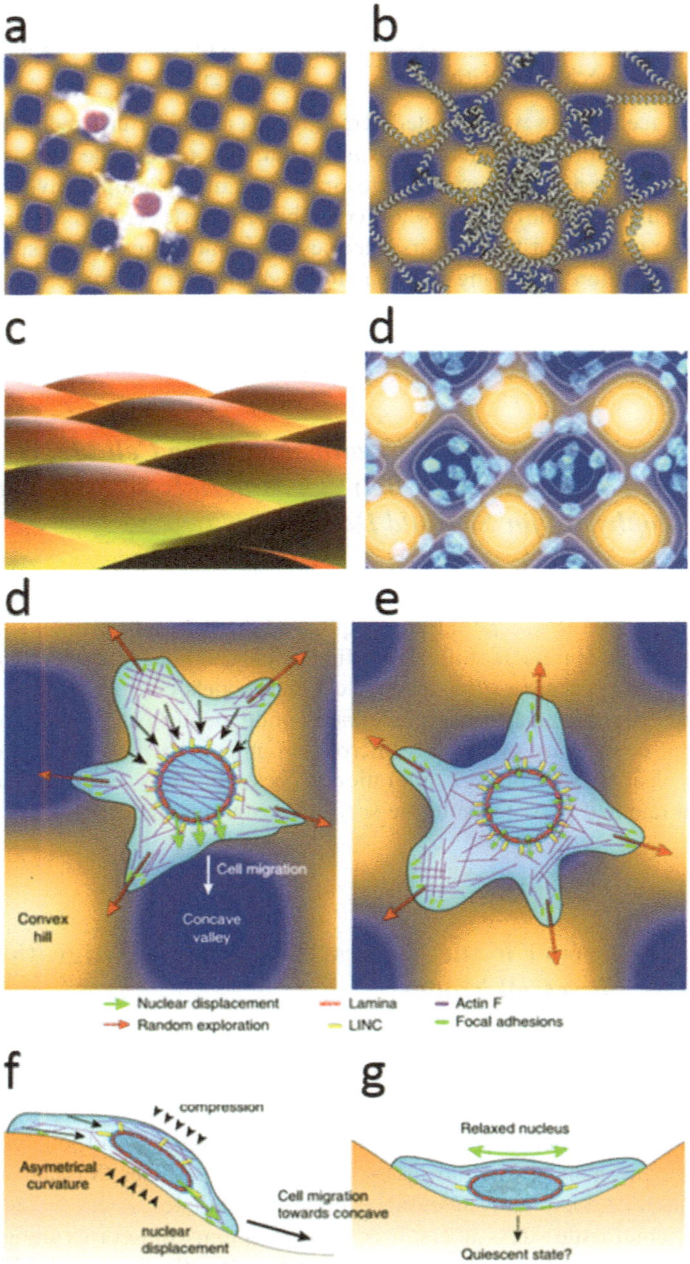

Fig. 12.1 In vitro observations of cell cultured on surfaces with sinusoidal topography (a). Cell migration trajectories remain in valleys and avoid peaks (b) of the sinusoidal topography viewed by side (c). Later, cells tend to stabilize their position in concave regions (d). Mechanical hypothesis to explain cell migration on sinus: curvature gradient breaks homogeneity in the compressive stress exerted by the cytoskeleton on the nucleus (f). The nucleus moves to lower pressure region. Then the cell migrates in the direction of nucleus movement so that the nucleus is in the center of the cell (g). Reproduced under the terms of the CC-BY 4.0 license (Pieuchot et al, 2018).

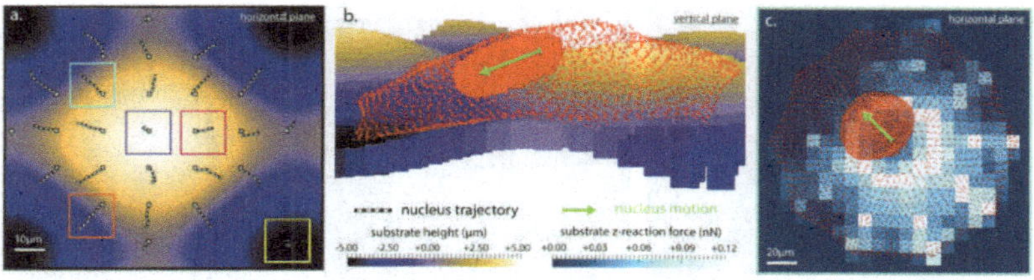

Fig. 12.2 Final displacements of the nucleus in the cell model depending on cell location on the sinus (a). Nucleus motion in cell model adhering on a peak (b) Nucleus motion in the opposite direction of greater traction force on the substrate. Reproduced under the terms of the CC-BY 4.0 license (Vassaux et al, 2019).

of the model as well as the calibration, validation, and adhesion simulation process can be found in Vassaux and Milan (2017). This mechanical cell model is able to capture realistic nucleus dynamics; the nucleus equilibrium is found at the center of the cell on a flat topography. These are governed by the coupled contribution of viscous, inertial (nucleus mass), and elastoplastic (conformational changes in the cytoskeleton) effects.

Simulations of cell adhesion follow a standardized procedure. In their initial configuration, the simulated cells display a spherical shape (Fig 12.4.b). Spreading is actioned after the displacement of the focal adhesions (FAs) away from the center of the cell following the topography of the substrate (Fig 12.4.c). This dynamic adhesion process, coupled with actomyosin contraction in stress fibers and the actin network, induces conformational changes in the cytoskeleton. At the end of the simulation (Fig. 12.5), cells are pulled onto the substrate and attached via a set number of focal points. FAs are distributed at the cell's periphery, regardless of the site of the cell adhesion in a concave, convex, or in the transitional areas.

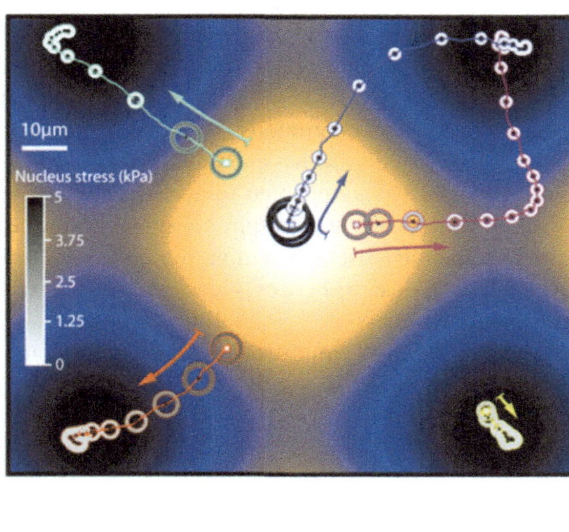

Fig. 12.3 Simulation of cell migration following the curvature-induced nucleus displacement. The cell model stabilizes when the nucleus stabilizes, the both in the center of the concave region. Reproduced under the terms of the CC-BY 4.0 license (Vassaux et al, 2019).

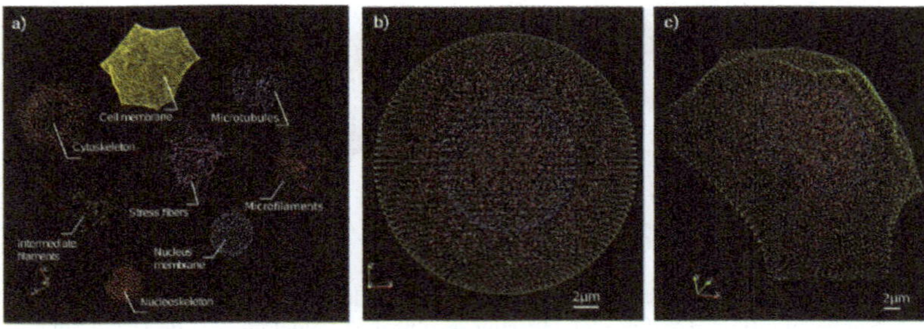

Fig. 12.4 Detailed structure of our in silico mechanical cell model. In the cell model, particles are interacting in one of three ways: repulsive contact, cable-like or spring-like. (a) The model encompasses a wide range of intracellular structures, for more details on what type of interaction is used for each structure and why see (Vassaux and Milan, 2017). (b) Initially, before adhesion, the cell is generated in a spherical shape, with all its constituents relaxed. (c) After adhesion, here on a convex substrate (not appearing), the cell finds its stretched configuration, with the filaments and stress fibres building up tension, and the microtubules and nucleus bearing compressive loads, ensuring structural stability of the cell. Reproduced under the terms of the CC-BY 4.0 licence (Vassaux and Milan, 2017).

Subsequently, the adhesion model has been extended to render migration tractable (Vassaux et al, 2019). The migration is simulated by reproducing in a simplified way the simultaneous protrusion of a lamellipodium at the front and the cell retraction at the back of the cell. The cell model migrates as new FAs are continuously assembled away from existing disassembling adhesions in the direction of motion. While the cytoskeleton connects the new FAs, the old ones are disassembled. We hypothesized that the lamellipodium forms in the direction of the topography-induced polarization of the nucleus and advances proportionally to nucleus internal motion. The internal displacement of the nucleus is computed as the vector directed from the cell barycenter to the nucleus barycenter. The spatial jump (amplitude, direction) from the disassembled adhesions to the assembled ones at a given step is equal to the internal displacement of the nucleus observed at the previous step. The simulation of cell migration ends when the nucleus displacement becomes negligible with respect to the cell dimensions; that is when the cell is assumed to have stabilized. Such a procedure renders a continuous migration of the cell.

The level of complexity encompassed in such *in silico* model is already high enough so that we are enabled to investigate the role of several intracellular structures, as well as the topography of substrate on cell mechanics, as well as on a hypothesized nucleosensitive migration mechanism. In comparison to *in vitro* models, the mechanical properties of each component of the model may easily be tuned and their role assessed on migration parameters such as rate and persistence. Acquiring data is also simplified, as in such computational models, mechanics are intrinsically quantified.

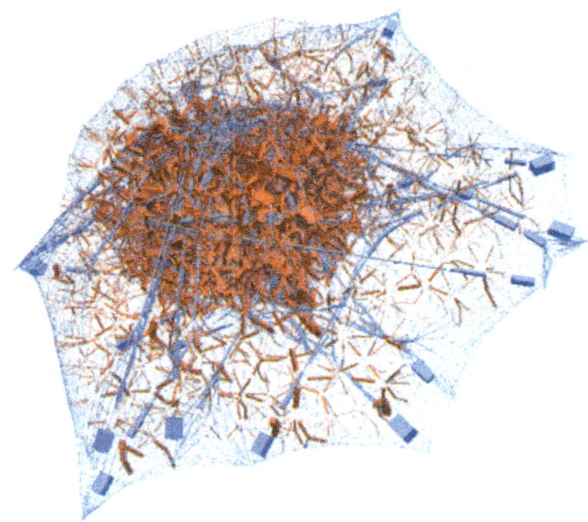

Fig. 12.5 Intracellular force network in the *in silico* cell model. Blue and red segments represent respectively tension and compression forces. Width of the segment is proportional to the magnitude of the force.

12.3.2 Cell-Scale Curvatures Optimize Migration Rates and Persistence

We hypothesized the importance of wavelength and amplitude for the nucleosensitivity guidance mechanism to occur. We made use of our *in silico* cell model whereby cell motility is induced by direction and the magnitude of the polarization of the nucleus to find optimal sinusoids to promote single mesenchymal stem cell migration rate. We demonstrated that on cell-scale curvatures an optimum of migration efficiency is reached. Cells were arbitrarily positioned in the neutral part of the sinusoid, that is in the middle of a flat portion of the sinusoid where the curvature is null. The adhesion and migration dynamics were simulated on three sinusoids, with a constant ratio of amplitude to wavelength: 3 μm / 30 μm, 10 μm / 100 μm, 30 μm / 300 μm (Fig. 12.5). The dynamics were observed until the cells stabilize and their motile behavior was considered inexistent. The efficiency of the cell model in finding the direction of the shortest path to the location of stabilization varied significantly with the sinusoid size. Radii of curvature approximately of the size of the cell led to the most straightforward to stabilization. On shorter and larger radii of curvature, cells exhibited curved trajectories (Fig. 12.6.a) or even sudden changes of direction (Fig. 12.6.c). In turn, migration rates were also much higher on cell-scale curvatures, reducing the time from the initiation of the dynamic migratory behavior to stabilization.

Simulations indicated that curvotaxis at small wavelength seems limited. Similarly, long-wave curvotaxis is also limited: large sinusoids are almost flat surfaces that offers almost no relaxation zone. The cell model cannot sense larger wavelength than its own spread diameter nor it senses smaller wavelength than the diameter of its nucleus. In other words, in this mechanism of cell migration induced by cur-

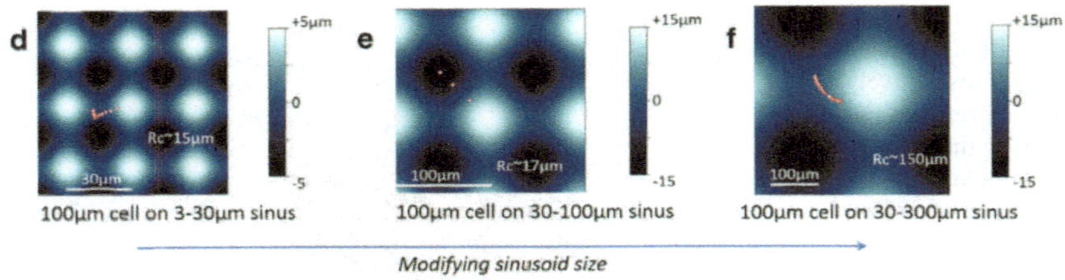

Fig. 12.6 Cell-scale curvatures optimize migration rates and persistence. Migration of a 100 μm cell on (d) 3-30 μm, (e) 30-100 μm, and (f) 30-300 μm sinusoids, the trajectories of the center of the cell model are indicated by the red/white data points. Cells dynamics are initiated on the flat part of the curvature (at the center of each map). Data points composing the trajectories are measured at identical time intervals, a large gap between two neighboring points indicates large migration velocity. Reproduced from Vassaux et al (2019) (CC-BY 4.0 license).

vature and intracellular movement of the nucleus, the diameter of the nucleus and the diameter of the cell constitute the minimum and the maximum of the spatial scale of the curvotaxis of the cell; cell curvotaxis is then related to its intrinsic dimensions. Extrapolating these results to scaffold design, topographies exhibiting cell-scale curvatures could be used for enhanced invasion and proliferation. As cells stabilize faster, they also enter more rapidly in growth and division phase.

12.3.3 Pieces of Evidence of a Will of the Cell to Relax

We led here additional simulation of cell migration decreasing drastically cortical tension or nuclear stiffness. In both cases, the greater is the decrease, the more the model lose the capability of sensing the curvature of the substrate and migration process stopped far away from the center of a concave region. Besides as a consequence the migration velocity dropped down. So, the cell model is able to sense the curvature and to reach concave region to relax only if it possesses full integrity in its cortical tension and nucleus stiffness. Simulation results are in good agreement with in vitro observations we reported in drugged cells obtained by either blocking F-polymerization or by knocking down nucleus lamina (Fig. 12.7).

The stiffness of the nucleus makes it an ideal topography sensor. Coupled with the actin cortex contractility, this renders a complex mechanism propelling the nucleus toward most relaxed locations inside the cell. Our *in silico* model enables to analyze and quantify directly the networks of forces established inside the cell throughout its migration (Fig. 12.8). This network of forces resulting from the interaction in the cytoskeleton and the nucleus is highly dynamic. Focusing on a 100 μm cell migrating on the 10 μm / 100 μm sinusoid, we observe the progressive relaxation of the forces as the cell migrates from convex to concave locations. In turn, the cell could be using its nucleus to find optimally relaxed and mechanically stable

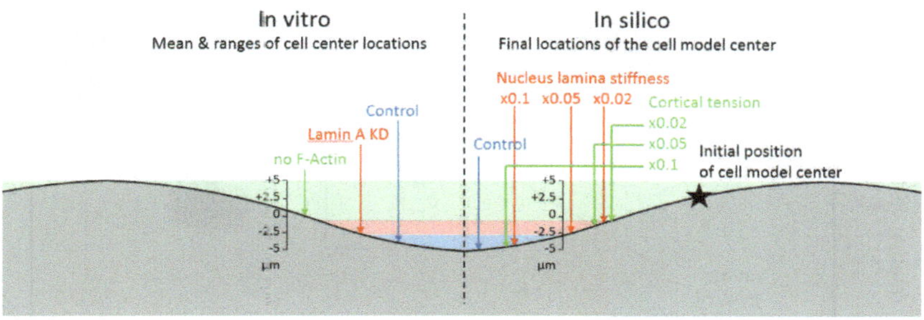

Fig. 12.7 *In silico* and *in vitro* results on final location of cells with altered properties of the cytoskeleton. In gray, the sinusoidal substrate. The bands in blue, red and green represent the location of the in vitro cells respectively, in control conditions, with low stiffness in the nucleus (lamin A knockdown) and without contractile cytoskeleton (no F-actin). For instance, cells in control condition located in concave area. For each in vitro condition, a vertical arrow indicates on the left side the mean position of the cells on the sinusoidal surface. On the right side, verticals arrows indicated the final position of the cell model at the end of migration in control conditions or with altered cell mechanical properties reproducing in vitro tests using drugs. *In silico* results are consistent with *in vitro* observations and lead to the same conclusion: the curvature-induced cell migration based on nucleus mechanosensitivity needs both nucleus stiffness and cytoskeleton contractility, and especially cytoskeleton contractility. Without one or both, and especially without the contractility of the cytoskeleton, cells lose their ability to detect curvature and can localize independently of the curvature gradient, whether convex or concave.

locations in its microenvironment. Concave locations in a sinusoid typically provide these two characteristics. In comparison, convex locations are highly unstable, small fluctuations in cell and nucleus centering on the topography could lead to large internal motions of organelles, highly damageable during mitosis. Flat locations are indeed more stable but do not enable the cell and its organelles to relax as much. Only cell-scale curvatures provide gradients of topography that can be perceived by the cell by impacting its mechanics. On smaller and larger radii of curvature, the topography is mostly integrated by the cell as a flat substrate, potentially not yielding a sufficiently strong mechanical signal.

12.3.4 Topography as a Parameter of Scaffold Design

Properties of the topography are a significant parameter in designing scaffolds and should be chosen depending on the type of cell, more precisely the size of the cell and its nucleus, for optimal invasion and proliferation. Our *in silico* cell model has served as a framework to integrate the hypothesized mechanism of migration called curvotaxis, whereby the cell motility is driven by the instability of its nucleus. We have been able to analyze the influence of the topography of the cell microenvironment on the motile behavior of mesenchymal stem cells. These primary results from our *in silico* stem cell model tend to show that curvotaxis could be an attempt to min-

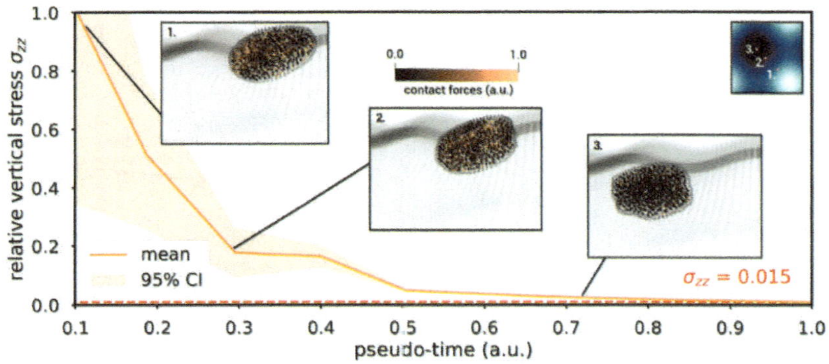

Fig. 12.8 Nucleus mechanical stress relax during curvotaxis. The mechanical vertical stress integrated over the nucleus relaxes during migration from convex to concave of a 100 µm-wide cell on a 30-100 µm sinusoid (Fig. 12.6.b). Snapshots of the nucleus shape and internal contact forces in the nucleoplasm are taken for three different cell positions during migration. Positions of the cell on the sinusoid in each snapshot are shown in the inset picture in the upper right corner. The vertical stress in the nucleus displays a relative decrease of 80% between onset of migration and stabilization in the nearest concave. Snapshots illustrate the simultaneous relaxation of nucleus shape, from elongated to rounded, and significant decrease of contact forces between particles constituting the nucleoplasm.

imize cells mechanical energy via relaxation, as well as a way to find mechanically more stable microenvironments. Such microenvironments are beneficial for a more robust cell growth and division. Our *in silico* cell model has also enabled to quantify optimal microenvironment topographies, that is sinusoid wavelength and amplitude. Curvotaxis is rendered more efficient by sinusoids displaying cell-scale curvatures. These results could inform the design of scaffolds used in tissue engineering to promote invasion and proliferation of mesenchymal stem cells. The methodology applied in this work could be repeated for different cells types, hence enabling to design cell-type specific scaffold topographies.

12.4 Design of a New Generation of Biomaterials of Dynamic Topography Aided by Silico Cell Models

Many *in vitro* studies exist on the influence of topography on cell migration, however in all these works the topography remains fixed (Caballero et al, 2015) (Fig. 12.9). We have shown that the concave regions attract the cells, but once these regions are reached, the cells stop their migration. To encourage the cells to migrate over a greater distance thanks only to the topography of the substrate and by using their curvotaxis capacity, we may propose a substrate of variable geometry, with changing topography, which can become alternately and locally concave then convex, and this cyclically. Some authors developed photochemical protocols to modulate in real time the local stiffness or strain of hydrogel substrates (Kloxin et al, 2010; Chandorkar et al, 2019). We have shown that cells cultured on sinusoidal surfaces, migrate naturally, with no other stimulus than the only curvature of the surface. What would be the migratory behavior of the cells on a sinusoidal surface animated by an undulatory movement (Figure 12.10). Would the cells follow the ripple? Would the cells start surfing the surface of the substrate following the ripple wave? What would be the influence of the ripple frequency? The optimal frequency of ripple, or in other words the speed of the wave front, should be *a priori* of the same order of magnitude as the migration speed of the cell. However, what could be the influence of a ripple at very high frequencies? In such a case, a displacement of the nucleus would be observed within the cell, a displacement going in the same direction as the wave front. As we showed that the displacement of the nucleus and its decentering is a signal for the cell to migrate to center again its nucleus, this would stimulate continuous cell migration.

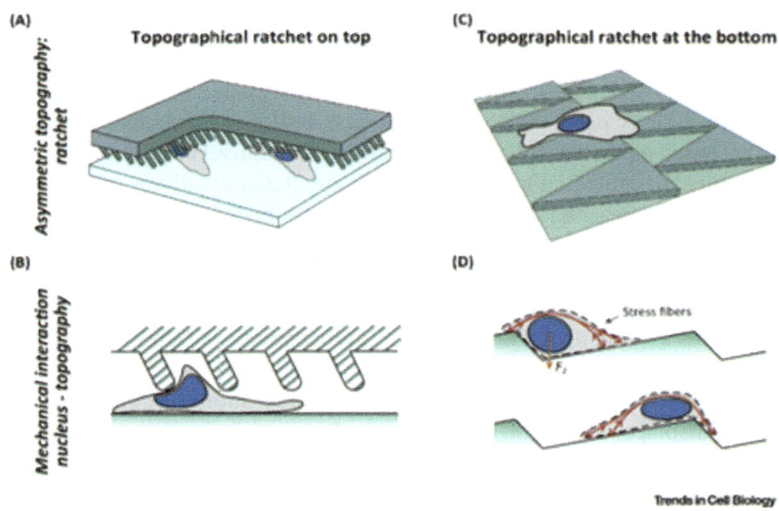

Fig. 12.9 Effect of the Cell Nucleus on Symmetry Breaking and Directional Migration. Cells move directionally in local asymmetric topographical ratchets imposed by confinement (A) or adhesion (C). A mechanical interaction between the cell nucleus and the tilted micropillars (B) or actomyosin stress fibers (D) guides cell polarization and motility. Reproduced from Caballero et al (2015) (CC-BY 4.0 licence).

As a perspective, we can imagine an evolution of the biomaterials and scaffolds with dynamic topography to induce cell migration and invasion. Typically, *in silico* modelling can play a role here. Indeed, *in silico* experiments can be pushed beyond what is technically feasible in vitro for the time being. We propose here to analyze *in silico* the influence of a dynamic topography on the migratory behavior of cells using our computational cell models we presented above. In the present study, we simulated

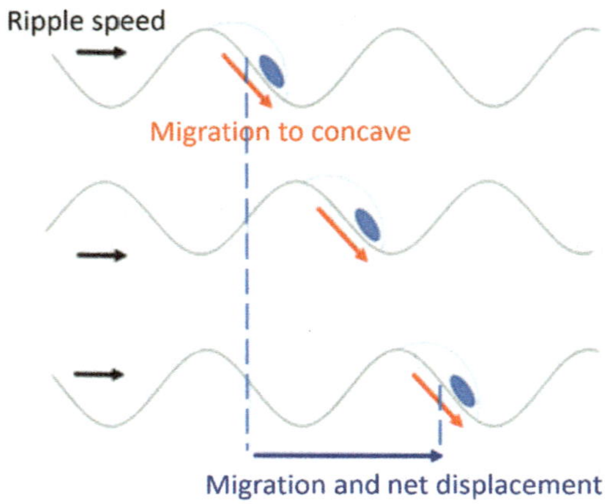

Fig. 12.10 Substrate animated by wave motion. Are migrating cells able to surf the wave?

curvature-guided cell migration on a deformable sinus animated by sequential ripple motion. We imposed at the location of the cell, a deformation of the substrate to reach a sinus morphology or micro-corrugated shape. Following the same process of cell migration based on the interplay between curvature-induced nucleus decentering and cell movement to center the nucleus again, we simulated iteratively the displacement of the cell until it reached the most concave region of the sinus (Fig. 12.11). Then the substrate deforms to become flat as at the beginning. The cell migrates with a net displacement of 45 μm. Then we deformed the substrate, a second time, with the same sinusoid morphology with a dephasing of 45 μm, inducing at the cell location a convex region. Following the same cell migration process, we simulate a second time cell displacement until it reached the new concave region. At the end, in imposing two deformations of the substrate, we induced cell migration in a controlled direction with a net displacement of 90 μm corresponding to 1.5 times the diameter of the cell. It is worth to be noted there is no gravitation here and the cell migrates only following the nucleus decentering induced by curvature.

In the same way we can imagine to study *in silico* the cellular migration in interaction with a dynamic substrate, micro-channels or micro-tubes able to be piloted in radial deformation by shrinkage movements or on the contrary of swelling. In such a case, are the cells able to migrate by accompanying the deformations of the micro-tubes?

We can also design *in silico*, a fibrous substrate whose fibers and their crosslinking could be driven dynamically to locally animate the fibrous matrix by contraction or extension. We could experiment with the potential of migrating cells and predict whether cells are able to take advantage of the movements of their environment to migrate.

This type of dynamic topography substrates continuously stimulating cell migration could help the colonization of porous biomaterials by the cells, a colonization that is still insufficient to obtain volume tissue regeneration. And this, proposing an

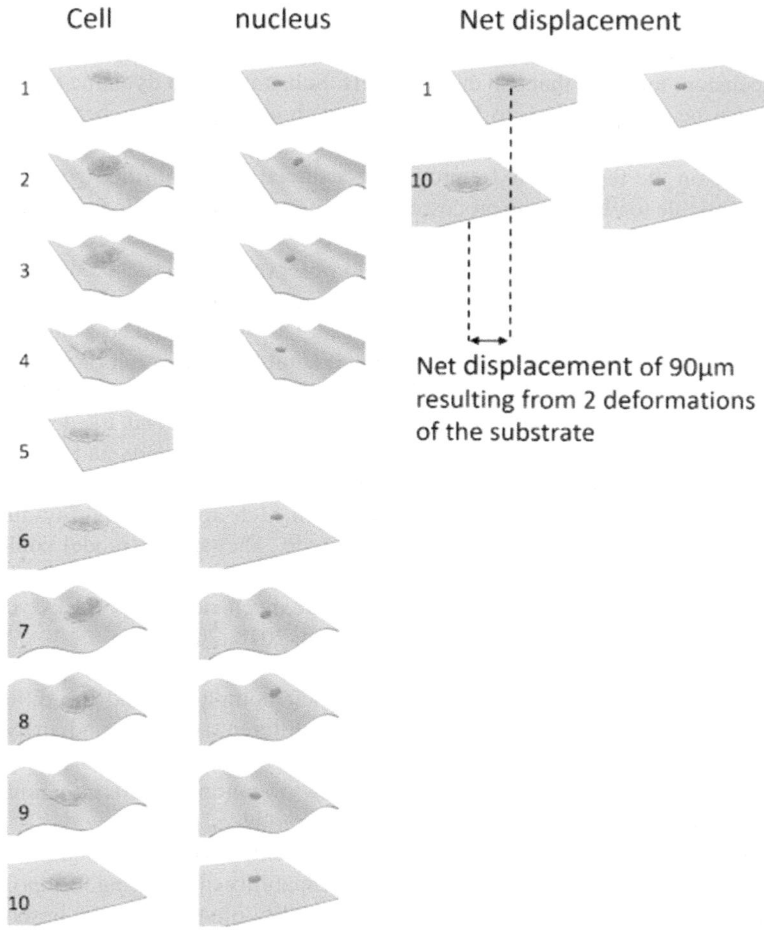

Fig. 12.11 Image series of the sequential cell migration on a flat substrate animated by 2 successive sinus deformations. The diameter of the adherent cell shape is 100 µm. After the first substrate deformation, the cell model migrated over 44.5 µm. As a result, the 2nd sinus deformation was imposed with a phase shift of 44.5 µm from the first one. After the two successive deformations of the substrate, the cell model migrate over 90 µm in a controlled direction. It is worth to be noted that there is no gravitation in this simulation.

original method, natural because based on the normal migration of cells, and alternative to conventional mechanical methods such as perfusion or pumping of cells in suspension, methods that can damage cells. Substrates with dynamic topography could also be an alternative to biochemical methods employing chemoattractants and which raise the question of the duration of release and the duration of action.

Playing on the topography by proposing artificial and controlled geometry can make it possible to identify the processes of setting up of the adhesion and migration, to identify the cellular preferences, the processes of optimization of their form and position, of their potential adaptation, to observe the emergence of alternative solutions when one is blocked. This work could provide a great deal of information on cellular functioning and adaptation resources. This work could also inform future improvements in the design of biomaterials to stimulate migration or proliferation or differentiation by time. This cellular model could be used for the design of scaffolds specifically dedicated to bone reconstruction. To this end, the design of the scaffold should promote the invasion of mesenchymal stem cells and osteoblastic differentiation. The scaffold should also stimulate the osteoblastic activity of bone tissue synthesis via mechanical stimuli based on high apparent rigidity allowing deformation of high frequency and low amplitude. The cellular model could be a complementary approach at the cellular level to those which are developed at the tissue level and which succeed in embracing bone mechanobiology (Lekszycki and dell'Isola, 2012; Giorgio et al, 2017; George et al, 2018, 2019).

In vitro experiments have their limits. While they do not fully reproduce the reality of in vivo conditions, but especially their complexity and the difficulties of producing biomaterials prevent testing many different solutions and analyze the cellular response in completely new conditions. The contribution of *in silico* or numerical simulation experiment, precisely allows to put in the cells situation under conditions impossible to consider *in vitro* and / or *in vivo*.

Virtually we can culture cells in a 3D environment, in contact with a material, a surface, or a fibrous matrix that would have the capability of changing its topography according to whether we are looking for the viability of stem cells by proposing a quiescent state or on the contrary a state of stress that will push them to migrate or differentiate. These controllable materials could adapt their conformation to the cellular time and specific cell function. Those smart materials are difficult or impossible to design for now. Nonetheless *in silico* experiments make it possible to overcome this problem by testing unrealistic conditions while identifying cellular behaviors never observed in vitro and dynamic microstructures and their associated deformation modes capable of stimulating cells. Based on these results, we would be able to imagine technical and feasible solutions to reproduced in vitro and in vivo the cellular response predicted by the model. The *in silico* approach can then join current developments in the field of intelligent materials such as meta-materials and nanomotors. For example, meta-materials thanks to their exotic electromagnetic or mechanical properties can modify their structural arrangement under the passage of electromagnetic waves or can have a negative Poisson's ratio, contracting transversely during compression (Barchiesi et al, 2019; Del Vescovo and Giorgio, 2014; dell'Isola et al, 2019). This type of material could be used to reproduce the optimal dynamic topography of the substrate identified by the cell model. Similarly, the properties of meta-materials could be modeled to predict and analyze the behavior of cells.

References

Adey W (1983) Molecular aspects of cell membranes as substrates for interaction with electromagnetic fields. In: Synergetics of the Brain, Springer, Berlin, Heidelberg, pp 201–211

Almonacid M, Ahmed WW, Bussonnier M, Mailly P, Betz T, Voituriez R, Gov NS, Verlhac MH (2015) Active diffusion positions the nucleus in mouse oocytes. Nature cell biology 17(4):470–479

Barchiesi E, Spagnuolo M, Placidi L (2019) Mechanical metamaterials: a state of the art. Mathematics and Mechanics of Solids 24(1):212–234

Caballero D, Comelles J, Piel M, Voituriez R, Riveline D (2015) Ratchetaxis: long-range directed cell migration by local cues. Trends in cell biology 25(12):815–827

Chandorkar Y, Nava AC, Schweizerhof S, Van Dongen M, Haraszti T, Köhler J, Zhang H, Windoffer R, Mourran A, Möller M, et al (2019) Cellular responses to beating hydrogels to investigate mechanotransduction. Nature communications 10(1):1–13

Chen T, Callan-Jones A, Fedorov E, Ravasio A, Brugués A, Ong HT, Toyama Y, Low BC, Trepat X, Shemesh T, et al (2019) Large-scale curvature sensing by directional actin flow drives cellular migration mode switching. Nature physics 15(4):393–402

Clark P, Connolly P, Curtis A, Dow J, Wilkinson C (1991) Cell guidance by ultrafine topography in vitro. Journal of cell science 99(1):73–77

Czeisler C, Short A, Nelson T, Gygli P, Ortiz C, Catacutan FP, Stocker B, Cronin J, Lannutti J, Winter J, et al (2016) Surface topography during neural stem cell differentiation regulates cell migration and cell morphology. Journal of Comparative Neurology 524(17):3485–3502

Darling NJ, Sideris E, Hamada N, Carmichael ST, Segura T (2018) Injectable and spatially patterned microporous annealed particle (MAP) hydrogels for tissue repair applications. Advanced Science 5(11):1801,046

Del Vescovo D, Giorgio I (2014) Dynamic problems for metamaterials: review of existing models and ideas for further research. International Journal of Engineering Science 80:153–172

dell'Isola F, Seppecher P, Alibert JJ, et al (2019) Pantographic metamaterials: an example of mathematically driven design and of its technological challenges. Continuum Mechanics and Thermodynamics 31(4):851–884

Dillon R, Fauci L, Gaver III D (1995) A microscale model of bacterial swimming, chemotaxis and substrate transport. Journal of theoretical biology 177(4):325–340

Doyle AD, Wang FW, Matsumoto K, Yamada KM (2009) One-dimensional topography underlies three-dimensional fibrillar cell migration. Journal of cell biology 184(4):481–490

George D, Allena R, Remond Y (2018) A multiphysics stimulus for continuum mechanics bone remodeling. Mathematics and Mechanics of Complex Systems 6(4):307–319

George D, Allena R, Remond Y (2019) Integrating molecular and cellular kinetics into a coupled continuum mechanobiological stimulus for bone reconstruction. Continuum Mechanics and Thermodynamics 31(3):725–740

Giorgio I, Andreaus U, dell'Isola F, Lekszycki T (2017) Viscous second gradient porous materials for bones reconstructed with bio-resorbable grafts. Extreme Mechanics Letters 13:141–147

Griffin DR, Weaver WM, Scumpia PO, Di Carlo D, Segura T (2015) Accelerated wound healing by injectable microporous gel scaffolds assembled from annealed building blocks. Nature materials 14(7):737–744

Isenberg BC, DiMilla PA, Walker M, Kim S, Wong JY (2009) Vascular smooth muscle cell durotaxis depends on substrate stiffness gradient strength. Biophysical journal 97(5):1313–1322

Kloxin AM, Benton JA, Anseth KS (2010) In situ elasticity modulation with dynamic substrates to direct cell phenotype. Biomaterials 31(1):1–8

Lekszycki T, dell'Isola F (2012) A mixture model with evolving mass densities for describing synthesis and resorption phenomena in bones reconstructed with bio-resorbable materials. ZAMM-Zeitschrift für Angewandte Mathematik und Mechanik 92(6):426–444

Milan JL, Manifacier I, Beussman KM, Han SJ, Sniadecki NJ, About I, Chabrand P (2016) In silico CDM model sheds light on force transmission in cell from focal adhesions to nucleus. Journal of biomechanics 49(13):2625–2634

Mogilner A (2009) Mathematics of cell motility: have we got its number? Journal of mathematical biology 58(1-2):105

Nickaeen M, Berro J, Pollard TD, Slepchenko BM (2019) Actin assembly produces sufficient forces for endocytosis in yeast. Molecular biology of the cell 30(16):2014–2024

Pieuchot L, Marteau J, Guignandon A, Dos Santos T, Brigaud I, Chauvy PF, Cloatre T, Ponche A, Petithory T, Rougerie P, et al (2018) Curvotaxis directs cell migration through cell-scale curvature landscapes. Nature communications 9(1):1–13

Rens EG, Merks RM (2017) Cell contractility facilitates alignment of cells and tissues to static uniaxial stretch. Biophysical journal 112(4):755–766

Rodriguez ML, McGarry PJ, Sniadecki NJ (2013) Review on cell mechanics: experimental and modeling approaches. Applied Mechanics Reviews 65(6)

Schakenraad K, Ravazzano L, Sarkar N, Wondergem JA, Merks RM, Giomi L (2019) Topotaxis of active Brownian particles. arXiv preprint arXiv:190806078

Vassaux M (2018) mvassaux/adhsc: Initial release of the adhesion cell model (version v1.0.0). URL https://zenodo.org/record/1187087

Vassaux M, Milan J (2017) Stem cell mechanical behaviour modelling: substrate's curvature influence during adhesion. Biomechanics and modeling in mechanobiology 16(4):1295–1308

Vassaux M, Pieuchot L, Anselme K, Bigerelle M, Milan JL (2019) A biophysical model for curvature-guided cell migration. Biophysical journal 117(6):1136–1144

Winkler B, Aranson IS, Ziebert F (2019) Confinement and substrate topography control cell migration in a 3D computational model. Communications Physics 2(1):1–11

Zigmond SH, Hirsch JG (1973) Leukocyte locomotion and chemotaxis: new methods for evaluation, and demonstration of a cell-derived chemotactic factor. The Journal of experimental medicine 137(2):387–410

Zink M, Szillat F, Allenstein U, Mayr SG (2013) Interaction of ferromagnetic shape memory alloys and RGD peptides for mechanical coupling to cells: from ab initio calculations to cell studies. Advanced Functional Materials 23(11):1383–1391

Chapter 13
Viscoelastic Characterization of Dacron Graft and Aortic Tissue

Christopher Zikry, Stewart McLennan, Gilles Soulez, Raymond Cartier, and Rosaire Mongrain

Abstract We present the elastic and viscoelastic characterization of aortic tissue and a synthetic material used for the fabrication of artificial vessels (Dacron). Using biaxial, high deformation and oscillating mechanical testing protocols, we assessed the hyperelastic and viscoelastic properties of both aortic tissue and Dacron. Energy loss is a viscous measure of energy absorbed by a material during deformation. It provides information of the materials time dependence and capacity to dissipate energy. Investigation of the correlation between smooth muscle cell (SMC) content and energy loss within healthy and aneurysmal aortic tissue was carried out via biaxial tensile testing of aortic tissue samples. The results of aortic tissue energy loss investigation show that an acceptable correlation exists between the presence of SMCs and the magnitude of energy loss. In addition, our data suggests that the condition and organization of SMCs may affect the viscous behaviour of tissue, instead of their mere presence. The results of the high deformation and oscillating mechanical testing show significant differences between the biological and the synthetic materials. Histologic examination of selected samples revealed healthy tissue was characterized by higher elastin content, 45.3±2.07% vs. 39.78±1.84%. Aneurysmal tissue was found to have greater SMC content in comparison to healthy tissue.

Keywords: Energy loss · Artificial vessel · Hyperelastic properties· Viscoelastic properties · Mechanical bio-integration · Graft design

C. Zikry, S. McLennan, R. Mongrain
Department of Biomedical Engineering, McGill University, Montreal, Quebec, Canada
e-mail: christopher.zikry@mail.mcgill.ca, stewart.mclennan@mail.mcgill.ca, rosaire.mongrain@mcgill.ca

G. Soulez
Centre Hospitalier de l'Université de Montréal, University of Montreal, Montreal, Quebec, Canada
e-mail: gilles.soulez@umontreal.ca

R. Cartier
Montreal Heart Institute, University of Montreal, Montreal, Quebec, Canada
e-mail: cartierr@icm.umontreal.ca

B. E. Abali and I. Giorgio (eds.), *Developments and Novel Approaches in Biomechanics and Metamaterials*, Advanced Structured Materials 132,
https://doi.org/10.1007/978-3-030-50464-9_13

13.1 Introduction

The aorta carries oxygenated blood and nutrients throughout the human body via systemic circulation (Maton et al, 1993). The aortic wall is a viscoelastic material that exhibits a measure of compliance and energy dissipation, by which it preserves pressured flow and protects itself against damage from cyclic stresses. Its J-shaped stress-strain curve has been linked to elastin and collagen fiber behaviour (Roach and Burton, 1957). Measures of viscosity along the aorta, such as circumferential relaxation, strength and energy loss, are thought to be correlated with the degree of SMC presence (Azuma and Hasegawa, 1971). Viscosity plays an important biomechanical role in the ascending aorta by dissipating arterial pulse waves, preserving the aortic wall and directing the precise timing of the Windkessel elastic function, where the aorta acts as an elastic buffering chamber behind the heart.

Prior investigation has suggested the 'elastic' arteries (conducting arteries) allow expansion and offer elastic recoil, while 'muscular' arteries limit expansion and promote vasoconstriction through increased SMC content (Rhodin, 1980). In addition it is believed that SMCs are preferentially aligned circumferentially to provide resistance to circumferential loads (Rhodin, 1980). Prior investigation has utilised energy loss as a robust biomechanical parameter to describe aortic pathological condition (Chung et al, 2014). Energy loss can be seen as a measure of how aortic tissue distributes energy and is thought to correlate with microstructural defects in the media (Chung et al, 2017).

During each cardiac cycle, the aortic wall receives energy from the pumping heart with each ventricular ejection of blood. The aortic wall absorbs this energy by distending. In accordance with its Windkessel elastic function, the wall returns to its resting circumference, transferring energy back to the blood. In this process, the aortic wall dissipates some energy into itself. While systolic pressure rises rapidly, diastolic pressure returns in a much slower fashion, as result of the reduced energy returned by the wall (Westerhof and Noordergraaf, 1970). Although the tissue compliance enables the Windkessel elastic function, the dissipative capacity of tissue directs the timing. This behaviour is highlighted in the stress-strain response hysteresis loop of aortic tissue and has been thoroughly observed *in vitro* and *in vivo*. Unlike the loading curve, the unloading curve is moderately unaffected by changes in the strain or strain rate (Goto and Kimoto, 1966). Therefore, the width of the hysteresis loop is determined by the magnitude of stretch or strain rate (Remington, 1954). When considering the behaviour of the aortic wall, this response is clear. As the tissue experiences higher levels of stress, it demands increasingly more relaxation to restore the tissue to its equilibrium state.

Pressure-area dynamics of the aorta can be obtained *in vivo* by mapping the stress and strain of the tissue under physiological loading cycles (Imura et al, 1990; Stefanadis et al, 1995). Armentano et al (1995) quantified the viscous component of the aortic wall by analysing *in vivo* hysteresis loops in conscious dogs. The study confirmed that the viscous modulus depends upon the arterial pressure and smooth muscle activation.

Understanding the biomechanical role of SMC content is important when designing artificial vessels. Vascular grafts are implanted in the human body in order to replace damaged or blocked vessels (Puskas and Chen, 2004). The development of synthetic grafts began in the 1940s (Puskas and Chen, 2004). In the 1950s, the concept of porous, fabric vascular grafts were introduced using polyvinyl chloride (PVC) and polyacrylonitrile (PAN). By the 1960s, most fabrics were abandoned except for Dacron and Teflon. At present, Prosthetic Dacron grafts are widely used as a synthetic substitute following aortic resection in cases of aortic aneurysms, aortic dilatations and aortic dissections. While these polyester grafts are readily available, very durable, and biocompatible. They exhibit mechanical properties inconsistent with native aortic tissue. Current Dacron graft implantation, despite its high patency rates, yields notable and potentially harmful consequences to the arterial circulation. This is a result of its disparate geometry and mechanical properties, with issues such as endoleaks and unwanted stress concentrations potentially arising.

From a bioengineering perspective, there are important geometrical and mechanical constraints that the synthetic replacements need to mimic to properly replace the aorta. This includes compliant functioning with the soft tissue environment and avoidance of damaging local stresses (pressure, friction). Furthermore, graft insertion creates discontinuity in the mechanical response of the entire vessel wall, associated with abnormal hemodynamics and shear stresses. Dacron grafts are approximately 75% thinner than the aortic wall and have well-defined corrugations along the longitudinal direction to increase axial distensibility and prevent buckling. The aortic wall is comprised of three distinct layers, while Dacron is a single-layered woven or knitted fabric. At physiological strain, Dacron is approximately 24 times stiffer than healthy ascending aorta (Tremblay et al, 2009). In accordance with its increased stiffness, Dacron exhibits reduced circumferential compliance (Puskas and Chen, 2004; Tremblay et al, 2009; Hasegawa and Azuma, 1979; Walden et al, 1980). The exceedingly high circumferential stiffness of Dacron grafts also results in its increased anisotropy (directional dependency) when compared to healthy ascending aorta (Tremblay et al, 2009) and iliac artery tissue (Lee and Wilson, 1986).

While Dacron grafts compromise non-physiological mechanical properties for long-term strength and durability, there is need to better characterize their viscoelastic behaviour in order to improve future graft design. *In vitro* planar biaxial tensile testing has been widely used to characterize arterial tissue (Azadani et al, 2013; Babu et al, 2015; Debes and Fung, 1995; Emmott et al, 2016; Geest et al, 2006; Mohan and Melvin, 1983; Okamoto et al, 2002). Biaxial tensile testing identifies the tensile behaviour of a material in two distinct directions under physiological multi-loading conditions. This reproducible testing method provides understanding of local tissue hyperelasticity, anisotropy and viscoelasticity.

13.2 Methods

13.2.1 Smooth Muscle Cell Content and Energy Loss Investigation of Aortic Tissue

Five aortic tissue samples were cut into 15×15 mm^2 patches. Quintuplicate thickness measurements were performed along the samples using a digital thickness calliper (Litematic VL-50A, Mitutoyo Corp., Japan). The samples were assumed to hold constant thickness at the average value for subsequent analysis. All samples were fastened with 4-0 prolene sutures to mount them onto the tensile arms. The no-load length was measured between sutures again using a calliper. The samples were then loaded in the ElectroForce Planar Biaxial TestBench (ELF 3200, TA Instruments, New Castle, DE, USA), as seen in Fig. 13.1, equipped with WinTest software (V8.0, Build 2011) linked to a displacement transducer and a 1kg load-cell (Model 31, Sensotec Honeywell).

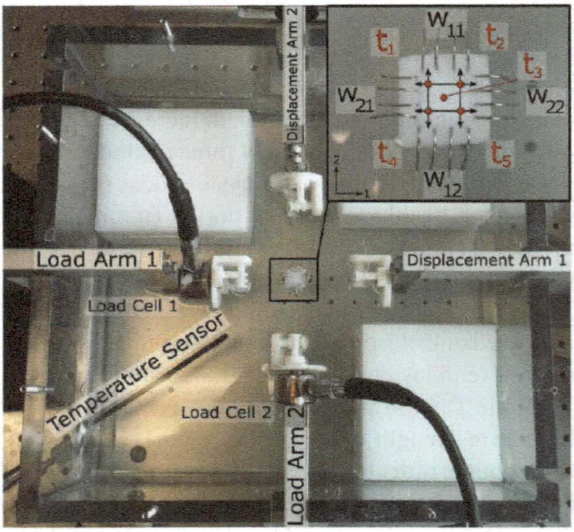

Fig. 13.1 ELF 3200 Biaxial Tensile Tester setup. Inset displays suture layout of a sample, including locations of thickness measurements (t) and definition of gauge lengths (w) between sutures.

All samples were immersed in buffered solution at 37°C for 15 minutes prior to testing to attain thermal equilibrium and subsequently remained there throughout all testing. Samples were preloaded to a force of 0.05 N in both directions before testing to ensure sutures were slightly tensioned. The zero displacement position was taken following this preloading. Samples then underwent 10 preconditioning loading-unloading cycles at a strain rate of 0.4mm/s to a displacement of 6mm and then back to the zero displacement position, in both perpendicular directions. Depending on suture location, this corresponded to a maximum strain of 60-65%.

Following preconditioning, 7 experimental cycles were performed at a strain rate of 0.1 mm/s to the same displacement.

Immediately following the last experimental cycle, the samples were pulled to a 6mm displacement at a rise time of 1s. Each sample was held at this displacement for 2000 s and the stress relaxation response was recorded. The sample was then slowly unloaded at a rate of 0.1 mm/s to its starting position and then quickly loaded in a similar fashion for a total of 5 cycles.

To investigate the correlation between SMC content and energy loss, healthy and aneurysmal human ascending aorta tissue samples were used. Viscous energy loss was calculated as the percentage of energy lost during the loading-unloading cycles, out of the total energy absorbed by the material during loading in the circumferential direction. A portion of tissue adjacent to each testing specimen was stained with Movat's pentachrome. Images were captured with a microscope and the percent micrograph field coverage of SMC content was measured.

13.2.2 Hyperelastic Characterization of Dacron Graft and Aortic Tissue

For hyperelastic characterization the same ELF 3200 Biaxial Tensile TesBench was used as for the SMC and energy loss investigation, with the setup and loading procedure remaining the same. This time, both Dacron and aortic tissues samples were tested.

A number of hyperelastic material models can be used to classify biological material properties. However, for simplicity, the elastic modulus at different strains (incremental elastic modulus) was measured in order to provide information on the overall hyperelastic behavior. Incremental elastic modulus was defined as the instantaneous slope at the low (15%) and high (45%) strain values, taken from the hyperelastic stress-strain plots for each sample. These values were determined using polynomial functions in MATLAB.

13.2.3 Viscoelastic Characterization of Dacron Graft and Aortic Tissue

The ELF 3200 also allows for frequency characterization (it can perform a frequency sweep from 0.1–100 Hz). Briefly, an oscillating force (load) was generated ($\sigma = \sigma_0 \sin(\omega t)$, with amplitude '$\sigma_0$' and frequency '$\omega$') which produced an out-of-phase displacement ($\varepsilon = \varepsilon_0 \sin(\omega t + \delta)$, with amplitude '$\varepsilon_0$' and phase lag '$\delta$'). From this, the complex viscoelastic stiffness was assessed using $E^* = E_s + iE_l$ where, $E_s = \sigma_0/\varepsilon_0 \cos(\delta)$ is the storage modulus and $E_l = \sigma_0/\varepsilon_0 \sin(\delta)$ is the loss modulus, from which $\tan(\delta) = E_l/E_s$ can be found.

It can be shown that for a Kelvin–Voigt model (Fig. 13.2), $E^* = k + i\omega\,\eta$ with 'k' and 'η' being the elasticity and viscosity, respectively (then $|E^*|^2 = k^2 + \omega^2\eta^2$ and $\tan(\delta) = \omega\,\eta/k$).

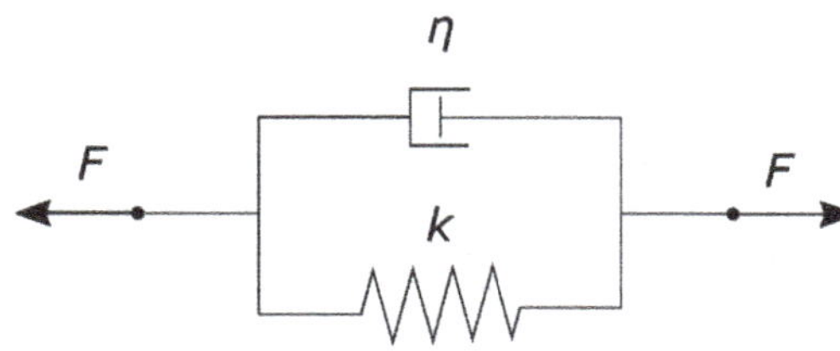

Fig. 13.2 Kelvin–Voigt Spring and Dashpot Model.

It can also be shown that the corresponding hysteresis is modelled as an ellipse (with an area $W_d = \pi\omega\epsilon^2$). From here, the Loss Factor (or energy loss) was calculated. This is defined as the ratio of the area of the hysteresis over the storage deformation energy, LF=Energy hysteresis/Storage energy or $LF = W_d/(W_e - W_d)$ with W_e being the elastic deformation energy (Chung et al, 2014; Shahmansouri et al, 2016; Zikry, 2018). The hysteresis loop calculation are illustrated in Fig. 13.3.

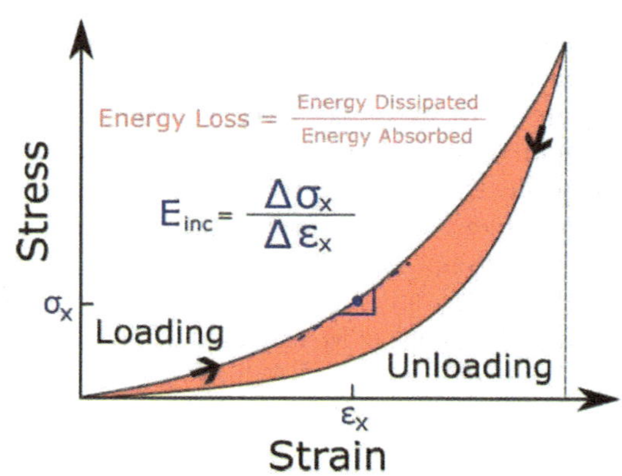

Fig. 13.3 Example hysteresis plot of typical viscoelastic material stress-strain response. Graphical representation of energy loss and incremental elastic modulus is depicted.

To investigate the applicability of the Kelvin–Voigt model for viscoelastic characterization of Dacron graft and aortic tissue we used the Fung Quasi-linear Viscoelasticity (QLV) model for comparison. Here, the QLV parameters were obtained using a custom made MATLAB script that fitted the biaxial testing data onto a Cauchy stress-time plot and extracted the QLV parameters via curve fitting of the constitutive equations. The constitutive equations of the Fung QLV material model were taken from Fung (1972, 1993).

13.3 Results

13.3.1 Smooth Muscle Cell Content and Energy Loss Investigation of Aortic Tissue

Histologic examination of the aortic tissue samples showed that healthy tissue was characterized by higher elastin content, 45.3±2.07% vs. 39.78±1.84%. In addition, thick, continuous elastinous sheets encapsulate single layers of SMCs inside healthy tissue. In aneurysmal tissue, several layers of SMCs are separated by weak, discontinuous elastin fibers, which vary in integrity across the thickness. Aneurysmal tissue was found to have a larger SMC content. An acceptable correlation ($R^2 = 0.3519$) exists between the presence of SMCs and the magnitude of viscous energy loss (Fig. 13.4), where viscous energy loss is defined as the area of dissipated energy over the area under the stress-strain loading curve and is given by

$$LF = \frac{W_L - W_U}{W_L} \tag{13.1}$$

where W_L and W_U refer to the energy absorbed during loading and energy released during unloading, respectively. More evident correlation ($R^2 = 0.9168$) was found between patient age and energy loss (Fig. 13.5).

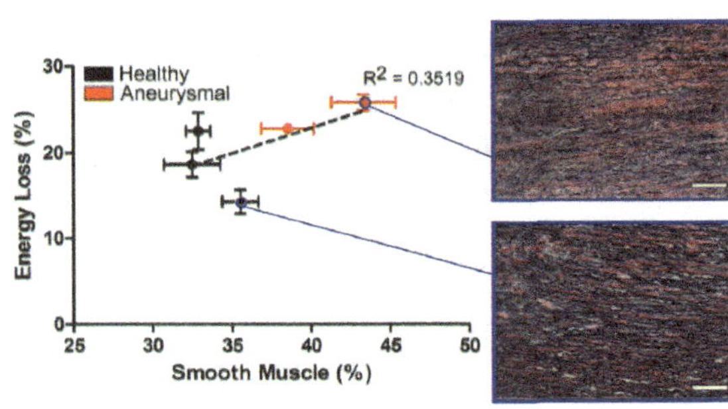

Fig. 13.4 Energy loss as a function of SMC presence in the aortic media. Scale bars represents 100 μm.

13.3.2 Hyperelastic and Viscoelastic Characterization of Dacron Graft and Aortic Tissue

The results for the incremental modulus (or tangent modulus) at 15% and 45% strain and the viscoelasticity characterization assessed with the energy loss are summarised in Fig. 13.6 and 13.7, respectively. Results are shown for healthy aortic

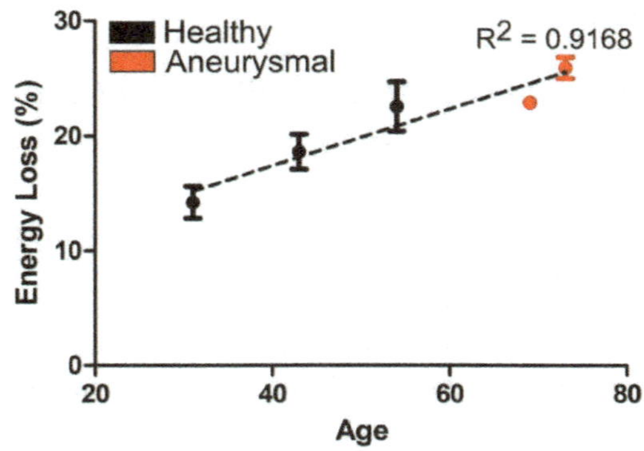

Fig. 13.5 Correlation between patient age and aortic tissue energy loss.

tissue, aneurysmal tissue and Dacron. Significant differences between the synthetic material and the biological tissues are observed.

Dacron grafts are approximately 5 and 11 times stiffer than healthy aortic tissue in the circumferential direction at 15% and 45% strain, respectively. In the longitudinal direction, this modulus disparity between Dacron and tissue is decreased. Slight differences were observed among samples considered healthy and aneurysmal. In Fig. 13.7, the energy loss of each sample is presented for each direction. Dacron grafts were found to exhibit twice as much energy loss as healthy ascending aortic tissue.

The solved Fung QLV parameters for healthy aortic tissue, aneurysmal aortic tissue, and Dacron graft are presented in Table 13.1. Here the C parameter can be thought of a measure of viscosity for each sample type and as such, can be compared against the prior energy loss values. The MATLAB QLV curve fitting for a typical aortic tissue sample is illustrated in Fig. 13.8.

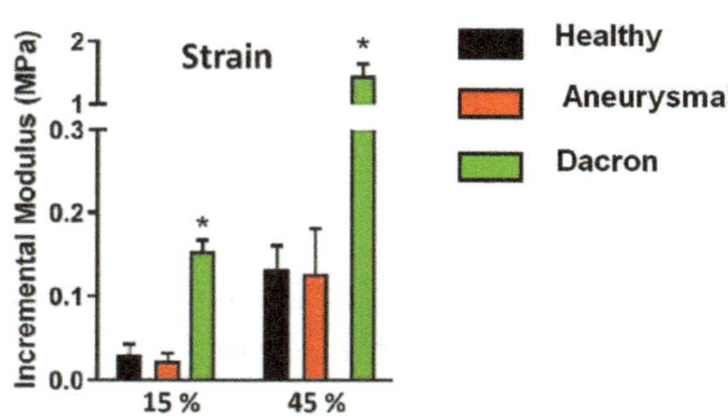

Fig. 13.6 Incremental modulus at 15% and 45% strain in the circumferential direction. Chart values represent mean ± SD. $*p < 0.05$ measured for same strain.

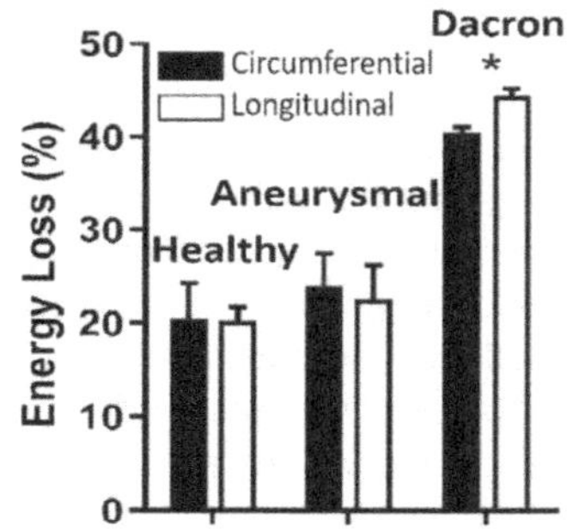

Fig. 13.7 Circumferential and longitudinal energy loss of each sample. Chart values represent mean±SD. $*p < 0.05$ measured for same strain.

Table 13.1 Fung QLV parameter values from circumferential biaxial testing presented as mean ± SD; $*p < 0.05$ for each parameter across all samples.

	A (MPa)	B (10^{-5})	C	τ_1 (s)	τ_2 (s)
Healthy	1984.62 ± 216.41	2.94 ± 0.64	0.1168 ± 0.0420	0.5444 ± 0.0852	825.25 ± 66.331
Aneurysmal	1830.55 ± 279.18	2.24 ± 0.52	0.1728 ± 0.0860	0.5836 ± 0.1818	382.78 ± 173.93
Dacron	5222.69 ± 200.83*	9.30 ± 1.18*	0.4845 ± 0.1166*	0.3445 ± 0.0727	195.05 ± 32.614*

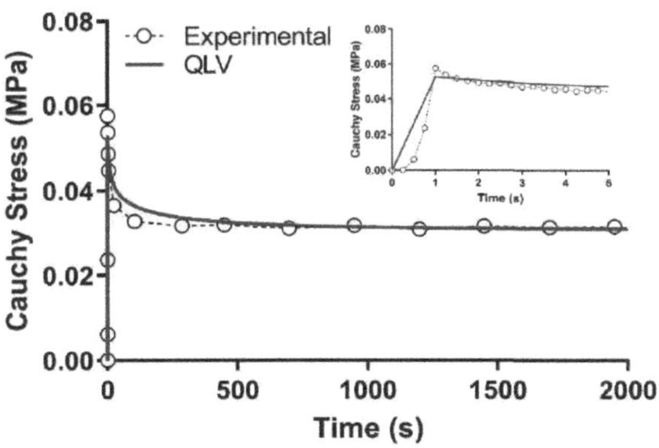

Fig. 13.8 Fung QLV model fit on typical stress relaxation response of aortic tissue. Inset displays first 5 seconds of the test.

13.4 Discussion

Important research into aortic tissue hyperelasticity characterization has been previously carried out (Avril, 2017; Avril et al, 2010). Avril et al (2010) established hyperelastic material parameters of human arteries using full-field experimental data from inflation/extension tests and the virtual fields method (VFM). The VFM was shown to be successful in identifying hyperelastic properties and as such may be considered an alternative to the incremental modulus method used in this study.

A seminal study by Holzapfel and Gasser (2001) presented a viscoelastic model for investigating three-dimensional stress and deformation response of fiber-reinforced composites, such as arterial walls. Here, using a particular anisotropic Helmholtz free-energy function, three components of the wall were modeled (a matrix material and two families of fibers), all of which were considered to have viscoelastic behaviour. Implementation of the model constitutive equations in numerical simulation showed good qualitative agreement. As such, future research efforts may benefit from using such a model opposed to the simplified viscoelastic characterisation used in this study. Additionally, It should be noted that other wall constituents can also affect the tissue viscoelasticity, however further studies are required to establish the extent of this.

While energy loss is a robust parameter and hysteresis loops are observed *in vivo*, it provides limited information of the time-dependent behaviour of aortic tissue under a peak load. In standard biaxial tests, unloading is performed at a controlled rate. In physiological loading conditions, the aortic wall unloading is determined by its previous distension, heart rate, and smooth muscle activation. For this reason, stress relaxation tests were performed in this study to assess viscoelastic behaviour over a time span, where stress decrease is achieved by the material alone. The correlation between SMC content and energy loss presented in this study ($R^2 = 0.3519$) was affected by the relatively large error bars on the data points, as seen in Figure 1.4. It should be noted, however, if using mean data values, a stronger correlation would be observed.

As previously mentioned, multiple stress relaxation cycles preconditioned the material to achieve repeatable curves. Holzapfel (2009) noted that Cauchy stress-stretch hysteresis loops during uniaxial preconditioning of muscular artery tissue shift toward a larger deformation, before converging to a certain deformation with smaller hysteresis. Here, Holzapfel (2009) found 5 loading-unloading cycles to be sufficient for convergence, which supports the 10 cycle approach used in our study. Conversely, Carew et al (2004) determined that repeatable stress relaxation curves in porcine aortic valves cannot be achieved by standard loading-unloading preconditioning. Zou and Zhang (2011) addressed this by compensating for drops in initial stress level (in aortic elastin), after each cycle by modifying the target stretch at each cycle. In the present study, the materials initial stress level dropped significantly after the first cycle, just as it does in standard preconditioning protocols. However, the maximum stretch was consistent through all cycles. Excised aorta achieved up to 50% relaxation after 5 cycles, which is substantially larger than the 20% relaxation observed by Zou and Zhang (2011). This may be due to the strain-rate dependence of viscoelasticity in soft tissues (Doehring et al, 2004). In the case of the present study, samples achieved target stretch in half of the total preconditioning time.

Histological examination revealed noticeable disparities between healthy and aneurysmal tissue. Although only 5 samples were used, increased elastin and energy loss were also observed to be correlated. At low stress levels, elastin fibers sustain more load than collagen fibers. Elastin viscoelasticity, while minimal, can contribute more to the overall viscous behaviour of the material at these stress levels. So combining SMC and elastin effects could reinforce the viscous assessment. It should be noted that this study examined tissue under controlled strain rate, as opposed to controlled stress value. It was also observed that SMCs in tissue with a lower collagen-to-elastin (C/E) ratio may hold higher integrity than those with higher C/E ratio. This may be the result of high stresses caused by increased collagen content causing degradation of SMC relaxation properties over time. Conversely, it may be a result of elastin degradation increasing load-bearing responsibility of SMCs and also inducing cell fatigue. However, it is important to note that SMC content is not believed to be the sole determinant of arterial viscoelastic properties.

Increasing the amount of tissue samples would help to account for variability in SMC content of aortic tissue. Furthermore, measuring the distance from the aortic valve and accounting for regional location of the tested sample may help illuminate reasons for potential variation across aortic tissue. A more sophisticated closed-loop system that accounts for transmural pressure, characteristic impedance, and physiological frequency can be replicated with a pulsatile blood pump flowing through a conduit.

It should be noted that this study revealed only normalized stress relaxation values. Circumferential stretch during the stress relaxation tests were 50% and 10% for aortic tissue and Dacron, respectively. These different stretch values need to be considered when interpreting the results. One study however, by Lee and Wilson (1986), determined stress relaxation behaviour of arterial tissue and vascular grafts by accounting for equal load, as opposed to strain. By doing so it was found that Dacron grafts displayed much more relaxation than iliac arteries.

Recent investigation of woven Dacron aims to explain the role of viscoelasticity on *in vivo* dynamics (Amabili et al, 2018; Tubaldi et al, 2018). Understanding of Dacron graft viscoelasticity and its unique effect on hemodynamics is limited and requires further study. Dacron grafts expand after initial placement in the thoracic aorta with a distinct early and late growth rate, mimicking a typical creep response of polymers. In the ascending aorta, grafts dilate 17% within a week following implantation, with smaller diameter grafts experiencing higher diameter increase (Etz et al, 2007). While the graft seems to adapt to the aorta geometry and hemodynamic demands through structural changes, its pre-implant circumferential mechanical properties are maintained. This was evidenced in a case study of a woven double-velour Dacron graft explanted after 27 years (Nagano et al, 2007).

From our obtained results, the measured aortic tissue and Dacron incremental modulus values were observed to be consistent with previous literature (Tremblay et al, 2009; Emmott et al, 2017). In addition, the incremental modulus of Dacron at 45% strain was observed to correlate with the physiological moduli at the transition between elastin and collagen load-bearing in rat aortic tissue (Danpinid et al, 2010). The dependence of viscosity on stress level, as seen in this study, may account for the increased energy loss in Dacron, which achieved supraphysiological loads at 60% strain. Dacron's large increase in stress levels at the same strain may explain our observed proportionately higher viscosity when compared to aortic tissue. Azuma and Hasegawa (1971) verified that Dacron grafts stretched to 10% matched the viscoelasticity of the proximal aorta when stretched to 50%. Furthermore, Amabili et al (2018) found that loss factor decreased with pre-stretching of Dacron. Their achieved magnitude and frequency response was similar to that achieved by the aortic tissue samples in this study. In their study, Amabili et al (2018) used a dynamic amplitude of 0.3% strain compared to the 25% strain used in this study.

A limitation of this study was the use of only one method for hyperelasticity characterization. It has been shown that classical hyperelasticity models produce different behaviour responses for living tissues (Holzapfel and Gasser, 2001; Avril, 2017; Avril et al, 2010) and as such future studies should investigate the behavioural differences between models for Dacron. In one case, a similar studies involving prosthetic material design for cardiovascular application found the Mooney–Rivlin hyperelastic model to be successful for characterizing a range of soft tissues (Mohammadi et al, 2009).

Another limitation is the use of the Kelvin–Voigt viscoelastic material model. The Kelvin–Voigt model is the simplest viscoelastic model capable to recover the creep and relaxation of soft tissue (1 spring and 1 dashpot in parallel). A more detailed alternative model would be the Zener model, which uses an additional spring. However, under low frequency (heart beat), the creep response of the Zener model has been shown to be very similar to that of the Kelvin–Voigt model (Bronshtein et al, 2013), but is slightly better at higher frequencies. As such, the use of the Kelvin–Voigt model is justified in this study, however, future studies using more complex loading assignments may wish to use a less simplified model. While we made use of the Fung QLV model, the Kelvin–Voigt model was found to be more practical in terms of practicality.

The parameter C of the QLV model represents a measure of viscosity within the material and τ_1 and τ_2 represent the immediate and long-term time constants, respectively. The QLV model results showed that Dacron holds the highest initial relaxation rate and the lowest normalized level of remaining stress, resulting in the largest estimation of C. Dacron also reaches its final value much quicker, confirmed by the shorter τ_2 value. While the QLV model has been widely used (Carew et al, 1999; Doehring et al, 2004; Kohandel et al, 2008), it improperly assumes that strain and time effects on stress are independent. For this reason, it insufficiently simulates the magnitude of initial stress relaxation during biaxial testing. In addition, the model is inadequate in predicting the stress levels during the ramp phase of biaxial testing, but is however able to recover in predicting the long-term relaxation. This is evidenced by the increasing residuals level during the first second, displayed in Fig. 13.8. The model overestimates the material's elasticity, resulting in an underestimation of viscosity. This effect is more pronounced in materials with high elastic modulus and viscosity, such as Dacron.

The established Dacron energy loss values do not resemble *in vivo* behaviour, but do however further attest to its significantly lower compliance in comparison to aortic tissue.

13.5 Conclusion

The SMC content of aortic tissue carries a certain fraction of the aortic walls viscosity as determined by energy loss mechanical testing.

In addition, an evident positive correlation was found between patient age and energy loss. An acceptable positive correlation was found between SMC content and energy loss, which suggests that SMC content alone, is not an effective predictor of the viscous parameter of aortic tissue.

Aortic tissue increases in stiffness as it is stretched. This hyperelastic characteristic is similar in Dacron, although its magnitude of stiffness is much higher. Furthermore, the compliance of this artificial tissue matches that of the aortic wall.

Current synthetic Dacron material has intrinsic stiffness and viscoelastic properties that are significantly different from biological tissue. Improvement in terms of biomechanical integration would require adjustments of the viscoelastic properties to achieve closer properties to native tissue.

Biaxial tensile tests are simple and reproducible, which help in providing a standard for material characterization. However they are limited by the fact that they do not mimic the conditions that the aorta endures *in vivo*. Stress relaxation and dynamic mechanical analysis incorporate time-dependent behaviours, but are still insufficient.

References

Amabili M, Balasubramanian P, Breslavsky I, Ferrari G, Tubaldi E (2018) Viscoelastic characterization of woven dacron for aortic grafts by using direction-dependent quasi-linear viscoelasticity. Journal of the Mechanical Behavior of Biomedical Materials 82:282–290

Armentano R, Barra J, Levenson J, Simon A, Pichel R (1995) Arterial wall mechanics in conscious dogs: Assessment of viscous, inertial, and elastic moduli to characterize aortic wall behavior. Circulation Research 76(3):468–478

Avril S (2017) Material Parameter Identification and Inverse Problems in Soft Tissue Biomechanics, vol 573, Springer, Saint-Etienne, chap Hyperelasticity of Soft Tissues and Related Inverse Problems, pp 37–66

Avril S, Badel P, Duprey A (2010) Anisotropic and hyperelastic identification of *in vitro* human arteries from full-field optical measurements. Journal of Biomechanics 43(15):2978–2985

Azadani A, Chitsaz S, Mannion A, Mookhoek A, Wisneski A, Guccione J, Hope M, Ge L, Tseng E (2013) Biomechanical properties of human ascending thoracic aortic aneurysms. The Annals of Thoracic Surgery 96(1):50–58

Azuma T, Hasegawa M (1971) A rheological approach to the architecture of arterial walls. The Japanese Journal of Physiology 21:27–47

Babu A, Byju A, Gundiah N (2015) Biomechanical properties of human ascending thoracic aortic dissections. Journal of Biomechanical Engineering 137(8):0810,131–0810,139

Bronshtein T, Au-Yeung G, Sarig U, Bao-Vi N, Mhaisalkar P, Yin C, Venkatraman S, Machluf M (2013) A mathematical model for analyzing the elasticity, viscosity, and failure of soft tissue: Comparison of native and decellularized porcine cardiac extracellular matrix for tissue engineering. Tissue Engineering: Part C 19(8):620–630

Carew E, Talman E, Boughner D, Vesely I (1999) Quasi-linear viscoelastic theory applied to internal shearing of porcine aortic valve leaflets. Journal of Biomechanical Engineering 121(4):386–392

Carew E, Garg A, Barber J, Vesely I (2004) Stress relaxation preconditioning of porcine aortic valves. Annals of Biomedical Engineering 32(4):563–572

Chung J, Lachapelle K, Wener E, Cartier R, Varennes B, Fraser R, Leask R (2014) Energy loss, a novel biomechanical parameter, correlates with aortic aneurysm size and histopathologic findings. The Journal of Thoracic and Cardiovascular Surgery 148(3):1082–1088

Chung J, Lachapelle K, Cartier R, Mongrain R, Leask R (2017) Loss of mechanical directional dependency of the ascending aorta with severe medial degeneration. Cardiovascular Pathology 226:45–50

Danpinid A, Luo J, Vappou J, Terdtoon P, Konofagou E (2010) *In Vivo* characterization of the aortic wall stress-strain relationship. Ultrasonics 50(7):654–665

Debes J, Fung Y (1995) Biaxial mechanics of excised canine pulmonary arteries. American Journal of Physiology-Heart and Circulatory Physiology 269(2):433–442

Doehring T, Carew E, Vesely I (2004) The effect of strain rate on the viscoelastic response of aortic valve tissue: A direct-fit approach. Annals of Biomedical Engineering 32(2):223–232

Emmott A, Garcia J, Chung J, Lachapelle K, El-Hamamsy I, Mongrain R, Cartier R, Leask R (2016) Biomechanics of the ascending thoracic aorta: A clinical perspective on engineering data. Canadian Journal of Cardiology 32(1):35–47

Emmott A, El-Hamamsy I, Leask R (2017) Case report: Histopathological and biomechanical properties of the aortic wall in 2 patients with chronic type a aortic dissection. Cardiovascular Pathology 29:48–52

Etz C, Homann T, Silovitz D, Bodian C, Luehr M, Di Luozzo G, Plestis K, Griepp R (2007) Vascular graft replacement of the ascending and descending aorta: Do dacron grafts grow? The Annals of Thoracic Surgery 84(4):1206–1213

Fung Y (1972) Stress-strain history relations of soft tissues in simple elongation. Englewood

Fung Y (1993) Biomechanics: mechanical properties of living tissues. Springer-Verlag, New York

Geest J, Sacks M, Vorp D (2006) The effects of aneurysm on the biaxial mechanical behavior of human abdominal aorta. Journal of Biomechanics 39(7):1324–1334

Goto M, Kimoto Y (1966) Hysteresis and stress-relaxation of the blood vessels studied by a universal tensile testing instrument. Japanese Journal of Physiology 16(2):169–184

Hasegawa M, Azuma T (1979) Mechanical properties of synthetic arterial grafts. Journal of Biomechanics 12(7):509–517

Holzapfel G (2009) Biomechanical Modelling at the Molecular, Cellular and Tissue Levels, vol 508, CISM International Centre for Mechanical Sciences, Springer, Vienna, chap Arterial Tissue in Health and Disease: Experimental Data, Collagen-Based Modeling and Simulation, Including Aortic Dissection, pp 259–344

Holzapfel G, Gasser T (2001) A viscoelasic model for fiber-reinforced composites at finite strains: Continuum basis, computational aspects and applications. Computer Methods in Applied Mechanics and Engineering 190(34):4379–4403

Imura T, Yamamoto K, Satoh T, Kanamori K, Mikami T, Yasuda H (1990) *In Vivo* viscoelastic behavior in the human aorta. Circulation Research 66(5):1413–1419

Kohandel M, Sivaloganathan S, Tenti G (2008) Estimation of the quasi-linear viscoelastic parameters using a genetic algorithm. Mathematical and Computer Modelling 47(3):266–270

Lee J, Wilson G (1986) Anisotropic tensile viscoelastic properties of vascular graft materials tested at low strain rates. Biomaterials 7(6):423–431

Maton A, Hopkins R, McLaughlin C, Johnson D, Warner S, LaHart D, Wright J (1993) Human Biology and Health. Prentice Hall, New Jersey, USA

Mohammadi H, Boughner D, Millon L, Wan W (2009) Design and simulation of a poly(vinyl alcohol) bacterial cellulose nanocomposite mechanical aortic heart valve prosthesis. Proceedings of the Institution of Mechanical Engineers, Part H: Journal of Engineering in Medicine 223(6):697–711

Mohan D, Melvin J (1983) Failure properties of passive human aortic tissue. ii-biaxial tension tests. Journal of Biomechanics 16(1):39–44

Nagano N, Cartier T, Zigras T, Mongrain R, Leask R (2007) Mechanical properties and microscopic findings of a dacron graft explanted 27 years after coarctation repair. The Journal of Thoracic and Cardiovascular Surgery 134(6):1577–1578

Okamoto R, Wagenseil J, DeLong W, Peterson S, Kouchoukos N, Sundt T (2002) Mechanical properties of dilated human ascending aorta. Annals of Biomedical Engineering 30(5):624–635

Puskas J, Chen Y (2004) Biomedical application of commercial polymers and novel polyisobutylene-based thermoplastic elastomers for soft tissue replacement. Biomacromolecules 5(4):1141–1154

Remington J (1954) Hysteresis loop behavior of the aorta and other extensible tissues. American Journal of Physiology 180(1):83–95

Rhodin J (1980) Architecture of the vessel wall, handbook of physiology, the cardiovascular system. The American Physiological Society 2:1–31

Roach M, Burton A (1957) The reason for the shape of the distensibility curves of arteries. Canadian Journal of Biochemistry and Physiology 35(8):681–690

Shahmansouri N, Alreshidan M, Emmott A, Lachapelle K, El-Hamamsy I, Cartier R, Leask R, Mongrain R (2016) Investigation on the regional loss factor and its anisotropy for aortic aneurysmns. Materials 9(11):1–16

Stefanadis C, Stratos C, Vlachopoulos C, Marakas S, Boudoulas H, Kallikazaros I, Tsiamis E, Toutouzas K, Sioros L, Toutouzas P (1995) Pressure-diameter relation of the human aorta. Circulation 92(8):2210–2219

Tremblay D, Zigras T, Cartier R, Leduc L, Butany J, Mongrain R, Leask R (2009) A comparison of mechanical properties of materials used in aortic arch reconstruction. The Annals of Thoracic Surgery 88(5):1484–1491

Tubaldi E, Païdoussis M, Amabili M (2018) Nonlinear dynamics of dacron aortic prostheses conveying pulsatile flow. Journal of Biomechanical Engineering 140(6):1–12

Walden R, Gilbert J, Megerman J, Abbott W (1980) Matched elastic properties and successful arterial grafting. Archives of Surgery 115(10):1166–1169

Westerhof N, Noordergraaf A (1970) Arterial viscoelasticity: A generalized model. effect on input impedance and wave travel in the systematic tree. Journal of Biomechanics 3(2):357–370

Zikry C (2018) Evaluation of poly(vinyl alcohol) cryogel as viscoelastic reconstruction graft for the ascending aorta. Master's thesis, Department of Bioengineering, McGill University, Montreal, Canada

Zou Y, Zhang Y (2011) The orthotropic viscoelastic behavior of aortic elastin. Biomechanics and Modeling in Mechanobiology 10(5):613–625

Part II
Metamaterials

Chapter 14
Plane Waves Transmission and Reflection at the Interface between Thermoelastic Continua in Absence of Dissipation: The Influence of Magnetic Field and Rotation

Abdelmoody M. Abd-Alla, Said M. Abo-Dahab, Abo-el-nour N. Abd-alla, and Mohamed Elsagheer

Abstract In the present contribution, it is addressed the problem of plane wave reflection and transmission at the imperfect interface between thermoelastic half-spaces having different properties when the effect of magnetic and rotation fields is relevant. Using Green and Naghdi theory, we formulate the needed governing equations for thermoelastic bodies occupying both an half space under the assumption that each of them is homogeneous and isotropic. We could solve in a closed form the problem of planar waves whose propagation is determined by postulated balance equations. The amplitude of transmitted and reflected waves are obtained so that we can prove that one can observe three kinds of waves, namely, $P-$waves, $T-$waves and $SV-$waves. Our closed form solution allows us to determine the amplitude ratios between reflected and refracted waves also in the case of imperfect boundary. In particular, we obtain that these amplitude ratios depend on the elastic stiffness, the magnetic field and on the thermal properties of the considered bodies. Also the case of stress-free boundary has been considered. The effect of the magnetic and rotation fields is investigated via suitable numerical computations. We observe that rotation and magnetic fields may induce relevant phenomenology.

Keywords: Magneto-thermoelasticity · Thermal stress · Energy dissipation · Reflection and Transmission · Magnetic field · G-N model

---

A. M. Abd-Alla, A. N. Abd-alla, M. Elsagheer
Math. Dept., Faculty of Science, Sohag University, Egypt
e-mail: mohmrr@yahoo.com, aboelnourabdalla@yahoo.com, elsagheer_mohamed@yahoo.com

S. M. Abo-Dahab
Math. Dept., Faculty of Science, Taif University, Saudi Arabia
Math. Dept., Faculty of Science, SVU, Qena, Egypt
e-mail: sdahb@yahoo.com

B. E. Abali and I. Giorgio (eds.), *Developments and Novel Approaches in Biomechanics and Metamaterials*, Advanced Structured Materials 132,
https://doi.org/10.1007/978-3-030-50464-9_14

Nomenclature

α_t is the coefficient of linear thermal expansion,
δ_{ij} is the kroneckar delta,
λ, μ are the Lame's constants (material constants),
μ_e is the magnetic permeability,
$\nu = \alpha_t\,(3\lambda + 2\mu)$ is the thermal parameter
ρ is the density of the medium,
σ_{ij} are the components of stress tensor,
τ_{ij} are the components of Maxwell's stress tensor,
∇^2 is the Laplace operator,
ω is the frequency,
b_c is heat transfer coefficient at the interface,
C_e is the specific heat per unit mass,
$\vec{E}$ is the electric intensity vector,
$\vec{H}$ is the magnetic field vector,
$\vec{J}$ is the electric current density vector,
K is the thermal conductivity,
K_c is the thermal contact conductance with dimension,
K_n, K_t are normal and transverse stiffness coefficients of a unit layer thickness,
t is the time,
T is the absolute temperature,
T_0 is the initial temperature, $\left|\dfrac{T - T_0}{T_0}\right| \ll 1$,
u, v are the components of displacement vector $\vec{u}$.

14.1 Introduction

The recent push towards the formulation of generalized continuum models (Altenbach et al, 2010; Eremeyev and dell'Isola, 2018; Piccardo et al, 2014; Andreaus et al, 2018; Barchiesi et al, 2019a; Eremeyev and Pietraszkiewicz, 2012; Eremeyev et al, 2018; Abdoul-Anziz and Seppecher, 2018) has been motivated by the need of supplying a solid theoretical ground for the design (dell'Isola et al, 2015b; Eugster et al, 2019; Pideri and Seppecher, 1997; Placidi, 2015; Alibert et al, 2003; Bouchitté et al, 2019; Camar-Eddine and Seppecher, 2001, 2003; dell'Isola and Placidi, 2011; Spagnuolo and Andreaus, 2019; Giorgio, 2020; Baroudi et al, 2019; Eremeyev et al, 2019) and development of multiphysics multiscale metamaterials (Barchiesi et al, 2019b; dell'Isola et al, 2019a,b; Barchiesi and Placidi, 2017; Carcaterra et al, 2015; De Angelo et al, 2019b; Abali and Zohdi, 2020). Generalized continua were formulated already by Piola (dell'Isola et al, 2015a, 2014; Turco et al, 2018), but only recently attracted again the interest of scholars in Mechanics (dell'Isola et al, 2015c; De Angelo et al, 2020; Del Vescovo and Giorgio, 2014; Eugster et al, 2019; Misra et al, 2018; Spagnuolo et al, 2019; Turco et al, 2017; Altenbach and Eremeyev, 2009, 2010, 2011; Altenbach et al, 2015; Eremeyev and Pietraszkiewicz, 2016). However up to now it has been neglected the possibility of exploiting multiphysics phenomena for getting exotic behaviors at macro level (Andreaus et al, 2004; Batra et al, 1998; dell'Isola and Rosa, 1997; dell'Isola et al, 2003, 2004; Maurini et al, 2004b,a; Giorgio et al, 2009, 2015; Eremeyev, 2016, 2018; Abali and Queiruga, 2019). Moreover, the recent theories of thermoelasticity, in which the thermal signal has a finite speed, have attracted more and more often intense research. These theories, based on the seminal paper by Cattaneo (1948), are formulated to avoid the paradox of infinite speed of heat signals, which instead, is present in the standard thermoelasticity theory (Biot, 1956). Lord and Shulman (1967) postulated a model where a finite speed is obtained using one relaxation time only. Green and Lindsay (1972) assumed in the constitutive equations a temperature rate, but at the cost of loosing the classical Fourier's law for heat conduction in the case when the material considered has a center of symmetry. This last theory manages to predict a finite speed of heat propagation but introducing two relaxation times. A specific thermoelastic problem in an infinite cylinder in presence of initial stress is considered in El-Naggar and Abd-Alla (1987). Green and Lindsay (1972); Green and Naghdi (1991, 1992, 1993); Chandrasekharaiah (1996a) have been formulated three other models using alternative approaches.

In (Chandrasekharaiah, 1996b) it is considered the propagation of plane harmonic waves in absence of dissipation. Moreover, in (Chandrasekharaiah and Srinath, 1997a,b) it is dealt with cylindrical/spherical waves originated by (i) loads applied at the boundary of considered cylindrical/spherical hole surrounded by an infinite body, (ii) line/point sources of heat in an infinite body. In the paper by Mukhopadhyay (2002) one finds the solution of a problem involving thermoelastic interactions in absence of dissipation when the considered body is infinite and a spherical cavity is present. A thermal shock is initiated in the cavity.

On the other hand, Lavrentyev and Rokhlin (1988) studied wave transmission and reflection in the case of imperfect boundary conditions at the interface dividing two elastic solid bodies occupying half spaces. Othman and Song (2007) have investigated a reflection phenomena of the plane waves from a homogeneous thermoelastic solid half-space with hydrostatic initial stress without energy dissipation. When the two half spaces are filled with metamaterials then the reflection and transmission problem deserves a particular interest especially because closed form solutions in the linear case allow for parametric optimization of metamaterial parameters (Placidi et al, 2018b,a, 2019; Placidi and Barchiesi, 2018; Placidi et al, 2017; De Angelo et al, 2019a; Scerrato and Giorgio, 2019; Turco et al, 2020). Of course the investigations about multiphysics metamaterials are not yet so advanced as those for the mechanical metamaterials (dell'Isola et al, 2003; Maurini et al, 2006; Porfiri et al, 2004; Rosi et al, 2010; Shen et al, 2010). We therefore believe that the present results may suggest new design solutions in metamaterial theory (dell'Isola and Steigmann, 2020; Spagnuolo et al, 2020a; Yildizdag et al, 2019; Abali and Reich, 2018, 2017).

In this context, magneto-thermo-elasticity, that models the coupling phenomena between deformation, heat conduction and electro-magnetism, may deserve the attention of those scholars studying the design of novel metamaterials (dell'Isola and Steigmann, 2020; Spagnuolo et al, 2020b; Spagnuolo, 2020; Spagnuolo et al, 2017). These novel applications may be added to those already considered in literature. A short list of them may include:

i. Geophysics: in particular, for the description of the interaction between seismic waves and earth magnetic field;
ii. Acoustics: in particular, for studying how magnetic fields may damp acoustic waves;
iii. Nuclear devices design: for studying the most efficient and highly sensitive superconducting magnetometer;
iv. Superconductivity: for studying the effects of deformation in superconductors;
v. Electrical power engineering: for obtaining design specification of high power transmission lines.

The present work develops some ideas already presented in the literature. In particular, the reflection of plane waves in magneto-thermo-viscoelasticity has been studied in Abd-Alla and Mahmoud (2013), while coupled magneto-thermo-viscoelastic phenomena have been studied, in an infinite body and in presence of a spherical cavity by Abd-Alla et al (2004), in presence of periodic loading the cavity. The seminal papers by Deresiewicz (1960, 1962) showed how to study the reflection and transmission of plane waves in thermoelasticity: his studies are a benchmark in the researches concerning the effects of interface properties on reflection and transmission. When one relaxation time is introduced in heat conduction the reflection and propagation results for planar waves were found by Sinha and Sinha (1974); Beevers and Bree (1975) . Instead, Sinha and Elasibai (1995, 1996) considered the case of two relaxation times. In (Singh and R., 1998; Singh and Kumar, 2003; Singh, 2003) the reflection and transmission at a plane boundary dividing micropolar and microstretched thermoelastic infinite half spaces has been investigated, in the same spirit as Placidi et al (2014); dell'Isola et al (2012). Finally, Singh (2005) studied the reflection of $P-$ and $SV-$waves at the free surface of a body where thermo diffusion occurs.

Recently, Kumar and Sarathi (2006) investigated the reflection and refraction of thermoelastic plane waves at an interface between two thermoelastic media without energy dissipation. Effect of rotation and imperfection on reflection and transmission of plane waves in an isotropic generalized thermoelastic media illustrated by Kumar and Singh (2009). Zhou et al (2012) studied the reflection and transmission of plane waves at the interface of pyroelectric bi-materials. Abd-Alla and Mahmoud (2010) investigated the magneto-thermoelastic problem in rotating non-homogeneous orthotropic hollow cylinder under the hyperbolic heat conduction model. Abd-Alla et al (2012b) studied the problem of transient coupled thermoelasticity of an annular fin. Abd-Alla et al (2011) discuss the generalized magneto-thermoelastic Rayleigh waves in a granular medium under influence of the gravity field and initial stress. Abd-Alla and Mahmoud (2013) investigated the problem of radial vibrations in non-homogeneity isotropic cylinder under influence of initial stress and magnetic field. Abd-Alla et al (2012a) studied the propagation of Rayleigh waves in magneto-thermo-elastic half–space of a homogeneous orthotropic material under the effect of the rotation, initial stress and gravity field.

In this paper, the reflection and transmission phenomena of thermoelastic plane waves at an imperfect interface between two dissimilar thermoelastic solid half-spaces with magnetic field and the rotation has been discussed. The basic governing equations for isotropic and homogeneous thermoelastic half-space are formulated in the context of the Green and Naghdi theory. These governing equations are solved analytically to obtain the amplitude of reflected waves in an $xz-$plane. It is shown that there exist three plane waves, namely, a $P-$wave, $T-$wave and $SV-$wave. The amplitude ratios of various reflected and refracted waves are investigated for an imperfect boundary. Some special cases of the problem are discussed. It is found that the amplitude ratios of various reflected and refracted waves are affected by the stiffness, magnetic field and thermal properties of the media. The amplitude ratios of reflected waves are also deduced for a special case of stress-free boundary. Numerical computations are carried out and comparisons made with the results predicted in the presence and absence of magnetic field and rotation. The results obtained calculated numerically. Some special cases have been deduced from this work. Finally, the results obtained are displayed graphically.

14.2 Formulation of the Problem

Le us consider two homogeneous isotropic thermoelastic solids without energy dissipation being in contact with each other at a plane surface, which we denote as the plane $z = 0$ of a rectangular coordinate system axis. We consider thermoelastic plane waves in the $xz-$plane with wave front parallel to $y - axis$ and all the field variables depend only on x, z and t taking into consideration Green and Naghdi (1993) theory, the field equations of the thermoelastic solid without energy dissipation in the absence of body forces and heat sources and the constitutive relations can be written as the following:

$$\sigma_{ij} = (\lambda e_{kk} - \nu T)\,\delta_{ij} + 2\mu e_{ij}, \quad (i, j = 1, 2, 3) \tag{14.1}$$

The field equation of the thermoelastic solid with the energy we can write the glop equation in elastic with body force and heat source and constitutive relation can be written as the following:

$$(\lambda + 2\mu)\,\nabla\left(\vec{\nabla}.\vec{u}\right) - \mu\vec{\nabla}\times\vec{\nabla}\times\vec{u} - \nu\vec{\nabla}T + \vec{F} = \rho\left[\frac{\partial^2 u}{\partial t^2} + \left(\vec{\Omega}\times\vec{\Omega}\times\vec{u}\right)\right] \tag{14.2}$$

where $\vec{\Omega}\times\vec{\Omega}\times\vec{u}$ is the centripetal acceleration due to the time varying motion only.

Let us consider the medium is a perfect electric conductor and the linearized Maxwell's equations governing the electromagnetic field, in the absence of the displacement current (SI) in the form:

$$\begin{aligned} &\vec{J} = \operatorname{curl} \vec{h}, -\mu_e \frac{\partial \vec{h}}{\partial t} = \operatorname{curl} \vec{E}, \\ &\operatorname{div} \vec{h} = 0, \ \operatorname{div} \vec{E} = 0, \\ &E = -\mu_e \left(\frac{\partial \vec{u}}{\partial t} \times \vec{H} \right). \end{aligned} \tag{14.3}$$

The Maxwell's electromagnetic stress tensor τ_{ij} is given by

$$\tau_{ij} = \mu_e \left[H_i h_j + H_j h_i - (H_k h_k) \delta_{ij} \right], \ \ (i, j = 1, 2, 3) \tag{14.4}$$

The Lorentz's force given by:

$$\vec{F} = \mu_e H_0 \vec{\nabla} \left(\vec{\nabla} . \vec{u} \right) \tag{14.5}$$

We can write equation 14.2 as

$$(\lambda + 2\mu + \mu_e H_0) \nabla \left(\vec{\nabla} . \vec{u} \right) - \mu \vec{\nabla} \times \vec{\nabla} \times \vec{u} - \nu \vec{\nabla} T = \rho \left[\frac{\partial^2 u}{\partial t^2} + \left(\vec{\Omega} \times \vec{\Omega} \times \vec{u} \right) \right] \tag{14.6}$$

The temperature equation takes the following:

$$K \nabla^2 T = \rho C_\nu \frac{\partial^2 T}{\partial t^2} + \nu T_0 \frac{\partial^2}{\partial t^2} \left(\vec{\nabla} . \vec{u} \right) \tag{14.7}$$

For a two dimensional problem, the displacement vector $\vec{u}$ is taken by

$$\vec{u} = (u, 0, w) \tag{14.8}$$

Let us consider the displacement vector is related with the potential function

$$\vec{u} = \operatorname{grad} \phi + \operatorname{curl} \psi, \qquad \operatorname{div} \psi = 0. \tag{14.9}$$

where, the displacement components u and w take the form:

$$u = \frac{\partial \phi}{\partial x} - \frac{\partial \psi}{\partial z}, \qquad w = \frac{\partial \phi}{\partial z} + \frac{\partial \psi}{\partial x} \tag{14.10}$$

From Eq. (14.9) we get

$$\begin{aligned} &\nabla \cdot u = \frac{\partial^2 \phi}{\partial x^2} + \frac{\partial^2 \phi}{\partial z^2} = \nabla^2 \phi, \qquad \nabla^2 u = \Delta_1 \left(\nabla^2 \phi \right) + \Delta_2 \left(\nabla^2 \psi \right), \\ &\frac{\partial^2 \vec{u}}{\partial t^2} = \Delta_1 \frac{\partial^2 \phi}{\partial t^2} + \Delta_2 \frac{\partial^2 \psi}{\partial t^2}. \end{aligned} \tag{14.11}$$

where

$$\Delta_1 = \left(\frac{\partial}{\partial x}, 0, \frac{\partial}{\partial z} \right), \qquad \Delta_2 = \left(-\frac{\partial}{\partial z}, 0, \frac{\partial}{\partial x} \right), \tag{14.12}$$

and we let the absolute temperature $T = \Theta$.

Substituting from Eqs. (14.10) and (14.11) into Eq. (14.6), we obtain

$$\begin{aligned}\left(\lambda+\mu+\mu_e H_0^2\right)\nabla\left(\nabla^2\phi\right)+\mu\left[\Delta_1\left(\nabla^2\phi\right)+\Delta_2\left(\nabla^2\psi\right)\right]-\nu\nabla\Theta\\ =\rho\left[\Delta_1\frac{\partial^2\phi}{\partial t^2}+\Delta_2\frac{\partial^2\psi}{\partial t^2}+\left(\vec{\Omega}\times\vec{\Omega}\times\vec{u}\right)\right]\end{aligned} \tag{14.13}$$

which reduces to the following two equations as

$$\begin{aligned}\left(\lambda+\mu+\mu_e H_0^2\right)\left(\nabla^2\phi\right)-\nu\Theta=\rho\left[\frac{\partial^2\phi}{\partial t^2}-\Omega^2\phi\right]\\ \mu\nabla^2\psi=\rho\left[\frac{\partial^2\psi}{\partial t^2}-\Omega^2\psi\right]\end{aligned} \tag{14.14}$$

and Eq. (14.7) reduce to

$$K\nabla^2\Theta=\rho C_\nu\frac{\partial^2\Theta}{\partial t^2}+\nu T_0\frac{\partial^2}{\partial t^2}\nabla.\phi \tag{14.15}$$

Equations (2.12)-(2.14) take the following forms:

$$\begin{aligned}c_1^2\nabla^2\phi-\overline{\nu}\Theta&=\frac{\partial^2\phi}{\partial t^2}-\Omega^2\phi,\\ c_2^2\nabla^2\psi&=\frac{\partial^2\psi}{\partial t^2}-\Omega^2\psi,\\ \overline{K}\nabla^2\Theta&=C_\nu\frac{\partial^2\Theta}{\partial t^2}+\overline{\nu}T_0\frac{\partial^2}{\partial t^2}\nabla^2\phi.\end{aligned} \tag{14.16}$$

where

$$c_1^2=\frac{\lambda+\mu+\mu_e H_0^2}{\rho},\qquad c_1^2=\frac{\mu}{\rho},\qquad \overline{\nu}=\frac{\nu}{\rho},\qquad \overline{K}=\frac{K}{\rho}. \tag{14.17}$$

Operating by ∇^2 on the Eq. (14.16) we get

$$\nabla^4\phi-\frac{\overline{\nu}}{c_1^2}\nabla^2\Theta=\frac{1}{c_1^2}\left[\frac{\partial^2}{\partial t^2}\nabla^2\phi-\Omega^2\nabla^2\phi\right] \tag{14.18}$$

Substituting from Eq. (**??**) into Eq. (14.18) we get

$$\nabla^4\phi-\frac{\overline{\nu}}{c_1^2}\nabla^2\left[\frac{C_\nu}{\overline{K}}\frac{\partial^2\Theta}{\partial t^2}+\frac{\overline{\nu}T_0}{\overline{K}}\frac{\partial^2}{\partial t^2}\nabla^2\phi\right]=\frac{1}{c_1^2}\left[\frac{\partial^2}{\partial t^2}\nabla^2\phi-\Omega^2\nabla^2\phi\right] \tag{14.19}$$

Simplify Eq. (14.19) we get

$$\nabla^4\phi-\nabla^2\left[\frac{1}{c_1^2}\frac{\partial^2\phi}{\partial t^2}-\frac{\Omega^2}{c_1^2}\phi+\frac{\overline{\nu}^2T_0}{c_1^2\overline{K}}\frac{\partial^2}{\partial t^2}\nabla^2\phi\right]=\frac{\overline{\nu}C_\nu}{c_1^2\overline{K}}\frac{\partial^2\Theta}{\partial t^2} \tag{14.20}$$

Operating by $\frac{\partial^2}{\partial t^2}$ on the Eq. (14.16) we get

$$\overline{\nu}\frac{\partial^2 \Theta}{\partial t^2} = c_1^2\frac{\partial^2}{\partial t^2}\nabla^2\phi - \frac{\partial^4\phi}{\partial t^4} + \Omega^2\frac{\partial^2\phi}{\partial t^2} \tag{14.21}$$

Multiply both sides by $\frac{C_\nu}{c_1^2\overline{K}}$, we get

$$\frac{\overline{\nu}C_\nu}{c_1^2\overline{K}}\frac{\partial^2\Theta}{\partial t^2} = \frac{C_\nu}{\overline{K}}\frac{\partial^2}{\partial t^2}\nabla^2\phi - \frac{C_\nu}{c_1^2\overline{K}}\frac{\partial^4\phi}{\partial t^4} + \frac{C_\nu\Omega^2}{c_1^2\overline{K}}\frac{\partial^2\phi}{\partial t^2} \tag{14.22}$$

Then Eq. (14.20) implying that

$$\begin{aligned}\nabla^4\phi - \nabla^2\left[\frac{1}{c_1^2}\frac{\partial^2\phi}{\partial t^2} - \frac{\Omega^2}{c_1^2}\phi + \frac{\overline{\nu}^2T_0}{c_1^2\overline{K}}\frac{\partial^2}{\partial t^2}\nabla^2\phi\right]\\ = \frac{C_\nu}{\overline{K}}\frac{\partial^2}{\partial t^2}\nabla^2\phi - \frac{C_\nu}{c_1^2\overline{K}}\frac{\partial^4\phi}{\partial t^4} + \frac{C_\nu\Omega^2}{c_1^2\overline{K}}\frac{\partial^2\phi}{\partial t^2}\end{aligned} \tag{14.23}$$

If we put $\varepsilon = \frac{\overline{\nu}^2T_0}{c_1^2C_\nu}$ then Eq. (14.23) becomes

$$\nabla^4\phi - \nabla^2\left[\left\{\frac{1}{c_1^2} + \left(1 + \varepsilon\nabla^2 - \frac{\Omega^2}{c_1^2}\right)\frac{C_\nu}{\overline{K}}\right\}\frac{\partial^2}{\partial t^2} - \frac{\Omega^2}{c_1^2}\right]\phi + \frac{C_\nu}{c_1^2\overline{K}}\frac{\partial^4\phi}{\partial t^4} = 0. \tag{14.24}$$

We assume now the solution takes the following form:

$$\phi\text{-}\phi e^{-i\omega t}, \quad \Theta\text{-}\Theta e^{-i\omega t}, \quad \psi\text{-}\psi e^{-i\omega t}. \tag{14.25}$$

We can define the constants as

$$A = \frac{1}{c_1^2} + \left(1 + \varepsilon\nabla^2 - \frac{\Omega^2}{c_1^2}\right)\frac{C_\nu}{\overline{K}}, \quad B = \frac{C_\nu}{c_1^2\overline{K}}, \quad C = \frac{\Omega^2}{c_1^2}. \tag{14.26}$$

and

$$\frac{\partial^2\phi}{\partial t^2} = -\omega^2\phi e^{-i\omega t}, \quad \frac{\partial^4\phi}{\partial t^4} = \omega^4\phi e^{-i\omega t} \tag{14.27}$$

From Eqs. (14.26) and (14.27) we can write Eq. (14.24) as the form:

$$\nabla^4\phi + \omega^2\left(A + \frac{C}{\omega^2}\right)\nabla^2\phi + B\omega^4\phi = 0. \tag{14.28}$$

by solving Eq. (14.28) we get

$$\delta_{1,2}^2 = \omega^2\lambda_{1,2}^2 \tag{14.29}$$

where

$$\lambda_{1,2}^2 = \frac{-\left(A + \frac{C}{\omega^2}\right) \pm \sqrt{\left(A + \frac{C}{\omega^2}\right)^2 - 4B}}{2} \tag{14.30}$$

14.3 Reflection and Transmission

Let us consider a magneto-thermoelastic plane wave ($P-$ or $T-$ or $SV-$wave) propagate through the medium M, which we identify as the region $z > 0$ and falling at the plane $z = 0$, with its direction of propagation making an angle θ_0 with the normal to the surface. Corresponding to each incident wave, we get the waves in medium M as reflected $P-$, $T-$ and $SV-$waves and transmitted $P-$, $T-$ and $SV-$waves in the medium $M^{'}$. We write all the variables without the primes in the region $z > 0$ (medium M) and attach a prime to denote the variables in the region $z < 0$ (medium $M^{'}$), as displayed in Fig. 14.1.

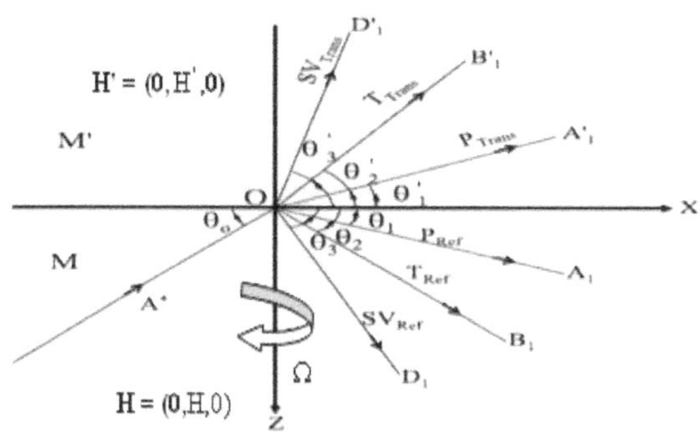

Fig. 14.1 Schematic of the problem

We consider a two-bonded homogenous isotropic thermoelastic solids in contact, as shown in Fig. 14.1. If the bonding is imperfect and the size and spacing between the imperfection is much smaller than the wavelength, then at the interface these can be described by using spring boundary condition (Lavrentyev and Rokhlin (1988)), i.e. at $z = 0$

$$\begin{aligned} &\sigma_{33}^{'} + \tau_{33}^{'} = K_n\left(\omega - \omega^{'}\right), \quad &&\sigma_{31}^{'} + \tau_{31}^{'} = K_t\left(u - u^{'}\right), \\ &K^{'}\frac{\partial T^{'}}{\partial z} = K_c\left(T - T^{'}\right), \quad &&K^{'}\frac{\partial T^{'}}{\partial z_{,}} = K\frac{\partial T}{\partial z}, \\ &\sigma_{33}^{'} + \tau_{33}^{'} = \sigma_{33} - \tau_{33}, \quad &&\sigma_{31}^{'} + \tau_{31}^{'} = \sigma_{31} + \tau_{31}. \end{aligned} \tag{14.31}$$

where K_n and K_t are normal transverse stiffness coefficients of a unit layer thickness and have dimension $\mathrm{N/m^2}$ and K_c is the thermal contact conductance with dimension $\mathrm{W/m^2K\,s}$.

The appropriate potentials satisfying the boundary conditions are:

Medium M:

$$\begin{aligned}\phi &= A_0 \exp\left\{i\delta_1\left(x\cos\theta_0 - z\sin\theta_0\right) - i\omega_1 t\right\}\\ &+ A_1 \exp\left\{i\delta_1\left(x\cos\theta_1 + z\sin\theta_1\right) - i\omega_1 t\right\}\\ &+ B_0 \exp\left\{i\delta_2\left(x\cos\theta_0 - z\sin\theta_0\right) - i\omega_2 t\right\}\\ &+ B_1 \exp\left\{i\delta_2\left(x\cos\theta_2 + z\sin\theta_2\right) - i\omega_2 t\right\},\end{aligned} \tag{14.32}$$

$$\begin{aligned}\Theta &= a_1 A_0 \exp\left\{i\delta_1\left(x\cos\theta_0 - z\sin\theta_0\right) - i\omega_1 t\right\}\\ &+ a_1 A_1 \exp\left\{i\delta_1\left(x\cos\theta_1 + z\sin\theta_1\right) - i\omega_1 t\right\}\\ &+ a_2 B_0 \exp\left\{i\delta_2\left(x\cos\theta_0 - z\sin\theta_0\right) - i\omega_2 t\right\}\\ &+ a_2 B_1 \exp\left\{i\delta_2\left(x\cos\theta_2 + z\sin\theta_2\right) - i\omega_2 t\right\},\end{aligned} \tag{14.33}$$

$$\begin{aligned}\psi &= D_0 \exp\left\{i\delta_3\left(x\cos\theta_0 - z\sin\theta_0\right) - i\omega_3 t\right\}\\ &+ D_1 \exp\left\{i\delta_3\left(x\cos\theta_3 + z\sin\theta_3\right) - i\omega_3 t\right\}.\end{aligned} \tag{14.34}$$

Medium M$'$:

$$\begin{aligned}\phi' &= A_1' \exp\left\{i\delta_1'\left(x\cos\theta_1' - z\sin\theta_1'\right) - i\omega_1' t\right\}\\ &+ B_1' \exp\left\{i\delta_2\left(x\cos\theta_2' - z\sin\theta_2'\right) - i\omega_2' t\right\},\end{aligned} \tag{14.35}$$

$$\begin{aligned}\Theta' &= a_1 A_0 \exp\left\{i\delta_1'\left(x\cos\theta_1' - z\sin\theta_1'\right) - i\omega_1' t\right\}\\ &+ a_1' A_1 \exp\left\{i\delta_2'\left(x\cos\theta_2' - z\sin\theta_2'\right) - i\omega_2' t\right\},\end{aligned} \tag{14.36}$$

$$\psi' = D_1' \exp\left\{i\delta_3'\left(x\cos\theta_3' - z\sin\theta_3'\right) - i\omega_3' t\right\}. \tag{14.37}$$

where

$$\begin{aligned}B_0 &= D_0 = 0 \text{ for incident } P - \text{wave},\\ A_0 &= D_0 = 0 \text{ for incident } T - \text{wave},\\ A_0 &= B_0 = 0 \text{ for incident } SV - \text{wave}.\end{aligned}$$

The Snell's law is given as

$$\frac{\cos\theta_0}{V_0} = \frac{\cos\theta_1}{\lambda_1^{-1}} = \frac{\cos\theta_2}{\lambda_2^{-1}} = \frac{\cos\theta_3}{\lambda_3^{-1}} = \frac{\cos\theta_1'}{\left(\lambda_1'\right)^{-1}} = \frac{\cos\theta_1'}{\left(\lambda_2'\right)^{-1}} = \frac{\cos\theta_1'}{\left(\lambda_3'\right)^{-1}} \tag{14.38}$$

where

$$\begin{aligned}\delta_1\left(\lambda_1^{-1}\right) &= \delta_2\left(\lambda_2^{-1}\right) = \delta_3\left(\lambda_3^{-1}\right) = \delta_1'\left[\left(\lambda_1'\right)^{-1}\right]\\ &= \delta_2'\left[\left(\lambda_2'\right)^{-1}\right] = \delta_3'\left[\left(\lambda_3'\right)^{-1}\right], \quad \text{at } z = 0\end{aligned} \tag{14.39}$$

and

$$V_0 = \begin{cases} \lambda_1^{-1} & \text{for incident } P-\text{ wave,} \\ \lambda_2^{-1} & \text{for incident } T-\text{ wave,} \\ \lambda_3^{-1} & \text{for incident } SV-\text{ wave.} \end{cases} \tag{14.40}$$

Using the potentials given by Eqs. (14.32)–(14.37) in boundary conditions (14.31) and using Eqs. (14.1), (14.4) and (14.10), we get a system of six non-homogeneous equations which can be written as:

$$\sum_{i=1}^{6} a_{ij} Z_j = Y_i, \quad (j = 1, 2, ..., 6) \tag{14.41}$$

where:

$$\begin{aligned}
a_{11} &= iK_n\delta_1\cos\theta_1, \quad a_{12} = iK_n\delta_2\cos\theta_2, \\
a_{14} &= iK_n\delta_1'\sin\theta_1' + \left(\lambda' + \mu_e' H_0'^2\right)\delta_1'^2 + \nu' a_1' + 2\mu'\delta_1'^2\sin^2\theta_1', \\
a_{15} &= iK_n\delta_2'\sin\theta_2' + \left(\lambda' + \mu_e' H_0'^2\right)\delta_2'^2 + \nu' a_2' + 2\mu'\delta_2'^2\sin^2\theta_2', \\
a_{13} &= iK_n\delta_3\cos\theta_3, \quad a_{16} = -\left(iK_n\delta_3'\cos\theta_3' + \mu'\delta_3'^2\sin^2\theta_3'\right), \\
a_{21} &= iK_t\delta_1\cos\theta_1, \quad a_{24} = -\left(\mu'\delta_1'^2\sin 2\theta_1' + i\delta_1' K_t\cos\theta_1'\right), \\
a_{22} &= iK_t\delta_2\cos\theta_2, \quad a_{25} = -\left(\mu'\delta_2'^2\sin 2\theta_2' + i\delta_2' K_t\cos\theta_2'\right), \\
a_{23} &= -iK_t\delta_3\sin\theta_3, \; a_{26} = -\left(\mu'\delta_3'^2\cos 2\theta_3' + i\delta_3' K_t\sin\theta_3'\right), \\
a_{31} &= K_c a_1, \qquad a_{34} = -\left(iK_c a_1' + i\delta_1' K' a_1'\sin\theta_1'\right), \\
a_{32} &= K_c a_2, \qquad a_{35} = -\left(iK_c a_2' + i\delta_2' K' a_2\sin\theta_2'\right), \\
a_{33} &= a_{36} = 0, \\
a_{41} &= -\left[\left(\lambda + \mu_e H_0^2\right) + 2\mu\sin^2\theta_1\right]\delta_1^2 - \nu a_1, \\
a_{44} &= \left[\left(\lambda' + \mu_e' H_0'^2\right) + 2\mu'\sin^2\theta_1'\right]\delta_1'^2 + \nu' a_1', \\
a_{42} &= -\left[\left(\lambda + \mu_e H_0^2\right) + 2\mu\sin^2\theta_2\right]\delta_2^2 - \nu a_2, \\
a_{45} &= \left[\left(\lambda' + \mu_e' H_0'^2\right) + 2\mu'\sin^2\theta_2'\right]\delta_2'^2 + \nu' a_2', \\
a_{43} &= -\mu\delta_3^2\sin 2\theta_3, \qquad a_{46} = -\mu'\delta_3'^2\sin 2\theta_3',
\end{aligned}$$

$$\begin{aligned}
a_{51} &= \mu\delta_1^2\sin 2\theta_1, \quad a_{52} = \mu\delta_2^2\sin 2\theta_2, \quad a_{53} = \mu\delta_3^2\cos 2\theta_3, \\
a_{54} &= \mu'\delta_1'^2\sin 2\theta_1', \quad a_{55} = \mu'\delta_2'^2\sin 2\theta_2', \quad a_{56} = -\mu'\delta_3'^2\cos 2\theta_3' \\
a_{61} &= iKa_1\delta_1\sin\theta_1, \quad a_{62} = iKa_2\delta_2\sin\theta_2, \quad a_{63} = a_{66} = 0, \\
a_{64} &= iK' a_1'\delta_1'\sin\theta_1', \quad a_{65} = iK' a_2'\delta_2'\sin\theta_2'.
\end{aligned}$$

and

$$Z_1 = \frac{A_1}{A^*}, \; Z_2 = \frac{B_1}{A^*}, \; Z_3 = \frac{D_1}{A^*}, \; Z_4 = \frac{A_1'}{A^*}, \; Z_5 = \frac{B_1'}{A^*}, \; Z_6 = \frac{D_1'}{A^*},$$

Taking into account the following cases:

(a) For incident $P-$ wave

$$A^* = A_0,\ Y_1 = a_{11},\ Y_2 = -a_{21},\ Y_3 = -a_{31},\ Y_4 = -a_{41},\ Y_5 = a_{51},\ Y_6 = a_{61}.$$

(b) For incident $T-$ wave

$$A^* = B_0,\ Y_1 = a_{12},\ Y_2 = -a_{22},\ Y_3 = -a_{32},\ Y_4 = -a_{42},\ Y_5 = a_{52},\ Y_6 = a_{62}.$$

(c) For incident $SV-$ wave

$$A^* = D_0,\ Y_1 = -a_{13},\ Y_2 = a_{23},\ Y_3 = a_{33},\ Y_4 = a_{43},\ Y_5 = -a_{53},\ Y_6 = -a_{63}.$$

where Z_1, Z_2 and Z_3 are the amplitude ratios of reflected $P-$, $T-$and $SV-$ waves and Z_4, Z_5 and Z_6 are the amplitude ratios of transmitted $P-$, $T-$and $SV-$ waves, respectively.

14.4 Cases

14.4.1 Case 1. Normal Stiffness

In the case of $K_n \neq 0,\ K_t \longrightarrow \infty,\ K_c \longrightarrow \infty$ we have a boundary with normal stiffness and obtain a system of six non-homogeneous equations as given by Eq. (14.41) with changed values of a_{ij} as

$$\begin{aligned} &a_{21} = i\delta_1 \cos\theta_1, \quad a_{22} = i\delta_2 \cos\theta_2, \quad a_{23} = -i\delta_3 \sin\theta_3, \\ &a_{24} = -i\delta_1^{'} \cos\theta_1^{'}, \quad a_{25} = -i\delta_2^{'} \cos\theta_2^{'}, \quad a_{26} = -i\delta_3^{'} \sin\theta_3^{'}, \\ &a_{31} = a_1, \quad a_{32} = a_2, \quad a_{34} = -a_1^{'}, \quad a_{35} = -a_2^{'}. \end{aligned} \tag{14.42}$$

14.4.2 Case 2. Transversal Stiffness

If we take into consideration $K_n \longrightarrow \infty,\ K_t \neq 0,\ K_c \longrightarrow \infty$ the imperfect boundary reduces to the transverse stiffness and we obtain a system of six non-homogeneous equations as given by Eq. (14.41) and the values of a_{ij} take the following form:

$$\begin{aligned} &a_{11} = i\delta_1 \cos\theta_1, \quad a_{12} = i\delta_2 \cos\theta_2, \quad a_{13} = -i\delta_3 \cos\theta_3, \\ &a_{14} = i\delta_1^{'} \sin\theta_1^{'}, \quad a_{15} = i\delta_1^{'} \sin\theta_1^{'}, \quad a_{16} = -i\delta_3^{'} \cos\theta_3^{'}, \\ &a_{31} = a_1, \quad a_{32} = a_2, \quad a_{34} = -a_1^{'}, \quad a_{35} = -a_2^{'}. \end{aligned} \tag{14.43}$$

14.4.3 Case 3. Thermal Contact Conductance

When $K_n \longrightarrow \infty$, $K_t \longrightarrow \infty$, $K_c \neq 0$ the imperfect boundary reduces to a thermally conducting imperfect surface, we get system of six non-homogeneous equations as given by Eq. (14.41) and the modified values of a_{ij} are

$$\begin{aligned}
&a_{11} = i\delta_1 \cos\theta_1, \quad a_{12} = i\delta_2 \cos\theta_2, \quad a_{13} = i\delta_3 \cos\theta_3,\\
&a_{14} = i\delta_1^{'} \sin\theta_1^{'}, \quad a_{15} = i\delta_2^{'} \sin\theta_2^{'}, \quad a_{16} = -i\delta_3^{'} \cos\theta_3^{'},\\
&a_{21} = i\delta_1 \cos\theta_1, \quad a_{22} = i\delta_2 \cos\theta_2, \quad a_{23} = -i\delta_3 \sin\theta_3,\\
&a_{24} = -i\delta_1^{'} \cos\theta_1^{'}, \quad a_{25} = -i\delta_2^{'} \cos\theta_2^{'}, \quad a_{26} = -i\delta_3^{'} \sin\theta_3^{'}.
\end{aligned} \tag{14.44}$$

14.4.4 Case 4. Welded Contact

In this case $K_n \longrightarrow \infty$, $K_t \longrightarrow \infty$, $K_n \longrightarrow \infty$ then we obtain a system of Eqs. (14.41) with changed values of a_{ij} as

$$\begin{aligned}
&a_{11} = i\delta_1 \cos\theta_1, \quad a_{14} = i\delta_1^{'} \sin\theta_1^{'}, \quad a_{21} = i\delta_1 \cos\theta_1, \quad a_{24} = -i\delta_1^{'} \cos\theta_1^{'},\\
&a_{12} = i\delta_2 \cos\theta_2, \quad a_{15} = i\delta_2^{'} \sin\theta_2^{'}, \quad a_{22} = i\delta_2 \cos\theta_2, \quad a_{25} = -i\delta_2^{'} \cos\theta_2^{'}\\
&a_{13} = i\delta_3 \cos\theta_3, \quad a_{16} = -i\delta_3^{'} \cos\theta_3^{'}, \quad a_{23} = -i\delta_3 \sin\theta_3, \quad a_{26} = -i\delta_3^{'} \sin\theta_3^{'},\\
&a_{31} = a_1, \quad a_{32} = a_2, \quad a_{34} = -a_1^{'}, \quad a_{35} = -a_2^{'}.
\end{aligned} \tag{14.45}$$

14.4.5 Case 5. Slip Boundary

If $K_n \longrightarrow \infty$, $K_t \longrightarrow 0$, $K_n \longrightarrow \infty$ then the imperfect boundary becomes a slip boundary and we obtain a system of six non-homogeneous equations as given by Eqs. (14.41) with modified values of a_{ij} as

$$\begin{aligned}
&a_{11} = i\delta_1 \cos\theta_1, \quad a_{12} = i\delta_2 \cos\theta_2, \quad a_{13} = i\delta_3 \cos\theta_3,\\
&a_{14} = i\delta_1^{'} \sin\theta_1^{'}, \quad a_{15} = i\delta_2^{'} \sin\theta_2^{'}, \quad a_{16} = -i\delta_3^{'} \cos\theta_3^{'},\\
&a_{21} = a_{22} = a_{23} = 0, \quad a_{25} = -\mu^{'} \delta_2^{'2} \sin 2\theta_2^{'},\\
&a_{24} = -\mu^{'} \delta_1^{'2} \sin 2\theta_1^{'}, \quad a_{26} = -\mu^{'} \delta_3^{'2} \cos 2\theta_3^{'},\\
&a_{31} = a_1, \quad a_{32} = a_2, \quad a_{34} = -a_1^{'}, \quad a_{35} = -a_2^{'},\\
&a_{44} = a_{45} = a_{46} = 0.
\end{aligned} \tag{14.46}$$

14.4.6 Special Case: Stress-Free Boundary

In this case when $K_n \longrightarrow 0,\ K_t \longrightarrow 0,\ K_n \longrightarrow \infty$ our results reduce to the stress free thermoelastic boundary and we obtain a system of three non-homogeneous equations:

$$\sum_{i=1}^{3} c_{ij} Z_j = Y_i, \quad (j = 1, 2, 3) \tag{14.47}$$

where

$$\begin{aligned}
c_{11} &= \left[\left(\lambda + \mu_e H_0^2\right) + 2\mu \sin^2\theta_1\right]\delta_1^2 + \nu a_1, \quad c_{21} = \mu\delta_1^2 \sin 2\theta_1,\\
c_{12} &= \left[\left(\lambda + \mu_e H_0^2\right) + 2\mu \sin^2\theta_2\right]\delta_2^2 + \nu a_2, \quad c_{22} = \mu\delta_2^2 \sin 2\theta_2,\\
c_{13} &= \mu\delta_3^2 \sin 2\theta_3, \qquad c_{23} = \mu\delta_3^2 \sin 2\theta_3,\\
c_{31} &= a_1, \quad c_{32} = a_2, \quad c_{33} = 0, \qquad \text{[Isothermal boundary]}
\end{aligned} \tag{14.48}$$

or

$$c_{31} = i a_1 \delta_1 \sin\theta_1, \quad c_{32} = i a_2 \delta_2 \sin\theta_2, \quad c_{33} = 0, \qquad \text{[Insulated boundary]} \tag{14.49}$$

with

$$Z_1 = \frac{A_1}{A^*}, \quad Z_2 = \frac{B_1}{A^*}, \quad Z_3 = \frac{D_1}{A^*} \tag{14.50}$$

Taking into account the following cases:

(a) For incident $P-$ wave; $A^* = A_0$

$$\begin{aligned}
&Y_1 = -c_{11}, \quad Y_2 = c_{21}, \quad Y_3 = -c_{31}, \qquad \text{(Isothermal)}\\
&Y_3 = c_{31} \qquad \text{(Insulated)}
\end{aligned} \tag{14.51}$$

(b) For incident $T-$ wave; $A^* = B_0$

$$\begin{aligned}
&Y_1 = -c_{12}, \quad Y_2 = c_{23}, \quad Y_3 = -c_{32}, \qquad \text{(Isothermal)}\\
&Y_3 = c_{32} \qquad \text{(Insulated)}
\end{aligned} \tag{14.52}$$

(c) For incident $SV-$ wave; $A^* = D_0$

$$Y_1 = c_{13}, \; Y_2 = -c_{23}, \; Y_3 = c_{33}, \tag{14.53}$$

where Z_1, Z_2 and Z_3 are the amplitude ratios of refracted $P-, T-$and $SV-$ waves, respectively.

14.5 Deductions

(A): In the absence of the thermal effect on the medium we obtain the elastic/ thermoelastic imperfect boundary. We obtain a system of five non-homogeneous equations, which can be written as

$$\sum_{i=1}^{5} a_{ij} Z_j^* = Y_j^*, \quad (j = 1, 2, ..., 5) \tag{14.54}$$

where

$$\begin{aligned}
a_{11} &= iK_n\delta_1 \cos\theta_1, \qquad a_{12} = iK_n\delta_2 \cos\theta_2, \qquad a_{13} = iK_n\delta_3 \cos\theta_3,\\
a_{14} &= iK_n\delta_1' \sin\theta_1' + \left(\lambda' + \mu_e' H_0'^2\right)\delta_1'^2 + 2\mu'\delta_1'^2 \sin^2\theta_1,\\
a_{15} &= -\left(iK_n\delta_3' \sin\theta_3' + \mu'\delta_3'^2 \sin^2\theta_3\right),\\
a_{21} &= -iK_t\delta_1 \cos\theta_1, \quad a_{22} = iK_t\delta_2 \cos\theta_2, \quad a_{23} = -iK_t\delta_3 \sin\theta_3,\\
a_{24} &= -\left(\mu'\delta_1'^2 \sin 2\theta_1' + i\delta_1' K_t \cos\theta_1\right),\\
a_{25} &= \mu'\delta_3'^2 \cos 2\theta_3' + i\delta_3' K_t \sin\theta_3,\\
a_{31} &= -\left[\left(\lambda + \mu_e H_0^2\right) + 2\mu \sin^2\theta_1\right]\delta_1^2 - \nu a_1,\\
a_{32} &= -\left[\left(\lambda + \mu_e H_0^2\right) + 2\mu \sin^2\theta_2\right]\delta_2^2 - \nu a_2,\\
a_{33} &= -\mu\delta_3^2 \sin 2\theta_3, \quad a_{34} = \left[\left(\lambda' + \mu_e' H_0'^2\right) + 2\mu' \sin^2\theta_1'\right]\delta_1'^2,\\
a_{35} &= -\mu'\delta_3'^2 \sin 2\theta_3',\\
a_{41} &= \mu\delta_1^2 \sin 2\theta_1, \quad a_{42} = \mu\delta_2^2 \sin 2\theta_2, \quad a_{43} = \mu\delta_3^2 \cos 2\theta_3,\\
a_{44} &= \mu'\delta_1'^2 \cos 2\theta_1', \quad a_{45} = -\mu'\delta_3'^2 \cos 2\theta_3',\\
a_{51} &= a_{52} = a_{53} = a_{54} = a_{55} = 0
\end{aligned} \tag{14.55}$$

with

$$\lambda_1'^2 = \frac{1}{c_1'^2}, \quad \lambda_2'^2 = 0 \tag{14.56}$$

and

$$Z_1^* = \frac{A_1}{A^*}, \quad Z_2^* = \frac{B_1}{A^*}, \quad Z_3^* = \frac{D_1}{A^*}, \quad Z_4^* = \frac{A_1'}{A^*}, \quad Z_5^* = \frac{D_1'}{A^*} \tag{14.57}$$

Taking into account the following cases:

(a) For incident $P-$wave; $A^* = A_0$

$$Y_1^* = a_{11}, \quad Y_2^* = -a_{21}, \quad Y_3^* = -a_{31}, \quad Y_4^* = a_{41}, \quad Y_5^* = -a_{51} \tag{14.58}$$

(b) For incident $T-$wave; $A^* = B_0$

$$Y_1^* = a_{12}, \quad Y_2^* = -a_{22}, \quad Y_3^* = -a_{32}, \quad Y_4^* = a_{42}, \quad Y_5^* = -a_{52} \tag{14.59}$$

(c) For incident $SV-$wave; $A^* = D_0$

$$Y_1^* = -a_{13}, \quad Y_2^* = a_{23}, \quad Y_3^* = a_{33}, \quad Y_4^* = -a_{43}, \quad Y_5^* = a_{53} \tag{14.60}$$

where Z_1^*, Z_2^* and Z_3^* are the amplitude ratios of refracted $P-$, $T-$and $SV-$waves and Z_4^* and Z_5^* are the amplitude ratios of transmitted respectively.

(B): Taking $\mu^{'} \longrightarrow 0$ in medium $M^{'}$, we obtain an interface of thermally conducting liquid/thermoelastic solid half-spaces, leading to a system of five non-homogeneous equations, which can be written as

$$\sum_{i=1}^{5} a_{ij} Z_j = Y_j, \quad (j = 1, 2, ..., 5) \tag{14.61}$$

where

$$\begin{aligned}
&a_{11} = iK_n\delta_1 \cos\theta_1, \quad a_{12} = iK_n\delta_2 \cos\theta_2, \quad a_{13} = iK_n\delta_3 \cos\theta_3, \\
&a_{14} = iK_n\delta_1^{'} \sin\theta_1^{'} + \lambda^{'}\delta_1^{'2} + \nu^{'} a_1^{'}, \\
&a_{15} = iK_n\delta_2^{'} \sin\theta_2^{'} + \lambda^{'}\delta_1^{'2} + \nu^{'} a_2^{'}, \\
&a_{21} = a_1, \quad a_{22} = a_2, \quad a_{23} = 0, \quad a_{24} = -a_1^{'}, \quad a_{25} = -a_2^{'}, \\
&a_{31} = -\left[\left(\lambda + \mu_e H_0^2\right) + 2\mu \sin^2\theta_1\right] \delta_1^2 - \nu a_1, \\
&a_{32} = -\left[\left(\lambda + \mu_e H_0^2\right) + 2\mu \sin^2\theta_2\right] \delta_2^2 - \nu a_2, \\
&a_{33} = -\mu\delta_3^2 \sin 2\theta_3, \quad a_{34} = \left(\lambda^{'} + \mu_e^{'} H_0^{'2}\right) \delta_1^{'2} + \nu^{'} a_1^{'}, \\
&a_{35} = \left(\lambda^{'} + \mu_e^{'} H_0^{'2}\right) \delta_2^{'2} + \nu^{'} a_2^{'}, \\
&a_{41} = \mu\delta_1^2 \sin 2\theta_1, \quad a_{42} = \mu\delta_2^2 \sin 2\theta_2, \quad a_{43} = \mu\delta_3^2 \cos 2\theta_3, \quad a_{44} = a_{45} = 0, \\
&a_{51} = a_1\left(iK^*\delta_1 \sin\theta_1 - b_c\right), \quad a_{52} = a_2\left(iK^*\delta_2 \sin\theta_2 - b_c\right), \quad a_{53} = 0, \\
&a_{54} = b_c a_1^{'}, \quad a_{53} = b_c a_2^{'},
\end{aligned} \tag{14.62}$$

where

$$\lambda_1^{'} = \left[\frac{1}{2}\left(\sqrt{A^{'2} - 4B^{'}} + A^{'}\right)\right]^{\frac{1}{2}}, \quad \lambda_2^{'} = \left[\frac{1}{2}\left(-\sqrt{A^{'2} - 4B^{'}} + A^{'}\right)\right]^{\frac{1}{2}}. \tag{14.63}$$

and

$$\begin{aligned}
&A^{'} = \frac{1}{c_1^{'2}} + (1+\varepsilon)^{'} \frac{C^{*'}}{\overline{K}^*}, \quad B^{'} = \frac{C_\nu^{'}}{c_1^{'2}\overline{K}}, \quad c_1^{'} = \frac{\lambda^{'}}{\rho^{'}} \\
&Z_1 = \frac{A_1}{A^*}, \quad Z_2 = \frac{B_1}{A^*}, \quad Z_3 = \frac{D_1}{A^*}, \quad Z_4 = \frac{A_1^{'}}{A^*}, \quad Z_5 = \frac{B_1^{'}}{A^*}
\end{aligned} \tag{14.64}$$

(a) For incident $P-$ wave; $A^* = A_0$

$$Y_1 = a_{11}, \quad Y_2 = -a_{21}, \quad Y_3 = -a_{31}, \quad Y_4 = a_{41}, \quad Y_5 = a_{51} \tag{14.65}$$

(b) For incident $T-$ wave; $A^* = B_0$

$$Y_1 = a_{22}, \quad Y_2 = -a_{23}, \quad Y_3 = -a_{32}, \quad Y_4^* = a_{42}, \quad Y_5 = a_{52} \tag{14.66}$$

(c) For incident $SV-$ wave; $A^* = D_0$

$$Y_1 = -a_{13}, \quad Y_2 = a_{23}, \quad Y_3 = a_{33}, \quad Y_4 = -a_{43}, \quad Y_5 = a_{53} \tag{14.67}$$

The amplitude ratios Z_1, Z_2 and Z_3 are for reflected $P-, T-$and $SV-$ waves and are for transmitted $P-$ and $T-$ waves respectively. If the magnetic field is neglected, the relevant previous results obtained are deduced as a special case from this study that discussed by Kumar and Sarthi (2006).

14.6 Particular Cases

We take into our consideration the following cases:

- For deduction (A) and (E):

  i. $K_n \neq 0,\ K_t \longrightarrow \infty$, corresponds to the case of normal stiffness boundary.
  ii. Corresponding $K_n \longrightarrow \infty,\ K_t \neq 0$, the results discussed above reduce to transverse stiffness boundary.
  iii. If $K_n \longrightarrow \infty,\ K_t \longrightarrow 0$, the corresponding results reduce to slip boundary.
  iv. When $K_n \longrightarrow \infty,\ K_c \neq 0$, then we obtain the corresponding results for welded boundary.

- For deduction (B), (C) and (D):

  i. $K_n \neq 0$, corresponding to the case of normal stiffness boundary.
  ii. If we take $K_n \longrightarrow \infty$, we obtain the corresponding results for a welded boundary.

14.7 Numerical Results and Discussion

With the view of illustrating results obtain in the preceding sections and comparing these in various cases, we now study some numerical results. The materials chosen for this purpose are Magnesium (M) and Zinc (M'). Physical data of these metals are given (Dhaliwal and Singh (1980)) below as follows:

Magnesium (M)

$$\lambda = 2.696 \times 10^{10}\ \mathrm{Nm}^{-2}, \quad \mu = 1.639 \times 10^{10}\ \mathrm{Nm}^{-2}, \quad \rho = 1.74 \times 10^{3}\ \mathrm{kg\,m}^{-3},$$
$$C^{*} = 1.70 \times 10^{3}\ J\,kg^{-1}\,\mathrm{deg}^{-1}, \quad \nu = 2.68 \times 10^{6}\ \mathrm{Nm}^{-2}\,\mathrm{deg}^{-1}, \quad T_0 = 298\ \mathrm{K}.$$

Zinc (M')

$$\lambda' = 8.58 \times 10^{10}\ \mathrm{Nm}^{-2}, \quad \mu' = 3.85 \times 10^{10}\ \mathrm{Nm}^{-2}, \quad \rho' = 1.74 \times 10^{3}\ \mathrm{kg\,m}^{-3}$$
$$C^{*'} = 7.14 \times 10^{3}\ \mathrm{J\,kg}^{-1}\,\mathrm{deg}^{-1}, \quad \nu' = 3.9 \times 10^{6}\ \mathrm{Nm}^{-2}\,\mathrm{deg}^{-1}, \quad T_0' = 296\ \mathrm{K}.$$

Figure 2 we observe the variation of the $|\delta_1|, |\delta_2|$ and $|\delta_3|$ of reflected of $P-$wave, $T-$wave and $SV-$wave, respectively, $\left|\acute{\delta}_1^{strain4}\right|, \left|\acute{\delta}_2\right|$ and $\left|\acute{\delta}_3\right|$ of transmission of

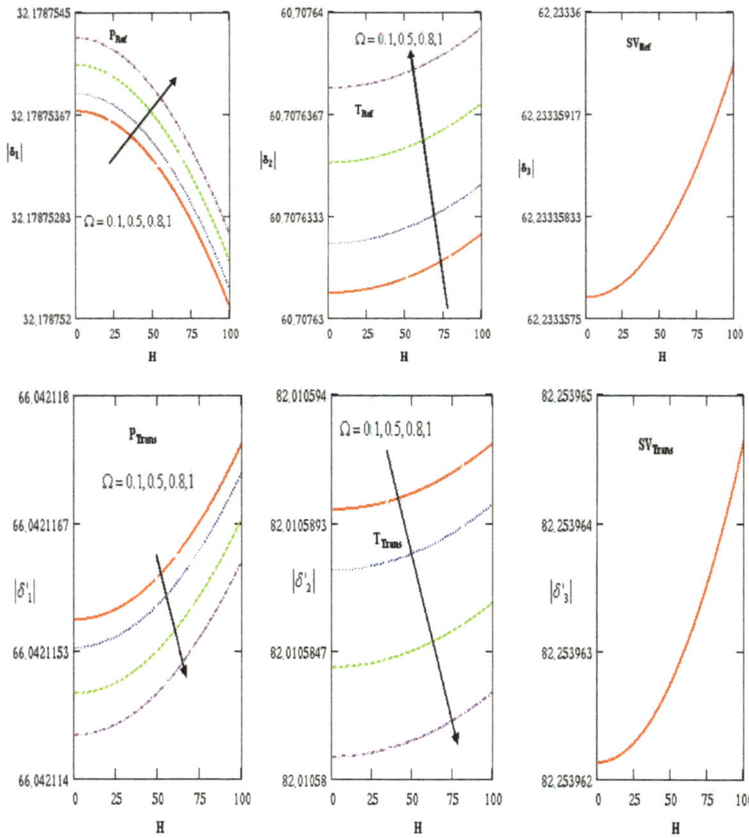

Fig. 14.2 Effects of the rotation on variations of $|\delta_1|, |\delta_2|, |\delta_3|$, $\left|\delta_1'\right|, \left|\delta_2'\right|$ and $\left|\delta_3'\right|$ of the waves respect the magnetic field

$P-$wave, $T-$wave and $SV-$wave respectively with respect to the magnetic field H, the reflected of $P-$wave $\left|\delta_1'\right|$ decreases with increasing of magnetic field, while it increases with increasing of rotation, as well the reflected of $T-$wave and $SV-$wave increases with increasing of magnetic field, while the reflected of $T-$wave increases with increasing of rotation , as well there is no effect of rotation on the reflected of $SV-$wave and the transmission of $P-$wave, $T-$wave and $SV-$wave increase with increasing of magnetic field, while the transmission of $P-$wave and $T-$wave decrease with increasing of rotation, while there is no effect of rotation on the transmission of $SV-$wave.

Figure 14.3 denotes the variation of the reflected of $P-$wave speed $\left|\lambda_1^{-1}\right|$, $T-$wave speed $\left|\lambda_2^{-1}\right|$ and $SV-$wave speed $\left|\lambda_3^{-1}\right|$, transmission of $P-$wave speed

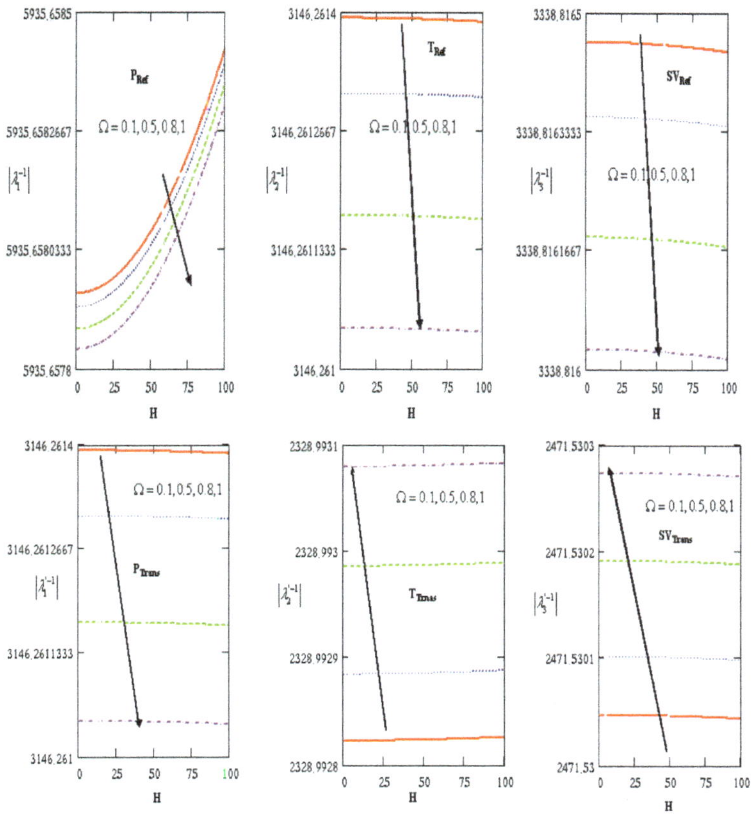

Fig. 14.3 Effects of the rotation on variations of waves speeds $|\lambda_1|, |\lambda_2|, |\lambda_3|$, $\left|\lambda_1'\right|$, $\left|\lambda_2'\right|$ and $\left|\lambda_3'\right|$ respect the magnetic field

$\left|\lambda_1^{'-1}\right|$, $T-$wave speed $\left|\lambda_2^{'-1}\right|$ and $SV-$wave speed $\left|\lambda_3^{'-1}\right|$ with respect to the magnetic field H, the reflected of $P-$wave speed increases with increasing of magnetic field, while it decreases with increasing of rotation, as well the reflected of $T-$wave speed and $SV-$wave speed decrease with increasing of magnetic field and rotation, the transmission of $P-$wave speed and $SV-$wave speed decrease with increasing of magnetic field, while the $T-$wave speed increases with the increasing of magnetic field, the $T-$wave speed and $SV-$wave speed increase with increasing of rotation, as well $P-$wave speed decreases with increasing of rotation.

Figure 14.4 shows the variation of the amplitude of reflected P_1-wave $|z_1|$, the amplitude of reflected P_2-wave $|z_2|$, the amplitude of reflected P_3-wave $|z_3|$, the amplitude of reflected P_4-wave $|z_4|$, the amplitude of reflected P_5-wave $|z_5|$ and the amplitude of the reflected P_6-wave $|z_6|$with respect to the angle of incident

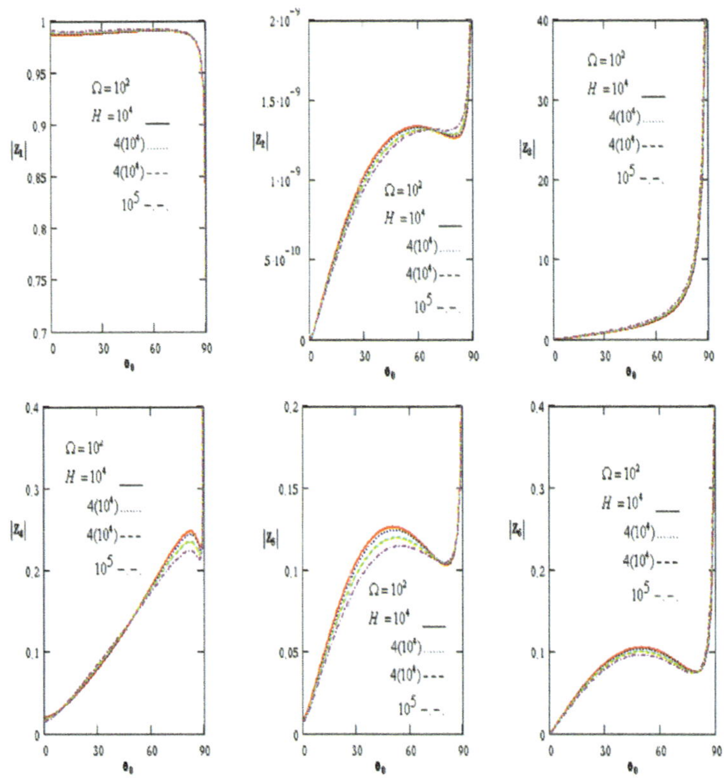

Fig. 14.4 Effects of the magnetic field on variations of amplitudes of P-waves respect the angle of incidence θ_0

θ_0,the amplitude of reflected P_1-wave decreases with increasing of the angle of the incident, while it increases with increasing of magnetic field, the amplitude of the reflected P_2-wave and P_3-wave increases with increasing of the angle of the incident, the amplitude of reflected P_2-wave decreases with increasing of magnetic field, as well the amplitude of the reflected P_3-wave increases with increasing the magnetic field, it is seen that the amplitude of reflected P_1-wave and P_2-wave and P_3-wave become maximum values at $\theta = 90$, the amplitude of the reflected P_4-wave, P_5-wave and P_6-wave increase until becomes maximum values at $\theta = 90$ with increasing of the and angle of incidence, while the amplitude of the reflected P_4-wave decreases with increasing of magnetic field, the amplitude of the reflected P_5-wave and the amplitude of the reflected P_6-wave decreases with increasing of magnetic field.

Figure 14.5 points to the variation of the amplitude of reflected P_1-wave $|z_1|$, the amplitude of reflected P_2-wave $|z_2|$, the amplitude of reflected P_3-wave $|z_3|$, the amplitude of reflected P_4-wave $|z_4|$, the amplitude of reflected P_5-wave $|z_5|$

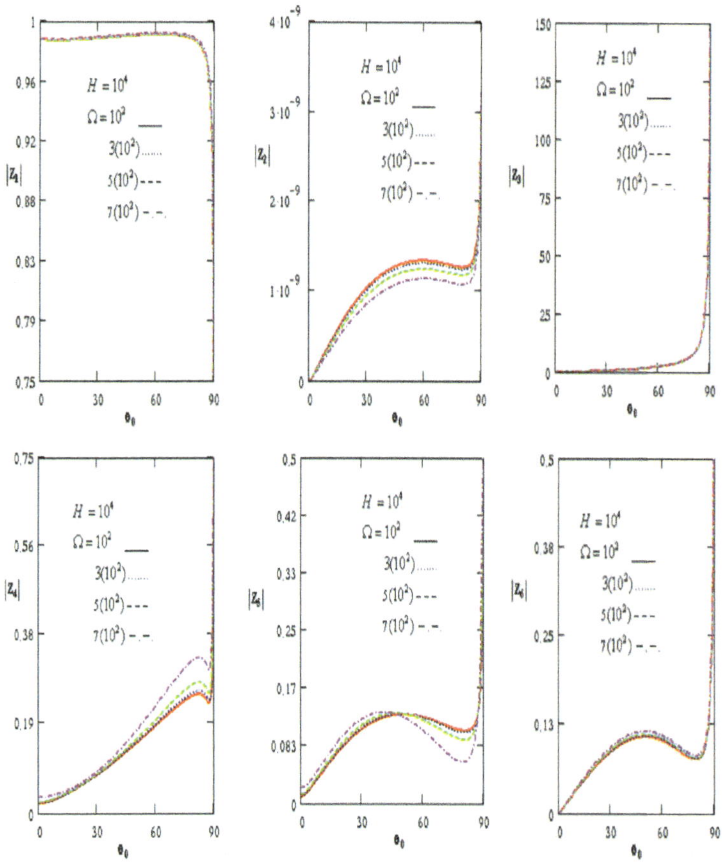

Fig. 14.5 Effects of the rotation on variations of amplitudes of P-waves respect the angle of incidence θ_0

and the amplitude of reflected P_6-wave $|z_6|$with respect to the angle of the incident θ_0,the amplitude of reflected P_1-wave decreases with increasing of the angle of incident, while the effect of rotation is very small on the amplitude of the reflected P_1-wave, the amplitude of reflected P_2-wave increases and then decreases and then increases with increasing of the angle of incident and the rotation until it reaches a maximum value at $\theta_0 = 90$, the amplitude of reflected P_3-wave increases with increasing of the angle of incident until it reaches a maximum value at $\theta_0 = 90$, while there is no effect of rotation on it, the amplitude of reflected P_4-wave increases and then decreases and then increases with increasing of the angle of incident and rotation until it reaches a maximum value at $\theta_0 = 90$, as well the amplitude of reflected P_5-wave increases and then decreases and then increases with increasing

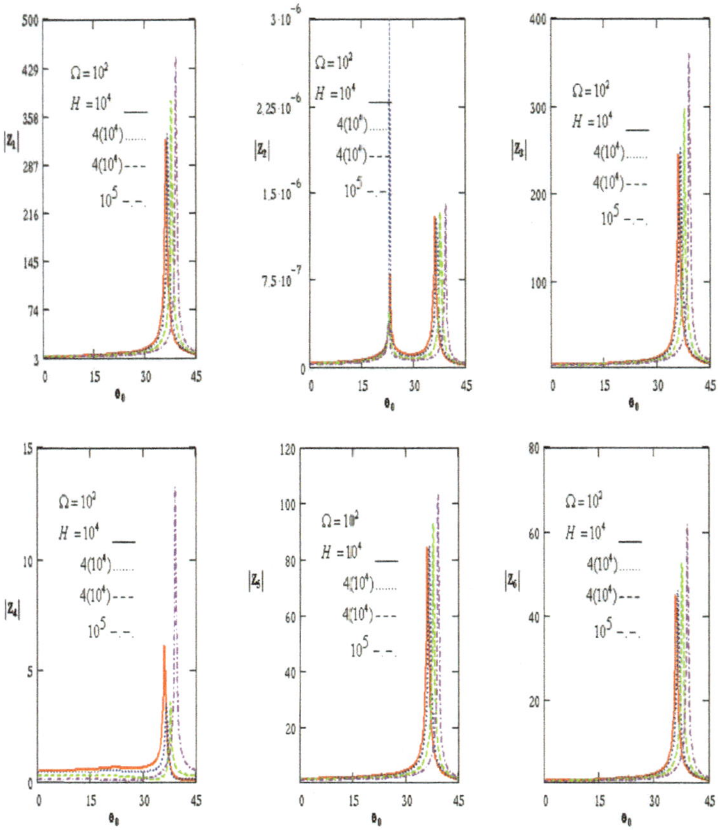

Fig. 14.6 Effects of the magnetic field on variations of P-waves respect the angle of incidence θ_0

of the angle of incident and rotation until it reaches a maximum value at $\theta_0 = 90$ of the angle of the incident, the amplitude of reflected P_6-wave increases and then decreases and then increases with increasing of the angle of incident and rotation until it reaches a maximum value at $\theta_0 = 90$,while there is no effect of rotation on it at $\theta_0 > 80$.

In Fig. 14.6 we see the variation of the amplitude of reflected T_1-wave $|z_1|$, the amplitude of reflected T_2-wave $|z_2|$, the amplitude of reflected T_3-wave $|z_3|$, the amplitude of reflected T_4-wave $|z_4|$, the amplitude of reflected T_5-wave $|z_5|$ and the amplitude of reflected T_6-wave $|z_6|$with respect to the angle of incident θ_0,the amplitude of the reflected T_1-wave increases until it reach a maximum value at $\theta_0 = 37.5, 40$ of the angle of incident and then decreases until vanish at $\theta_0 = 45$,while it decreases with increasing of magnetic field, the amplitude of reflected T_2-wave increases until it reaches a maximum value at $\theta_0 = 22.5, 37.5$ of

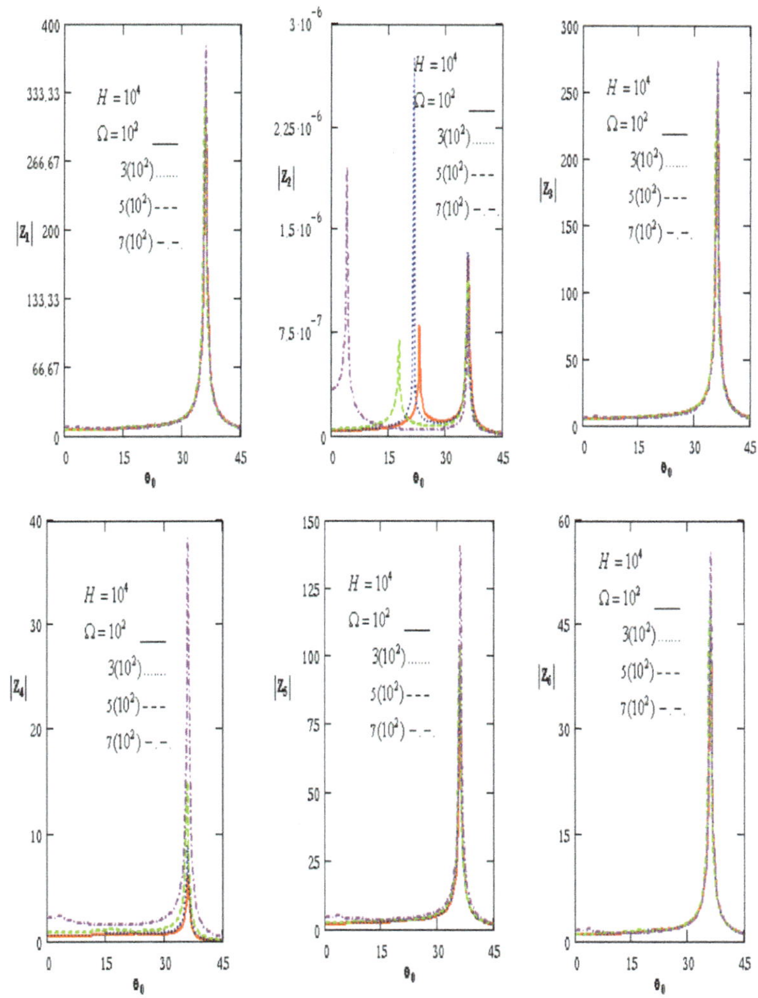

Fig. 14.7 Effects of the rotation on variations of amplitudes of T-waves respect the angle of incidence θ_0

the angle of incident and then decreases and increases until vanish at $\theta_0 = 45$,while it decreases with increasing of magnetic field, the amplitude of the reflected T_3-wave increases until it reaches a maximum value at $\theta_0 = 37.5, 40$ of the angle of incident and then decreases until vanish at $\theta_0 = 45$,while it decreases with increasing of magnetic field, as well the amplitude of reflected T_4-wave, T_5-wave and T_6-wave increase until it reach a maximum value at $\theta_0 = 37.5, 40$ of the angle of incident and

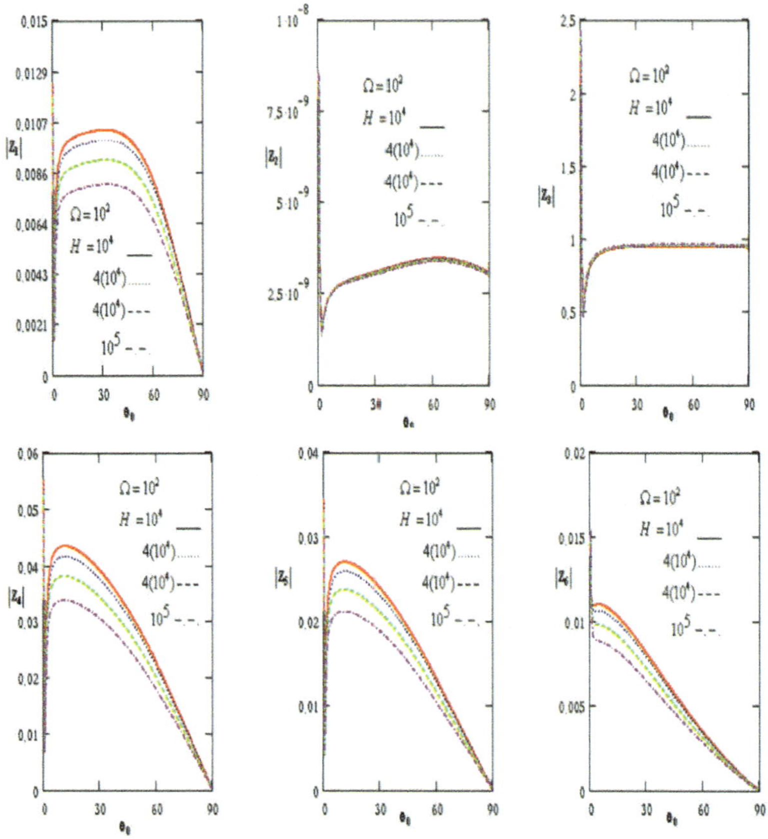

Fig. 14.8 Effects of the magnetic field on variations of amplitudes of SV-waves respect the angle of incidence θ_0

then decreases until vanish at $\theta_0 = 45$,while it decrease with increasing of magnetic field.

Figure 14.7 shows the variation of the amplitude of reflected T_1-wave $|z_1|$, the amplitude of reflected T_2-wave $|z_2|$, the amplitude of reflected T_3-wave $|z_3|$, the amplitude of reflected T_4-wave $|z_4|$, the amplitude of reflected T_5-wave $|z_5|$ and the amplitude of reflected T_6-wave $|z_6|$with respect to the angle of the incident θ_0,the amplitude of the reflected T_1-wave increases until it reach a maximum value at $\theta_0 = 37.5, 40$ of the angle of incident and then decreases until vanish at $\theta_0 = 45$,while it increases with increasing of magnetic field, the amplitude of reflected T_2-wave increases until it reach a maximum value at $\theta_0 = 22.5, 37.5$ of the angle of incident and then decreases and increases until vanish at $\theta_0 = 45$,while it increases with increasing of rotation, the amplitude of reflected T_3-wave increases

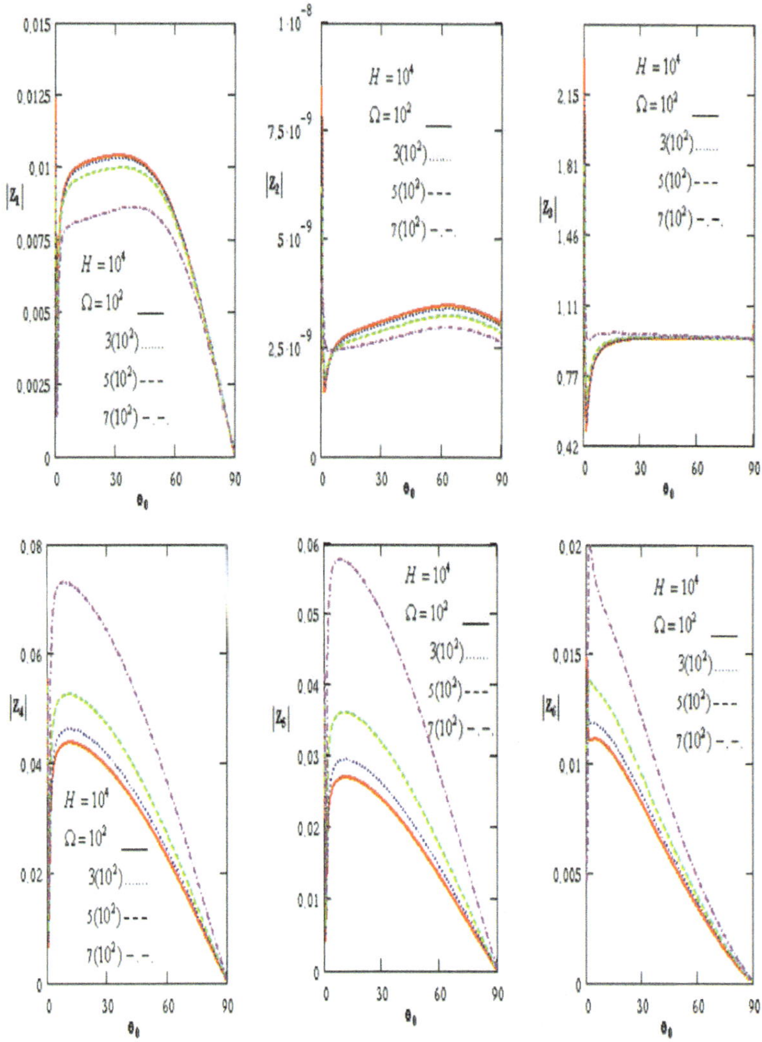

Fig. 14.9 Effects of the rotation on variations of amplitudes of amplitudes of SV-waves respect the angle of incidence θ_0

until it reach a maximum value at $\theta_0 = 37.5, 40$ of the angle of incident and then decreases until vanish at $\theta_0 = 45$,while it increases with increasing of magnetic field, as well the amplitude of the reflected T_4-wave, T_5-wave and T_6-wave increase until it reaches a maximum value at $\theta_0 = 37.5$ of the angle of incident and then decreases until vanish at $\theta_0 = 45$,while it increase with increasing of magnetic field.

In Fig. 14.8 the variation is seen because of the amplitude of reflected SV_1-wave $|z_1|$, the amplitude of reflected SV_2-wave $|z_2|$, the amplitude of reflected SV_3-wave $|z_3|$, the amplitude of reflected SV_4-wave $|z_4|$, the amplitude of reflected SV_5-wave $|z_5|$ and amplitude of reflected SV_6-wave $|z_6|$ with respect to the angle of incident θ_0,the amplitude of reflected SV_1-wave increases and then decreases until it vanish at $\theta_0 = 90$ with increasing of the angle of incident, while it decreases with increasing of magnetic field, the amplitude of reflected SV_2-wave and SV_3-wave increase with the increasing of the angle of the incident, while it decreases with increasing of magnetic field, the amplitude of reflected SV_4-wave and SV_5-wave increase and the decrease with increasing of the angle of incident until vanish at $\theta_0 = 90$,while it decrease with increasing of magnetic field, and the amplitude of reflected SV_6-wave decreases with increasing of the angle of incident until vanish at $\theta_0 = 90$, while it decreases with increasing of magnetic field.

Figure 14.9 denotes the variation of the amplitude of reflected SV_1-wave $|z_1|$, the amplitude of reflected SV_2-wave $|z_2|$, the amplitude of reflected SV_3-wave $|z_3|$, the amplitude of reflected SV_4-wave $|z_4|$, theamplitude of reflected SV_5-wave $|z_5|$ and the amplitude of reflected SV_6-wave $|z_6|$with respect to the angle of the incident θ_0,the amplitude of reflected SV_1-wave increases and then decreases with increasing of the angle of incident until it vanish at $\theta_0 = 90$, while it decreases with increasing of rotation, the amplitude of reflected SV_2-wave and SV_3-wave decrease and then increase with increasing of angle of incident, while SV_2-wave decreases with the increasing of rotation, as well the SV_3-wave increases with increasing of rotation, the amplitude of reflected SV_4-wave and SV_5-wave increase and then decrease with increasing of the angle of incident until vanish at $\theta_0 = 90$,while it increases with the increasing of rotation, and the amplitude of the reflected SV_6-wave decreases with increasing of the angle of incident until vanish at $\theta_0 = 90$, while it increases with increasing of rotation.

14.8 Conclusions

We model Maxwell's stresses and rotation effects on reflection and transmission of plane waves between two thermoelastic media without energy dissipation. The reflected waves velocity with the magnetic field, the rotation and amplitude of the reflected wave with the angle of incidence are obtained in the framework of dynamical coupling theory, Green and Naghdi theory. The effects of applied magnetic field and rotation are discussed numerically and illustrated graphically.

The following conclusions can be made:

i. the reflected waves velocity and amplitude of the reflected wave depend on the angle of incidence, rotation and magnetic field, the nature of this dependence is different for different reflected waves;
ii. the rotation and magnetic field play a significant role and the two effects have the inverse trend for the reflected waves velocity and amplitude of the reflected wave;

iii. the rotation and magnetic field have a strong effect on the reflected waves velocity and amplitude of the reflected wave.

It is observed that the reflected waves velocity and amplitude of reflected wave change their values in the presence of rotation and magnetic field. Hence, the presence of rotation and magnetic field affect the reflection phenomena significantly.

References

Abali BE, Queiruga AF (2019) Theory and computation of electromagnetic fields and thermomechanical structure interaction for systems undergoing large deformations. Journal of Computational Physics 394:200–231

Abali BE, Reich FA (2017) Thermodynamically consistent derivation and computation of electro-thermo-mechanical systems for solid bodies. Computer Methods in Applied Mechanics and Engineering 319:567–595

Abali BE, Reich FA (2018) Verification of deforming polarized structure computation by using a closed-form solution. Continuum Mechanics and Thermodynamics pp 1–16

Abali BE, Zohdi TI (2020) Multiphysics computation of thermal tissue damage as a consequence of electric power absorption. Computational Mechanics 65:149–158

Abd-Alla AM, Mahmoud SR (2010) Magneto-thermoelastic problem in rotating non-homogeneous orthotropic hollow cylinder under the hyperbolic heat conduction model. Meccanica 45:451–462

Abd-Alla AM, Mahmoud SR (2013) On the problem of radial vibrations in non-homogeneity isotropic cylinder under influence of initial stress and magnetic field. Journal of Vibration and Control, (In press)

Abd-Alla AM, Hammad HA, Abo-Dahab SM (2004) Magneto-thermo-viscoelastic interactions in an unbounded body with a spherical cavity subjected to a periodic loading. Applied Mathematics and Computation 155:235–248

Abd-Alla AM, Abo-Dahab SM, Mahmoud SR, Hammad HA (2011) On generalized magneto-thermoelastic rayleigh waves in a granular medium under influence of gravity field and initial stress. Journal of Vibration and Control 40:451–372

Abd-Alla AM, Abo-Dahab SM, Bayones FS (2012a) Propagation of rayleigh waves in magneto-thermo-elastic half–space of a homogeneous orthotropic material under the effect of the rotation, initial stress and gravity field. Journal of Vibration and Control, (In press)

Abd-Alla AM, Mahmoud SR, Abo-Dahab SM (2012b) On problem of transient coupled thermoelasticity of an annular fin. Meccanica 47:1295–1306

Abdoul-Anziz H, Seppecher P (2018) Strain gradient and generalized continua obtained by homogenizing frame lattices. Mathematics and mechanics of complex systems 6(3):213–250

Alibert JJ, Seppecher P, dell'Isola F (2003) Truss modular beams with deformation energy depending on higher displacement gradients. Mathematics and Mechanics of Solids 8(1):51–73

Altenbach H, Eremeyev V (2009) On the linear theory of micropolar plates. ZAMM-Journal of Applied Mathematics and Mechanics/Zeitschrift für Angewandte Mathematik und Mechanik 89(4):242–256

Altenbach H, Eremeyev VA (2010) Thin-walled structures made of foams. In: Cellular and Porous Materials in Structures and Processes, Springer, pp 167–242

Altenbach H, Eremeyev VA (2011) Shell-like structures: non-classical theories and applications, vol 15. Springer Science & Business Media

Altenbach H, Eremeyev VA, Naumenko K (2015) On the use of the first order shear deformation plate theory for the analysis of three-layer plates with thin soft core layer. ZAMM-Journal of

Applied Mathematics and Mechanics/Zeitschrift für Angewandte Mathematik und Mechanik 95(10):1004–1011

Altenbach J, Altenbach H, Eremeyev VA (2010) On generalized cosserat-type theories of plates and shells: a short review and bibliography. Archive of Applied Mechanics 80(1):73–92

Andreaus U, dell'Isola F, Porfiri M (2004) Piezoelectric passive distributed controllers for beam flexural vibrations. Journal of Vibration and Control 10(5):625–659

Andreaus U, Spagnuolo M, Lekszycki T, Eugster SR (2018) A Ritz approach for the static analysis of planar pantographic structures modeled with nonlinear Euler–Bernoulli beams. Continuum Mechanics and Thermodynamics 30(5):1103–1123

Barchiesi E, Placidi L (2017) A review on models for the 3D statics and 2D dynamics of pantographic fabrics. In: Wave Dynamics and Composite Mechanics for Microstructured Materials and Metamaterials, Springer, pp 239–258

Barchiesi E, Ganzosch G, Liebold C, Placidi L, Grygoruk R, Müller WH (2019a) Out-of-plane buckling of pantographic fabrics in displacement-controlled shear tests: experimental results and model validation. Continuum Mechanics and Thermodynamics 31(1):33–45

Barchiesi E, Spagnuolo M, Placidi L (2019b) Mechanical metamaterials: a state of the art. Mathematics and Mechanics of Solids 24(1):212–234

Baroudi D, Giorgio I, Battista A, Turco E, Igumnov LA (2019) Nonlinear dynamics of uniformly loaded elastica: Experimental and numerical evidence of motion around curled stable equilibrium configurations. ZAMM-Zeitschrift für Angewandte Mathematik und Mechanik 99(7):e201800,121

Batra RC, dell'Isola F, Vidoli S (1998) A second-order solution of Saint-Venant's problem for a piezoelectric circular bar using signorini's perturbation method. Journal of elasticity 52(1):75–90

Beevers CE, Bree J (1975) A note on wave reflection problems in linear thermoelasticity, j. J Math Phys Sci 9:355–362

Biot MA (1956) Thermoelasticity and irreversible thermodynamics, j. Appl Phys 27:240–253

Bouchitté G, Mattei O, Milton GW, Seppecher P (2019) On the forces that cable webs under tension can support and how to design cable webs to channel stresses. Proceedings of the Royal Society A 475(2223):20180,781

Camar-Eddine M, Seppecher P (2001) Non-local interactions resulting from the homogenization of a linear diffusive medium. Comptes Rendus de l'Académie des Sciences-Series I-Mathematics 332(5):485–490

Camar-Eddine M, Seppecher P (2003) Determination of the closure of the set of elasticity functionals. Archive for rational mechanics and analysis 170(3):211–245

Carcaterra A, dell'Isola F, Esposito R, Pulvirenti M (2015) Macroscopic description of microscopically strongly inhomogeneous systems: A mathematical basis for the synthesis of higher gradients metamaterials. Archive for Rational Mechanics and Analysis 218(3):1239–1262

Cattaneo C (1948) Sulla conduzione del calore. Atti Sem Mat Fis Univ Modena 3:83–101

Chandrasekharaiah DS (1996a) One-dimensional wave propagation in the linear theory of thermoelasticity with energy dissipation, j. Thermal Stresses 19:695–710

Chandrasekharaiah DS (1996b) Thermoelastic plane waves without energy dissipation. Mech Res Commun 23:549–555

Chandrasekharaiah DS, Srinath DS (1997a) Axisymmetric thermoelastic interactions without energy dissipation in an unbounded body with cylindrical cavity. J Elasticity 46((1)):19–31

Chandrasekharaiah DS, Srinath DS (1997b) Thermoelastic plane waves without energy dissipation in a rotating body. Mech Res Commun 24((5)):551–560

De Angelo M, Barchiesi E, Giorgio I, Abali BE (2019a) Numerical identification of constitutive parameters in reduced-order bi-dimensional models for pantographic structures: application to out-of-plane buckling. Archive of Applied Mechanics 89(7):1333–1358

De Angelo M, Spagnuolo M, D'Annibale F, Pfaff A, Hoschke K, Misra A, Dupuy C, Peyre P, Dirrenberger J, Pawlikowski M (2019b) The macroscopic behavior of pantographic sheets depends mainly on their microstructure: experimental evidence and qualitative analysis of damage in metallic specimens. Continuum Mechanics and Thermodynamics pp 1–23

De Angelo M, Placidi L, Nejadsadeghi N, Misra A (2020) Non-standard timoshenko beam model for chiral metamaterial: Identification of stiffness parameters. Mechanics Research Communications 103:103,462

Del Vescovo D, Giorgio I (2014) Dynamic problems for metamaterials: review of existing models and ideas for further research. International Journal of Engineering Science 80:153–172

dell'Isola F, Placidi L (2011) Variational principles are a powerful tool also for formulating field theories. In: Variational models and methods in solid and fluid mechanics, Springer, pp 1–15

dell'Isola F, Rosa L (1997) Almansi-type boundary conditions for electric potential inducing flexure in linear piezoelectric beams. Continuum Mechanics and Thermodynamics 9(2):115–125

dell'Isola F, Steigmann DJ (2020) Discrete and continuum models for complex metamaterials. Cambridge University Press

dell'Isola F, Porfiri M, Vidoli S (2003) Piezo-electromechanical (pem) structures: passive vibration control using distributed piezoelectric transducers. Comptes Rendus Mecanique 331(1):69–76

dell'Isola F, Maurini C, Porfiri M (2004) Passive damping of beam vibrations through distributed electric networks and piezoelectric transducers: prototype design and experimental validation. Smart Materials and Structures 13(2):299

dell'Isola F, Madeo A, Placidi L (2012) Linear plane wave propagation and normal transmission and reflection at discontinuity surfaces in second gradient 3D continua. ZAMM-Journal of Applied Mathematics and Mechanics/Zeitschrift für Angewandte Mathematik und Mechanik 92(1):52–71

dell'Isola F, Maier G, Perego U, et al (2014) The complete works of Gabrio Piola: Volume I. Advanced Structured Materials, vol. 38

dell'Isola F, Andreaus U, Placidi L (2015a) At the origins and in the vanguard of peridynamics, non-local and higher-gradient continuum mechanics: An underestimated and still topical contribution of Gabrio Piola. Mathematics and Mechanics of Solids 20(8):887–928

dell'Isola F, Lekszycki T, Pawlikowski M, Grygoruk R, Greco L (2015b) Designing a light fabric metamaterial being highly macroscopically tough under directional extension: first experimental evidence. Zeitschrift für angewandte Mathematik und Physik 66:3473–3498

dell'Isola F, Seppecher P, Della Corte A (2015c) The postulations á la D'Alembert and á la Cauchy for higher gradient continuum theories are equivalent: a review of existing results. Proc R Soc A 471(2183):20150,415

dell'Isola F, Seppecher P, Alibert JJ, et al (2019a) Pantographic metamaterials: an example of mathematically driven design and of its technological challenges. Continuum Mechanics and Thermodynamics 31(4):851–884

dell'Isola F, Seppecher P, Spagnuolo M, et al (2019b) Advances in pantographic structures: design, manufacturing, models, experiments and image analyses. Continuum Mechanics and Thermodynamics 31:1231–1282

Deresiewicz H (1960) Effect of boundaries on waves in a thermoelastic solid: Reflexion of plane waves from a plane boundary. J Mech Phys Solids 8:164–172

Deresiewicz H (1962) Corrections and additions: Effect of boundaries on waves in a thermoelastic solid. J Mech Phys Solids 10:179–181

Dhaliwal RS, Singh A (1980) Dynamic coupled thermoelasticity. Hindustan Publ Corp, New Delhi, India 726

El-Naggar AM, Abd-Alla AM (1987) On a generalized thermoelastic problem in an infinite cylinder under initial stress. Earths, Moon& planets 37:213–223

Eremeyev VA (2016) On effective properties of materials at the nano-and microscales considering surface effects. Acta Mechanica 227(1):29–42

Eremeyev VA (2018) On the material symmetry group for micromorphic media with applications to granular materials. Mechanics Research Communications 94:8–12

Eremeyev VA, dell'Isola F (2018) A note on reduced strain gradient elasticity. In: Generalized Models and Non-classical Approaches in Complex Materials 1, Springer, pp 301–310

Eremeyev VA, Pietraszkiewicz W (2012) Material symmetry group of the non-linear polar-elastic continuum. International Journal of Solids and Structures 49(14):1993–2005

Eremeyev VA, Pietraszkiewicz W (2016) Material symmetry group and constitutive equations of micropolar anisotropic elastic solids. Mathematics and Mechanics of Solids 21(2):210–221

Eremeyev VA, dell'Isola F, Boutin C, Steigmann D (2018) Linear pantographic sheets: existence and uniqueness of weak solutions. Journal of Elasticity 132(2):175–196

Eremeyev VA, Rosi G, Naili S (2019) Comparison of anti-plane surface waves in strain-gradient materials and materials with surface stresses. Mathematics and mechanics of solids 24(8):2526–2535

Eugster S, Steigmann D, et al (2019) Continuum theory for mechanical metamaterials with a cubic lattice substructure. Mathematics and Mechanics of Complex Systems 7(1):75–98

Giorgio I (2020) A discrete formulation of Kirchhoff rods in large-motion dynamics. Mathematics and Mechanics of Solids 25(5):1081–1100

Giorgio I, Culla A, Del Vescovo D (2009) Multimode vibration control using several piezoelectric transducers shunted with a multiterminal network. Archive of Applied Mechanics 79(9):859

Giorgio I, Galantucci L, Della Corte A, Del Vescovo D (2015) Piezo-electromechanical smart materials with distributed arrays of piezoelectric transducers: current and upcoming applications. International Journal of Applied Electromagnetics and Mechanics 47(4):1051–1084

Green AE, Lindsay A (1972) Thermoelasticity. J Elasticity 2:1–7

Green AE, Naghdi PM (1991) A re-examination of the basic postulate of thermo mechanics. In: Proc. Roy. Soc. London 432, pp 171–194

Green AE, Naghdi PM (1992) On undamped heat waves in an elastic solid. J Thermal Stresses 15:253–264

Green AE, Naghdi PM (1993) The theory of thermoelasticity without energy dissipation. J Elasticity 31:189–208

Kumar R, Sarathi P (2006) Reflection and refraction of thermoelastic plane waves at an interface between two thermoelastic media without energy dissipation. Arch Mech 58((1)):155–185

Kumar R, Singh M (2009) Effect of rotation and imperfection on reflection and transmission of plane waves in anisotropic generalized thermoelastic media. Journal of Sound and Vibration 324:773–797

Lavrentyev AI, Rokhlin SI (1988) Ultrasonic spectroscopy of imperfect contact interfaces between a layer and two solids. J Acoust Soc Am 103((2)):657–664

Lord HW, Shulman Y (1967) A generalized dynamical theory of thermoelasticity. J Mech Phys Solids 15:299–309

Maurini C, dell'Isola F, Pouget J (2004a) On models of layered piezoelectric beams for passive vibration control. In: Journal de Physique IV (Proceedings), EDP sciences, vol 115, pp 307–316

Maurini C, Pouget J, dell'Isola F (2004b) On a model of layered piezoelectric beams including transverse stress effect. International journal of solids and structures 41(16-17):4473–4502

Maurini C, Pouget J, dell'Isola F (2006) Extension of the Euler–Bernoulli model of piezoelectric laminates to include 3D effects via a mixed approach. Computers & structures 84(22-23):1438–1458

Misra A, Lekszycki T, Giorgio I, Ganzosch G, Müller WH, dell'Isola F (2018) Pantographic metamaterials show atypical Poynting effect reversal. Mechanics Research Communications 89:6–10

Mukhopadhyay S (2002) Thermoelastic interactions without energy dissipation in an unbounded medium with a spherical cavity due to a thermal shock at the boundary. J Thermal Stresses 25:877–887

Othman MIA, Song Y (2007) Reflection of plane waves from an elastic solid half-space under hydrostatic initial stress without energy dissipation. International Journal of Solids and Structures 44:5651–5664

Piccardo G, Ranzi G, Luongo A (2014) A complete dynamic approach to the generalized beam theory cross-section analysis including extension and shear modes. Mathematics and Mechanics of Solids 19(8):900–924

Pideri C, Seppecher P (1997) A second gradient material resulting from the homogenization of an heterogeneous linear elastic medium. Continuum Mechanics and Thermodynamics 9(5):241–257

Placidi L (2015) A variational approach for a nonlinear 1-dimensional second gradient continuum damage model. Continuum Mechanics and Thermodynamics 27(4-5):623

Placidi L, Barchiesi E (2018) Energy approach to brittle fracture in strain-gradient modelling. Proc R Soc A 474(2210):20170,878

Placidi L, Rosi G, Giorgio I, Madeo A (2014) Reflection and transmission of plane waves at surfaces carrying material properties and embedded in second-gradient materials. Mathematics and Mechanics of Solids 19(5):555–578

Placidi L, Andreaus U, Giorgio I (2017) Identification of two-dimensional pantographic structure via a linear D4 orthotropic second gradient elastic model. Journal of Engineering Mathematics 103(1):1–21

Placidi L, Barchiesi E, Misra A (2018a) A strain gradient variational approach to damage: a comparison with damage gradient models and numerical results. Mathematics and Mechanics of Complex Systems 6(2):77–100

Placidi L, Misra A, Barchiesi E (2018b) Two-dimensional strain gradient damage modeling: a variational approach. Zeitschrift für angewandte Mathematik und Physik 69(3):56

Placidi L, Misra A, Barchiesi E (2019) Simulation results for damage with evolving microstructure and growing strain gradient moduli. Continuum Mechanics and Thermodynamics 31(4):1143–1163

Porfiri M, dell'Isola F, Frattale Mascioli F (2004) Circuit analog of a beam and its application to multimodal vibration damping, using piezoelectric transducers. International Journal of Circuit Theory and Applications 32(4):167–198

Rosi G, Pouget J, dell'Isola F (2010) Control of sound radiation and transmission by a piezoelectric plate with an optimized resistive electrode. European Journal of Mechanics-A/Solids 29(5):859–870

Scerrato D, Giorgio I (2019) Equilibrium of two-dimensional cycloidal pantographic metamaterials in three-dimensional deformations. Symmetry 11(12):1523

Shen H, Qiu J, Ji H, Zhu K, Balsi M, Giorgio I, dell'Isola F (2010) A low-power circuit for piezoelectric vibration control by synchronized switching on voltage sources. Sensors and actuators A: Physical 161(1-2):245–255

Singh B (2003) Wave propagation in an anisotropic generalized thermoelastic solid. Indian J pure Applied Math 34(10):1479–1485

Singh B (2005) Reflection of p and sv waves from free surface of an elastic solid with generalized thermodiffusion. J Earth Syst Sci 114(2):1–10

Singh B, Kumar R (2003) Wave propagation in a generalized thermo-microstretch elastic solid. Int J Engng Sci 36:891–912

Singh B, R K (1998) Reflection of plane waves from the flat boundary of a micropolar generalized thermoelastic half-space with stretch. Indian J Pure Appl Math 29(6):657–669

Sinha AN, Sinha SB (1974) Reflection of thermoelastic waves at a solid half-space with thermal relaxation. J Phys Earth 22:237–244

Sinha SB, Elasibai KA (1995) Reflection and refraction of thermoelastic waves at an interface of two semi-infinite media with two relaxation times. J Thermal Stresses 20:129–145

Sinha SB, Elasibai KA (1996) Reflection of thermoelastic waves at a solid half-space with two relaxation times. J Thermal Stresses 19:749–762

Spagnuolo M (2020) Circuit analogies in the search for new metamaterials: Phenomenology of a mechanical diode. In: Nonlinear Wave Dynamics of Materials and Structures, Springer, pp 411–422

Spagnuolo M, Andreaus U (2019) A targeted review on large deformations of planar elastic beams: extensibility, distributed loads, buckling and post-buckling. Mathematics and Mechanics of Solids 24(1):258–280

Spagnuolo M, Barcz K, Pfaff A, dell'Isola F, Franciosi P (2017) Qualitative pivot damage analysis in aluminum printed pantographic sheets: Numerics and experiments. Mechanics Research Communications 83:47–52

Spagnuolo M, Peyre P, Dupuy C (2019) Phenomenological aspects of quasi-perfect pivots in metallic pantographic structures. Mechanics Research Communications 101:103,415

Spagnuolo M, Franciosi P, dell'Isola F (2020a) A Green operator-based elastic modeling for two-phase pantographic-inspired bi-continuous materials. International Journal of Solids and Structures 188:282–308

Spagnuolo M, Yildizdag EM, Andreaus U, Cazzani AM (2020b) Are higher gradient models capable of predicting the mechanical behavior also in case of wide-knit pantographic structures? Mathematics and Mechanics of Solids

Turco E, Giorgio I, Misra A, dell'Isola F (2017) King post truss as a motif for internal structure of (meta) material with controlled elastic properties. Royal Society open science 4(10):171,153

Turco E, Misra A, Pawlikowski M, dell'Isola F, Hild F (2018) Enhanced Piola–Hencky discrete models for pantographic sheets with pivots without deformation energy: Numerics and experiments. International Journal of Solids and Structures 147:94–109

Turco E, Barchiesi E, Giorgio I, dell'Isola F (2020) A Lagrangian Hencky-type non-linear model suitable for metamaterials design of shearable and extensible slender deformable bodies alternative to Timoshenko theory. International Journal of Non-Linear Mechanics 123:103,481

Yildizdag ME, Tran CA, Barchiesi E, Spagnuolo M, dell'Isola F, Hild F (2019) A multi-disciplinary approach for mechanical metamaterial synthesis: A hierarchical modular multiscale cellular structure paradigm. In: State of the Art and Future Trends in Material Modeling, Springer, pp 485–505

Zhou ZD, Yang FP, Kuang ZB (2012) Reflection and transmission of plane waves at the interface of pyroelectric bi-materials. Journal of Sound and Vibration 331:3558–3566

Chapter 15
Casimir Effect on Amplitude-Frequency Response of Parametric Resonance of Electrostatically Actuated NEMS Cantilever Resonators

Dumitru I. Caruntu & Christian A. Reyes

Abstract This paper deals with the effect of Casimir force on the amplitude-frequency response of parametric resonance of electrostatically actuated nano-resonators. The resonator is actuated by using an electrostatic force to include a first order fringe correction. Casimir force and viscous damping force are included in the model, as well. Both electrostatic and Casimir forces are nonlinear. The behavior of the resonator is investigated using two methods, the Method of Multiple Scales (MMS) for a Reduced Order Model (ROM) using one mode of vibration, and numerical integration of ROMs using up to five modes of vibration. ROM is based on the application of a Galerkin procedure that uses the undamped mode shapes of the cantilevered beam as the basis of functions. The amplitude-frequency response consists of two bifurcations, namely subcritical and supercritical. The increase of Casimir effect shows an increase of the interval of frequencies of the unstable zero steady-state solutions, and a larger range of frequencies for which the system has stable steady-state solutions for amplitudes larger than 0.5 of the gap.

Keywords: NEMS · Non-linear · Amplitude-frequency · Parametric resonance · Casimir effect

15.1 Introduction

Nano-electromechanical systems (NEMS) are of great interest in the development of miniaturized sensors (Zhang and Turner, 2005; Zhu et al, 2007; Caruntu et al, 2019; Cheng et al, 2007; Cimalla et al, 2007; Zhang et al, 2014), filters (Rhoads et al, 2005; Lamoreaux, 2004), resonators (Caruntu and Oyervides, 2017; Nayfeh

D. I. Caruntu, C. A. Reyes
University of Texas Rio Grande Valley,
Mechanical Engineering Department, Edinburg, Texas 78539, U.S.A.
e-mail: caruntud2@asme.org, c.reyes47@gmail.com

B. E. Abali and I. Giorgio (eds.), *Developments and Novel Approaches in Biomechanics and Metamaterials*, Advanced Structured Materials 132,
https://doi.org/10.1007/978-3-030-50464-9_15

et al, 2007; Nayfeh and Younis, 2005; Caruntu and Juarez, 2019; Younis and Nayfeh, 2003; Alsaleem et al, 2009; Zhu et al, 2007; Ke, 2009; Blom et al, 1992; Caruntu and Knecht, 2011, 2015; Caruntu and Taylor, 2014), actuators (Zand and Ahmadian, 2009; Hu et al, 2004; Younis et al, 2003; Daqaq et al, 2009; Batra et al, 2006; Krylov, 2008; Caruntu et al, 2013b), and switches, motors and relays (Lamoreaux, 2004). NEMS have numerous applications due to their low power consumption, ease of fabrication, high efficiency, simple structure, and quick response. These applications can be achieved via a variety of element configurations and actuation methods (Zhang et al, 2014). Electrostatic actuation could be used for such NEMS configurations (Caruntu and Knecht, 2011). One of the critical effects in electrostatically actuated devices is the presence of pull-in instability which arises due to nonlinearities in the system. Pull-in occurs when a moving element contacts and "sticks" to another element within the system. In many systems, this is a major limitation as it can significantly limit the range of motion of the device (Zhang et al, 2014; Caruntu et al, 2016). Hence, it is of great interest to predict how system parameters influence pull-in in order to control or mitigate its effects.

Nonlinearities can arise from a number of sources such as geometric nonlinearities due to large deflections (Spagnuolo and Andreaus, 2019; Baroudi et al, 2019), squeeze-film damping effects, and actuating forces. In particular, capacitive electrostatic forces are commonly used as the actuating force in nano devices and introduce nonlinear effects. Additionally, at submicron scales, intermolecular surface forces, such as Casimir or van der Waals, can affect the behavior of micro- and nano-beams as well (Ramezani et al, 2007; Caruntu and Juarez, 2019).

The Casimir force and a first order fringe effect models that are considered here are reported by Ramezani et al (2008). The electrostatic excitation produced by a fluctuating voltage across the length of the nano-beam parametrically excites the system. Specifically, a parametric term arises in both linear and nonlinear terms of the equation of motion. This is similar to Rhoads et al (2006) who investigated a parametrically excited comb drive system. It was found that such an excitation does not create just a single defined type of nonlinear effect for the system, but a variety of effects depending on system parameters in addition to the frequency and amplitude of excitation. Various bifurcations were found in the system in which the frequency of excitation was used as the bifurcation parameter. These bifurcations created mixed nonlinearities in addition to the familiar hardening and/or softening effects. In this paper, a parametrically excited cantilevered nano-beam is investigated and similar frequency dependant bifurcations are found.

Understanding the effect of parametric excitations is of general interest. The stability of these systems and the types of nonlinearities that occur are highly sensitive to physical parameters as well as frequency and initial amplitude (Nayfeh et al, 2007; Nayfeh and Younis, 2005; Younis and Nayfeh, 2003; Alsaleem et al, 2009; Rhoads et al, 2006; Caruntu et al, 2016). It is then important to identify bifurcation parameters and bifurcation points in order to design and control systems under parametric excitation. Bifurcation phenomena have been investigated in literature for such parametrically excited systems, mainly for discrete comb drive systems (Rhoads et al, 2006; DeMartini et al, 2007), but not for cantilevered beam elements. Most of the analysis in literature investigated pull-in stability, amplitude-frequency responses or found limit cycles and time responses of such systems (Zand and Ahmadian, 2009; Ramezani et al, 2008). Daqaq et al (2009) discussed how parametric excitations in a cantilevered beam coupled to an electrical system via a piezoelectric patch can be used for energy harvesting. They found that there is an optimal value for the electromechanical coupling terms for maximizing the output voltage of the harvester. They also discussed the sensitivity for the harvester in that there is a critical value for excitation forces below which oscillations will not occur. The value of the critical excitation force is dependent on the electromechanical coupling term.

Ke (2009) investigated a double-sided electromechanically driven nanotube resonator taking into account van der Waals forces. An energy based method was used to find analytical relationships for the steady state amplitude of the nanotube as a function of driving frequency and excitation voltage. An analytical relationship for the resonant pull-in voltage was also developed. The analytical results where then verified numerically using a reduced order Galerkin method carried out to 5th order. Sedighi et al (2014) investigated electrostatically actuated nano cantilever including the Casmir and van der Waals force using the Parameter Expansion Method.

In this paper, an investigation of the effect of the Casimir forces on the amplitude-frequency response of parametric resonance of electrostatically actuated NEMS cantilever resonators is conducted. Forces acting on the resonator are 1) electrostatic forces to include fringe effect, given by an AC voltage of frequency near natural frequency of the cantilever, 2) Casimir force due to the gap between the resonator and a parallel ground plate less than 1 μm, and 3) viscous damping force. All forces are in the category of soft excitation, and weak nonlinearities and damping. In the case of soft excitation the structure experiences very small amplitudes if not in a resonant zone.

To the best of our knowledge, this is the first time when 1) the Casimir forces effect on the amplitude-frequency response is investigated using 2) ROMs up to five modes of vibration. 3) ROM with one mode of vibration, and all nonlinear terms expanded in Taylor series with all terms up to cubic power retained is solved using the Method of Multiple Scales (MMS). 4) ROMs with two, three, four, and five modes of vibration are numerically integrated using a MATLAB ode solver, namely ode15s, in order to predict time response of the resonator. 5) ROMs with up to five modes of vibration are used to perform a continuation and bifurcation analysis using AUTO 07p software package. 6) This work shows that ROM with five modes of vibration accurately predict the amplitude-frequency bifurcation diagram (response) in all amplitudes up to the gap. 7) Casimir forces effect on the amplitude-frequency response shows that as Casimir forces increase, the softening effect increases and the bifurcation points (as well as the entire steady-state amplitudes) are shifted to lower frequencies.

15.2 System Model

In the present model of the NEMS cantilever resonator, the Euler–Bernoulli theory of thin beams is used. The resonator is a uniform cantilever.

Electrostatic force is a common source of actuation in Nano-electromechanical system (NEMS) devices. On micro- and nano-scales, electrostatic actuation is able to provide sizable forces with relatively low voltages and power consumption. The Palmer approximate formula (a first order fringing correction) gives the electrostatic force per unit length as follows:

$$F_e = \frac{\varepsilon_0 W}{2} \frac{V(t)^2}{(g-w)^2} \left[1 + 0.65 \frac{(g-w)}{W} \right] \tag{15.1}$$

where ε_0 is the permittivity of free space, W is the beam width, g is the gap between the beam and the ground plate, w is the deflection of the beam, and $V(t)$ is the applied voltage.

Between the beam and the ground plate is a dielectric material such as air. The actuation forces are given by the Casimir effect and electrostatic force produced by a potential difference across the upper beam and underlying ground conducting plate.

The Casimir force is significant for nano-scale systems, and it accounts for the dispersion forces which arise between closely spaced, uncharged conducting surfaces. In the context of the system used in this paper, plate separations are large enough when pair interactions between atoms and molecules (referred to as the van der Waals interaction following a $1/d^3$ relation) are considered non-significant, and the force between plates depends on only bulk material properties (this is sometimes referred to as the retarded van der Waals interaction following a $1/d^4$ relation). This occurs when the separation of surfaces is significantly larger than the atomic spacing. The transition between the van der Waals and Casimir regime occurs at a distance of approximately 20 nm (Ramezani et al, 2008) and hence a larger gap distance will be considered in this investigation. On the other hand, in order for the Casimir effect to be significant, surface separations should be less than approximately 1000 nm. The Casimir force per unit length along the beam is (Lamoreaux, 2004)

$$F_c = \frac{\pi^2 \hbar c W}{240(g-w)^4} \tag{15.2}$$

where $\hbar = 1.055 \times 10^{-34}$ Js is Planck's constant divided by 2π, and $c = 2.998 \times 10^8$ ms^{-1} is the speed of light. The source of this effect is a matter of debate but is typically attributed to a zero-point energy in the electromagnetic field. Casimir's original derivation of Eq. (15.2) is based on this assertion, and with the experimental verification of this effect has been argued as proof of zero-point energy. Other approaches, however, have been used to explain the Casimir force completely removed from zero-point energy resulting in the identical results of Eq. (15.2). One of these alternative approaches was proposed by Lifshitz who attributed the phenomena to charge and current fluctuations due to Johnson noise. These fluctuations within a material produce a field that can extend beyond its surface resulting in an attractive force at very small distances (Lamoreaux, 2004).

When modeling structures at the micro and nano scale, air damping effects are significant and impact the behavior of the system. Viscous flow as the beam moves through the air is one of the main components of air damping. The force due to viscous damping is described by

$$F_d = b\frac{\partial w}{\partial t} \tag{15.3}$$

where b is the coefficient of viscous damping per unit length.

15.3 Partial-Differential Equation of Motion

The flexible silicon nano cantilever resonator suspended over a grounded substrate (underlying plate), Fig. 15.1, is electrostatically actuated by applying a potential difference between the cantilever and the ground plate. In addition to the electrostatic force, Casimir and viscous damping forces are included in the model. The length of the beam is considered to be relatively long compared to the underlying gap and hence the beam will experience only small to moderate deflections, i.e. the slope of the beam will be relatively small, so the Euler–Bernoulli theory is suitable. This is important since the model used to describe the electrostatic and Casimir forces assumes the upper and lower plates to be locally parallel.

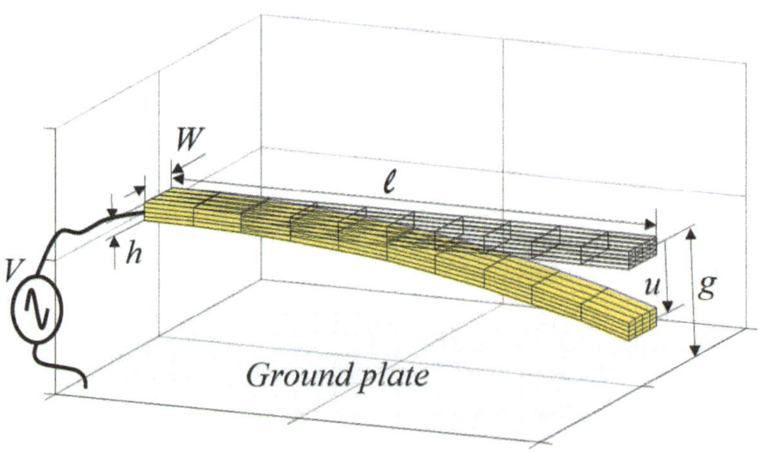

Fig. 15.1 Electrostatically Actuated Uniform NEMS Resonator

The dimensionless equation of motion of the resonator (Caruntu and Knecht, 2011; Caruntu and Martinez, 2014) to include boundary and initial conditions, is given by

$$\begin{cases} \dfrac{\partial^4 u(\tau,z)}{\partial z^4} + \dfrac{\partial^2 u(\tau,z)}{\partial \tau^2} = -\, b^* \dfrac{\partial u(\tau,z)}{\partial \tau} + \dfrac{\alpha}{[1-u(\tau,z)]^4} + \\ \qquad \dfrac{\delta V^2}{[1-u(\tau,z)]^2} + \dfrac{f\delta V^2}{1-u(\tau,z)} \\ u(\tau,0) = \dfrac{\partial u}{\partial z}(\tau,0) = \dfrac{\partial^2 u}{\partial z^2}(\tau,1) = \dfrac{\partial^3 u}{\partial z^3}(\tau,1) = 0 \\ u(0,z) = f(z), \dfrac{\partial u}{\partial z}(0,z) = p(z) \end{cases} \tag{15.4}$$

One can see at the right-hand side of the differential equation of motion the dimensionless forces acting on the cantilever; from left to right, they are damping, Casimir, electrostatic, and fringe effect forces. Variables z, τ and $u(\tau, z)$ are dimensionless longitudinal coordinate, dimensionless time, and dimensionless deflection, respectively, and they are related to their corresponding dimensional quantities x, t, and $w(t, x)$, respectively, as follows:

$$u = w/g\,, \quad z = x/\ell\,, \quad \tau = t \cdot \frac{1}{\ell^2}\sqrt{\frac{EI_0}{\rho A_0}}\,, \tag{15.5}$$

where ℓ , g, E, A_0, I_0, and ρ are the beam's length, initial gap between cantilever and ground plate, Young's modulus, cross section area, cross section moment of inertia, and material density, respectively. The dimensionless parameters α, δ, f and b^* in Eqs. (15.4) track the effects of Casimir forces, voltage (or electrostatic excitation amplitude), fringe, and damping, respectively, and they are given by

$$\alpha = \frac{\pi^2 \hbar c W \ell^4}{240 g^5 E I_0}, \; \delta = \frac{\varepsilon_0 W \ell^4}{2g^3 E I_0} V_0^2, \; f = \frac{0.65g}{W}, \; b^* = b\frac{\ell^2}{\sqrt{\rho A_0 E I_0}} \tag{15.6}$$

where b is the dimensional damping, and V_0 the voltage amplitude. In this investigation, the dimensionless voltage $V(\tau)$, dimensionless frequency of excitation Ω, and the dimensionless natural frequency ω are as follows:

$$V(\tau) = \cos(\Omega\tau), \quad \Omega = \Omega^* \ell^2 \sqrt{\frac{\rho A_0}{EI_0}}, \quad \omega = \omega^* \ell^2 \sqrt{\frac{\rho A_0}{EI_0}} \tag{15.7}$$

where Ω^* is the dimensional frequency of excitation, and ω^* is the dimensional natural frequency of the resonator. The quality factor is related to dimensionless damping (Caruntu and Knecht, 2011).

15.4 Method of Multiple Scales

In what follows, the case of weak nonlinear forces and damping is considered, i.e. the Casimir, electrostatic, fringe, and damping parameters $\alpha,\ \delta,\ f$, and b^* in Eq. (15.4) are small. The Casimir, electrostatic, and fringe effect terms in Eq. (15.4) are expanded in Taylor series around $u = 0$ and all terms up to cubic power are retained. Then MMS is applied. A small dimensionless bookkeeping parameter ε is introduced as factor of all small terms in the resulting equation

$$\begin{aligned}\frac{\partial^2 u}{\partial \tau^2} + \frac{\partial^4 u}{\partial z^4} &= -\varepsilon b^* \frac{\partial u}{\partial \tau} + \varepsilon\alpha\left[1 + 4u + 10u^2 + 20u^3\right] \\ &\quad + \varepsilon\delta\left[(1 + f) + (2 + f)u + (3 + f)u^2 + (4 + f)u^3\right]V^2(\tau) \end{aligned} \tag{15.8}$$

A first-order expansion of the dimensionless transverse displacement u is considered. This is given by

$$u(z, \tau, \varepsilon) = u_0(z, T_0, T_1) + \varepsilon \cdot u_1(z, T_0, T_1) \tag{15.9}$$

where $T_0 = \tau$ is fast time scale and $T_1 = \varepsilon \cdot \tau$ is slow time scale. The time derivatives become

$$\partial/\partial\tau = D_0 + \varepsilon \cdot D_1,\ D_n = \partial/\partial T_n,\ n = 0, 1 \tag{15.10}$$

where $D_n,\ n = 0, 1$ represent partial derivatives with respect to the fast and slow time scales. Substituting Eq. (15.9) and Eq. (15.10) into Eq. (15.8), and equating coefficients of like powers of the bookkeeping parameter, the following two approximation problems, namely zero-order and first-order, result

$$\text{Order } \varepsilon^0 : D_0^2 u_0 + \frac{\partial^4 u_0}{\partial z^4} = 0 \tag{15.11}$$

$$\begin{aligned}\text{Order } \varepsilon^1 : D_0^2 u_1 + \frac{\partial^4 u_1}{\partial z^4} &= -2D_0 D_1 u_0 - b^* D_0 u_0 + \alpha\left[1 + 4u_0 + 10u_0^2 + 20u_0^3\right] \\ &\quad + \delta\left[(1 + f) + (2 + f)u_0 + (3 + f)u_0^2 + (4 + f)u_0^3\right]V^2(T_0) \end{aligned} \tag{15.12}$$

The solution u_0 of Eq. (15.11) is given by

$$u_0(z, T_0, T_1) = \varphi(z)\left[A(T_1)e^{i\omega T_0} + \overline{A}(T_1)e^{-i\omega T_0}\right] \tag{15.13}$$

where A and $\overline{A}$ are complex conjugate coefficients depending on the slow time scale. Enforcing the boundary conditions given by Eq. (15.4), the mode shapes $\varphi_k(z)$ and the corresponding natural frequency ω_k are obtained. Natural frequencies and mode shapes for uniform cantilevers are reported in the literature Alsaleem et al (2009) and Zhu et al (2007); Caruntu and Knecht (2011). Natural frequencies and mode shapes of nonuniform structures as well as mathematical methods for finding them can be found in the literature Caruntu (2007, 2005, 2013). The mode shapes $\varphi_k(z)$ of a uniform cantilever form an orthonormal set, satisfying

$$\langle\varphi_k(z),\varphi_n(z)\rangle=\int_0^1\varphi_k(z)\varphi_n(z)dz=\delta_{kn}\,,\ \frac{d^4\varphi_k(z)}{dz^4}=\omega_k^2\varphi_k(z)$$
$$\varphi_k(0)=0,\ \varphi_k'(0)=0,\ \varphi_k''(1)=0,\ \varphi_k'''(1)=0\,. \tag{15.14}$$

where δ_{kn} is Kronecker's delta. The first-order approximation can be found by solving the nonhomogeneous Eq. (15.12). Substituting Eq. (15.13) into Eq. (15.12) , it results

$$\begin{aligned}
D_0^2u_1+\frac{\partial^4u_1}{\partial z^4}&=-2D_0D_1\varphi_k\left[A_k(T_1)e^{i\omega_kT_0}+\overline{A_k}(T_1)e^{-i\omega_kT_0}\right]\\
&-b^*D_0\varphi_k\left[A_k(T_1)e^{i\omega_kT_0}+\overline{A_k}(T_1)e^{-i\omega_kT_0}\right]+\alpha\Big\{1+4\varphi_k\left[A_k(T_1)e^{i\omega_kT_0}\right.\\
&\left.+\overline{A_k}(T_1)e^{-i\omega_kT_0}\right]+10\varphi_k^2\left[A_k(T_1)e^{i\omega_kT_0}+\overline{A_k}(T_1)e^{-i\omega_kT_0}\right]^2\\
&+20\varphi_k^3\left[A_k(T_1)e^{i\omega_kT_0}+\overline{A_k}(T_1)e^{-i\omega_kT_0}\right]^3\Big\}+\delta\Big\{(1+f)+\\
&(2+f)\varphi_k\left[A_k(T_1)e^{i\omega_kT_0}+\overline{A_k}(T_1)e^{-i\omega_kT_0}\right]+(3+f)\varphi_k^2\left[A_k(T_1)e^{i\omega_kT_0}+\right.\\
&\left.\overline{A_k}(T_1)e^{-i\omega_kT_0}\right]^2+(4+f)\varphi_k^3\left[A_k(T_1)e^{i\omega_kT_0}+\overline{A_k}(T_1)e^{-i\omega_kT_0}\right]^3\Big\}V^2(T_0)^2
\end{aligned} \tag{15.15}$$

In this investigation the AC frequency of excitation is considered near natural frequency $\Omega\approx\omega_k$. This nearness is showed by a small detuning parameter σ as follows:

$$\Omega=\omega_k+\varepsilon\sigma \tag{15.16}$$

Equation (15.15) is then expanded. The square of the dimensionless voltage V is given by

$$V^2(T_0)=\frac{1}{2}+\frac{1}{4}\left(e^{2i\Omega T_0}+e^{-2i\Omega T_0}\right) \tag{15.17}$$

After substituting Eq. (15.16) and Eq. (15.17) into Eq. (15.15), the secular terms containing $\left(e^{i\omega_kT_0}\right)$ are collected and their sum set equal to zero. In addition, the non-homogeneous Eq. (15.15) has to be orthogonal to every solution of the homogeneous problem Eq. (15.11), so the equation of secular terms becomes

$$\begin{aligned}
&-2i\omega_kg_{1kk}A_k'-i\omega_kb^*g_{1kk}A_k+\left[4\alpha+C_2\right]g_{1kk}A_k+3\left[20\alpha+C_4\right]g_{3kk}A_k^2\overline{A_k}+\\
&\frac{1}{2}C_2g_{1kk}\overline{A_k}e^{2i\sigma T_1}+\frac{3}{2}C_4g_{3kk}A_k\overline{A_k}^2e^{2i\sigma T_1}+\frac{1}{2}C_4g_{3kk}A_k^3e^{-2i\sigma T_1}=0
\end{aligned} \tag{15.18}$$

where

$$C_m=\frac{1}{2}(m+f)\delta,\qquad m=1,2,3,4\,,\qquad g_{nkk}=\langle\varphi_k^n,\varphi_k\rangle=\int_0^1\varphi_k^n\varphi_kdz \tag{15.19}$$

and n is greater than or equal to zero. A_k' is the derivative of A_k with respect to the slow time scale T_1. Express A_k in polar form

$$A_k=\frac{1}{2}a_ke^{i\beta_k} \tag{15.20}$$

where a_k and β_k are real and represent the amplitude of the beam and the phase of the system relative to the excitation frequency, respectively. Substituting Eq. (15.20) into Eq. (15.18) and separating the real and imaginary parts, the following amplitude-phase system of equations results

$$a_k' = a_k\left[-\frac{b^*}{2} + \left(C_2 + \frac{C_4}{2}\frac{g_{3kk}}{g_{1kk}}a_k^2\right)\frac{\sin 2\gamma_k}{4\omega_k}\right] \tag{15.21}$$

$$\begin{aligned} a_k\gamma_k' &= a_k\sigma + \frac{4\alpha + C_2}{2\omega_k}a_k + \frac{3(20\alpha + C_4)}{8\omega_k}\frac{g_{3kk}}{g_{1kk}}a_k^3 \\ &+ a_k\left(C_2 + C_4\frac{g_{3kk}}{g_{1kk}}a_k^2\right)\frac{\cos 2\gamma_k}{4\omega_k} \end{aligned} \tag{15.22}$$

where

$$\gamma_k = \sigma T_1 - \beta_k \tag{15.23}$$

The steady-state solutions result by substituting $a_k' = \gamma_k' = 0$ into Eqs. (15.21) and (15.22). One of the steady-state solutions is the trivial solution $a_k = 0$ for all values of the detuning parameter σ. The non-trivial steady state solution case consists of a set of parametric equations describing the amplitude-frequency bifurcation diagram (response) as follows:

$$a_k^2 = \frac{2g_{1kk}}{C_4 g_{3kk}}\left(\frac{2\omega_k b^*}{\sin 2\gamma_k} - C_2\right) \tag{15.24}$$

$$\sigma = -\frac{4\alpha + C_2}{2\omega_k} - \frac{3(20\alpha + C_4)}{8\omega_k}\frac{g_{3kk}}{g_{1kk}}\alpha_k^2 - \left(C_2 + C_4\frac{g_{3kk}}{g_{1kk}}a_k^2\right)\frac{\cos 2\gamma_k}{4\omega_k} \tag{15.25}$$

The MMS steady-state amplitudes of the amplitude-frequency response, Eqs. (15.24) and (15.25), are afterwards presented in Fig. 15.2 for a uniform beam. It is seen that a softening type of behavior occurs with two branches, lower and upper, which are unstable and stable, respectively.

15.5 Reduced Order Model of Uniform NEMS Cantilevers

The analytical results based on steady-state amplitudes which were obtained using MMS are compared to numerical solutions of Eq. (15.4), in the case of uniform resonators. The system is considered at nano scale, where the Casimir force is significant.

To numerically investigate the system, a reduced order model (ROM) is constructed (Alsaleem et al, 2009; Younis et al, 2003; Caruntu et al, 2013a). This is done by utilizing a Galerkin procedure in which the solution is

$$u(z,\tau) = \sum_{i=1}^{N} u_i(\tau)\varphi_i(z) \tag{15.26}$$

where $u_i(\tau)$ are time dependent functions to be determined, N the number modes of vibrations used, and $\varphi_i(z)$ the mode shape functions of the uniform cantilever. The mode shapes satisfy Eqs. (15.14).

It is important to note that when constructing ROM, the treatment of the excitation force is very important. It was reported by Younis et al (2003) that the exact form of the forcing function must be used to numerically solve the equations of motion accurately. It was shown that when the forcing function was Taylor expanded out to third order before solving, that erroneous results were predicted in amplitudes larger than 0.5 of the gap. In addition, for the solutions to converge, the number of terms in Eq. (15.26) must be at least $N = 5$ (Caruntu et al, 2013b,c).

To implement the ROM, Eq. (15.4) is first multiplied by $(1-u)^4$ in order to eliminate any displacement terms from appearing in the denominator. Using Eq. (15.26) and (15.14), multiplying the resulting equation by $\varphi_n(z)$ and integrating from $z = 0$ to 1, the following system of equations results

$$\begin{aligned}
&\sum_{i=1}^{N} \ddot{u}_i h_{ni} - 4\sum_{ij=1}^{N} \ddot{u}_i u_j h_{nij} + 6\sum_{ijk=1}^{N} \ddot{u}_i u_j u_k h_{nijk} - 4\sum_{ijkl=1}^{N} \ddot{u}_i u_j u_k u_l h_{nijkl} \\
&+ \sum_{ijklm=1}^{N} \ddot{u}_i u_j u_k u_l u_m h_{nijklm} + b^* \sum_{i=1}^{N} \dot{u}_i h_{ni} - 4b^* \sum_{ij=1}^{N} \dot{u}_i u_j h_{nij} \\
&+ 6b^* \sum_{ijk=1}^{N} \dot{u}_i u_j u_k h_{nijk} - 4b^* \sum_{ijkl=1}^{N} \dot{u}_i u_j u_k u_l h_{nijkl} \\
&+ b^* \sum_{ijklm=1}^{N} \dot{u}_i u_j u_k u_l u_m h_{nijklm} + \sum_{i=1}^{N} \omega_i^2 u_i h_{ni} - 4\sum_{ij=1}^{N} \omega_i^2 u_i u_j h_{nij} \\
&+ 6\sum_{ijk=1}^{N} \omega_i^2 u_i u_j u_k h_{nijk} - 4\sum_{ijkl=1}^{N} \omega_i^2 u_i u_j u_k u_l h_{nijkl} \\
&+ \sum_{ijklm=1}^{N} \omega_i^2 u_i u_j u_k u_l u_m h_{nijklm} = \delta V^2 \Bigg[(1+f)\sum_{i=1}^{N} h_n \\
&- (2+3f)\sum_{i=1}^{N} u_i h_{ni} + (1+3f)\sum_{ij=1}^{N} u_i u_j h_{nij} + f\sum_{ijk=1}^{N} u_i u_j u_k h_{nijk} \Bigg] + \alpha h_n
\end{aligned} \tag{15.27}$$

where $n = 1, 2, \ldots, N$, and $i, j, k, l, m = 1, 2, \ldots, N$, and

$$h_n = \int_0^1 \varphi_n dz,\ h_{ni} = \int_0^1 \varphi_n \varphi_i dz,\ h_{nij} = \int_0^1 \varphi_n \varphi_i \varphi_j dz,$$
$$h_{nijk} = \int_0^1 \varphi_n \varphi_i \varphi_j \varphi_k dz,\ h_{nijkl} = \int_0^1 \varphi_n \varphi_i \varphi_j \varphi_k \varphi_l dz, \tag{15.28}$$
$$h_{nijklm} = \int_0^1 \varphi_n \varphi_i \varphi_j \varphi_k \varphi_l \varphi_m dz$$

Equations (15.27) form a system of N non-explicit coupled, nonlinear ordinary-differential equations. A finite number of terms N are used in Eq. (15.27). $N = 2, 3, 4$, and 5 were the ROMs considered.

15.6 Numerical Simulations

The case of uniform cantilever beams is considered. The dimensionless cantilevers mode shapes, Eqs. (15.13) and (15.14) are reported by Weaver Jr et al (1990) and given by

$$\varphi_k(z) = -\left\{\cos\left(\sqrt{\omega_k} z\right) - \cosh\left(\sqrt{\omega_k} z\right) + C_k\left[\sin\left(\sqrt{\omega_k} z\right) - \sinh\left(\sqrt{\omega_k} z\right)\right]\right\} \tag{15.29}$$

where ω_k are the dimensionless natural frequencies. These frequencies and the constant coefficients C_k of the first five modes of vibration are given in Table 15.1 (Weaver Jr et al, 1990). Substituting Eq. (15.29) into Eq. (15.19) the coefficients g_{1kk}

Table 15.1 First five natural frequencies and mode shape coefficients for uniform cantilever

	$k = 1$	$k = 2$	$k = 3$	$k = 4$	$k = 5$
ω_k	3.51602	22.0344	61.6972	120.9019	199.8595
C_k	-0.734	-1.0185	-0.9992	-1.00003	0.99999

and g_{3kk} can be obtained. The first mode $k = 1$ is considered. The g coefficients in this case are

$$g_{011} = 0.7830,\ g_{111} = 1.0000,\ g_{211} = 1.4778,\ g_{311} = 2.3488 \tag{15.30}$$

Similarly, substituting Eq. (15.29) into Eq. (15.28), the h coefficients of Eq. (15.27) are calculated. Table 15.2 gives the constants of the system. Table 15.3 shows values of physical characteristics of a typical nano-beam. This leads to realistic values of the dimensionless parameters given by Eqs. (15.6), and shown in Table 15.4. Substituting the values given in Table 15.4 and Eq. (15.30) into the steady-state Eqs. (15.24) and (15.25), the amplitude-frequency relationships are obtained.

Table 15.2 System Constants

Planck's constant/2π	$\hbar$	1.055×10^{-34} J s
Speed of light	c	2.998×10^{8} m/s
Permittivity of free space	ε_0	8.854×10^{-12} C^2/N/m^2

Table 15.3 Dimensional System Parameters

Beam width	W	125 nm
Beam length	l	132 μm
Beam thickness	h	165 nm
Initial gap distance	g	500 nm
Material density	ρ	2330 kg/m^3
Young's modulus	E	169 GPa
Quality factor	Q	350
Voltage	V_0	0.02 V

Table 15.4 Dimensionless System Parameters

Casimir parameter	α	0.01
Voltage parameter	δ	0.10
Fringe parameter	f	0.26
Damping parameter	b^*	0.01

Figure 15.2 shows the amplitude-frequency response using three different methods: MMS, 5T ROM AUTO, and 5T Time Response. In the horizontal axis is the detuning frequency σ, and in the vertical axis U_{max} the amplitude of the tip of the cantilever. MMS is a perturbation method used to solve a ROM with one mode of vibration and predict the amplitude-frequency response, Eqs. (15.24) and (15.25). 5TROM AUTO is a continuation and bifurcation analysis of ROM with five modes of vibration, Eqs. (15.27) with $N = 5$, by using the software package AUTO 07p in order to predict the amplitude-frequency response. 5T Time Response is a numerical integration of ROM with five modes of vibration, Eqs. (15.27) with $N = 5$, using ode15s, a MATLAB solver of ordinary differential equations, in order to obtain time responses of the structure. The three methods can be seen to be in good agreement at amplitudes lower than 0.4 of the gap. However, at higher amplitudes, MMS overestimates the amplitudes. At larger amplitudes 5T ROM AUTO and 5T Time responses are in good agreement.

The amplitude-frequency response consists of zero-amplitude steady-states, and two non-zero steady-state amplitude branches. The stable and unstable steady-state solutions are represented by solid and dashed lines, respectively. Nontrivial amplitudes resulting from Eqs. (15.24) and (15.25) are shown in Fig. 15.2. The solid

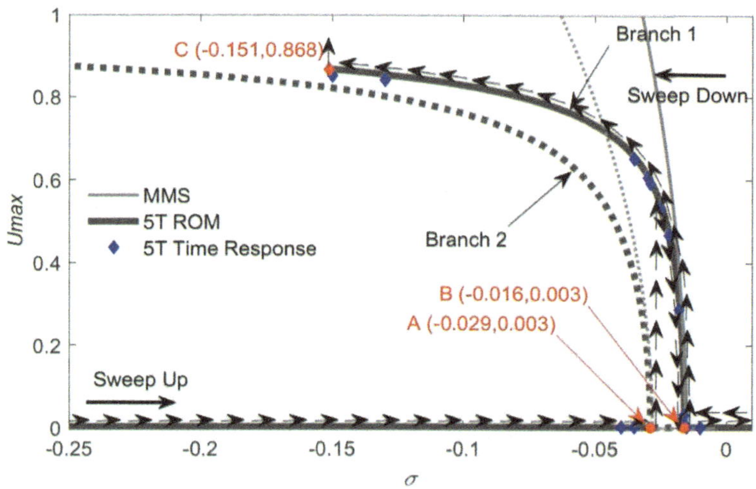

Fig. 15.2 Comparison between MMS, 5T ROM AUTO, and 5T Time Responses. $\alpha = 0.01$, $\delta = 0.1$, $b^* = 0.01$, and $f = 0.26$

branch 1, shows the stable steady-state amplitudes. The dashed branch 2, shows the unstable steady-state amplitudes. Zero amplitude solutions exist for all frequencies. The zero-amplitude solution is stable except for the detuning parameter values of $\sigma_A \leq \sigma \leq \sigma_B$. Two distinct bifurcations, one subcritical and the other one supercritical, are shown. Bifurcation points A and B are subcritical and supercritical bifurcation points, respectively. The results shown in Fig. 15.2 are similar to those reported by Rhoads et al (2006) for a parametrically excited comb drive. Zhu et al (2007) also obtained similar results in nonlinearly coupled micromechanical oscillators where a double pitchfork bifurcation was found with softening like characteristics. One can notice a softening nonlinear behavior of the system, i.e. the non-zero amplitude branches are bent towards lower frequencies. As the frequency is swept up, the system has zero steady-state amplitude until reaching the subcritical bifurcation point A. At this point the system loses stability and the system jumps to larger amplitudes located on branch 1. As the sweeping of frequency continues, the amplitude decreases along branch 1 until it reaches the supercritical bifurcation point B. After this point, the response continues with zero amplitudes. When the frequency is swept down, the system has zero steady-state amplitude until it reaches bifurcation point B. The amplitude starts increasing along branch 1 until it reaches the end of the branch, point C. After this point the system loses stability and pull-in occurs.

It should be noted that the results of Fig. 15.2 will never exceed unity since the beam displacement of the tip is being considered relative to the initial gap between the beam and ground electrode. If the beam's displacement does reach or approach unity it will experience a pull-in phenomenon. The only stable non-zero amplitudes are found in a band around $-0.151 < \sigma < -0.016$. For frequencies to the right of the stable branch, $\sigma > -0.016$, all steady-state amplitudes are zero. For any given frequency to the left of the unstable branch, $\sigma < -0.029$, depending on the initial amplitude, the nano-cantilever settles either to zero steady-state amplitude, or larger steady-state amplitude on branch 1, or experiences pull-in.

Figure 15.3 shows the effect of the Casimir parameter on the amplitude-frequency response. As the Casimir force increases the response is shifted to lower frequencies. Both branches, along with the bifurcation points, are shifted as a whole. While the bifurcation points A and B are shifted to lower frequencies, the gap between them does not significantly change. However, the softening effect increases, i.e. the non-zero amplitude branches 1 and 2 are bent to a larger degree towards lower frequencies. Therefore the system starts experiencing lower amplitudes with greater Casimir parameter. The unstable branch experiences less bending from the influence of the Casimir force than the stable branch. This makes the stable branch get closer to the unstable branch, reducing the gap between both branches. The end point C of the stable branch 1 is shifted to lower frequencies as the Casimir force parameter increases.

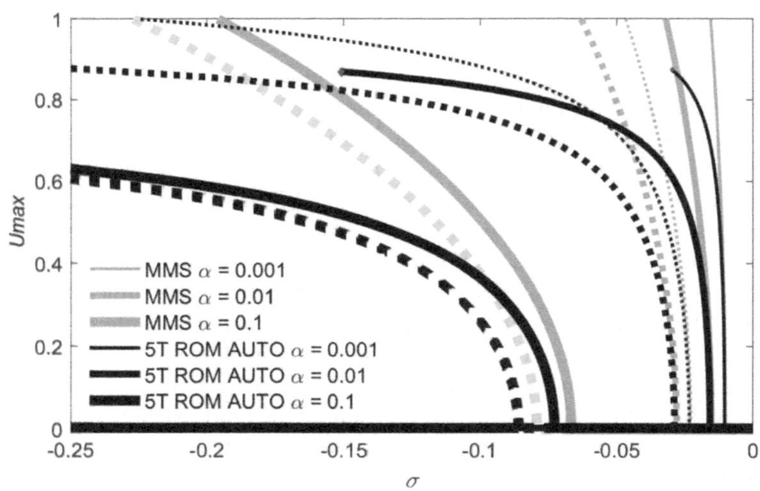

Fig. 15.3 Casimir influence on the frequency response using MMS and 5T ROM AUTO. $\delta = 0.1$, $b^* = 0.01$, and $f = 0.26$

Figures 15.4 through 15.6 show the time responses for specific frequencies and initial amplitudes. These time responses are in agreement with 5T ROM AUTO branches in Fig. 15.2. Time responses in Figures 15.4 and 15.5 settle to steady-state amplitudes on stable branches. For initial amplitudes below and above the unstable branch one can see that different steady-state amplitudes are reached for the same frequency, Figure 15.5 c) and d). Figure 15.6 shows two time responses in agreement with the predicted last stable amplitude, point C, achieved at high amplitudes. A slight change in the frequency in Figure 15.6 shows that the nano-cantilever either settles to a steady-state amplitude on branch BC or experiences pull-in.

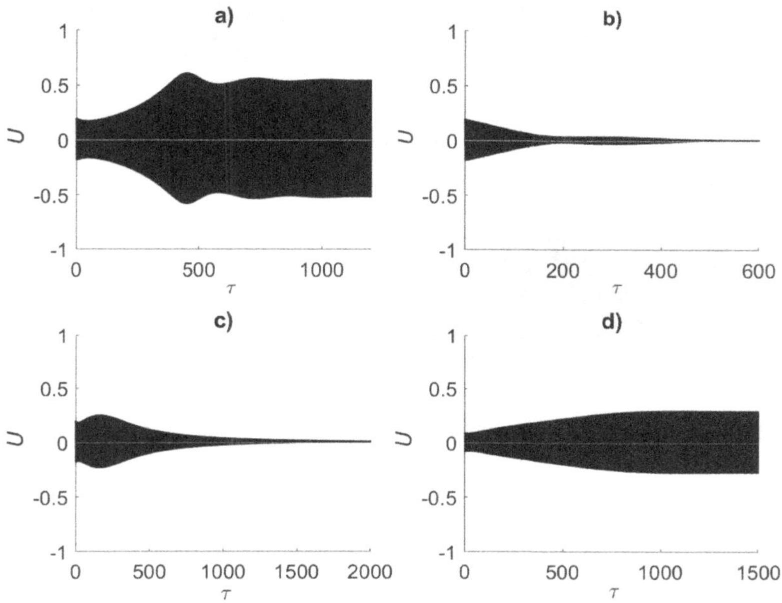

Fig. 15.4 Time Responses using five term (5T) ROM. $b^* = 0.01$, $f = 0.26$, $\delta = 0.1$, $\alpha = 0.01$ and a) $\sigma = -0.025$ with initial amplitude $U_0 = 0.2$, b) $\sigma = -0.035$ with initial amplitude $U_0 = 0.2$, c) $\sigma = -0.015$ with initial amplitude $U_0 = 0.2$, d) $\sigma = -0.018$ with initial amplitude $U_0 = 0.1$.

Figure 15.7 shows the convergence of the amplitude-frequency response when using MMS, 2 terms, 3 terms, 4 terms, and 5 terms ROM AUTO. When increasing the number of terms used in the ROM, the softening effect experienced in the system is better captured. The stable branch 1 with the end point C can be seen bending as the number of terms increases. For 4 and 5 term ROM AUTO the branches do not drastically change, so 5 term ROM AUTO is used in this research. The end point C of the stable branch can also be seen moving towards lower frequency as the number of terms increases. 5T ROM AUTO predicts the response and pull-in accurately.

Figure 15.8 and 15.9 show the convergence of the bifurcation points A and B with the number of terms of ROM. Both figures show that there is no significant difference between 4 and 5 term ROM AUTO, the responses overlapping each other.

Figure 15.10 shows the effect of the voltage parameter δ on the amplitude-frequency response of the resonator under Casimir force. One noticeable impact on the amplitude-frequency response is the frequency gap between the stable and unstable branches. As the voltage parameter increases, the frequency gap between

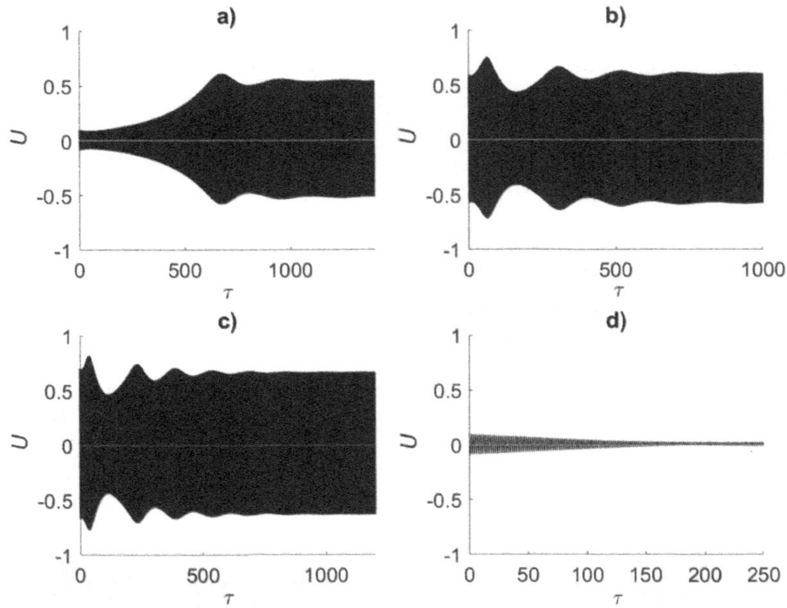

Fig. 15.5 Time Responses using five term (5T) ROM. $b^* = 0.01$, $f = 0.26$, $\delta = 0.1$, $\alpha = 0.01$ and a) $\sigma = -0.025$ with initial amplitude $U_0 = 0.1$, b) $\sigma = -0.029$ with initial amplitude $U_0 = 0.6$, c) $\sigma = -0.035$ with initial amplitude $U_0 = 0.7$, d) $\sigma = -0.035$ with initial amplitude $U_0 = 0.1$.

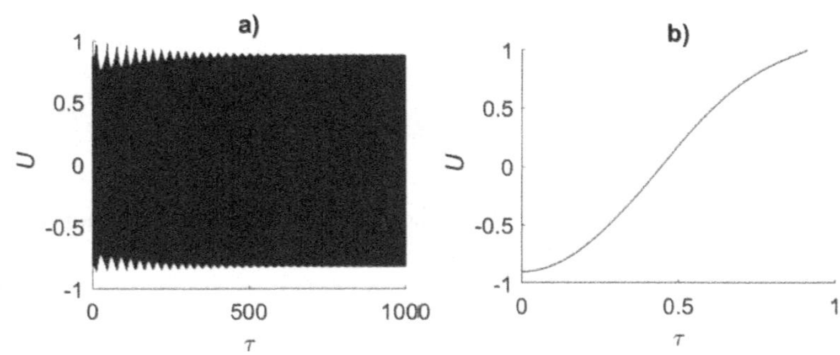

Fig. 15.6 Time Responses using five term (5T) ROM. $b^* = 0.01$, $f = 0.26$, $\delta = 0.1$, $\alpha = 0.01$ and a) $\sigma = -0.15$ with initial amplitude $U_0 = 0.87$, b) $\sigma = -0.16$ with initial amplitude $U_0 = 0.9$

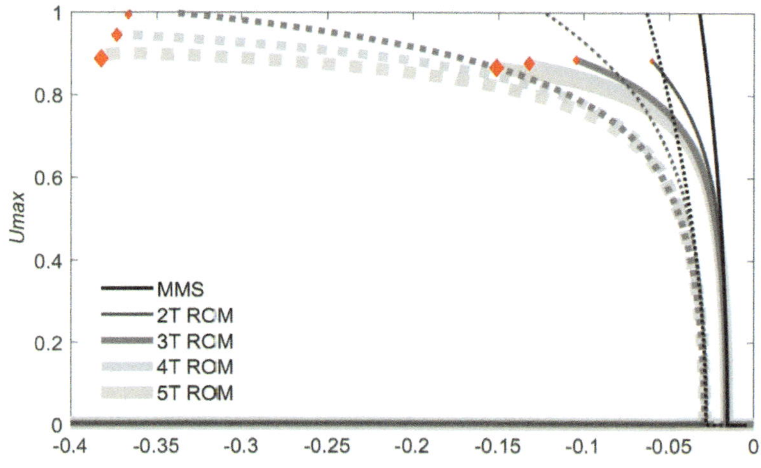

Fig. 15.7 Convergence of frequency response using MMS, 2 terms, 3 terms, 4 terms, and 5 terms ROM AUTO. $\alpha = 0.01$, $\delta = 0.1$, $b^* = 0.01$, and $f = 0.26$

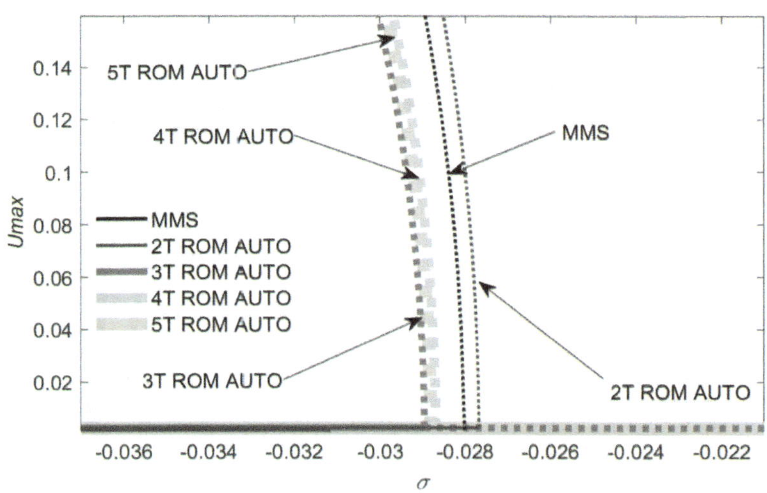

Fig. 15.8 Zoom in showing the convergence of the subcritical bifurcation point A with the number of terms in ROM.

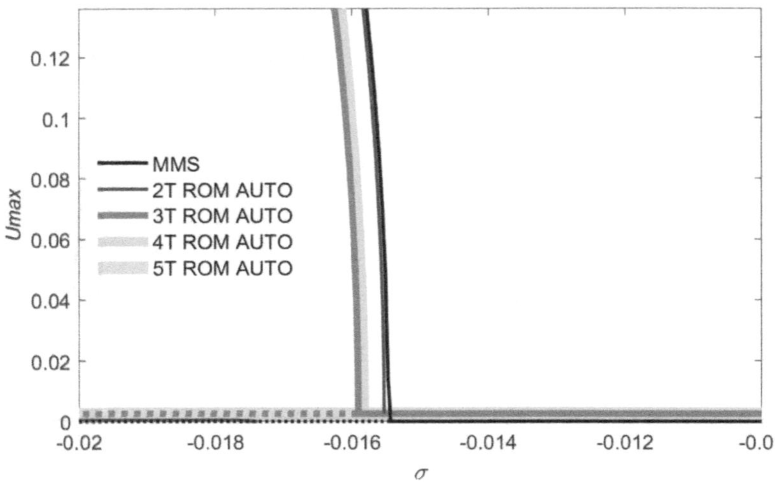

Fig. 15.9 Zoom in showing the convergence of the supercritical bifurcation point B with the numbers of terms in ROM.

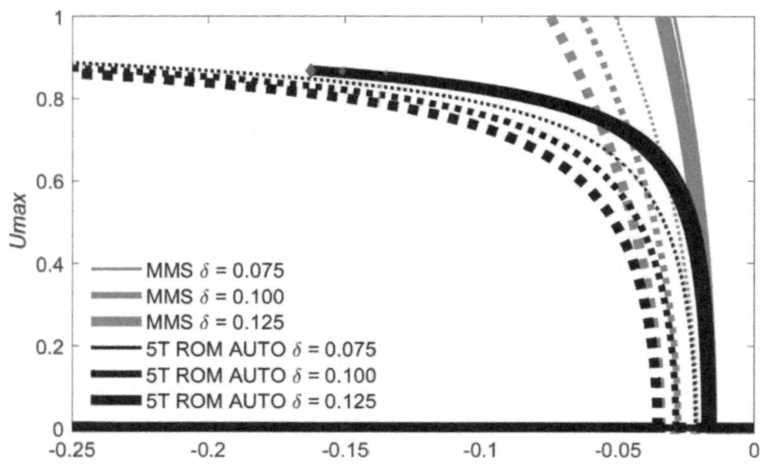

Fig. 15.10 Voltage influence on the frequency response using MMS and 5T ROM AUTO. $\alpha = 0.01$, $b^* = 0.01$, and $f = 0.26$

the stable and unstable branch increases due to shifting of the unstable branch, and the subcritical bifurcation point A, to lower frequencies. The stable branch and the supercritical bifurcation point B does not have a significant shifting. So, the zero-amplitude region between branches widens with an increase in the voltage parameter. Therefore, a larger voltage parameter δ increases the range of values of frequencies (σ_A, σ_B) for which the resonator experiences non-zero steady-state amplitudes. The end point C of the stable branch is shifted to lower frequencies as the voltage parameter is increased. This causes the range of values of resonant frequencies (σ_B, σ_C) for which the resonator can reach non-zero steady-state amplitudes to increase.

Figure 15.11 shows the effect of fringing parameter f on the amplitude-frequency response of the resonator under Casimir force. It is seen that as the fringe parameter increases, the stable and unstable branches are shifted to lower frequencies. Similar to the voltage effect, the unstable branch along with the subcritical bifurcation point A are significantly shifted towards lower frequencies with the increase of the fringe parameter. The supercritical bifurcation point B located on the stable branch is shifted towards lower frequencies as well, but the shifting is not significant. The end point C on the stable branch is shown to keep the same amplitude but is shifted towards lower frequencies. The unstable region (σ_A, σ_B) between branches widens.

Since the value of the fringe effect parameter depends on gap to width ratio, resonators using narrow beam elements relative to the gap size should pay particular attention to the fringing that arises in the electrostatic field. The fringe effect enhances the electrostatic force.

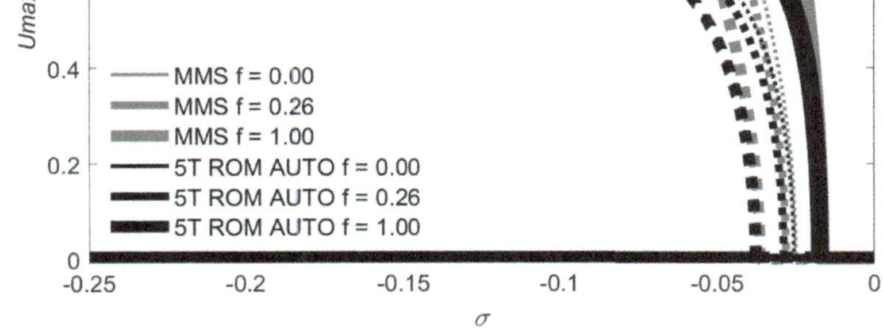

Fig. 15.11 Fringe effect on the frequency response using MMS and 5T ROM AUTO. $\alpha = 0.01$, $\delta = 0.1$, and $b^* = 0.01$

Figure 15.12 shows the effect of damping parameter on the amplitude-frequency response of the resonator under Casimir force. Increasing the damping parameter b^* the subcritical bifurcation point A is shifted to higher frequencies, while the supercritical bifurcation point B is shifted to lower frequencies. Therefore, as the damping is increased, the frequency gap between the stable and unstable branches, between points A and B as well, is reduced. At higher amplitudes, the end point C of the stable branch is shifted to higher frequencies. As for the unstable branch, there is minor change at higher amplitudes.

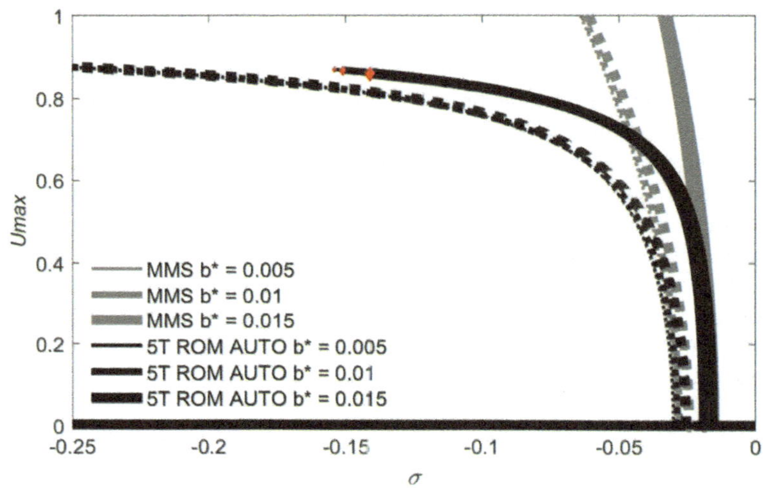

Fig. 15.12 Damping influence on frequency response using MMS and 5T ROM AUTO. $\alpha = 0.01$, $\delta = 0.1$, and $f = 0.26$

15.7 Discussion and Conclusions

This paper investigates the Casimir effect on the behavior of electrostatically actuated NEMS cantilever resonators under parametric resonance. Euler–Bernoulli beam theory was used for modeling the cantilevered beam under electrostatic actuation. AC voltage of frequency near natural frequency of the resonator was used to actuate the cantilever. This led the system into parametric resonance. The forces included in the model consisted of electrostatic force to include fringe effect, Casimir and damping forces. The effects of the dimensionless parameters on the amplitude-frequency response were investigated and reported.

After nondimensionalizing the equation of motion, two methods were used to solve the equation. The Method of Multiple Scales (MMS) was used in a direct approach to find the amplitude-frequency relationship of the system. One should mention that MMS solved a ROM with one mode of vibration. The second method used was the Reduced Order Model method with up to five modes of vibration. ROM was based on a Galerkin procedure. ROM is accurate for strong nonlinearities. The amplitude-frequency responses from these two methods are compared. Although MMS captures the system's behavior quite well, it is limited to weak nonlinearities and small amplitudes. Nayfeh et al (2007); Nayfeh and Younis (2005) reported the use of ROM to predict periodic motions. Using ROM method with five modes of vibration, the response at higher amplitudes is better captured when compared to MMS. ROM captured also the behavior of the resonator for larger amplitudes including the pull-in instability. A similar ROM was used for the cantilevered resonator considered here using up to five modes of vibration. Using four or more modes guarantees the convergence of the steady state amplitude (Nayfeh et al, 2007; Younis et al, 2003). The results of the ROM where compared with the direct (Nayfeh et al, 2007) approach using the Method of Multiple Scales (MMS) for all cases.

It is important to note than both methods are in agreement for amplitudes less than 0.4 of the gap. The ROM is more accurate for amplitudes larger than 0.4 of the gap. However, the increased accuracy comes at a cost of increased computational time.

Acknowledgements We would like to acknowledge Martin W. Knecth and Roberto J. Zapata for their contributions to this work.

References

Alsaleem FM, Younis MI, Ouakad HM (2009) On the nonlinear resonances and dynamic pull-in of electrostatically actuated resonators. Journal of Micromechanics and Microengineering 19(4):045,013

Baroudi D, Giorgio I, Battista A, Turco E, Igumnov LA (2019) Nonlinear dynamics of uniformly loaded elastica: Experimental and numerical evidence of motion around curled stable equilibrium configurations. ZAMM-Zeitschrift für Angewandte Mathematik und Mechanik 99(7):1–20

Batra RC, Porfiri M, Spinello D (2006) Electromechanical model of electrically actuated narrow microbeams. Journal of Microelectromechanical systems 15(5):1175–1189

Blom F, Bouwstra S, Elwenspoek M, Fluitman J (1992) Dependence of the quality factor of micromachined silicon beam resonators on pressure and geometry. Journal of Vacuum Science & Technology B: Microelectronics and Nanometer Structures Processing, Measurement, and Phenomena 10(1):19–26

Caruntu D, Martinez I, Knecht M (2013a) Rom analysis of frequency response of AC near half natural frequency electrostatically actuated MEMS cantilevers. ASME J Comput Nonlinear Dyn 8(3):031,011

Caruntu DI (2005) Self-adjoint differential equations for classical orthogonal polynomials. Journal of Computational and Applied Mathematics 180(1):107–118

Caruntu DI (2007) Classical Jacobi polynomials, closed-form solutions for transverse vibrations. Journal of Sound and Vibration 306(3-5):467–494

Caruntu DI (2013) Factorization of self-adjoint ordinary differential equations. Applied Mathematics and Computation 219(14):7622–7631

Caruntu DI, Juarez E (2019) Voltage effect on amplitude–frequency response of parametric resonance of electrostatically actuated double-walled carbon nanotube resonators. Nonlinear Dynamics 98(4):3095–3112

Caruntu DI, Knecht M (2011) On nonlinear response near-half natural frequency of electrostatically actuated microresonators. International Journal of Structural Stability and Dynamics 11(04):641–672

Caruntu DI, Knecht MW (2015) Microelectromechanical systems cantilever resonators under soft alternating current voltage of frequency near natural frequency. Journal of Dynamic Systems, Measurement, and Control 137(4):041,016

Caruntu DI, Martinez I (2014) Reduced order model of parametric resonance of electrostatically actuated MEMS cantilever resonators. International Journal of Non-Linear Mechanics 66:28–32

Caruntu DI, Oyervides R (2017) Frequency response reduced order model of primary resonance of electrostatically actuated MEMS circular plate resonators. Communications in Nonlinear Science and Numerical Simulation 43:261–270

Caruntu DI, Taylor KN (2014) Bifurcation type change of AC electrostatically actuated MEMS resonators due to DC bias. Shock and Vibration 2014

Caruntu DI, Martinez I, Knecht MW (2013b) Reduced order model analysis of frequency response of alternating current near half natural frequency electrostatically actuated MEMS cantilevers. Journal of Computational and Nonlinear Dynamics 8(3):031,011

Caruntu DI, Martinez I, Taylor KN (2013c) Voltage–amplitude response of alternating current near half natural frequency electrostatically actuated MEMS resonators. Mechanics Research Communications 52:25–31

Caruntu DI, Martinez I, Knecht MW (2016) Parametric resonance voltage response of electrostatically actuated micro-electro-mechanical systems cantilever resonators. Journal of Sound and Vibration 362:203–213

Caruntu DI, Botello MA, Reyes CA, Beatriz JS (2019) Voltage–amplitude response of superharmonic resonance of second order of electrostatically actuated MEMS cantilever resonators. Journal of Computational and Nonlinear Dynamics 14(3):031,005

Cheng K, Weng Z, Oliver D, Thomson D, Bridges G (2007) Microelectromechanical resonator characterization using noncontact parametric electrostatic excitation and probing. Journal of Microelectromechanical Systems 16(5):1054–1060

Cimalla V, Niebelschütz F, Tonisch K, Foerster C, Brueckner K, Cimalla I, Friedrich T, Pezoldt J, Stephan R, Hein M, et al (2007) Nanoelectromechanical devices for sensing applications. Sensors and Actuators B: Chemical 126(1):24–34

Daqaq MF, Stabler C, Qaroush Y, Seuaciuc-Osório T (2009) Investigation of power harvesting via parametric excitations. Journal of Intelligent Material Systems and Structures 20(5):545–557

DeMartini BE, Butterfield HE, Moehlis J, Turner KL (2007) Chaos for a microelectromechanical oscillator governed by the nonlinear Mathieu equation. Journal of Microelectromechanical Systems 16(6):1314–1323

Hu YC, Chang C, Huang S (2004) Some design considerations on the electrostatically actuated microstructures. Sensors and Actuators A: Physical 112(1):155–161

Ke C (2009) Resonant pull-in of a double-sided driven nanotube-based electromechanical resonator. Journal of Applied Physics 105(2):024,301

Krylov S (2008) Parametric excitation and stabilization of electrostatically actuated microstructures. International Journal for Multiscale Computational Engineering 6(6):563–584

Lamoreaux SK (2004) The Casimir force: background, experiments, and applications. Reports on progress in Physics 68(1):201

Nayfeh AH, Younis MI (2005) Dynamics of MEMS resonators under superharmonic and subharmonic excitations. Journal of Micromechanics and Microengineering 15(10):1840

Nayfeh AH, Younis MI, Abdel-Rahman EM (2007) Dynamic pull-in phenomenon in MEMS resonators. Nonlinear dynamics 48(1-2):153–163

Ramezani A, Alasty A, Akbari J (2007) Closed-form solutions of the pull-in instability in nanocantilevers under electrostatic and intermolecular surface forces. International Journal of Solids and Structures 44(14-15):4925–4941

Ramezani A, Alasty A, Akbari J (2008) Analytical investigation and numerical verification of Casimir effect on electrostatic nano-cantilevers. Microsystem Technologies 14(2):145–157

Rhoads JF, Shaw SW, Turner KL, Baskaran R (2005) Tunable microelectromechanical filters that exploit parametric resonance. Journal of Vibration and Acoustics 127(5):423–430

Rhoads JF, Shaw SW, Turner KL, Moehlis J, DeMartini BE, Zhang W (2006) Generalized parametric resonance in electrostatically actuated microelectromechanical oscillators. Journal of Sound and Vibration 296(4-5):797–829

Sedighi HM, Daneshmand F, Zare J (2014) The influence of dispersion forces on the dynamic pull-in behavior of vibrating nano-cantilever based NEMS including fringing field effect. Archives of Civil and Mechanical Engineering 14(4):766–775

Spagnuolo M, Andreaus U (2019) A targeted review on large deformations of planar elastic beams: extensibility, distributed loads, buckling and post-buckling. Mathematics and Mechanics of Solids 24(1):258–280

Weaver Jr W, Timoshenko SP, Young DH (1990) Vibration problems in engineering. John Wiley & Sons

Younis MI, Nayfeh A (2003) A study of the nonlinear response of a resonant microbeam to an electric actuation. Nonlinear Dynamics 31(1):91–117

Younis MI, Abdel-Rahman EM, Nayfeh A (2003) A reduced-order model for electrically actuated microbeam-based MEMS. Journal of Microelectromechanical systems 12(5):672–680

Zand MM, Ahmadian M (2009) Vibrational analysis of electrostatically actuated microstructures considering nonlinear effects. Communications in Nonlinear Science and Numerical Simulation 14(4):1664–1678

Zhang W, Turner KL (2005) Application of parametric resonance amplification in a single-crystal silicon micro-oscillator based mass sensor. Sensors and Actuators A: Physical 122(1):23–30

Zhang WM, Yan H, Peng ZK, Meng G (2014) Electrostatic pull-in instability in MEMS/NEMS: A review. Sensors and Actuators A: Physical 214:187–218

Zhu J, Ru C, Mioduchowski A (2007) High-order subharmonic parametric resonance of nonlinearly coupled micromechanical oscillators. The European Physical Journal B 58(4):411–421

Chapter 16
Scintillating Crystals as Continua with Microstructure

Fabrizio Daví

Abstract A scintillator material converts ionizing radiations into visible light. The process, a microscopic one related to the band gap energy in the atoms, is mediated by excited charge carriers which evolve and recombine in photons. Such recombination process evolves at a scale between the microscopic scale and the macroscopic scale of the bulk crystals. Here we show how this evolution process can be modeled with the mechanics of a continua with microstructure. By the means of thermodynamics we arrive at constitutive relations which lead to a reaction-diffusion-drift coupled system. Such a system resembles those already obtained, by starting from a different approach, for semiconductors. The mathematical study of these equations gives an insight about some properties of scintillator crystal which are confirmed by known experimental results.

Keywords: Continua with microstructure · Maxwell equations · Reaction-diffusion-drift equations · Scintillating crystals

16.1 Introduction

A scintillator is a material that converts ionizing radiations like γ- or X-rays into photons which in turn can be collected into light rays and hence plays a fundamental role in detectors for high-energy physic or medical imaging. The scintillation is a phenomena well understood, even if not completely, at the microscopic atomic scale where it depends on the band structure of the atoms (Lecoq et al, 2017; McGregor, 2018): the energy which hits the crystal generates a population of excitation carriers (charged particles) which recombine in a dissipative and non linear manner to yield photons in the visible range. The scales of observable phenomena (mainly, the

F. Daví
DICEA & ICRYS, Universitá Politecnica delle Marche, via Brecce Bianche, Ancona, Italy
e-mail: davi@univpm.it

B. E. Abali and I. Giorgio (eds.), *Developments and Novel Approaches in Biomechanics and Metamaterials*, Advanced Structured Materials 132,
https://doi.org/10.1007/978-3-030-50464-9_16

light collection, the light yield and the decay-time or timing) is the macroscopic one whereas the recombination of excitation carriers and the light generation are phenomena which take place at a mesoscopic scale which is between the microscopic one of excitation carriers production and the macroscopic one of light propagation (*vid. e.g.* Vasil'ev and Getkin, 2014).

Scintillation is an intrinsically non-linear problem that received a lot of attention from an experimental point of view (*vid. e.g.* the references in Lecoq et al (2017)): from the theoretical side what is missing is a model which links the microscopic scale with the global behavior of the bulk crystals. The problem was partially overset by two numerical tools, namely the GEANT-4 (Agostinelli et al, 2003) and LITRANI (Gentit, 2002) programs: moreover, some evolution models borrowed from the kinetics of the chemical reaction are widely used (*vid. e.g.* amongst the others Vasil'ev, 2008; Singh and Koblov, 2015; Williams et al, 2015, and the references quoted therein). However, there is the need for a consistent mechanical model, based on a limited set of parameters, to link the world of experimental results with the results of the numerical analysis performed on bulk crystals.

In order to attain such a goal, here we assume that the whole process of charge carriers recombination and photon production can be described in terms of the mechanics of a continuum with a suitable microstructure, in the sense of Capriz (1989). Indeed, provided by a suitable scaling procedure we can to define a state variable which represents at the mesoscopic scale the microscopic generation phenomena, by a judicious selection of the director we obtain a balance of microforce equation which represents, in mechanical language, the conservation of charge for a non-deformable and non-isothermal scintillator. In Daví (2019a) such an equation was postulated directly by adding an evolutionary term to the balance of microforce equation: here we show instead that, provided we introduce the notion of *self-energy* associated to the scintillation process, such a balance law follows directly by the invariance of total power as proposed in Mariano (2001). We first obtain constitutive relations from thermodynamics and then we show how the excitation carriers generation and recombination process can be described by a reaction-diffusion-drift equation (Daví, 2019a), coupled with a heat equation which admits heat sources generated by the scintillation process and with the equation of electrostatic for non-polar bodies. This result closely resembles the results obtained, by starting from different hypotheses, in semiconductors (Albinus et al, 2002).

For this set of coupled equations we obtained results about the existence, uniqueness and asymptotic behavior of the solution which are related to meaningful physical quantities like *decay time* and *light yield*, a measure of the scintillator resolution and efficiency (Daví, 2019b). In another paper (Daví, 2019c) we showed how with these results we can well estimate the measured decay time for some well-studied scintillators.

The present treatment differs from those presented into (Daví, 2019a) for the central role represented by the self-energy associated with the scintillation process: moreover we extend the model previously obtained to encompass within such an approach other phenomena like persistent luminescence. The extension to deformable continua and mechanoluminescence (light production by deformation) will be the object of a forthcoming paper (*vid.* some partial results in Daví, 2019d).

16.2 Excitation Carriers Density and Scintillation Self-Power

Let $\mathcal{B}$ a region of the three-dimensional euclidean space which we assume comprised of a dielectric, non-magnetizable and rigid scintillating crystal. When incoming ionizing radiation hits $\mathcal{B}$ at given point x and time t, the radiation energy generates a distribution of charged particles (electrons, holes, bounded electron-hole pairs called exciton) referred as *excitation carriers*. These excitation carriers which evolves in the matter according to the Bethe–Bloch equation (Inokuti, 1971; Ziegler, 1999), follow a complicated path with many kinks and bends for about 10 microns until either they lose their energy, excites other particles or recombines, generating a shower of charged particle within a region $\mathcal{P} \subset \mathcal{B}$ whose diameter is about 100 microns (Jaffe, 2007). We call the region $\mathcal{P}$ about $(x\,,t)$ the *Scintillation region*: it is within this region that the charged carriers recombine, some of them generating photons.

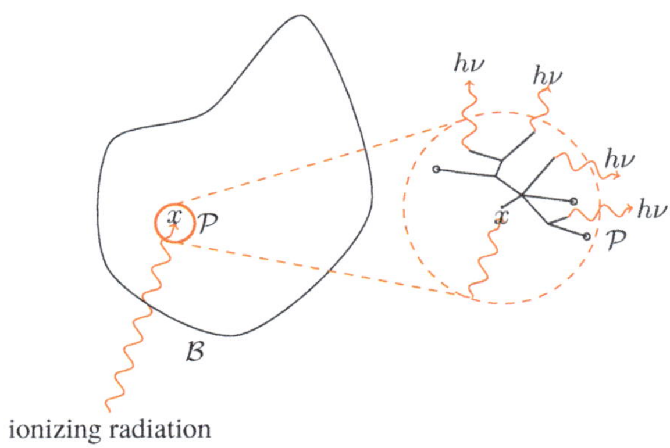

Fig. 16.1 The scintillation region $\mathcal{P}$.

In the initial stages the particles follow a straight path of few nanometers: in (Daví, 2019a) we showed how, by the means of a suitable scaling procedure of an approximate solution of the Bethe–Bloch equation (Ulmer, 2007), we may obtain a mesoscopic state variable $N = N(x\,,t)$, the *excitation carrier density*. Such a variable, which represents the particle density in the initial stage, brings to the mesoscopic scale the material properties and the ray-energy which impinges the material at $(x\,,t)$. We henceforth assume N a continuous field on the scintillation region

$$N : \mathcal{P} \times [0, \tau) \to \mathbb{R} \cup \{0\}, \quad (x, t) \mapsto N(x, t) > 0. \tag{16.1}$$

Since the charged particles may have different charge sign and also exhibits different recombination and dissipation mechanics it is useful to differentiate them by introducing an *excitation carrier vector*, the $m-$dimensional array:

$$\boldsymbol{n} \equiv (n_1, n_2, \dots, n_m), \quad \sum_{k=1}^{m} n_k(x, t) = N(x, t); \tag{16.2}$$

the simplest non-trivial case is for $m = 2$ with say n_1 representing the electrons density (which is equal to the holes density) and n_2 the exciton density, *i.e.* bounded electron-hole pairs which evolve together. The case with $m = 3$ was proposed in Williams et al (2011); Moses et al (2012) whereas in Gridin et al (2015) it was assumed $m = 7$, whereas the model proposed in Vasil'ev (2008) it is $m \geq 11$.

For e the elementary charge, excitation carriers (16.2) induce a free-charge density ρ_f within the scintillation volume:

$$\rho_f = e\boldsymbol{q} \cdot \boldsymbol{n}, \quad \text{in } \mathcal{P}, \tag{16.3}$$

where $\boldsymbol{q} = (q_1, q_2, \dots, q_m)$, $q_k \in \mathbb{Z}$ is the charge number vector.

We assume that within the scintillation region $\mathcal{P}$ there are no bound charges and hence, by the Maxwell–Lorentz equations in absence of magnetic fields (Wang, 1979), the excitation carrier densitiy induces an electric potential φ:

$$\begin{aligned} -\epsilon\Delta\varphi &= \rho^*, \quad && \text{in } \mathbb{R}^3 \\ [\![\nabla\varphi]\!] \cdot \mathbf{m} &= 0, && \text{on } \partial\mathcal{P}, \end{aligned} \tag{16.4}$$

where $\mathbf{m}$ is the outward unit normal to $\partial\mathcal{P}$ and:

$$\rho^* = \begin{cases} e\boldsymbol{q} \cdot \boldsymbol{n}, & \text{in } \mathcal{P}, \\ 0, & \text{in } \mathbb{R}^3/\mathcal{P}. \end{cases} \tag{16.5}$$

We define the energy associated with the excitation carriers as the sum of the electrostatic energy with an entropic term $\mathfrak{F}(\boldsymbol{n})$ depending on the excitation carrier density

$$S(\mathcal{P}) = \int_{\mathbb{R}^3} \frac{1}{2}\epsilon\|\nabla\varphi\|^2 \,\mathrm{d}V + \int_{\mathcal{P}} \mathfrak{F}(\boldsymbol{n}) \,\mathrm{d}V, \tag{16.6}$$

were $\mathrm{d}V$ denotes the volume element: we leave at this stage the entropic term unspecified, its nature shall be made precise later on.

Let $\Pi_{\text{self}}(\mathcal{P})$ be the *Scintillation self-power*, that is the power expended by the charge carriers in $\mathcal{P}$ on $\mathcal{P}$ itself: then, by a constitutive assumption introduced in (DeSimone and Podio Guidugli, 1996) we set:

$$\Pi_{\text{self}}(\mathcal{P}) + \dot{S}(\mathcal{P}) = 0; \tag{16.7}$$

by (16.4), from (16.7) we arrive at:

$$\Pi_{\text{self}}(\mathcal{P}) = -\int_{\mathcal{P}} (e\boldsymbol{q}\varphi + \frac{\partial \mathfrak{F}}{\partial \boldsymbol{n}}) \cdot \dot{\boldsymbol{n}} \, \mathrm{d}V \,. \tag{16.8}$$

We define *Scintillation potential*, the $m-$dimensional array whose components have the dimension of an energy

$$\boldsymbol{g}(\boldsymbol{n}) = e\boldsymbol{q}\varphi(\boldsymbol{n}) + \frac{\partial \mathfrak{F}}{\partial \boldsymbol{n}}(\boldsymbol{n}) \,, \tag{16.9}$$

and which expends power for the rate-of-change of the excitation carriers density.

16.3 Balance Laws

We assume that the scintillation region $\mathcal{P}$ is comprised of a macroscopic continuum superposed to a continuum with microstructure which, according to (Capriz, 1989) we represent with a director $\boldsymbol{d}$ defined on $(x\,,t)$ and which belongs to a manifold $\mathcal{M}$: at this stage we didn't specify nor the manifold neither the physical meaning of $\boldsymbol{d}$. The *volume microforce* $\boldsymbol{b}$ and the *surface microforce* $\boldsymbol{s}$ represent the external actions on the microstructure in $\mathcal{P}$, whereas the *microstress* $\boldsymbol{T}$ represents the internal action and the *interactive microforce* $\boldsymbol{k}$ accounts for the interaction between the microstructure and the macroscopic continua.

As far as the macroscopic state variables here we already assume that the scintillator crystal is rigid and then the only macroscopic descriptor is the *absolute temperature* $\theta = \theta(x\,,t) > 0$: accordingly we assume that on $\mathcal{P}$ be defined a *heat flux* h and a *heat supply* r; the more general case of deformable crystal will be treated into a forthcoming paper (*vid. also* Daví, 2019d).

We define the *external power* expended on $\mathcal{P}$ by the microstructure as:

$$\Pi_{\text{ext}}(\mathcal{P}) = \int_{\mathcal{P}} \boldsymbol{b} \cdot \dot{\boldsymbol{d}} \, \mathrm{d}V + \int_{\partial\mathcal{P}} \boldsymbol{s} \cdot \dot{\boldsymbol{d}} \, \mathrm{d}A \,, \tag{16.10}$$

where $\mathrm{d}A$ denote the surface element, and likewise, the *internal power*

$$\Pi_{\text{in.}}(\mathcal{P}) = \int_{\mathcal{P}} (\boldsymbol{T} \cdot \nabla\dot{\boldsymbol{d}} + \boldsymbol{k} \cdot \dot{\boldsymbol{d}}) \, \mathrm{d}V \,. \tag{16.11}$$

In order to led physical significance to (16.10) and (16.11) we identify the scintillation potential with the director velocity:

$$\boldsymbol{g} = \dot{\boldsymbol{d}} \,; \tag{16.12}$$

accordingly, $\mathcal{M} \equiv \mathbb{R}^m$, $\boldsymbol{s}$ and $\boldsymbol{T}$ represent electric currents (normalized with respect to the elementary charge e) whereas $\boldsymbol{b}$ and $\boldsymbol{k}$ represent the rates of change of charge density (again normalized with respect to e) due respectively to external and internal charges and the *Total power* expended on $\mathcal{P}$:

$$\Pi(\mathcal{P}) = \Pi_{ext}(\mathcal{P}) + \Pi_{\text{self}}(\mathcal{P}) - \Pi_{\text{int}}(\mathcal{P})\,, \tag{16.13}$$

represents the electromagnetic power associated to scintillation.

By (16.8), (16.10), (16.11), with (16.9) and by the divergence theorem we get:

$$\Pi(\mathcal{P}\,,\boldsymbol{g}) = \int_{\mathcal{P}} (\operatorname{div}\boldsymbol{T} - \boldsymbol{k} + \boldsymbol{b} - \dot{\boldsymbol{n}}) \cdot \boldsymbol{g}\,\mathrm{d}V + \int_{\partial\mathcal{P}} (\boldsymbol{s} - \boldsymbol{T}\mathbf{m}) \cdot \boldsymbol{g}\,\mathrm{d}A\,; \tag{16.14}$$

we follow (Mariano, 2001) and require that the total power (16.14) be invariant under the change of observer

$$\Pi(\mathcal{P}\,,\boldsymbol{g}) = \Pi(\mathcal{P}\,,\boldsymbol{g}^*)\,, \quad \boldsymbol{g}^* = \boldsymbol{g} + \boldsymbol{g}_o\,, \quad \boldsymbol{g}_o = \text{const.}\,, \tag{16.15}$$

to obtain the balance laws:

$$\begin{aligned} \operatorname{div}\boldsymbol{T} - \boldsymbol{k} + \boldsymbol{b} = \dot{\boldsymbol{n}}\,, &\quad \text{in } \mathcal{P} \times [0\,,\tau)\,, \\ \boldsymbol{T}\mathbf{m} = \boldsymbol{s}\,, &\quad \text{on } \partial\mathcal{P} \times [0\,,\tau)\,. \end{aligned} \tag{16.16}$$

We remark that since the director is a collection of $m-$ scalar fields there is no rotational invariance, as pointed out in Mariano (2001): moreover, the change of observer (16.14) means that the total power is invariant for scintillation potentials which differs by a constant.

As far as the director $\boldsymbol{d}$ is concerned, by (16.9) and (16.12) its components represent what in classical mechanics is called the *Action*:

$$d_k(x\,,t) = d_k(x\,,0) + \int_0^t g_k(x\,,\tau)\mathrm{d}\tau\,, \quad k = 1\,,2\,,\ldots\,,m\,. \tag{16.17}$$

16.4 Thermodynamics. Constitutive Relations

We assume that within the scintillation region $\mathcal{P}$ an *Internal energy* $\varepsilon = \varepsilon(x\,,t)$ and an *Entropy* $\eta = \eta(x\,,t)$ are well-defined. Accordingly we assume that we can write the *balance of energy* as:

$$\frac{\mathrm{d}}{\mathrm{d}t}\int_{\mathcal{P}} \varepsilon\,\mathrm{d}V = -\int_{\partial\mathcal{P}} \mathbf{h}\cdot\mathbf{m}\,\mathrm{d}A + \int_{\mathcal{P}} r\,\mathrm{d}V + \Pi_{\text{ext}}(\mathcal{P}) + \Pi_{\text{self}}(\mathcal{P})\,, \tag{16.18}$$

and the *entropy inequality* as

$$\frac{\mathrm{d}}{\mathrm{d}t}\int_{\mathcal{P}} \eta \geq \mathrm{d}V - \int_{\partial\mathcal{P}} \frac{\mathbf{h}\cdot\mathbf{m}}{\theta}\,\mathrm{d}A + \int_{\mathcal{P}} \frac{r}{\theta}\,\mathrm{d}V\,; \tag{16.19}$$

by the divergence theorem and the balance laws (16.16) the local forms of (16.18) and (16.19) read:

$$\dot{\varepsilon} = -\operatorname{div}\mathbf{h} + r + \boldsymbol{T}\cdot\nabla\dot{\boldsymbol{d}} + \boldsymbol{k}\cdot\dot{\boldsymbol{d}}\,,$$
$$\dot{\eta} \geq \frac{1}{\theta}(-\operatorname{div}\mathbf{h} + r) + \frac{1}{\theta^2}\mathbf{h}\cdot\nabla\theta\,. \tag{16.20}$$

If we introduce a *Gibbs free-energy*:

$$\psi = \varepsilon - \theta\eta\,, \tag{16.21}$$

then from (16.20) we arrive at the *reduced dissipation inequality*:

$$\dot{\psi} + \eta\dot{\theta} - \boldsymbol{T}\cdot\nabla\dot{\boldsymbol{d}} - \boldsymbol{k}\cdot\dot{\boldsymbol{d}} + \frac{1}{\theta}\mathbf{h}\cdot\nabla\theta \leq 0\,. \tag{16.22}$$

We assume as constitutive hypotheses for the scintillator crystal that:

- the Gibbs free-energy depends at most on θ, $\boldsymbol{d}$ and $\nabla\boldsymbol{d} = \boldsymbol{D}$:

$$\psi = \hat{\psi}(\theta\,,\boldsymbol{d}\,,\boldsymbol{D})\,; \tag{16.23}$$

- the microstress and the interactive microforce can be split additively into a conservative and a non-conservative (dissipative) part:

$$\boldsymbol{T} = \boldsymbol{T}^c + \boldsymbol{T}^d\,, \quad \boldsymbol{k} = \boldsymbol{k}^c + \boldsymbol{k}^d\,. \tag{16.24}$$

In force of these constitutive hypotheses, from the reduced dissipation inequality (16.22) a possible separation into a conservative and a dissipative parts leads to:

$$(\frac{\partial\hat{\psi}}{\partial\boldsymbol{D}} - \boldsymbol{T}^c)\cdot\dot{\boldsymbol{D}} + (\frac{\partial\hat{\psi}}{\partial\boldsymbol{d}} - \boldsymbol{k}^c)\cdot\dot{\boldsymbol{k}} + (\frac{\partial\hat{\psi}}{\partial\theta} + \eta)\dot{\theta}$$
$$-\boldsymbol{T}^d\cdot\nabla\boldsymbol{g} - \boldsymbol{k}^d\cdot\boldsymbol{g} + \frac{1}{\theta}\mathbf{h}\cdot\nabla\theta \leq 0\,, \tag{16.25}$$

which by customary assumptions leads to the constitutive relations:

$$\boldsymbol{T}^c(\theta\,,\boldsymbol{d}\,,\boldsymbol{D}) = \frac{\partial\hat{\psi}}{\partial\boldsymbol{D}}\,,$$
$$\boldsymbol{k}^c(\theta\,,\boldsymbol{d}\,,\boldsymbol{D}) = \frac{\partial\hat{\psi}}{\partial\boldsymbol{d}}\,, \tag{16.26}$$
$$\eta(\theta\,,\boldsymbol{d}\,,\boldsymbol{D}) = -\frac{\partial\hat{\psi}}{\partial\theta}\,;$$

and to a possible representation of the dissipative terms

$$\boldsymbol{T}^d(\theta\,,\boldsymbol{d}\,,\boldsymbol{D}\,,\boldsymbol{g}\,,\nabla\boldsymbol{g}) = \mathbb{S}(\theta\,,\boldsymbol{d}\,,\boldsymbol{D}\,,\boldsymbol{g})[\nabla\boldsymbol{g}]\,,$$
$$\boldsymbol{k}^d(\theta\,,\boldsymbol{d}\,,\boldsymbol{D}\,,\boldsymbol{g}\,,\nabla\boldsymbol{g}) = \mathbb{H}(\theta\,,\boldsymbol{d}\,,\boldsymbol{D}\,,\nabla\boldsymbol{g})\boldsymbol{g}\,, \tag{16.27}$$
$$\mathbf{h}(\theta\,,\boldsymbol{d}\,,\boldsymbol{D}\,,\boldsymbol{g}\,,\nabla\boldsymbol{g}) = -\kappa(\theta\,,\boldsymbol{d}\,,\boldsymbol{D}\,,\boldsymbol{g}\,,\nabla\boldsymbol{g})[\nabla\theta]\,,$$

where $\mathbb{S}$ and $\mathbb{H}$ are two $m \times m$ positive definite matrices and $\kappa > 0$ (the *thermal conductivity*). We remark that since (16.9), (16.12) and (16.17) $\boldsymbol{d}$, $\boldsymbol{D}$, $\boldsymbol{g}$ and $\nabla\boldsymbol{g}$ depends implicitly on $\boldsymbol{n}$, then we can formally rewrite (16.27) as in (Daví, 2019a):

$$\begin{aligned} \boldsymbol{T}^d(\theta\,,\boldsymbol{n}\,,\nabla\boldsymbol{g}) &= \mathbb{S}(\theta\,,\boldsymbol{n})[\nabla\boldsymbol{g}]\,, \\ \boldsymbol{k}^d(\theta\,,\boldsymbol{n}\,,\boldsymbol{g}) &= \mathbb{H}(\theta\,,\boldsymbol{n})\boldsymbol{g}\,, \\ \mathrm{h}(\theta\,,\boldsymbol{n}\,,\nabla\theta) &= -\kappa(\theta\,,\boldsymbol{n})[\nabla\theta]\,. \end{aligned} \tag{16.28}$$

By (16.21), (16.26) and (16.28) the local form $(16.20)_1$ of the energy balance reduces to:

$$\theta\dot{\eta} = \kappa\Delta\theta + r + \delta(\boldsymbol{g}\,,\nabla\boldsymbol{g})\,, \tag{16.29}$$

where the *dissipation* $\delta(\boldsymbol{g}\,,\nabla\boldsymbol{g})$ is defined as:

$$\delta(\boldsymbol{g}\,,\nabla\boldsymbol{g}) = \mathbb{S}[\nabla\boldsymbol{g}]\cdot\nabla\boldsymbol{g} + \mathbb{H}\boldsymbol{g}\cdot\boldsymbol{g} > 0\,; \tag{16.30}$$

if we assume a temperature-dependent entropy of the Boltzmann type, $\eta(\theta) = K\log\theta$, $K > 0$, then from (16.29) we obtain the heat equation with an additional heat supply provided by the scintillation:

$$K\dot{\theta} = \kappa\Delta\theta + r + \delta(\boldsymbol{g}\,,\nabla\boldsymbol{g})\,. \tag{16.31}$$

By using (16.24), (16.26) and (16.28) into (16.16) we arrive at the evolution equations for the excitation carrier density:

$$\begin{aligned} &\mathrm{div}(\frac{\partial\hat{\psi}}{\partial\boldsymbol{D}} + \mathbb{S}[\nabla\boldsymbol{g}]) - (\frac{\partial\hat{\psi}}{\partial\boldsymbol{d}} + \mathbb{H}\boldsymbol{g}) + \boldsymbol{b} = \dot{\boldsymbol{n}}\,, \quad \boldsymbol{g} = \dot{\boldsymbol{d}}\,, \quad \text{in } \mathcal{P}\times[0\,,\tau)\,, \\ &(\frac{\partial\hat{\psi}}{\partial\boldsymbol{D}} + \mathbb{S}[\nabla\boldsymbol{g}])\mathrm{m} = \boldsymbol{s} \quad \text{in } \partial\mathcal{P}\times[0\,,\tau)\,, \end{aligned} \tag{16.32}$$

which are coupled with the Laplace equation (16.4) and the heat equation (16.31).

In the evolution equation $(16.32)_1$ we have two different regimes: one which depends on the pair $(\boldsymbol{g}\,,\nabla\boldsymbol{g})$ and which is purely dissipative, the other which depends on the pair $(\boldsymbol{d}\,,\boldsymbol{D})$: the first regime describes the generation and recombination of excitons into photons, a process whose decay in time is fast. With the second regime, the equation describes phenomena which decays in longer times, like *e.g.* persistent luminescence and related phenomenologies.

16.5 Reaction-Diffusion-Drift Equations for Scintillators

If we assume for the Gibbs free-energy the restricted form:

$$\psi = \tilde{\psi}(\theta)\,, \tag{16.33}$$

then from $(16.26)_{1,2}$ the conservative part of the stress and interactive microforce both vanish and $\boldsymbol{T} = \boldsymbol{T}^d$, $\boldsymbol{k} = \boldsymbol{k}^d$, with (16.28) still holding. Then, from (16.16) we obtain the evolution equations in term of the scintillation potential:

$$\begin{aligned} \operatorname{div} \mathbb{S}[\nabla \boldsymbol{g}] - \mathbb{H}\boldsymbol{g} = \dot{\boldsymbol{n}}\,, &\quad \text{in } \mathcal{P} \times [0\,,\tau)\,, \\ \mathbb{S}[\nabla \boldsymbol{g}]\mathbf{m} = \mathbf{0}\,, &\quad \text{on } \partial\mathcal{P} \times [0\,,\tau)\,, \end{aligned} \tag{16.34}$$

where without loss of generality we assumed $\boldsymbol{b} = \mathbf{0}\,,\, \boldsymbol{s} = \mathbf{0}$ and which can be put in the gradient-flow form

$$D \int_{\mathcal{P}} \frac{1}{2} \delta(\boldsymbol{g}\,, \nabla \boldsymbol{g})\, \mathrm{d}V = \dot{\boldsymbol{n}}\,, \tag{16.35}$$

where D denotes the Frechet derivative and the dissipation is given by (16.30).

From (16.9) we have:

$$\boldsymbol{T} = \mathbb{S}(\theta\,, \boldsymbol{n})[\nabla \boldsymbol{g}] = \mathbb{S}(\theta\,, \boldsymbol{n})[e\boldsymbol{q} \otimes \nabla\varphi + \frac{\partial^2 \mathfrak{F}}{\partial \boldsymbol{n}^2} \nabla \boldsymbol{n}]\,; \tag{16.36}$$

let $\mathbb{N}(\boldsymbol{n})$ be the $m \times m$ matrix:

$$\mathbb{N}(\boldsymbol{n}) \equiv \begin{bmatrix} n_1 & 0 & \dots & 0 \\ 0 & n_2 & \dots & 0 \\ \vdots & \vdots & \ddots & \vdots \\ 0 & 0 & \dots & n_m \end{bmatrix}, \tag{16.37}$$

and let $\mathbb{N}^{-1}(\boldsymbol{n})$ be the $m \times m$ matrix whose entries are $1/n_k$ for $n_k \neq 0$ and 0 for $n_k = 0$. We define the $m \times m$ positive semi-definite *Mobility* matrix as

$$\mathbb{M}(\theta) = e\mathbb{S}(\theta\,, \boldsymbol{n})\mathbb{N}^{-1}(\boldsymbol{n})\,. \tag{16.38}$$

Likewise we define the positive semi-definite $m \times m$ *Diffusivity* matrix:

$$\mathbb{D}(\theta) = e^{-1}\mathbb{M}(\theta)\mathbb{N}(\boldsymbol{n})\frac{\partial^2 \mathfrak{F}}{\partial \boldsymbol{n}^2}(\boldsymbol{n})\,, \tag{16.39}$$

to arrive from (16.36) to the Nernst–Planck type relation:

$$\boldsymbol{T} = \mathbb{D}(\theta)[\nabla \boldsymbol{n}] + \mathbb{M}(\theta)[\mathbb{N}(\boldsymbol{n})\boldsymbol{q} \otimes \nabla\varphi]\,. \tag{16.40}$$

As far as the dissipative microforce is concerned, as in Daví (2019a), by following and idea of Mielke (2011) we set:

$$\boldsymbol{k} = \mathbb{H}(\theta\,, \boldsymbol{n})\boldsymbol{g} = \mathbb{K}(\theta\,, \boldsymbol{n})\boldsymbol{n}\,, \tag{16.41}$$

where the $m \times m$ *Recombination* matrix $\mathbb{K}(\theta\,, \boldsymbol{n})$ has a polynomial dependence on $\boldsymbol{n}$.

Provided these hypotheses, then from the evolution equation $(16.32)_1$ we arrive at the same reaction-diffusion-drift equation obtained in (Daví, 2019a)

$$\operatorname{div}(\mathbb{D}(\theta)[\nabla \boldsymbol{n}] + \mathbb{M}(\theta)[\mathbb{N}(\boldsymbol{n})\boldsymbol{q} \otimes \nabla\varphi]) - \mathbb{K}(\boldsymbol{n})\boldsymbol{n} = \dot{\boldsymbol{n}}\,, \qquad \text{in } \mathcal{P} \times [0\,,\tau)\,,$$

$$\mathbb{D}(\theta)[\nabla \boldsymbol{n}]\mathbf{m} = \mathbf{0}\,, \qquad \text{in } \partial\mathcal{P} \times [0\,,\tau)\,. \tag{16.42}$$

where we took into account the boundary condition $(16.4)_2$: also, from (16.16) we are able to write the dissipation (16.30) as

$$\int_{\mathcal{P}} \delta\, \mathrm{d}V = -\int_{\mathcal{P}} \boldsymbol{g} \cdot \dot{\boldsymbol{n}}\, \mathrm{d}V = \Pi_{\text{self}}(\mathcal{P}) > 0\,, \tag{16.43}$$

which shows that in this case the self-power has a dissipative nature, that the evolution equation is the gradient flow of the self power and that the scintillation potential opposes itself to the rate-of-change of excitation carriers densitry.

Equation $(16.42)_1$ is the same as proposed into (Vasil'ev, 2008) by following ideas given by (Fok, 1964) and (Antonov-Romanovskiy, 1966) and which was widely used in its *Kinetic* or *Diffusive* approximation (disregarding respectively the microstress or the interactive microforce) in many experimental and theoretical papers like *e.g* Bizzarri and Dorenbos (2007); Bizzarri et al (2009b,a); Li et al (2011); Williams et al (2011); Singh (2011); Moses et al (2012); Khodyuk and Dorenbos (2012); Grim et al (2012); Khodyuk et al (2012); Singh and Koblov (2015); Williams et al (2015); Gridin et al (2015); Lu et al (2015); Vasil'ev (2017); Lu et al (2017).

The same result was obtained, by starting from a different approach, in Albinus et al (2002); Mielke (2011) for semiconductors.

We look now in some details to the entropic term $\mathfrak{F}(\boldsymbol{n})$: in Daví (2019a) it was assumed to represent the Gibbs entropy of the excitation carriers density :

$$\mathfrak{F}(\boldsymbol{n}) = \theta k_B \sum_{k=1}^{m} n_k(\log(C_k n_k) - 1)\,, \tag{16.44}$$

where C_k are normalizing constants and k_B is the Boltzmann constant. It easy to show that:

$$\frac{\partial^2 \mathfrak{F}}{\partial \boldsymbol{n}^2} = \theta k_B \mathbb{N}^{-1}\,, \tag{16.45}$$

and then from our definition (16.39) of Diffusivity matrix we are able to recover the Einstein-Smoluchowsky relation:

$$\mathbb{D} = \frac{\theta k_B}{e}\mathbb{M}\,. \tag{16.46}$$

Under such an hypothesis into (Daví, 2019b) we show how, by adapting to the results obtained by Desvillettes and Fellner (2006); Fischer (2017); Chen and Jüngel (2017); Fellner and Kniely (2018); Chen and Jüngel (2019) in the case $m = 2$ we may proof the existence of renormalized and weak solutions for (16.42) as well as we may obtain upper-bound estimates for the asymptotic behaviour of the solution. This last result is very important because is an estimate of one of the most important properties of scintillators, the *decay time*. To this regard, into Daví (2019c) we took the experimental data available for four well-studied scintillating crystals and applied the result obtained into Daví (2019b) to estimate the decay time. The results obtained were a very good estimate of the decay times experimentally measured for some scintillators. The method would also be helpful, together with computer-aided methods, to design crystals with faster decay times.

16.6 Conclusion

We obtained a mechanical thermodynamically-consistent model for the evolution of charge carriers in scintillators, much has it was done in semiconductors by (Albinus et al, 2002). The novelty of our approach is that the results were obtained within the mechanics of continua with microstructure (Capriz, 1989) and that the relevant balance law was obtained by the requirement of invariance of the total power, taking into account the electronic self-power generated by the excitation carriers population.

The coupled reaction-diffusion-drift systems which we obtained has obtained a great degree of attention in these recent years (Fischer, 2017; Chen and Jüngel, 2017; Fellner and Kniely, 2018; Chen and Jüngel, 2019), an instance which allows to obtain a good mathematical estimate of the scintillator decay time, besides other results concerning *e.g.* the existence of solutions. These results are in accordance with the existing experimental data.

However, there are still many points which deserve further investigations:

- The relevant mathematical result concerning the decay time holds for $m = 2$ (which is enough for classical semiconductor models): it has to be extended to the case $m > 2$;
- the purely dissipative reaction-diffusion-drift system (16.42) is a special case of a more general system (16.32) which contains conservative terms. The insight is that these terms should describe phenomena like the persistent luminescence which is in some cases associated to scintillation. A complete constitutive theory which can explain these behaviors, which are experimentally well documented, is still missing;
- the model has to be extended to the deformable case (as in Xiao and Bhattacharya (2008) for semiconductors) to account also for defects associated with the residual stress due to crystal growth process and to account for radiation damage and viscous-like phenomena due to transient stress. Some partial results are presented into Daví (2019d);
- in (Mielke, 2011), the advantages and limits of the choice (16.44) are discussed in some details. Indeed it would be more appropriate to choose for the term $\mathfrak{F}(\boldsymbol{n})$ an expression based on the Fermi-Dirac statistics proposed in (Albinus et al, 2002):

$$\mathfrak{F}(\boldsymbol{n}) = k_B\theta \sum_k^m C_k\Big(z f_{\gamma-1}^{-1}(z) - f_\gamma(f_{\gamma-1}^{-1}(z))\Big), \quad z = C_k n_k\,, \tag{16.47}$$

where

$$f_{\gamma}(z) = \frac{1}{\Gamma(\gamma+1)} \int_0^{\infty} \frac{\xi^{\gamma}}{1+\exp(\xi - z)} \mathrm{d}\xi\,, \quad \gamma > -1\,. \tag{16.48}$$

Acknowledgements This research topic, which is within the scope of CERN R&D Experiment 18 "Crystal Clear Collaboration" and the PANDA Collaboration at GSI-Darmstadt, was funded by the European Research Council under the COST action TD-1401 "FAST-Fast Advanced Scintillation Timing" and from the Progetto Strategico di Ateneo 2017 "Scintillating crystals: an interdisciplinary, applications-oriented approach aimed to the scientific knowledge and process control for application concerning life-quality improvement."

References

Agostinelli S, Allison J, Amako K, Apostolakis J, Araujo H, Arce P, Asai M, Axen D, Banerjee S, Barrand G, Behner F, Bellagamba L, Boudreau J, Broglia L, Brunengo A, Burkhardt H, Chauvie S, Chuma J, Chytracek R, Cooperman G, Cosmo G, Degtyarenko P, Dell'Acqua A, Depaola G, Dietrich D, Enami R, Feliciello A, Ferguson C, Fesefeldt H, Folger G, Foppiano F, Forti A, Garelli S, Giani S, Giannitrapani R, Gibin D, Gómez Cadenas J, González I, Gracia Abril G, Greeniaus G, Greiner W, Grichine V, Grossheim A, Guatelli S, Gumplinger P, Hamatsu R, Hashimoto K, Hasui H, Heikkinen A, Howard A, Ivanchenko V, Johnson A, Jones F, Kallenbach J, Kanaya N, Kawabata M, Kawabata Y, Kawaguti M, Kelner S, Kent P, Kimura A, Kodama T, Kokoulin R, Kossov M, Kurashige H, Lamanna E, Lampén T, Lara V, Lefebure V, Lei F, Liendl M, Lockman W, Longo F, Magni S, Maire M, Medernach E, Minamimoto K, Mora de Freitas P, Morita Y, Murakami K, Nagamatu M, Nartallo R, Nieminen P, Nishimura T, Ohtsubo K, Okamura M, O'Neale S, Oohata Y, Paech K, Perl J, Pfeiffer A, Pia M, Ranjard F, Rybin A, Sadilov S, Di Salvo E, Santin G, Sasaki T, Savvas N, Sawada Y, Scherer S, Sei S, Sirotenko V, Smith D, Starkov N, Stoecker H, Sulkimo J, Takahata M, Tanaka S, Tcherniaev E, Safai Tehrani E, Tropeano M, Truscott P, Uno H, Urban L, Urban P, Verderi M, Walkden A, Wander W, Weber H, Wellisch J, Wenaus T, Williams D, Wright D, Yamada T, Yoshida H, Zschiesche D (2003) Geant4 - a simulation toolkit. Nuclear Instruments and Methods in Physics Research A506(4):250–303

Albinus G, Gajewski H, Hünlich R (2002) Thermodynamic design of energy models of semiconductor devices. Nonlinearity 15:367–383

Antonov-Romanovskiy V (1966) Photoluminescence Kinetics of Crystal Phosphors. Nauka, Moscow, (in Russian)

Bizzarri G, Dorenbos P (2007) Charge carriers and exciton dynamics in $LaBr_3$:Ce^{3+} scintillators: Experiment and model. Physical Review B 75:184,302

Bizzarri G, Cherepy N, Chong W, Hull G, Moses W, Payne S, Singh J, Valentine J, Vasil'ev A, Williams R (2009a) Progress in studying scintillator proportionality: Phenomenological model. IEEE Transaction on Nuclear Science 56:2313–2316

Bizzarri G, Moses W, Singh J, Vasil'ev A, Williams R (2009b) An analytical model of nonproportional scintillator light yield in terms of recombination rates. J of Applied Physics 105:044,507

Capriz G (1989) Continua with microstructure. Springer Tracts in Natural Philosophy, Springer Verlag, Berlin

Chen X, Jüngel A (2017) Global renormalized solutions to reaction-cross diffusion systems. preprint ArXiv177101463v1 pp 1–30

Chen X, Jüngel A (2019) Weak-strong uniqueness of renormalized solutions to reaction-cross-diffusion systems. Mathematical Models and Methods in Applied Sciences 29(02):237–270

Daví F (2019a) A Continuum Theory of Scintillation in Inorganic Scintillating Crystals. European Physical Journal B 92(1):16

Daví F (2019b) Decay time estimates by a Continuum model for Inorganic Scintillators. Crystals 9(1):4

Daví F (2019c) Light yield, decay time and reaction diffusion-drift equation in scintillators'. To appear in: Proceedings of INDAM Meeting *Harnack's inequalities and nonlinear operators*

Daví F (2019d) Mechanoluminescence in scintillators. In: Proceedings of AIMETA 2019 Conference

DeSimone A, Podio Guidugli P (1996) On the continuum theory of deformable ferromagnetic solids. Arch Rat Mech Anal 136:201–233

Desvillettes L, Fellner K (2006) Exponential decay toward equilibrium via entropy methods for reaction-diffusion equations. J Math Anal Appl 319:157–176

Fellner K, Kniely M (2018) On the entropy method and exponential convergence to equilibrium for a recombination-drift-diffusion system with self-consistent potential. Applied Mathematics Letters 79:196–204

Fischer J (2017) Weak-strong uniqueness of solutions to entropy-dissipating reaction-diffusion equations. Nonlinear Analysis 159:181–207

Fok M (1964) Introduction to Luminescence Kinetics of Crystal Phosphors. Nauka, Moscow, (in Russian)

Gentit FX (2002) Litrani: a general purpose Monte-Carlo program simulating light propagation in isotropic or anisotropic media. Nuclear Instruments and Methods in Physics Research A486(1-2):35–39

Gridin S, Belsky A, Dujardin C, Getkin A, Shiran N, Vasil'ev A (2015) Kinetic Model of Energy Relaxation in CsI:A (A=Tl and In) Scintillators. The Journal of Physical Chemistry C 119:20,578–20,590

Grim J, Li Q, Ucer K, Burger A, Bizarri G, Moses W, Williams R (2012) The roles of thermalized and hot carrier diffusion in determining light yield and proportionality of scintillators. Phys Status Solidi A 209:2421–2426

Inokuti M (1971) Inelastic Collisions of Fast Charged Particles with Atoms and Molecules - The Bethe Theory Revisited. Reviews of Modern Phys 43(3):297–347

Jaffe J (2007) Energy and length scales in scintillator nonproportionality. Nuclear Instruments and Methods in Physics Research A570:72–83

Khodyuk I, Dorenbos P (2012) Trends and patterns of scintillator nonproportionality. IEEE Transaction on Nuclear Science 59(6):3320–3331

Khodyuk I, Quarati F, Alekhin M, Dorenbos P (2012) Charge carrier mobility and non proportionality of $LaBr_3$:Ce scintillators. arXiv preprint arXiv:12095278

Lecoq P, Gektin A, Korzhik M (2017) Inorganic Scintillators for Detector Systems: Physical Principles and Crystal Engineering. Particle Acceleration and Detection, Springer International Publishing, Switzerland, second Edition

Li Q, Grim J, Williams R, Bizarri G, Moses W (2011) A transport-based model of material trends in nonproportionality of scintillators. J of Applied Physics 109:123,716

Lu X, Li Q, Bizarri G, Yang K, Mayhugh M, Menge P, Williams R (2015) Coupled rate and transport equations modeling proportionality of light yield in high-energy electron tracks: CsI at 295 K and 100 K; CsI:tl at 295 K. Physical Review B92:115,207

Lu X, Gridin S, Williams R, Mayhugh M, Gektin A, Syntfeld-Kazuch A, Swiderski L, Moszynski M (2017) Energy-Dependent Scintillation Pulse Shape and Proportionality of Decay Components for CsI:Tl: Modeling with Transport and Rate Equations. Physical Review Applied 7:014,007

Mariano P (2001) Multifield theories in mechanics of materials. Adv. Appl. Mech., Academic Press, Orlando

McGregor D (2018) Materials for Gamma-Ray Spectrometers: Inorganic Scintillators. Annual Review Material Research 48(13):1–33

Mielke A (2011) A gradient structure for reaction-diffusion systems and for energy-drift diffusion systems. Nonlinearity 24:1329–1346

Moses W, Bizzarri G, Williams R, Payne S, Vasil'ev A, Singh J, Li Q, Grim J, Chong W (2012) The Origins of Scintillator Non-Proportionality. IEEE Transaction on Nuclear Science 59(5):2038–2044

Singh J (2011) Study of nonproportionality in the light yield of inorganic scintillators. J of Applied Physics 110:024,503

Singh J, Koblov A (2015) Influence of Excitonic Processes in the Energy Resolution of Scintillators. In: Singh J, Williams R (eds) Excitonic and Photonic Processes in Materials, Springer Series in Materials Science, vol 203, Springer, Heidelberg

Ulmer W (2007) Theoretical aspects of energy-range relations, stopping power and energy straggling of protons. Radiation Physics and Chemistry 76:1089–1107

Vasil'ev A (2008) From Luminescence Non-Linearity to Scintillation Non-Proportionality. IEEE Transaction on Nuclear Science 55(3):1054–1061

Vasil'ev A (2017) Microtheory of Scintillation in Crystalline Materials. In: Korzhik M, Gektin A (eds) Engineering of Scintillation Materials and Radiation Technologies, Springer Proceedings in Physics, vol 200, Springer, Heidelberg, pp 1–32

Vasil'ev A, Getkin A (2014) Multiscale Approach to Estimation of Scintillation Characteristics. IEEE Transaction on Nuclear Science 61(1):235–245

Wang C (1979) Mathematical Principles of Mechanics and Electromagnetism. Part B: Electromagnetism and Gravitation. Springer Science, New York

Williams R, Grim J, Li Q, Ucer K, Moses W (2011) Excitation density, diffusion-drift, and proportionality in scintillators. Phys Status Solidi B 248:426–438

Williams R, Grim G, Li Q, Ucer K, Bizarri A, Burger A (2015) Scintillation Detectors of Radiation: Excitations at High Densities and Strong Gradients. In: Singh J, Williams R (eds) Excitonic and Photonic Processes in Materials, Springer Series in Materials Science, vol 203, Springer, Heidelberg

Xiao Y, Bhattacharya K (2008) A Continuum Theory of Deformable, Semiconducting Ferroelectrics. Arch Rat Mech Anal 189:59–95

Ziegler J (1999) The Stopping of Energetic Light Ions in Elemental Matter. J Appl Phys/Rev Applied Phys 85:1249–1272

Chapter 17
Modeling of Pyramidal Lattice Structures Compared to Tomographic Analysis

Valentin Hassaine-Daouadji, Rui-Pedro Carreira, Jean-François Witz, and Mathias Brieu

Abstract Architectured materials can now be easily produced thanks to recent developments in additive manufacturing processes. Such mesostructures have great potential to supplement the classical materials used for shock absorption in multiple protective applications (with expanded polystyrene (EPS) or ethylene-vinyl acetate (EVA) components), conferring security, comfort and lightness. Beyond the selected raw material, the choice of the lattice pattern and the way it is repeated directly affects the macroscopic mechanical response of the manufactured structure under compressive loading. With respect to the colossal amount of lattice shapes, a cost effectiveness design of experiment is needed in order to efficiently find the right compromise between geometries, material, and applications. Based on identified mechanical properties of a thermoplastic polyurethane (TPU) material selective laser sintered, and the characterization of several lattice structures, we propose a relevant numerical modeling tool of pyramidal lattice structures and validate its reliability and robustness.The finite element model (FEM) is based on a beam design of trusses with parameterized stiffeness at the vertices (beam intersections). The patterns studied are octet-structures of 1mm diameter beams with 35°, 45° and 55° angles. In parallel, X-Ray computed tomographic analysis performed during compressive tests provided the macroscopic static behavior and kinematic behavior of the considered structures. The tomographic images are analyzed and directly confronted to the FEM results, which enables us to improve and assess our model.

V. Hassaine-Daouadji, J. F. Witz
Univ. Lille, CNRS, Centrale Lille, FRE 2016 - LaMcube - Laboratoire de mécanique multiphysique multiéchelle, F-59000, Lille, France
e-mail: valentin.hassainedaouadji@centralelille.fr, jean-francois.witz@centralelille.fr

R. P. Carreira
DECATHLON Technical Office, DECATHLON SA, Lille, France
e-mail: ruipedro.carreira@decathlon.com

M. Brieu
California State University, Los Angeles, ECST College, Dept. Mechanical Engineering, USA
e-mail: mbrieu@calstatela.edu

B. E. Abali and I. Giorgio (eds.), *Developments and Novel Approaches in Biomechanics and Metamaterials*, Advanced Structured Materials 132,
https://doi.org/10.1007/978-3-030-50464-9_17

Keywords: Lattice structure modeling · Micro-tomographic analysis · Hessian matrix's invariants of a 3D image · Beam finite element method

17.1 Introduction

Additive manufacturing is a method for generating parts by adding successive layers of matter. The entry into the public domain of patents related to printing technologies, like the fused deposition modeling (FDM) (Crump and Muir, 1992), or the selective laser sintering (SLS) (Deckard, 1986), has been accompanied by a rapid development of increasingly accurate and innovative new printers. Today, a real effervescence, both in industry and in the general public, results from the drastic drop in machine prices. This context, which is conducive to the emergence of additive manufacturing as a means of production in its own right, encourages organizations to change their traditional design schemes and rethink conventional technical solutions (Pradel et al, 2018).

In particular, additive manufacturing makes it possible to produce complex parts and create micro-architectural forms that were not previously manufacturable. Cellular materials as well as architectural materials and lattice structures are renowned for their good shock absorption properties (Avalle et al, 2001)(Barchiesi et al, 2019), and caught the attention of sports and leisure products manufacturers (Soe et al, 2015b), (Robinson et al, 2017; Fusco and Nike, 2004). Indeed, the possibilities offered by these types of geometries in terms of design, absorption performance, lightness and breathability seem uncommon. These architectural materials could replace part of the current offer of foams conventionally used in protective and footwear products. The injection manufacturing processes for helmet EPS foams as well as the EVA foams for soles have many defects, including poor control of local foam densities and considerable edge effects (Arefmanesh et al, 1990; Gosselin and Rodrigue, 2005).

Lattice structures will have to meet, like conventional foams, requirements in terms of conformity and standards (AFNOR, 2013; CPSC, 1998), guaranteeing the safety of users. These product validations consist mainly of laboratory tests on time-consuming and resource-intensive prototypes. However, these tests have their digital counterparts which make it possible to iterate more quickly on the design of products. It is therefore necessary to understand and represent the effective mechanical behavior of these structures. An accurate numerical representation of the behavior of architectural structures then becomes essential to assess the relevance of these components as substitutes for foams.

A huge diversity of lattice patterns is generable , such as body center cubic models (Yuan et al, 2017), octet-truss structures (Deshpande et al, 2001b), Kelvin patterns and other geometries (Luxner et al, 2007; Soe et al, 2015a). In addition, there is a large number of additive materials and manufacturing processes (Yuan et al, 2018). A choice was made to limit the study; the material used is a TPU. The pattern studied is the octet-truss pattern in its bending dominated form (Deshpande et al, 2001a; Hammetter et al, 2013). This particular shape can be stiffened by simply adding horizontal beams to the pattern. This ease of behavior change could be interesting in the sizing of sports products to locally stiffen the part and thus adapt the structure if required.

The first modeling approach to represent the behavior is the raw simulation of the structure with 3D finite elements (Tancogne-Dejean et al, 2016). The modeling of a complete part and its immeasurable facets of lattice structures is too costly in terms of computation time to iterate quickly designs in a industrial development of sports products context.

A second possible method is homogenization of lattice structures. A costless modeling could be implemented by only representing the homogeneous material equivalent to a representative elementary volume of void and lattice structures. Several authors have studied the question (Bohm, 2017; Hoang, 2012) and many industrial tools are emerging to propose ergonomic solutions to the homogenization of additive manufactured structures (Omairey et al, 2019; Adam et al, 2017; Lejeunes and Bourgeois, 2011). In the current configuration, the technology and targeted sports applications does not allow more than 3 to 4 repetitions of patterns in the thickness of a helmet or sole ($2-3\,\text{cm}$ maximum). This low number of pattern replication does not allow a representative elementary volume to emerge and homogenization is therefore unsuitable for our study. Avoiding homogeneisation and directly calculating the corresponding elasticity tensor of the whole structure is a method proposed by (Milton et al, 2017).

The approach finally considered as the most realistic is the following: the modeling of lattice structures as one-dimensional beam elements structure. Severals athors decided to represent structural mechanisms in this way (Turco et al, 2016, 2018). The beam simulation seems ideal to represent a lattice network. With this structural approach the computation time is very low, and accurately represents the lattice geometry.

In order to validate the relevance of the beam modeling, different TPU lattices structures from an SLS process have been studied in quasi-static compressions monitored by tomography.

The exact strain fields and macroscopic force signals have been recorded and considered as references to assess the relevance of the subsequent modeling. In a third part, a modeling with the finite element software ABAQUS™ of these same beam structures is presented. The comparisons between the kinematics and statics of the real tests and modelings are presented to validate the beam element modelings.

As a result, we will show this approach is able to model the mechanical behavior of octet-truss lattice structures with beam elements.

17.2 Materials and Prints

The SLS process is selected for this study. Such powder bed fusion technology builds objects by melting thin layers of powder feedstock. In our case, the laser is used to melt cross sections of the truss, a layer at a time. The EOS P380 printer is used to process a TPU material with a powder layer thickness of 0.15 mm. The scan speed was 5 m/s, with a CO2 laser type and a F-theta lens.

The TPU is well known for its mechanical performances close to those of plastics and its elastic properties adjoining those of elastomeric materials. More specifically, the TPU 92A-1 used in this study has a density of 1.2 g/cm^3, it is flexible, known for its durability and its high resistance to tearing. Three different octet lattice structures

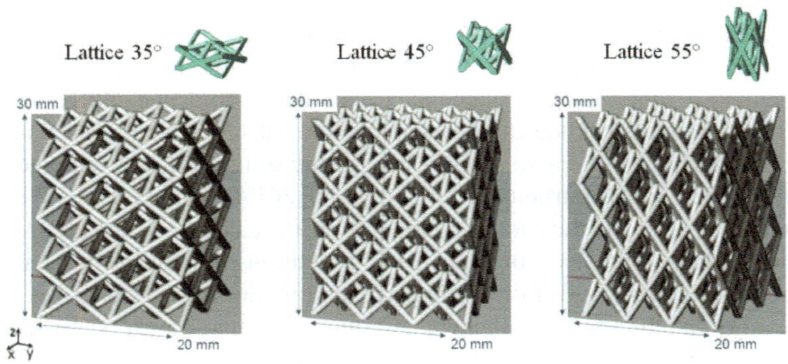

Fig. 17.1 CAD of the three lattice structures studied

in Fig.17.1 with three different angles were printed for this study. Octet patterns at 35°, 45° and 55° are repeated so that the lattice structure can match the maximum size admissible in the tomographic chamber ($20 \times 20 \times 30\,\text{mm}^3$). Each structures where designed with beams of 1 mm diameter (which is the minimal feature size admissible with this SLS technology). The resulting slenderness ratio of the 35°, 45° and 55° structures are respectively 6.10, 4.95 and 6.10 and has a volumetric density of 0.03, 0.04, and 0.04. These volumetric densities applied to the TPU material of 1200 g/L are coherent with the classical foams used by the sport and leisure manufacturers.

17.3 Mesoscopic Testings

17.3.1 Tomographic Device

The X-ray tomographic analysis (Baruchel et al, 2000; Buffiere et al, 2010) is a non-destructive characterizing technique revealing the internal composition and geometry of the tested sample. The platform Isis4D, partner of the LaMCube, propose a tungsten tomographic system which has the particularity of being added to a testing device inside the measurement chamber. This uni-axial electro-mechanical testing system enables compression and tensile testings on maximum 40 mm wide samples inside a transparent plexiglass cell. Loadings are monitored by a stepper motor gearbox added to a prestressed ball screws piston type with a displacement controlled motor. The centering of the cell on the rotation plate is ensured by a precision automated biaxial translation stage.

The whole system enables 3D images acquisition during stresses tests. The voxel size is $30\mu m^3$.

17.3.2 Compression Tests Method

The lattice structures are submitted to uniaxial compression beetween two flat plexiglass plates inside the tomographic device. The displacement apparatus is driven at very specified compressive strain rate. The upper and lower surfaces parts of the samples in contact with the plates are allowed to slide laterally. The other four faces of the samples (the lateral faces) are load-free. 3D images are taken during two loadings/unloadings cycles in order to observe if the cellular material is damaged during the first loadings. Five images per sample are performed.

- One image at the initial state with a 1 N preloading to ensure contact
- Second image at 10% compression
- Third image back to the initial state with at least 1 N, for contact preservation
- Fourth image at 25% compression
- Fifth image back to the initial state with at least 1 N, for contact preservation

Compressions are performed at 0.033 mm/s.

17.3.3 3D Images Reconstruction Method

The volume observed is obtained by digital reconstruction using a filtered back-projection algorithm (Kak and Slaney, 1987). The reconstructed volume is then a tridimensional mapping of the attenuation coefficient g_i of the tested sample (Baruchel et al, 2000). A grays segmentation method is used on these 3D images. The void zones are identified by thresholding and only the voxel corresponding to the material are considered. Many authors have already analysed lattice structures under tomographic studies (Bernal Ostos et al, 2012; Tancogne-Dejean et al, 2016; Liu et al, 2017).

17.3.4 3D Images Observations

The following figure 17.2 shows a 3D image of a lattice structure analyzed by tomography. The 3D images during the 25% compressions are reconstructed using the same method as presented just before. The image reconstruction is efficient. We are

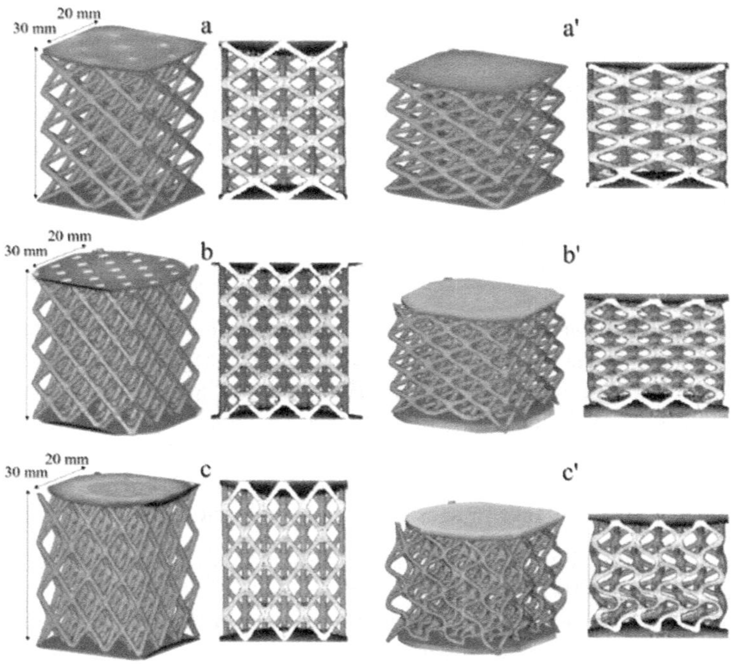

Fig. 17.2 Tomographic 3D images of the 35°, 45° and 55° structures before (left a,b,c) and during the 25% compression (right a',b',c') tests

able to recreate the lattice geometry thanks to the retro-projection filtered algorithm. The results of the tomographic pictures were noisy, especially near the compressive upper and lower plates. In fact the absorption coefficient of the constitutive material of the compressive plates (plexiglass) is very close to the one of the TPU constitutive of the studied lattice structures. The meticulous thresholding required to accurately distinguish the upper and lower vertexes from the plates induced some noisy voxels in this area. Successive erosions and dilations image treatments added with a Gaussian filter were applied to remedy this. The structures are then clearly visible with clean extremities Fig.17.2. No internal porosity can be observed. There are no apparent defects on the 3D pictures, no stratification or lamination, no impurities and no significant density variation.

On the tomographic images, we can observe on the lattice larger areas than the rest of the structure fig. 17.4. These are the areas where the beams intersect with each other that will be named vertices or vertexes in this study. These vertices are voluminous for two reasons: first, the mathematical intersection of two, three or four cylinders is indeed a three-dimensional space whose volume will depend on the angle between the bodies. A formulation of this zone is also proposed by some author (Hammetter et al, 2013). Secondly, the resolution of the printing machine does not allow these cylinder intersections to be perfectly represented and matters agglutinate in these

recesses. The selected technology of the printing machine makes the vertices larger than expected.
A very qualitative observation of those 3D images fig. 17.2 shows three different displacements behaviors for the three lattice structures. For the 35° structure, the lattice is deformed in the manner of a pantograph; the beams keep their rectilinear apparence. The trusses moved because of local bending at the vertices such that the structure seems to be folded. For the 45° structure, we can notice inflexion points for the truss and a moment appearance near the vertexes. For the 55° structure, the moments at the vertexes are dominating, there are rigid body-like rotations at the vertices and the truss are highly bent. For the three structures, the macroscopic Poisson's ratio is very small, and there is no horizontal convexity or "barrel effect."

17.3.5 3D Images Analysis

The resulting 3D images are 8bit grayscale images of approximately 1000 × 1000 × 1000 voxels. Those images are processed in order to extract some quantifiable informations useful for our study.

17.3.5.1 Beams Stability

The first useful information that can be extracted from these 3D images is the dimensional stability of the structure.

The diameters of the beams of the three structures were checked using the skeletonize and medial axis plugin from Python-skimage (Vanderesse et al, 2016). Averages analysis of several areas of the 3D images of the structures were performed. The equivalent average diameters and the associated standard deviations have been represented in the Fig. 17.3. We notice that diameter of 1 mm is well respected. The

Lattice structures	35°	45°	55°
Average beam section radius (mm)	0.466	0.479	0.419
Standard Deviation	0.054	0.066	0.037

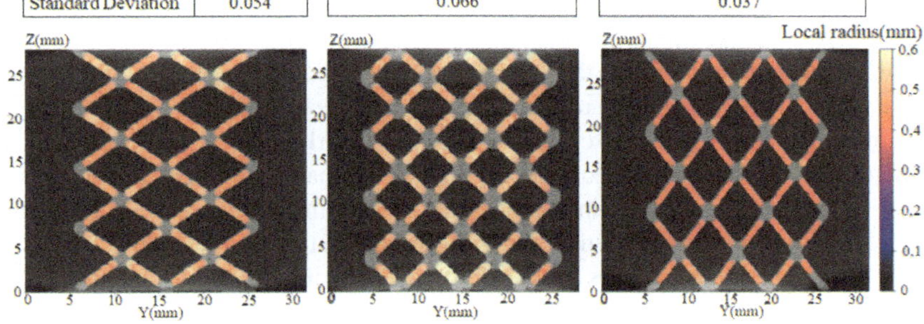

Fig. 17.3 Beam diameter analysis by Python-skeletonize-skimage plugin

overall diameter of the beam is 1 mm as drawn on the CAD.

The stability of the radius compiled in Fig. 17.3 for all the beams reassures us with the idea of modeling them with a uniform cross-section elements. The first simulation hypothesis consisted in proposing a Timoshenko beam approach with a homogeneous cross-section of 0.5 mm radius to represent the lattice.

17.3.5.2 Matter Caking at the Vertices

A second useful information is the size of the vertices areas. The size and shapes of theses interesting zones can lead to a better understanding of the mechanics of the whole structure. The way the vertices are constituted can help us to accurately represent the structure kinematics.

A cross section analysis approximate the average size of the vertices of each structures. This succession of 2D planes analysis was performed by a Python code based on thresholdings added to the labelization of wide cross section areas. Three

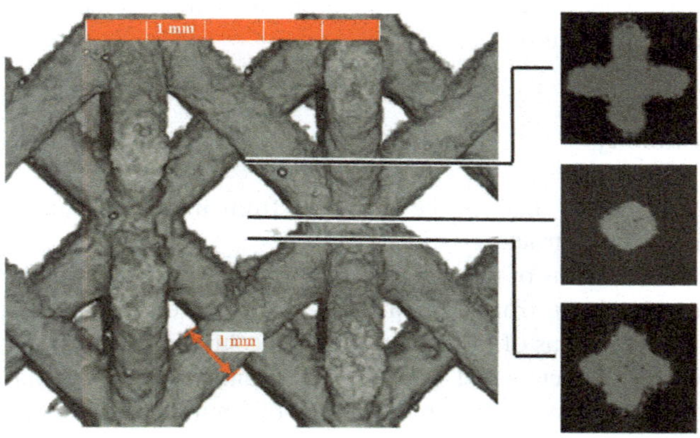

Fig. 17.4 Zoom of the tomographic picture of the lattice 45°

different vertices cross sections are presented Fig. 17.4. We consider that the vertice stops where the four beams constitutive of the vertex are clearly distinct and separated.

Three joining moment of the beams are presented by the three cutting views of the Fig. 17.4. On these 2D reduced images, the section size S can be extracted. Assuming that the section is circular (which is clearly false at some place of the vertice), we can estimate the equivalent radius $\sim R$ matching the measured section. This $\sim R$ is considered as an indicator of the stiffenning occuring near the vertices.

As a result of this cross section analysis, we understood that the vertices are very larges aeras where the stiffness must be increased by at least two to five times.

17.3.5.3 3D Images Analysis Method

A last useful information is the positions of the intersection areas between the beams i.e the coordinates of the center of the vertices.

To do so, a Gaussian filter is applied on the 3D images with :

$$G(x,y,z) = \frac{1}{\sqrt{2\pi}\sigma} e^{-\frac{x^2+y^2+z^2}{2\sigma^2}} \tag{17.1}$$

where σ is the windows of the filter in the x,y,z 3D images coordinates.

The local curvature of the 3D images can be calculated with :

$$\boldsymbol{H}(g_i) = \begin{bmatrix} \frac{\partial g_i}{\partial x^2} & \frac{\partial g_i}{\partial x \partial y} & \frac{\partial g_i}{\partial x \partial z} \\ \frac{\partial g_i}{\partial y \partial x} & \frac{\partial g_i}{\partial y^2} & \frac{\partial g_i}{\partial y \partial z} \\ \frac{\partial g_i}{\partial z \partial x} & \frac{\partial g_i}{\partial z \partial y} & \frac{\partial g_i}{\partial z^2} \end{bmatrix} \tag{17.2}$$

where g_i is the grays level intensity of the 3D images from the tomographic reconstruction, and $\boldsymbol{H}$ the Hessian matrix operator.

The results of the convolution of the gaussian filter G with the hessian $\boldsymbol{H}(g_i)$ is named H.

$$H = G(x,y,z) \circledast \boldsymbol{H}(g_i) \tag{17.3}$$

Each coefficient of H will be then identify as H_{ij}. Thanks to the Eq. (17.2) we can calculate the second invariant I_2 of H as follows:

$$\begin{aligned} I_2 &= \mathrm{tr}(\mathrm{com}(H_{ij})) \\ &= [\mathrm{tr}(H_{ij})^2 - \mathrm{tr}(H_{ij}^2)] \\ &= H_{11}H_{22} + H_{22}H_{33} + H_{33}H_{11} + H_{12}^2 + H_{23}^2 + H_{31}^2 \end{aligned} \tag{17.4}$$

The sign of the I_2 enables to distinguish the zones corresponding to the beams from the zones considered as vertices Fig. 17.5. The coordinates of the gravity center of the vertices zones are identified and recorded.

17.3.5.4 Vertices Displacements: Kinematic Reference

The analysis of this second invariant of the Hessian matrix applied to the intensity gradients of the 3D image from the tomograph enables the identification of the coordinates of each vertices of the structure. From these coordinates, a reconstruction of the lattice skeleton is performed by a Python code. The lattice skeleton as observed on the tomograph is thus reconstructed and serves as the basis for the mesh implemented in the Abaqus simulation Fig. 17.5. Similarly, the image processing

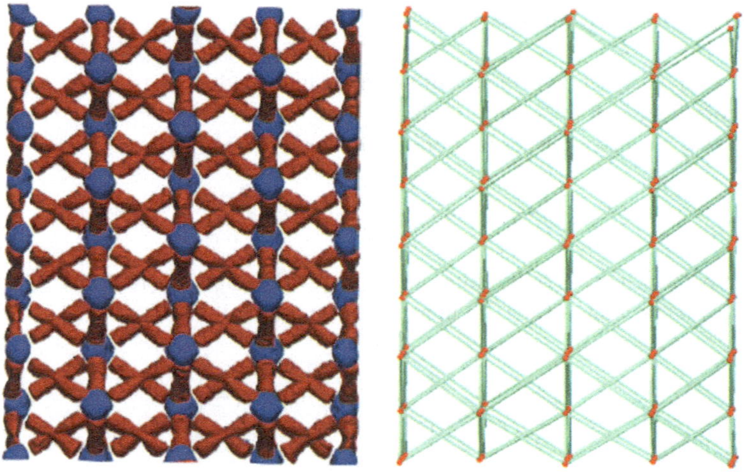

Fig. 17.5 Beam/Vertex distinction by the I_2 of the Hessian (left), Skeleton reconstruction for finite element analysis (right)

with the second invariant of the Hessian allows to distinguish the vertices from the beams on the deformed structures. Vertexes coordinates of the deformed structures are recorded. The displacement of each nodes from the upper and lower layers are imposed as boundary conditions in the Abaqus modeling. The positions of every vertices of the compressed structures will also serve as a reference for the kinematics. In fact, the upcoming modelings will be compared to the displacement field observed. The accuracies of each modeling will be quantified by a deviation criterion to those kinematics references.

17.3.5.5 Macroscopic Efforts Signals: Static Reference

The uni-axial cell registers force during the compressive loadings. For each of the three structures, the maximum load appends at the maximum strains (25%) and are synthesized in the Table 17.1.

Table 17.1 Cyclic compressive loadings measurements

Lattice Structures	35° compressive strain 0%	35° compressive strain 25%	45° compressive strain 0%	45° compressive strain 25%	55° compressive strain 0%	55° compressive strain 25%
Displacements(mm)	0.485	7.158	0.492	7.157	0.486	7.453
Efforts(N)	1	8	2	38	2	29

17.4 Simple Finite Elements Modelings

17.4.1 Simple Modelings Method

An Abaqus explicit model is implemented with the three considered architectures (at 35°, 45° and 55°). The lattice structures bases of modeling for the numerical investigation are made of Timoshenko linear beams. The exact geometry of the three lattice structures, and the exact nodal coordinates of each vertex, are determined by the tomographic analyses of the mesostructures. Beam elements are designed between the rightful nodes and the lattice structures are generated as observed. Each beam is segmented in ten elements, which is more than enough to achieve mesh convergence (Luxner et al, 2007). The 3D images of the deformed structures Fig. 17.2 shows large displacements and deflections due to rotations. We can thus assume the mechanical behavior can be modeled as an elastic behavior under large transformation. The transformation is large only because of rotation, but the strain remains low and the linear elastic Hooke's law remains relevant. The constitutive material of the beams is determined by macroscopic tensile tests on bulk samples printed with SLS. We choose a perfectly elastic linear material with $E = 30\,\text{MPa}$. The beams are considered as perfectly cylindrical with a $0.5\,\text{mm}$ radius circular

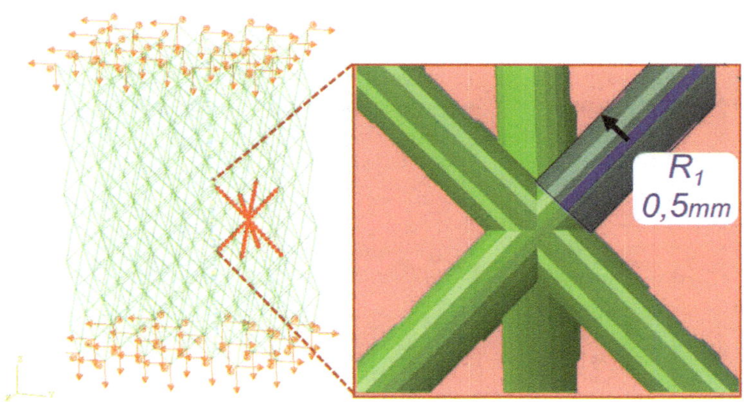

Fig. 17.6 Finite element modeling of the lattice 45° with homogeneous Timoshenko beams

section Fig. 17.6. The modeling consists in a 25% compression of the 30 mm high mesh. Node displacements are applied on the upper and lower layers while the rest of the structure is totally free. The boundary conditions of each node of the upper and lower layers are specified equal to the displacement observed during tomographic analyses of the mesostructured tests.

The modelings will be compared and validated firstly in the kinematic approach. An accuracy criterion is developed to do so: the average kinematic deviation Δ.

Table 17.2 Kinematic deviation results

Lattice structures	35°	45°	55°
Average Kinematic Deviation Δ (mm)	0.18	0.24	0.71
Standard Deviation	0.11	0.18	0.63

Table 17.3 Static deviation results

Lattice structures	35°	45°	55°
Efforts(N) Tomographe	8	38	29
Efforts(N) simulations	4.2	19.5	21.5

Very simply, Δ represents the average distance for each vertex from the observed tomographic picture (at 25% compression) to the deformed modeling result.

17.4.2 Simple Modelings Results

The 25% compression Abaqus modelings for the three structures converged well. The deformed networks are presented Fig. 17.7 for the three structures. The average kinematic deviation criteria and the resulting macroscopic forces (calculated as the sum of the forces at the upper layer nodes) are reported in Table 17.2 and Table 17.3 As observed on Fig. 17.7 the three simulations presented do not have the right

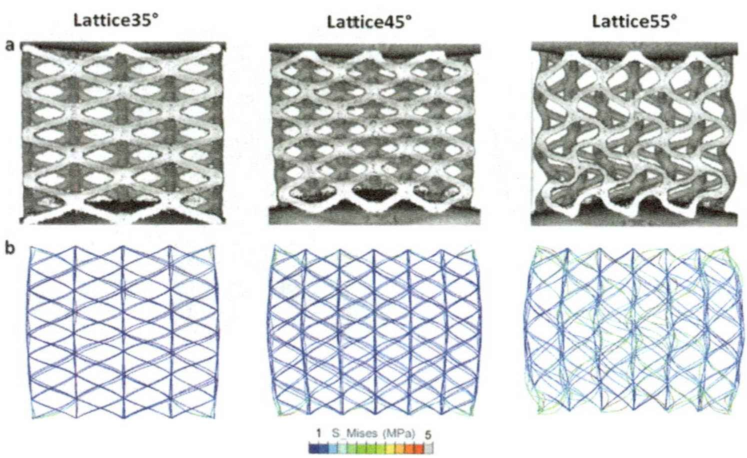

Fig. 17.7 Comparison of the kinematics of the modeling results for the three lattice structures. *Top*: tomographic pictures (a). *Bottom*: simulations results (b).

kinematics or statics behavior. In particular, the overall displacement field of the structure is not acceptable since a barrel effect, i.e. a horizontal convex swelling at half height on the structure in compression, is shown. This deviation from the reference kinematics is reflected in the value of the kinematic average deviation criterion Δ in Table 17.2, increasing with the angularity of the structures. The strength signals are also not acceptable since they are significantly lower than experimental measurements for all the three structures tab. 17.3. The volume envelope of the simulated structure does not correspond correctly. We cannot trust the stresses responses or the internal deformations of those modelings. Reviewing our approach is therefore essential.

In fact, the vertexes observed Fig. 17.4 are not nodal points, but volume areas. Those intersections localizations are zones where stresses are distributed, with each nearby beam-endings contributing to the diffusion of forces. Vertices cannot be represented only by one-dimensional beams. Our modeling approach with homogeneous beams at each central node of the vertices could not work. Despite all that, beam modeling of the filiform parts seems appropriate.

17.5 Improved Finite Elements Modeling

17.5.1 Improved Modelings Method

Finite element modelings with beam element are computationally cheap. But these straightforward modelings of the vertices by beam elements are limited by two main realities:

- These modelings do not take into account the volume area overlapping of the truss cylinders.
- The solid body existing where the beams intersect is not represented and the vertex is simply considered as a nodal point where constraints cannot exist.

Thereby, a local stiffening must be introduced in order to accurately depict the stiffness and density of the material distribution in the intersection. Authors such as Luxner et al (2005) noticed that a local stiffening is necessary to accurately represent the cohesion of the vertex with a beam finite element modeling. Their approach was to multiply the Young's modulus by 1000 in the vicinity of the vertices within a spherical domain with a radius equal to the strut radius. Our approach consists in keeping the same material and stiffening the part by changing the virtual section of the beams near the vertex. The idea is to find the size of the spherical domain around the vertex where the beam must be stiffened. We must also find the associated intensity of stiffening that will match the kinematic and static behavior of the whole structure. Two parameters are introduced: first the size of the vertex, that is to say the number of elements around the node intersections of beams that will be stiffened. Secondly, the stiffness of the vertex by increasing the circular section of these specified elements Fig. 17.8.

A design of experiment is proposed to find the good compromise between size of the vertexes and stiffness of these vertexes. The size of the vertex varies from $1/10^{\text{th}}$ of the strut length to $4/10^{\text{th}}$. The radius of the circular section of the beam

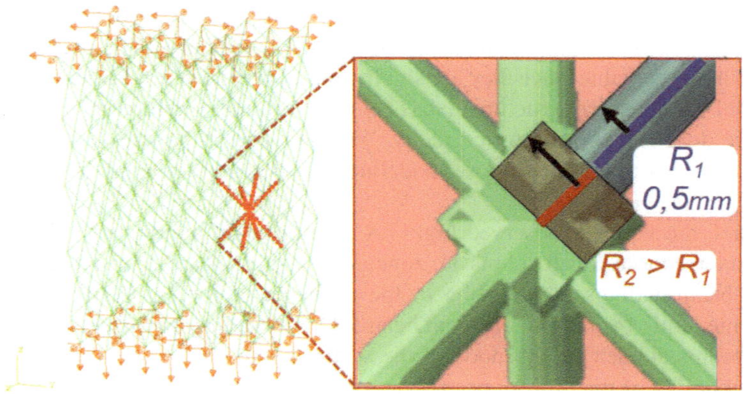

Fig. 17.8 Vertex stiffening of the lattice structure modeling by local section increase

considered as vertex varies from $0.6mm$ to $2.0mm$. For the three structures, the input Abaqus modeling files where written in each of these conditions. All calculation were computed in the same configuration of the simple homogeneous beam modeling previously presented.

From this design of experiment, several optimums emerged. The optimums chosen correspond to the condition minimizing the kinematic deviation criterion, but correspond also to the conditions closest to the real size observed in the tomographic pictures Fig. 17.4. The final parameters selected are a stiffening envelope area of $2/10^{\text{th}}$ of the length between vertices, and a radius of 1.2 mm for the cross-section of the vertices. The virtual section increasing is thereby considered as a physical reality and improves the validity of our model.

17.5.2 Improved Modelings Results

Simulations under these conditions are presented in Fig. 17.9. The calculations of the kinematic deviation criterion and the resulting forces are specified in tab. 17.4 and tab. 17.5.

We can see in Table 17.4 that we have reduced the kinematics error for the three structures by nearly 50%. More generally, the kinematics is much better since the "barrel effect" has almost disappeared as seen in Fig. 17.9. The buckling behavior of the $55°$ structure is well represented.

However, the static is still not acceptable. This time, the macroscopic efforts of the simulations are too high. This result seems logical since we are stiffening our structure in many ways. Several leads can explain these errors on the statics of our model:

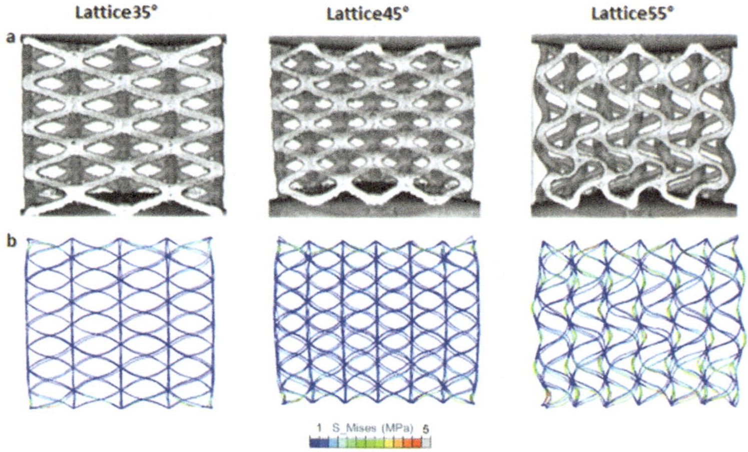

Fig. 17.9 Comparison of the kinematics of the modeling results for the three lattice structures with optimised vertices size and stiffness parameters. *Top*: tomographic pictures (a). *Bottom*: radius 1.2 mm, stiffening on $2/10^{th}$ of the beam length (b).

Table 17.4 Kinematic deviation of the stiffened simulation results

Lattice structures	35°	45°	55°
Average Kinematic Deviation(mm)	0.09	0.13	0.42
Standard Deviation	0.07	0.09	0.39

Table 17.5 Static deviation of the stiffened simulation results

Lattice structures	35°	45°	55°
Efforts(N) Tomographe	8	38	29
Efforts(N) simulations	12.4	44.6	42.3

- It already seems that the beam diameters were slightly overestimated during the simulations. Dimensional stability measurements of beams showed that the printed beams had diameters slightly smaller than 1 mm. By reducing the diameters in the abaqus modelings, the associated forces could be corrected and the static errors reduced.
- Moreover, the material proposed for the simulations is a 30 MPa Young's modulus linear elastic material. Experiments analyses have indeed shown that the TPU has this kind of behavior until a certain threshold (about 4 MPa). Once this threshold is reached, the material is damaged, there is softening and its equivalent modulus gradually decreases. Also, modeling the material by damaging behavior with modulus softening would also reduce the static response and thus reduce the current error.
- Finally, the TPU material was studied on macroscopic samples of the ISO-527 standard type. The associated macroscopic material modeling may not be representative of the material of the microbeams of the lattice structures.

With this proposed vertex stiffened model, it would be necessary to take a beam Young's modulus E_{Beam} lower than the Young's modulus of the TPU ($E_{\text{Bulk}} = 30\,\text{MPa}$) identified on macroscopic specimens. Table 17.6 summarizes the conversion to eliminate the static error of the simulations.

Table 17.6 Young's modulus reduction for static fitting

Lattice structures	35°	45°	55°
E_{Beam}	$0.65 E_{\text{Bulk}}$	$0.85 E_{\text{Bulk}}$	$0.68 E_{\text{Bulk}}$

17.6 Conclusion

Three lattice structures were printed and tested in compression under a tomographic device and compression testings were carried out. The structural behavior was described thanks to 3D images reconstructions. The places where the beams intersect, called vertices, were identified by an image analysis using the second invariant of the Hessian matrix applied to the grayscale images for these three structures. Vertexes centers movements were thus recorded. From these displacements, boundary conditions were implemented for a Timoshenko beam compression modeling. This first simulation involved only homogeneous beams with a circular cross-section of 1 mm diameter. These results were not conclusive, neither statically nor kinematically. In order to more accurately represent the agglomeration of material occurring at the intersections of the beams of the printed structures, local stiffening is carried out. This virtual stiffening consists in increasing the section to 1.2 mm radius at the two beam elements directly closest to each vertex. Under these conditions, the simulations are much better kinematically and accurately represent the overall deformation behavior for all three structures. However, an error on the static response persists. Incorrect adjustment of beam diameters, material damage and a wrong definition of Young's modulus are some of the possible explanations for these deviations. A correction of these modules has been proposed to remove the static error.

To continue this work, a possible further investigation would be to identify a hyper-visco-elastic material model for the TPU used. Transcription of these numerical models into dynamics could be used to characterize the shock absorption behavior of lattice structures. Work on the identification of contact laws would extend the analyses to more than 25% of deformation. Finally, after the calibration of these finite element modelings, we will be able to generate and digitally test a wide spectrum of geometries and thus obtain a "catalogue" of lattice mechanical behaviors. Sports and leisure manufacturers seeking to replace their shock absorbing components with this type of structure will then be able to choose from this database.

Acknowledgements The tomographic analysis were managed thanks to the ISIS4D X-Ray CT platform. This platform has been funded by International Campus on Safety and Intermodality in Transportation(CISIT), the Nord-Pas-de-Calais Region, the European Community and the National Center for Scientific Research. The authors gratefully acknowledge the support of these institutions.

References

Adam L, Assaker R, Hébert P, Lietaer O, Mathieu S (2017) Additive Manufacturing Simulation Solution as Enabler for Confident Lightweight Automotive Design. e-Xstream engineering Digimat Users' Manual(Release 2017.1.):22

AFNOR (2013) NF EN 1078+A1 Février 2013 Casques pour cyclistes et pour utilisateurs de planches à roulettes et de patins à roulettes

Arefmanesh A, Advani SG, Michaelides EE (1990) A numerical study of bubble growth during low pressure structural foam molding process. Polymer Engineering and Science 30(20):1330–1337

Avalle M, Belingardi G, Montanini R (2001) Characterization of polymeric structural foams under compressive impact loading by means of energy-absorption diagram. International Journal of Impact Engineering 25(5):455–472

Barchiesi E, Spagnuolo M, Placidi L (2019) Mechanical metamaterials: a state of the art. Mathematics and Mechanics of Solids 24(1):212–234

Baruchel J, Buffiere JY, Maire E, Merle P, Peix G (2000) X-Ray Tomography in Material Science, hermes science publications edn

Bernal Ostos J, Rinaldi R, Hammetter C, Stucky G, Zok F, Jacobsen A (2012) Deformation stabilization of lattice structures via foam addition. Acta Materialia 60(19):6476–6485

Bohm HJ (2017) A Short Introduction to Basic Aspects of Continuum Micromechanics. ILSB Report / ILSB-Arbeitsbericht 206 supersedes CDL FMD Report 3 1998(Institute of Lightweight Design and Structural Biomechanics (ILSB) Vienna University of Technology):183

Buffiere JY, Maire E, Adrien J, Masse JP, Boller E (2010) In Situ Experiments with X ray Tomography: an Attractive Tool for Experimental Mechanics. Experimental Mechanics 50(3):289–305

CPSC (1998) US CPSC 1203 : Safety Standard for Bicycle Helmets

Crump SS, Muir AEPD (1992) Apparatus and Method for Creating Three-Dimensional Objects

Deckard CR (1986) Apparatus for producing parts by selective sintering

Deshpande V, Ashby M, Fleck N (2001a) Foam topology: bending versus stretching dominated architectures. Acta Materialia 49(6):1035–1040

Deshpande VS, Fleck NA, Ashby MF (2001b) Effective properties of the octet-truss lattice material. J Mech Phys Solids 49(1747–1769):23

Fusco C, Nike I (2004) Footwear Sole Incorporating a Lattice Structure

Gosselin R, Rodrigue D (2005) Cell morphology analysis of high density polymer foams. Polymer Testing 24(8):1027–1035

Hammetter CI, Rinaldi RG, Zok FW (2013) Pyramidal Lattice Structures for High Strength and Energy Absorption. Journal of Applied Mechanics 80(4):041,015

Hoang DH (2012) Contribution à l'homogénéisation de matériaux hétérogènes viscoélastiques : milieux aléatoires et périodiques et prise en compte des interfaces. PhD thesis, Université Paris-Est Marne la Vallée, Laboratoire Modélisation et Simulation Multi Echelle (UMR CNRS 8208)

Kak AC, Slaney M (1987) Principles of Computerized Tomographic Imaging, ieee press edn. The Instbte of Electrical and Electronics Engineers, Inc., New York

Lejeunes S, Bourgeois S (2011) Une Toolbox Abaqus pour le calcul de propriétés effectives de milieux hétérogènes. 10e Colloque National en Calcul des Structures CSMA 2011:8

Liu L, Kamm P, Garcia-Moreno F, Banhart J, Pasini D (2017) Elastic and failure response of imperfect three-dimensional metallic lattices: the role of geometric defects induced by Selective Laser Melting. Journal of the Mechanics and Physics of Solids 107:160–184

Luxner MH, Stampfl J, Pettermann HE (2005) Finite element modeling concepts and linear analyses of 3d regular open cell structures. Journal of Materials Science 40(22):5859–5866

Luxner MH, Stampfl J, Pettermann HE (2007) Numerical simulations of 3d open cell structures - influence of structural irregularities on elasto-plasticity and deformation localization. International Journal of Solids and Structures 44(9):2990–3003

Milton G, Briane M, Harutyunyan D (2017) On the possible effective elasticity tensors of 2-dimensional and 3-dimensional printed materials. Mathematics and Mechanics of Complex Systems 5(1):41–94

Omairey SL, Dunning PD, Sriramula S (2019) Development of an ABAQUS plugin tool for periodic RVE homogenisation. Engineering with Computers 35(2):567–577

Pradel P, Zhu Z, Bibb R, Moultrie J (2018) Investigation of design for additive manufacturing in professional design practice. Journal of Engineering Design 29(4-5):165–200

Robinson M, Soe S, McShane G, Celeghini R, Burek R, Alves M, Hanna B, Theobald P (2017) Developing Elastomeric Cellular Structures for Multiple Head Impacts. Head Health Challenge III

Soe S, Ryan M, McShane G, Theobald P (2015a) Energy Absorbing Characteristics of Additively Manufactured TPE Cellular Structures. The Journal of Innovation Impact KES Transactions on Sustainable Design and Manufacturing II(ISSN 2051-6002):14

Soe SP, Martin P, Jones M, Robinson M, Theobald P (2015b) Feasibility of optimising bicycle helmet design safety through the use of additive manufactured TPE cellular structures. The International Journal of Advanced Manufacturing Technology 79(9-12):1975–1982

Tancogne-Dejean T, Spierings AB, Mohr D (2016) Additively-manufactured metallic micro-lattice materials for high specific energy absorption under static and dynamic loading. Acta Materialia 116:14–28

Turco E, dell'Isola F, Cazzani A, Rizzi NL (2016) Hencky-type discrete model for pantographic structures: numerical comparison with second gradient continuum models. Zeitschrift für angewandte Mathematik und Physik 67(4)

Turco E, Misra A, Pawlikowski M, dell'Isola F, Hild F (2018) Enhanced Piola–Hencky discrete models for pantographic sheets with pivots without deformation energy: Numerics and experiments. International Journal of Solids and Structures 147:94–109

Vanderesse N, Ky I, Quevedo Gonzalez F, Nuno N, Bocher P (2016) Image analysis characterization of periodic porous materials produced by additive manufacturing. Materials & Design 92:767–778

Yuan S, Shen F, Bai J, Chua CK, Wei J, Zhou K (2017) 3d soft auxetic lattice structures fabricated by selective laser sintering: TPU powder evaluation and process optimization. Materials & Design 120:317–327

Yuan S, Shen F, Chua CK, Zhou K (2018) Polymeric composites for powder-based additive manufacturing: Materials and applications. Progress in Polymer Science JPPS 1112

Chapter 18
Derivation of Imperfect Interface Laws for Multi-Physic Composites by a Multiscale Approach: Theoretical and Numerical Studies

Serge Dumont, Frederic Lebon, Raffaella Rizzoni, and Michele Serpilli

Abstract In the present study, we focus our attention to a specific type of composite, constituted by two media, called the adherents, bonded together with a thin interphase layer, called the adhesive. We assume that the composite constituents are made of different multi-physic materials with highly contrasted constitutive properties. The study considers a generic multi-physic coupling in a very general framework and can be adapted to well-known multi-physic behaviors, such as piezoelectricity, thermo-elasticity, as well as to multifield microstructural theories, such as micropolar elasticity.

Keywords: Composites, Interfaces, Multi-physic materials, Asymptotic analysis

---

S. Dumont
University of Nîmes, IMAG, CNRS UMR 5149 Pl. E. Bataillon, 34095 Montpellier cedex 5, France
e-mail: serge.dumont@unimes.fr

F. Lebon
Aix-Marseille University, CNRS, Centrale Marseille, LMA Impasse Nikola Tesla 4, Marseille, France
e-mail: lebon@lma.cnrs-mrs.fr

R. Rizzoni
Department of Engineering, University of Ferrara, Via Saragat 1, 44122, Ferrara, Italy
e-mail: raffaella.rizzoni@unife.it

M. Serpilli
Department of Civil and Building Engineering, and Architecture, Università Politecnica delle Marche, Via brecce bianche, 60131 Ancona, Italy
e-mail: m.serpilli@univpm.it

B. E. Abali and I. Giorgio (eds.), *Developments and Novel Approaches in Biomechanics and Metamaterials*, Advanced Structured Materials 132,
https://doi.org/10.1007/978-3-030-50464-9_18

18.1 Introduction

Structural bonding assembly has become an important technological solution over the past few years and is increasingly replacing bolting assembly (Ascione et al, 2017) as shown in Fig. 18.1). The resulting structure has many advantages, such as weight savings or the elimination of stress concentration. Similarly, in nature there are many living or natural structures that are composed of substructures, cells or soils for example, glued together. There are also many other examples of glued structures in the field of bioengineering (Breschi et al, 2008) as illustrated in Fig. 18.2. Understanding and modelling the bonding process then becomes a necessity.

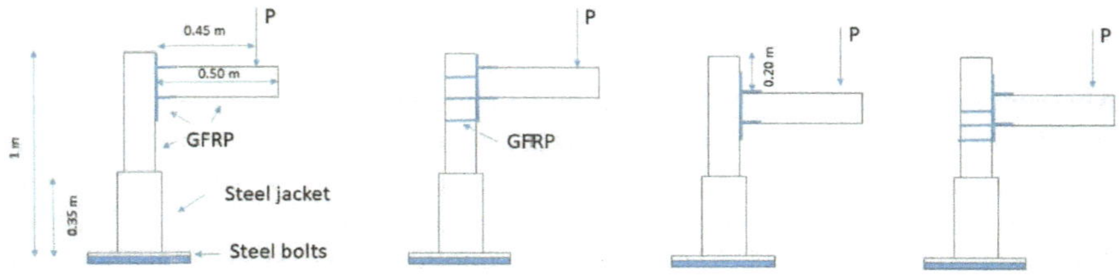

Fig. 18.1 An example of bonding, for civil engineering structures.

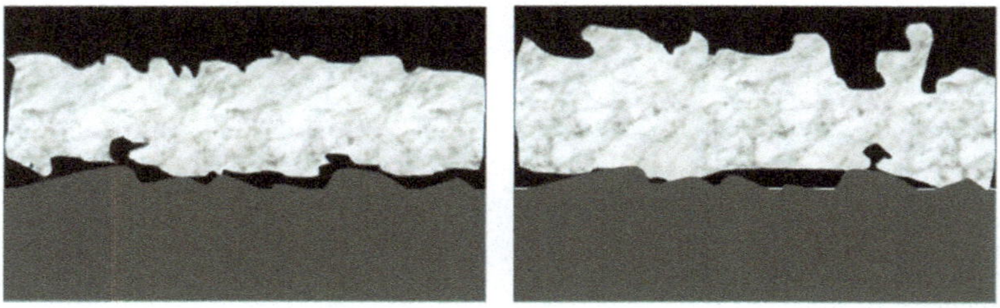

Fig. 18.2 An example of adhesion, for dental structures.

An obvious common point between all these bonded composite structures is the thinness of the adhesive compared to those of the substrates or adherents. This is true for both industrial structures and living or natural structures. This thinness will obviously lead to numerical modelling difficulties. Indeed, the mesh size of the glue will mechanically lead to computations with a very large number of degrees of freedom and therefore very expensive computations. These costs will increase further if the adhesive surface is irregular and has a high roughness. Similarly, in the presence of kinematic or behaviour non-linearities, in the presence of cracks, etc., the costs become prohibitive. "Direct" calculations are then limited to academic cases

(Dumont et al, 2014). There are two very different possibilities, either to develop suitable numerical methods (Alart and Lebon, 1995; Alart et al, 1997; Barbie et al, 2015) or to set up macroscopic models of the adhesive's behaviour. In this chapter, we will focus on the second family of solutions.

There are at least two methodologies in the literature for obtaining constitutive laws of adhesive (interface) behaviour. The most classic is the introduction of phenomenological models, usually based on experimental results, such as Coulomb models, compliances, etc. In this chapter, we will prefer to focus on deductive models. The idea of this methodology is to start from a micromechanical study of the adhesive (interphase) and to deduce, using mathematical methods, an equivalent macroscopic behaviour (interface).

In this chapter, we focus our attention to a bonded composite, constituted by two adherents, bonded together with a thin adhesive. We assume that the composite constituents are made of different multi-physic materials with highly contrasted constitutive properties. The study considers a generic multi-physic coupling in a very general framework and can be adapted to well-known multi-physic behaviors, such as piezoelectricity, thermo-elasticity, as well as to multifield microstructural theories, such as micropolar elasticity (see, e.g. Chatzigeorgiou et al, 2015). Several works have suggested a generalization of the classical interface models, including the effects of other physical (thermal, piezoelectric, etc.) interactions (dell'Isola and Romano, 1987; Chen, 2008; Wang et al, 2017; Firooz and Javili, 2019; Saeb et al, 2019), and within the framework of linear multifield theories, such as higher order continua theories (Placidi et al, 2014; Eremeyev, 2019).

The analysis has been carried out by means of the asymptotic expansions method, using the thickness as a small parameter. This technique is based on the fact that the thickness of the adhesive can be considered as a small parameter (intended to tend towards zero) and denoted by ε in the following. The asymptotic analysis has been applied to the rigorous derivation of simplified models for complex assemblies, presenting thin interphases, in the field of linear elasticity (Lebon and Rizzoni (2010); Dumont et al (2018); Rizzoni et al (2014); Serpilli and Lenci (2016)) as well as in piezoelectricity, taking into account other physical interactions, micropolar elasticity and poroelasticity (Serpilli et al (2013); Serpilli (2015, 2017, 2018, 2019)). As mentioned above, the asymptotic methods allow to replace the adhesive layer with a two-dimensional surface, the so-called imperfect interface, with non-classical transmission conditions between the two adherents. By defining the small parameter and constitutive properties of the middle layer, we perform an asymptotic analysis. We assume that the multi-physic stiffness ratios between the adherents and the adhesive depend on ε^p. As proposed by Caillerie (1970), we identify three critical exponents p, corresponding to different imperfect interface models: $p = 1$, the soft (also called lowly-conducting) multi-physic interface model; $p = 0$, the hard (also called moderately-conducting) multi-physic interface model; $p = -1$, the rigid (also called highly-conducting) multi-physic interface model. Following the approach introduced by Rizzoni et al (2014), we characterize the order zero and the order one transmission problems. Finally, a general multi-physic interface model is developed, and numerically tested through the finite element method. In particular, in the framework of piezoelectricity, we compare the results obtained by modeling the adhesive as an interphase, having a thin finite thickness, with the results obtained with the general multi-physic interface model.

18.2 Statement of the Problem

We consider the composite assembly constituted of two solids $\Omega^\varepsilon_\pm \subset \mathbb{R}^3$, called the adherents, bonded together by an intermediate thin layer $B^\varepsilon := S \times (-\frac{\varepsilon}{2}, \frac{\varepsilon}{2})$ of thickness ε, called the adhesive, with cross-section $S \subset \mathbb{R}^2$. In the following B^ε and S will be called interphase and interface, respectively. Let $S^\varepsilon_\pm$ be the plane contact surfaces between the adhesive and the adherents and let $\Omega^\varepsilon := \Omega^\varepsilon_+ \cup B^\varepsilon \cup \Omega^\varepsilon_-$ denote the composite system comprising the interphase and the adherents (cf. Fig. 1.3a).

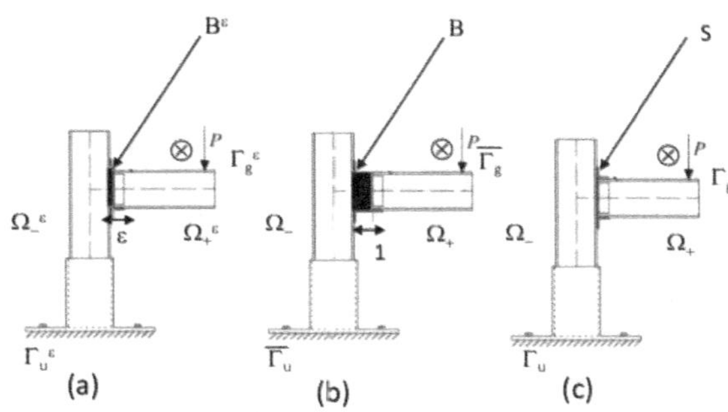

Fig. 18.3 Initial (a), rescaled (b) and limit (c) configurations of the composite.

We suppose that the composite is constituted by a multi-physic material, in which different physical behaviors interact together, such as in piezoelectricity. Its equilibrium state is determined by a collection of order parameters $\mathbf{s}^\varepsilon := (\mathbf{u}^\varepsilon_1, \ldots, \mathbf{u}^\varepsilon_N, \varphi^\varepsilon_1, \ldots, \varphi^\varepsilon_M)$: N vector state variables, namely $\mathbf{u}^\varepsilon_i$, and M scalar state variables, namely φ^ε_k. With the multi-physic state $\mathbf{s}^\varepsilon$, we associate its conjugated physical quantity $\mathbf{t}^\varepsilon = \mathbf{t}^\varepsilon(\nabla^\varepsilon \mathbf{s}^\varepsilon)$, where $\nabla^\varepsilon \mathbf{s}^\varepsilon$ denotes the gradient of $\mathbf{s}^\varepsilon$. The vector field $\mathbf{t}^\varepsilon := (\boldsymbol{\sigma}^\varepsilon_1, \ldots, \boldsymbol{\sigma}^\varepsilon_N, \mathbf{D}^\varepsilon_1, \ldots, \mathbf{D}^\varepsilon_M)$ represents a generalized stress field. We also consider the following homogeneous and linear constitutive law:

$$\mathbf{t}^\varepsilon = \mathbb{K}^\varepsilon \nabla^\varepsilon \mathbf{s}^\varepsilon,$$

where $\mathbb{K}^\varepsilon$ is a generalized linear constitutive matrix. The constitutive tensor $\mathbb{K}^\varepsilon$ satisfies suitable symmetry and positivity properties.

We assume that the adherents are subject to a generalized system of body forces $\mathbf{F}^\varepsilon : \Omega^\varepsilon_\pm \to \mathbb{R}^{3N\times M}$ and surface forces $\mathbf{G}^\varepsilon : \Gamma^\varepsilon_g \to \mathbb{R}^{3N\times M}$, where $\Gamma^\varepsilon_g \subset (\partial\Omega^\varepsilon_+ \setminus S^\varepsilon_+) \cup (\partial\Omega^\varepsilon_- \setminus S^\varepsilon_-)$. Body and surface forces are neglected in adhesive layer. On $\Gamma_u^\varepsilon := (\partial\Omega_+^\varepsilon \setminus S_+^\varepsilon) \cup (\partial\Omega_-^\varepsilon \setminus S_-^\varepsilon) \setminus \Gamma_g^\varepsilon$ homogeneous boundary conditions are prescribed, so that $\mathbf{s}^\varepsilon = \mathbf{0}$ on Γ_u^ε. We assume that on $\Gamma_{lat} := \partial S \times (-\frac{\varepsilon}{2}, \frac{\varepsilon}{2})$ homogeneous Neumann boundary conditions are applied. The differential formulation of the governing equations has the following structure:

$$\begin{cases} -\text{div }\mathbf{t}^\varepsilon = \mathbf{F}^\varepsilon \text{ in } \Omega^\varepsilon, \\ \mathbf{t}^\varepsilon \mathbf{n}^\varepsilon = \mathbf{G}^\varepsilon \quad \text{on } \Gamma_g^\varepsilon, \\ \mathbf{s}^\varepsilon = \mathbf{0} \quad\quad \text{on } \Gamma_u^\varepsilon, \end{cases} \tag{18.1}$$

where $\mathbf{t}^\varepsilon \mathbf{n}^\varepsilon := (\boldsymbol{\sigma}_1^\varepsilon \mathbf{n}^\varepsilon, \ldots, \boldsymbol{\sigma}_N^\varepsilon \mathbf{n}^\varepsilon, \mathbf{D}_1^\varepsilon \cdot \mathbf{n}^\varepsilon, \ldots, \mathbf{D}_M^\varepsilon \cdot \mathbf{n}^\varepsilon)$ represents the generalized traction vector on the boundary Γ_g^ε and $\mathbf{n}^\varepsilon$ the outer normal unit vector to Γ_g^ε. Let us introduce the functional space $V(\Omega^\varepsilon) := \{\mathbf{s}^\varepsilon \in H^1(\Omega^\varepsilon; \mathbb{R}^{3N\times M});\ \mathbf{s}^\varepsilon = \mathbf{0} \text{ on } \Gamma_u^\varepsilon\}$. The variational formulation of problem (18.1) defined on the variable domain Ω^ε can be written as follows:

$$\begin{cases} \text{Find } \mathbf{s}^\varepsilon \in V(\Omega^\varepsilon) \text{ such that} \\ \bar{A}_-^\varepsilon(\mathbf{s}^\varepsilon, \mathbf{r}^\varepsilon) + \bar{A}_+^\varepsilon(\mathbf{s}^\varepsilon, \mathbf{r}^\varepsilon) + \hat{A}^\varepsilon(\mathbf{s}^\varepsilon, \mathbf{r}^\varepsilon) = L^\varepsilon(\mathbf{r}^\varepsilon), \end{cases} \tag{18.2}$$

for all $\mathbf{r}^\varepsilon := (\mathbf{v}_1^\varepsilon, \ldots, \mathbf{v}_N^\varepsilon, \psi_1^\varepsilon, \ldots, \psi_M^\varepsilon) \in V(\Omega^\varepsilon)$, where defined by

$$\bar{A}_\pm^\varepsilon(\mathbf{s}^\varepsilon, \mathbf{r}^\varepsilon) := \int_{\Omega_\pm^\varepsilon} \bar{\mathbb{K}}^\varepsilon \nabla^\varepsilon \mathbf{s}^\varepsilon \cdot \nabla^\varepsilon \mathbf{r}^\varepsilon dx^\varepsilon, \quad \hat{A}^\varepsilon(\mathbf{s}^\varepsilon, \mathbf{r}^\varepsilon) := \int_{B^\varepsilon} \hat{\mathbb{K}}^\varepsilon \nabla^\varepsilon \mathbf{s}^\varepsilon \cdot \nabla^\varepsilon \mathbf{r}^\varepsilon dx^\varepsilon,$$

$$L^\varepsilon(\mathbf{r}^\varepsilon) := \int_{\Omega_\pm^\varepsilon} \mathbf{F}^\varepsilon \cdot \mathbf{r}^\varepsilon dx^\varepsilon + \int_{\Gamma_g^\varepsilon} \mathbf{G}^\varepsilon \cdot \mathbf{r}^\varepsilon d\Gamma^\varepsilon.$$

18.3 Method of Asymptotic Expansion

In order to perform an asymptotic analysis of problem (18.2) when ε tends to zero, we rewrite the problem on a fixed domain Ω independent of ε. By using the approach of Ciarlet (1997), we consider the change of variables $\pi^\varepsilon : x \in \overline{\Omega} \mapsto x^\varepsilon \in \overline{\Omega}^\varepsilon$ given by

$$\pi^\varepsilon : \begin{cases} \bar{\pi}^\varepsilon(x_1, x_2, x_3) = (x_1, x_2, x_3 \mp \frac{1}{2}(1-\varepsilon)), \text{ for all } x \in \overline{\Omega}_\pm, \\ \hat{\pi}^\varepsilon(x_1, x_2, x_3) = (x_1, x_2, \varepsilon x_3), \quad\quad\quad \text{for all } x \in \overline{B}, \end{cases}$$

where, after the change of variables, the adherents occupy $\Omega_\pm := \Omega_\pm^\varepsilon \pm \frac{1}{2}(1-\varepsilon)\mathbf{e}_3$ and the interphase $B = \{x \in \mathbb{R}^3 : (x_1, x_2) \in S,\ |x_3| < \frac{1}{2}\}$. The sets $S_\pm = \{x \in \mathbb{R}^3 : (x_1, x_2) \in S,\ x_3 = \pm\frac{1}{2}\}$ denote the interfaces between B and $\Omega_\pm$ and $\Omega = \Omega_+ \cup \Omega_- \cup B$ is the rescaled configuration of the composite. Lastly, Γ_g and Γ_u indicate the images through π^ε of Γ_g^ε and Γ_u^ε (cf. Fig. 1.3b). Consequently, $\frac{\partial}{\partial x_\alpha^\varepsilon} = \frac{\partial}{\partial x_\alpha}$ and $\frac{\partial}{\partial x_3^\varepsilon} = \frac{\partial}{\partial x_3}$ in $\Omega_\pm$, $\frac{\partial}{\partial x_\alpha^\varepsilon} = \frac{\partial}{\partial x_\alpha}$ and $\frac{\partial}{\partial x_3^\varepsilon} = \frac{1}{\varepsilon}\frac{\partial}{\partial x_3}$ in B.

We assume that the constitutive coefficients of $\Omega_\pm^\varepsilon$ are independent of ε, $\bar{\mathbb{K}}^\varepsilon = \bar{\mathbb{K}}$, while the constitutive coefficients of B^ε depend on ε, $\hat{\mathbb{K}}^\varepsilon = \varepsilon^p \hat{\mathbb{K}}$, with $p \in \{-1, 0, 1\}$. Finally, we assume that the forces are such that $L^\varepsilon(\mathbf{r}^\varepsilon) = L(\mathbf{r})$. By virtue of the previous hypothesis, the rescaled problem can be written in the following form:

$$\begin{cases} \text{Find } \mathbf{s}^\varepsilon \in V(\Omega), \text{ such that} \\ \bar{A}_-(\mathbf{s}^\varepsilon, \mathbf{r}) + \bar{A}_+(\mathbf{s}^\varepsilon, \mathbf{r}) + \varepsilon^{p-1}\hat{a}(\mathbf{s}^\varepsilon, \mathbf{r}) + \varepsilon^p \hat{b}(\mathbf{s}^\varepsilon, \mathbf{r}) + \varepsilon^{p+1}\hat{c}(\mathbf{s}^\varepsilon, \mathbf{r}) = L(\mathbf{r}), \end{cases} \tag{18.3}$$

for all $\mathbf{r} \in V(\Omega) := \{\mathbf{s} \in H^1(\Omega; \mathbb{R}^{3N\times M});\ \mathbf{s} = \mathbf{0} \text{ on } \Gamma_u\}$, where

$$\bar{A}_\pm(\mathbf{s}^\varepsilon, \mathbf{r}) := \int_{\Omega_\pm} \bar{\mathbb{K}} \nabla \mathbf{s}^\varepsilon \cdot \nabla \mathbf{r} dx, \quad \hat{a}(\mathbf{s}^\varepsilon, \mathbf{r}) := \int_B \hat{\mathbb{K}}_{33} \mathbf{s}^\varepsilon_{,3} \cdot \mathbf{r}_{,3} dx,$$

$$\hat{b}(\mathbf{s}^\varepsilon, \mathbf{r}) := \int_B \left\{ \hat{\mathbb{K}}_{3\alpha} \mathbf{s}^\varepsilon_{,3} \cdot \mathbf{r}_{,\alpha} + \hat{\mathbb{K}}_{\alpha 3} \mathbf{s}^\varepsilon_{,\alpha} \cdot \mathbf{r}_{,3} \right\} dx, \quad \hat{c}(\mathbf{s}^\varepsilon, \mathbf{r}) := \int_B \hat{\mathbb{K}}_{\alpha\beta} \mathbf{s}^\varepsilon_{,\beta} \cdot \mathbf{r}_{,\alpha} dx,$$

and $\hat{\mathbb{K}}_{ij}$ denote the sub-matrices of $\hat{\mathbb{K}}$, defined by

$$\hat{\mathbb{K}} = \begin{bmatrix} \hat{\mathbb{K}}_{\alpha\beta} & \hat{\mathbb{K}}_{\alpha 3} \\ \hat{\mathbb{K}}_{3\alpha} & \hat{\mathbb{K}}_{33} \end{bmatrix}, \quad (\hat{\mathbb{K}}_{ij})^T = \hat{\mathbb{K}}_{ji}.$$

We can now apply the asymptotic expansions method to the rescaled problem (18.3), whose fundamental assumption relies in considering the solution $\mathbf{s}^\varepsilon$ of the problem as a series of powers of ε:

$$\begin{aligned} \mathbf{s}^\varepsilon &= \mathbf{s}^0 + \varepsilon \mathbf{s}^1 + \varepsilon^2 \mathbf{s}^2 + \ldots, \\ \bar{\mathbf{s}}^\varepsilon &= \bar{\mathbf{s}}^0 + \varepsilon \bar{\mathbf{s}}^1 + \varepsilon^2 \bar{\mathbf{s}}^2 + \ldots, \\ \hat{\mathbf{s}}^\varepsilon &= \hat{\mathbf{s}}^0 + \varepsilon \hat{\mathbf{s}}^1 + \varepsilon^2 \hat{\mathbf{s}}^2 + \ldots. \end{aligned} \tag{18.4}$$

where $\bar{\mathbf{s}}^\varepsilon = \mathbf{s}^\varepsilon \circ \bar{\pi}^\varepsilon$ and $\hat{\mathbf{s}}^\varepsilon = \mathbf{s}^\varepsilon \circ \hat{\pi}^\varepsilon$. By injecting (18.4) into the rescaled problem (18.3), and by identifying the terms with identical power of ε, we obtain, as customary, a set of variational problems to be solved in order to characterize the limit multi-physic state $\mathbf{s}^0$, the first order corrector term $\mathbf{s}^1$ and their associated limit problem, for $p \in \{-1, 0, 1\}$.

Following the approach described in Rizzoni et al (2014); Dumont et al (2018), we introduce the matching conditions based on the continuity of the generalized traction $\mathbf{t}^\varepsilon \mathbf{e}_3$ and multiphyisic state $\mathbf{s}^\varepsilon$ at the interfaces $S^\varepsilon_\pm$ in the initial configuration and on the continuity of the traction and state $\bar{\mathbf{t}}^\varepsilon \mathbf{e}_3$, $\bar{\mathbf{s}}^\varepsilon$, $\hat{\mathbf{t}}^\varepsilon \mathbf{e}_3$, $\hat{\mathbf{s}}^\varepsilon$ at the interfaces $S_\pm$ in the rescaled configuration. Hence, one has

$$\begin{aligned} &[[\mathbf{s}^\varepsilon]] = [\bar{\mathbf{s}}^\varepsilon] - \varepsilon \langle\langle \mathbf{s}^\varepsilon_{,3} \rangle\rangle + o(\varepsilon), && \langle\langle \mathbf{s}^\varepsilon \rangle\rangle = \langle \bar{\mathbf{s}}^\varepsilon \rangle - \tfrac{\varepsilon}{4}[[\mathbf{s}^\varepsilon_{,3}]], \\ &[[\mathbf{t}^\varepsilon \mathbf{e}_3]] = [\bar{\mathbf{t}}^\varepsilon \mathbf{e}_3] - \varepsilon \langle\langle \mathbf{t}^\varepsilon_{,3} \mathbf{e}_3 \rangle\rangle + o(\varepsilon), && \langle\langle \mathbf{t}^\varepsilon \mathbf{e}_3 \rangle\rangle = \langle \bar{\mathbf{t}}^\varepsilon \mathbf{e}_3 \rangle - \tfrac{\varepsilon}{4}[[\mathbf{t}^\varepsilon_{,3} \mathbf{e}_3]], \end{aligned} \tag{18.5}$$

where

$$\begin{aligned} &\langle f \rangle(\tilde{x}) := \tfrac{1}{2}(f(\tilde{x}, (1/2)^+) + f(\tilde{x}, -(1/2)^-), \quad \tilde{x} := (x_\alpha) \in S, \\ &[f]\,(\tilde{x}) := f(\tilde{x}, (1/2)^+) - f(\tilde{x}, -(1/2)^-), \\ &\langle\langle f \rangle\rangle(\tilde{x}) := \tfrac{1}{2}(f(\tilde{x}, 0^+) + f(\tilde{x}, 0^-)), \\ &[[f]](\tilde{x}) := f(\tilde{x}, 0^+) - f(\tilde{x}, 0^-), \end{aligned}$$

denote, respectively, the mean value and the jump functions at the interfaces.

18.4 Multi-Physic Interface Models

In this section we present the asymptotic models for multi-physic interfaces obtained for the soft, hard and rigid cases at order 0 and order 1. For the sake of brevity, we will skip all the mathematical computations carried out in the deduction of the limit models.

18.4.1 The Soft Multi-Physic Interface

The transmission problems at order 0 and order 1 can be summarized as follows:

- Order 0

Governing equations

$$\begin{cases} -\mathrm{div}\, \bar{\mathbf{t}}^0 = \mathbf{F} & \text{in } \Omega_\pm, \\ \bar{\mathbf{t}}^0\mathbf{n} = \mathbf{G} & \text{on } \Gamma_g, \\ \bar{\mathbf{s}}^0 = \mathbf{0} & \text{on } \Gamma_u, \end{cases}$$

Transmission conditions on $S_\pm$

$$\begin{cases} [\bar{\mathbf{s}}^0] = (\hat{\mathbb{K}}_{33})^{-1}\langle \bar{\mathbf{t}}^0\mathbf{e}_3\rangle, \\ [\bar{\mathbf{t}}^0\mathbf{e}_3] = \mathbf{0}. \end{cases}$$

- Order 1

Governing equations

$$\begin{cases} -\mathrm{div}\, \bar{\mathbf{t}}^1 = \mathbf{0} & \text{in } \Omega_\pm, \\ \bar{\mathbf{t}}^1\mathbf{n} = \mathbf{0} & \text{on } \Gamma_g, \\ \bar{\mathbf{s}}^1 = \mathbf{0} & \text{on } \Gamma_u, \end{cases}$$

Transmission conditions on $S_\pm$

$$\begin{cases} [\bar{\mathbf{s}}^1] = (\hat{\mathbb{K}}_{33})^{-1}\left(\langle \bar{\mathbf{t}}^1\mathbf{e}_3\rangle - \hat{\mathbb{K}}_{\alpha 3}\langle \bar{\mathbf{s}}^0\rangle_{,\alpha}\right), \\ [\bar{\mathbf{t}}^1\mathbf{e}_3] = -\hat{\mathbb{K}}_{3\alpha}[\bar{\mathbf{s}}^0]_{,\alpha}. \end{cases}$$

The transmission problems for a soft multi-physic interface at order 0 and order 1 represent a formal generalization of the soft interface models obtained by means of the asymptotic methods in linear elasticity (see, e.g., Rizzoni et al, 2014; Dumont et al, 2018) and in other multifield frameworks, such as poroelasticity (see Serpilli, 2019). At order 0, the interface behaves from a mechanical point of view as a series of linear springs, reacting to the discontinuity of the multi-physic state between the upper and bottom faces, while the generalized traction vector remains continuous. At order 1, the interface conditions maintain a similar structure, but both the multi-physic state and the traction vector are discontinuous through the interface. Moreover, they depend on the in-plane derivatives of the jump and mean values of $\bar{\mathbf{s}}^0$, which can be considered a known source term, identified in the order 0 problem.

18.4.2 The Hard Multi-Physic Interface

The hard interface transmission problems at order 0 and order 1 take the following expressions:

- Order 0

Governing equations

$$\begin{cases} -\text{div}\,\bar{\mathbf{t}}^0 = \mathbf{F} & \text{in } \Omega_\pm, \\ \bar{\mathbf{t}}^0\mathbf{n} = \mathbf{G} & \text{on } \Gamma_g, \\ \bar{\mathbf{s}}^0 = \mathbf{0} & \text{on } \Gamma_u, \end{cases}$$

Transmission conditions on $S_\pm$

$$\begin{cases} [\bar{\mathbf{s}}^0] = \mathbf{0}, \\ [\bar{\mathbf{t}}^0\mathbf{e}_3] = \mathbf{0}. \end{cases}$$

- Order 1

Governing equations

$$\begin{cases} -\text{div}\,\bar{\mathbf{t}}^1 = \mathbf{0} & \text{in } \Omega_\pm, \\ \bar{\mathbf{t}}^1\mathbf{n} = \mathbf{0} & \text{on } \Gamma_g, \\ \bar{\mathbf{s}}^1 = \mathbf{0} & \text{on } \Gamma_u, \end{cases}$$

Transmission conditions on $S_\pm$

$$\begin{cases} [\bar{\mathbf{s}}^1] = (\hat{\mathbb{K}}_{33})^{-1}\left(\langle\bar{\mathbf{t}}^0\mathbf{e}_3\rangle - \hat{\mathbb{K}}_{\alpha 3}\langle\bar{\mathbf{s}}^0\rangle_{,\alpha}\right), \\ [\bar{\mathbf{t}}^1\mathbf{e}_3] = -\left(\hat{\mathbb{K}}_{3\alpha}[\bar{\mathbf{s}}^1]_{,\alpha} + \hat{\mathbb{K}}_{\alpha\beta}\langle\bar{\mathbf{s}}^0\rangle_{,\alpha\beta}\right). \end{cases}$$

It is interesting to notice that the hard multi-physic interface problems is equivalent to the ones derived in the case of linear elasticity in Lebon and Rizzoni (2010); Rizzoni et al (2014); Dumont et al (2018). At order 0, we recover the classical continuity conditions for both the multi-physic state and generalized traction vector. Thus, the adherents are perfectly bonded together. At order 1, we encounter a mixed interface model with a jump of the state and traction vector depending on the values of the multi-physic state and traction vector at order 0. These order 0 terms, being known from the solution of the previous problem, can be viewed as external source terms.

18.4.3 The Rigid Multi-Physic Interface

The differential formulations of the rigid interface problems at order 0 and order 1 take the following form:

- Order 0

Governing equations

$$\begin{cases} -\text{div}\,\bar{\mathbf{t}}^0 = \mathbf{F} & \text{in } \Omega_\pm, \\ \bar{\mathbf{t}}^0\mathbf{n} = \mathbf{G} & \text{on } \Gamma_g, \\ \bar{\mathbf{s}}^0 = \mathbf{0} & \text{on } \Gamma_u, \end{cases}$$

Transmission conditions on $S_\pm$

$$\begin{cases} [\bar{\mathbf{s}}^0] = \mathbf{0}, \\ [\bar{\mathbf{t}}^0\mathbf{e}_3] = -\hat{\mathbb{L}}_{\alpha\beta}\langle\bar{\mathbf{s}}^0\rangle_{,\alpha\beta}. \end{cases}$$

- Order 1

Governing equations

$$\begin{cases} -\mathrm{div}\,\bar{\mathbf{t}}^1 = \mathbf{0} & \text{in } \Omega_\pm, \\ \bar{\mathbf{t}}^1\mathbf{n} = \mathbf{0} & \text{on } \Gamma_g, \\ \bar{\mathbf{s}}^1 = \mathbf{0} & \text{on } \Gamma_u, \end{cases}$$

Transmission conditions on $S_\pm$

$$\begin{cases} [\bar{\mathbf{s}}^1] = -(\hat{\mathbb{K}}_{33})^{-1}\hat{\mathbb{K}}_{\alpha 3}\langle\bar{\mathbf{s}}^0\rangle_{,\alpha}, \\ [\bar{\mathbf{t}}^1\mathbf{e}_3] = -\hat{\mathbb{K}}_{3\alpha}(\hat{\mathbb{K}}_{33})^{-1}\langle\bar{\mathbf{t}}^0\mathbf{e}_3\rangle_{,\alpha} - \hat{\mathbb{L}}_{\alpha\beta}\langle\bar{\mathbf{s}}^1\rangle_{,\alpha\beta}, \end{cases}$$

where $\hat{\mathbb{L}}_{\alpha\beta} := \hat{\mathbb{K}}_{\alpha\beta} - \hat{\mathbb{K}}_{3\alpha}(\hat{\mathbb{K}}_{33})^{-1}\hat{\mathbb{K}}_{\beta 3}$. The rigid multi-physic interface problems show the same features of the rigid interface asymptotic models obtained in different frameworks in Bessoud et al (2009); Serpilli (2015, 2017, 2018, 2019). Concerning the order 0 model, we obtain a continuity of the multi-physic state at the interface level, while the traction vector is discontinuous and depends on the divergence of a generalized membrane stress vector $\mathbf{N}^0_\alpha := \hat{\mathbb{L}}_{\alpha\beta}\langle\bar{\mathbf{s}}^0\rangle_{,\beta}$. The interface behaves as a multi-physic membrane. The order 1 presents a discontinuity on both the multi-physic state and traction vector. Analogously to the order 0 model, we obtain a generalized equilibrium membrane problem defined on the plane of the interface.

18.4.4 The General Multi-Physic Interface

The approach of Rizzoni et al (2014) can be extended in order to obtain a condensed form of the transmission conditions summarizing both the orders 0 and 1 of the soft, hard and rigid cases in only one couple of equations in terms of the jump of the multi-physic state and generalized tractions at the interface.

Therefore, we denote by $\tilde{\mathbf{s}}^\varepsilon := \bar{\mathbf{s}}^0 + \varepsilon\bar{\mathbf{s}}^1 + \varepsilon^2\bar{\mathbf{s}}^2$ and $\tilde{\mathbf{t}}^\varepsilon := \bar{\mathbf{t}}^0 + \varepsilon\bar{\mathbf{t}}^1$, two suitable approximations for $\bar{\mathbf{s}}^\varepsilon$ and $\bar{\mathbf{t}}^\varepsilon$. Let us consider the rigid multi-physic interface conditions, as starting point. After rescaling back the constitutive coefficients $\hat{\mathbb{K}} = \varepsilon\hat{\mathbb{K}}^\varepsilon$ in B^ε, we can write $[\tilde{\mathbf{s}}^\varepsilon]$ and $[\tilde{\mathbf{t}}^\varepsilon\mathbf{e}_3]$. Indeed, one has

$$\begin{aligned} [\tilde{\mathbf{s}}^\varepsilon] &:= [\bar{\mathbf{s}}^0] + \varepsilon[\bar{\mathbf{s}}^1] + \varepsilon^2[\bar{\mathbf{s}}^2] = -\varepsilon(\hat{\mathbb{K}}^\varepsilon_{33})^{-1}\left(\hat{\mathbb{K}}^\varepsilon_{\alpha 3}\langle\tilde{\mathbf{s}}^\varepsilon\rangle_{,\alpha} - \langle\tilde{\mathbf{t}}^\varepsilon\mathbf{e}_3\rangle\right) + o(\varepsilon^2), \\ [\tilde{\mathbf{t}}^\varepsilon\mathbf{e}_3] &:= [\bar{\mathbf{t}}^0\mathbf{e}_3] + \varepsilon[\bar{\mathbf{t}}^1\mathbf{e}_3] = -\varepsilon\hat{\mathbb{K}}^\varepsilon_{3\alpha}(\hat{\mathbb{K}}^\varepsilon_{33})^{-1}\langle\tilde{\mathbf{t}}^\varepsilon\mathbf{e}_3\rangle_{,\alpha} - \varepsilon\hat{\mathbb{L}}^\varepsilon_{\alpha\beta}\langle\tilde{\mathbf{s}}^\varepsilon\rangle_{,\alpha\beta} + o(\varepsilon^2). \end{aligned}$$

An alternative expression of the above transmission conditions can be given in terms of $\langle\tilde{\mathbf{t}}^\varepsilon\mathbf{e}_3\rangle$ and $[\tilde{\mathbf{t}}^\varepsilon\mathbf{e}_3]$, which will be useful to write the variational formulation of the interface multi-physic problem:

$$\begin{aligned} \langle\tilde{\mathbf{t}}^\varepsilon\mathbf{e}_3\rangle &= \tfrac{1}{\varepsilon}\hat{\mathbb{K}}^\varepsilon_{33}[\tilde{\mathbf{s}}^\varepsilon] + \hat{\mathbb{K}}^\varepsilon_{\alpha 3}\langle\tilde{\mathbf{s}}^\varepsilon\rangle_{,\alpha} + o(\varepsilon^2), \\ [\tilde{\mathbf{t}}^\varepsilon\mathbf{e}_3] &= -\hat{\mathbb{K}}^\varepsilon_{3\alpha}[\tilde{\mathbf{s}}^\varepsilon]_{,\alpha} - \varepsilon\hat{\mathbb{K}}^\varepsilon_{\alpha\beta}\langle\tilde{\mathbf{s}}^\varepsilon\rangle_{,\alpha\beta} + o(\varepsilon^2). \end{aligned} \tag{18.6}$$

It is easy to prove that this interface law is general enough to describe the interface laws at order 0 and order 1 prescribing the multi-physic state jump and traction jump in the cases of the soft and hard interfaces, by choosing the following scalings for the constitutive matrices: $\hat{\mathbb{K}}^\varepsilon = \varepsilon\hat{\mathbb{K}}$, for the soft case, and $\hat{\mathbb{K}}^\varepsilon = \hat{\mathbb{K}}$, for the hard case.

The relations (18.6) can be transformed into interface equations defined on S, by making use of the matching relations (18.5), up to higher order terms:

$$
\begin{aligned}
\langle\langle \mathbf{t}\mathbf{e}_3 \rangle\rangle &= \tfrac{1}{\varepsilon}\hat{\mathbb{K}}_{33}[[\mathbf{s}]] + \hat{\mathbb{K}}_{\alpha 3}\langle\langle \mathbf{s} \rangle\rangle_{,\alpha}, \\
[[\mathbf{t}\mathbf{e}_3]] &= -\hat{\mathbb{K}}_{3\alpha}[[\mathbf{s}]]_{,\alpha} - \varepsilon\hat{\mathbb{K}}_{\alpha\beta}\langle\langle \mathbf{s} \rangle\rangle_{,\alpha\beta}.
\end{aligned}
\tag{18.7}
$$

18.5 Finite Element Implementation and Numerical Test

In order to derive the variational form of the multi-physic problem, which will be numerically tested through a finite element procedure, one can write the variational form of the equilibrium problem on each sub-domain Ω_+ and Ω_-:

$$
\int_{\Omega_\pm} \bar{\mathbb{K}}\boldsymbol{\nabla}\mathbf{s}\cdot\boldsymbol{\nabla}\mathbf{r}dx - \int_S \mathbf{t}(\tilde{x},0^+)\mathbf{n}(\tilde{x},0^+)\cdot\mathbf{r}d\Gamma - \int_S \mathbf{t}(\tilde{x},0^-)\mathbf{n}(\tilde{x},0^-)\cdot\mathbf{r}d\Gamma =
$$
$$
= \int_{\Omega_\pm}\mathbf{F}\cdot\mathbf{r}dx + \int_{\Gamma_g}\mathbf{G}\cdot\mathbf{r}d\Gamma,
$$

which can be rewritten as

$$
\int_{\Omega_\pm} \bar{\mathbb{K}}\boldsymbol{\nabla}\mathbf{s}\cdot\boldsymbol{\nabla}\mathbf{r}dx + \int_S [[\mathbf{t}\mathbf{e}_3\cdot\mathbf{r}]]d\tilde{x} = L(\mathbf{r}),
$$

letting $\mathbf{e}_3 = \mathbf{n}(\tilde{x},0^-) = -\mathbf{n}(\tilde{x},0^+)$ and $d\Gamma = d\tilde{x}$. Then, using the property $[[ab]] = \langle\langle a \rangle\rangle[[b]] + [[a]]\langle\langle b \rangle\rangle$, relations (18.7) and after an integration by parts, we obtain

$$
\begin{cases}
\text{Find } \mathbf{s} \in W(\tilde{\Omega}). \text{ such that} \\
\bar{A}_-(\mathbf{s},\mathbf{r}) + \bar{A}_+(\mathbf{s},\mathbf{r}) + \mathcal{A}(\mathbf{s},\mathbf{r}) = L(\mathbf{r}),
\end{cases}
\tag{18.8}
$$

for all $\mathbf{r} \in W(\tilde{\Omega}) := \{\mathbf{s} \in H^1(\tilde{\Omega};\mathbb{R}^{3N\times M}),\ \mathbf{s}|_S \in H^1(S;\mathbb{R}^{3N\times M}), \mathbf{s} = \mathbf{0} \text{ on } \Gamma_u\}$, with $\tilde{\Omega} := \Omega_+ \cup S \cup \Omega_-$ and

$$
\mathcal{A}(\mathbf{s},\mathbf{r}) := \int_S \Big(\frac{1}{\varepsilon}\hat{\mathbb{K}}_{33}[[\mathbf{s}]]\cdot[[\mathbf{r}]] + \hat{\mathbb{K}}_{\alpha 3}\langle\langle \mathbf{s} \rangle\rangle_{,\alpha}\cdot[[\mathbf{r}]] + \hat{\mathbb{K}}_{3\alpha}[[\mathbf{s}]]\cdot\langle\langle \mathbf{r} \rangle\rangle_{,\alpha} +
$$
$$
+\varepsilon\hat{\mathbb{K}}_{\alpha\beta}\langle\langle \mathbf{s} \rangle\rangle_{,\alpha}\cdot\langle\langle \mathbf{r} \rangle\rangle_{,\beta}\Big)\, d\tilde{x}.
$$

A standard finite element method is employed to solve this equation. In order to take into account the jumps in the displacements across the interface, a ‘flat" finite element is considered on the interface S that has all nodes on S, the first ones related to Ω_-, and the other ones related to Ω_+. It is then possible to write a stiffness matrix of this problem that is invertible and with standard error estimates (for more details, see for example Nairn, 2007).

Numerical Study: The Piezoelectric Composite Plate

The aim of this study is to numerically test the general interface law, expressed in (18.8), comparing it to a three-dimensional analysis of the problem. As preliminary test, we consider the piezoelectric case. The piezoelectric state at the equilibrium is determined by the pair $\mathbf{s} := (\mathbf{u}, \varphi)$, where $\mathbf{u}$ and φ represent the displacement field and the electric potential. The generalized stress vector is given by $\mathbf{t} := (\boldsymbol{\sigma}, \mathbf{D})$, where $\boldsymbol{\sigma}$ and $\mathbf{D}$ denote, respectively, the Cauchy stress tensor and the electric displacement.

Let us consider a piezoelectric three-phases composite plate, which occupies a 3D domain defined by $\Omega = [0, L_1] \times [0, L_2] \times [-h/2, h/2]$, with $L_1 = 10h$ and $L_2 = 5h$ (see Fig. 1.4). The adhesive thickness is set to be ε.

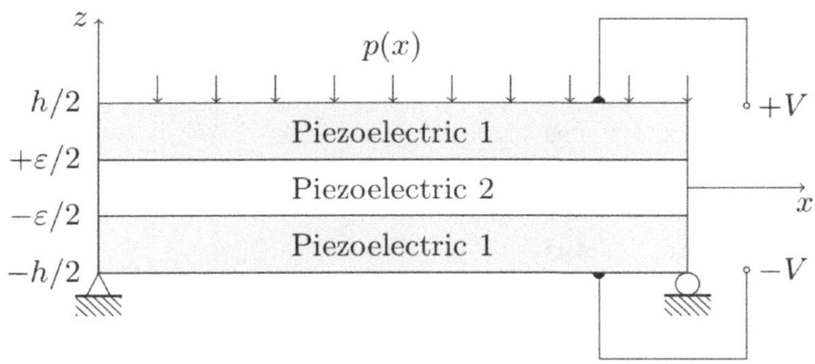

Fig. 18.4 The geometry of the piezoelectric composite plate in the plane (x, z).

The adherents are constituted by PVDF (Polyvinylidene fluoride), a monoclinic piezoelectric material with poling axis $\mathbf{e}_3$, while the adhesive is made of PZT-4, a transversally isotropic piezoelectric material with poling axis $\mathbf{e}_3$. This constitutive sub-matrices $(\mathbb{K}_{ij})$ are defined as follows:

$$\mathbb{K}_{33} = \begin{pmatrix} 2c_{55} & 0 & 0 & 0 \\ 0 & 2c_{44} & 0 & 0 \\ 0 & 0 & c_{33} & e_{33} \\ 0 & 0 & -e_{33} & H_{33} \end{pmatrix}, \quad \mathbb{K}_{12} = \begin{pmatrix} 0 & 2c_{66} + c_{12} & 0 & 0 \\ 0 & 0 & 0 & 0 \\ 0 & 0 & 0 & 0 \\ 0 & 0 & 0 & 0 \end{pmatrix},$$

$$\mathbb{K}_{13} = \begin{pmatrix} 0 & 0 & 2c_{55} & e_{15} \\ 0 & 0 & 0 & 0 \\ c_{13} & 0 & 0 & 0 \\ -e_{31} & 0 & 0 & 0 \end{pmatrix}, \quad \mathbb{K}_{23} = \begin{pmatrix} 0 & 0 & 0 & 0 \\ 0 & 0 & 2c_{44} & e_{24} \\ 0 & c_{23} & 0 & 0 \\ 0 & -e_{32} & 0 & 0 \end{pmatrix},$$

$$\mathbb{K}_{11} = \begin{pmatrix} c_{11} & 0 & 0 & 0 \\ 0 & 2c_{66} & 0 & 0 \\ 0 & 0 & 2c_{55} & e_{15} \\ 0 & 0 & -e_{15} & H_{11} \end{pmatrix}, \quad \mathbb{K}_{22} = \begin{pmatrix} 2c_{66} & 0 & 0 & 0 \\ 0 & c_{22} & 0 & 0 \\ 0 & 0 & 2c_{44} & e_{24} \\ 0 & 0 & -e_{24} & H_{22} \end{pmatrix}.$$

For the transversally isotropic material with poling axis $\mathbf{e}_3$, one has $c_{11} = c_{22}$, $c_{13} = c_{23}$, $c_{55} = c_{44}$, $c_{66} = (c_{11} - c_{12})/2$, $e_{15} = e_{24}$, $e_{31} = e_{32}$ and $H_{11} = H_{22}$. The elastic, dielectric and piezoelectric coefficients are listed in Table 18.1. The

Table 18.1 Piezoelectric material properties

Moduli	PZT-4	PVDF
c_{11}, GPa	139	238.24
c_{22}, GPa	139	23.6
c_{33}, GPa	115	10.64
c_{12}, GPa	77.8	3.98
c_{13}, GPa	74.3	2.19
c_{23}, GPa	74.3	1.92
$2c_{44}$, GPa	25.6	2.15
$2c_{55}$, GPa	25.6	4.4
$2c_{66}$, GPa	30.6	6.43
e_{31}, C/m^2	−5.2	−0.13
e_{32}, C/m^2	−5.2	−0.145
e_{33}, C/m^2	15.1	−0.276
e_{24}, C/m^2	12.7	−0.009
e_{15}, C/m^2	12.7	−0.135
H_{11}, nF/m	13.06	0.111
H_{22}, nF/m	13.06	0.106
H_{33}, nF/m	11.51	0.106

piezoelectric composite plate is subject to surface uniform load equal to $p = 1$ kN/m^2 on the top face, as shown in Fig. ??. We assume that no voltage is applied on the upper and lower faces. In this case, the composite plates behaves as a sensor (see Bonaldi et al, 2017).

The finite element simulations (made with GetFEM) are carried out using Q1 elements (linear on a cube), with 7280 nodes (29203 degrees of freedom) for the three phases problem and 5824 nodes (23379 degrees of freedom) for the problem with the interface law.

First, the influence of the relative thickness of the interphase $\frac{\varepsilon}{L}$ is investigated in order to evaluate the accuracy of the various modeling proposed in the previous sections. In particular, the quality of the solutions is evaluated considering the L^2-relative error $e = \frac{\|\mathbf{s}^\varepsilon - \mathbf{s}_{model}\|}{\|\mathbf{s}^\varepsilon\|}$, where $\mathbf{s}^\varepsilon$ denotes the reference solution computed using the three-phases problem with a fine finite element mesh, while $\mathbf{s}_{model}$ indicates the solution of the interface model (18.8). The convergence of the general interface model towards the three-phases one with respect to the thickness ratio $\frac{\varepsilon}{L}$ is presented in Fig. 18.5. From the plot, it can be observed that, by reducing the thickness of

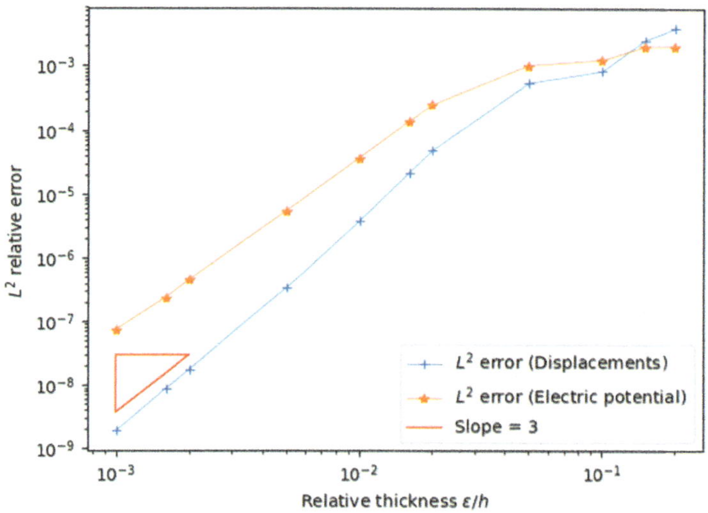

Fig. 18.5 Convergence results with respect to the thickness $\frac{\varepsilon}{L}$.

the adhesive, the relative error has a drastic reduction and so, the proposed general interface model provides an acceptable solution and it is able to correctly approximate the solution $\mathbf{s}^{\varepsilon}$. The convergence rate is ε^3. Besides, even if the relative thickness is of 1%, the relative error is equal to $7.65 \cdot 10^{-2}\%$ for the displacement field and $9.06 \cdot 10^{-4}\%$ for the electric potential, meaning that the general interface model can also be used for moderately thick adhesive layers.

Now, let us fix the relative thickness $\frac{\varepsilon}{L} = 0.02$. The numerical results for the variables are provided using the dimensionless units. We set:

$$(U_i, \Phi) = \frac{E_0}{V}(u_i, \frac{\varphi}{E_0}) \qquad (T_{ij}, \mathcal{D}_k) = \frac{hE_0}{C_{00}V}(\sigma_{ij}, E_0 D_k),$$

where we have chosen, for numerical convenience, $V = 50V$, $E_0 = 10^9 Vm^{-1}$ and $C_{00} = 1GPa$. The results are represented in Fig. 18.6, 18.7 and 18.8.

Figure 18.6 represents the trend of the displacement field and electric potential, evaluated in $x = L_1/2, y = L_2/2, z/h \in [-0.5, 0.5]$. The plot shows a good agreement between the solution of the general interface problem (dotted line) and the solution of the three-phases problem (solid line). The composite plate behaves

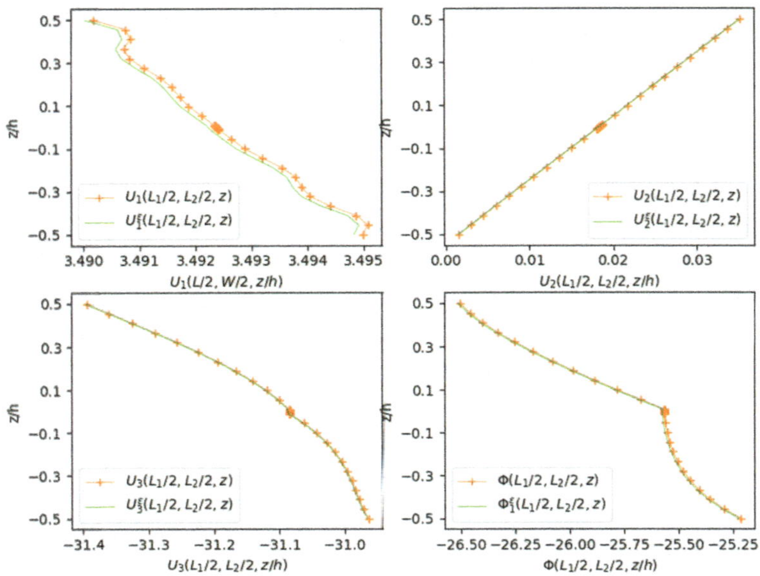

Fig. 18.6 Representation of the displacements and the electric potential on a section along the z-axis.

mostly as a Kirchhoff-Love single-layer plate, taking also into account the transversal deformation of the adhesive. From the electric point of view, the electric potential is constant through the adhesive: this is due to the fact that the intermediate layer (PZT-4) has a higher electrical conductivity with respect to upper and lower bodies (PVDF), see Table 1, and, hence, it behaves as a highly conducting interface.

Figure 18.7 and Fig. 18.8 represent the trend of the jumps of the displacement and electric potential and the jumps of the stress vector and normal electric displacement along the x-axis, namely $(x \in [0, L_1], y = L_2/2, z = 0)$, and y-axis, namely $(x = L_1/2, y =\in [0, L_2], z = 0)$. The numerical simulations highlight that the proposed model is able to describe the mechanical behavior of the composite. Few solution oscillations can be found close to the lateral boundaries, due to the presence of edges, which produce expected stress concentrations and singularities.

18.6 Concluding Remarks

General imperfect contact conditions have been proposed, simulating the behavior of a thin interphase undergoing linear coupled multi-physic phenomena. These con-

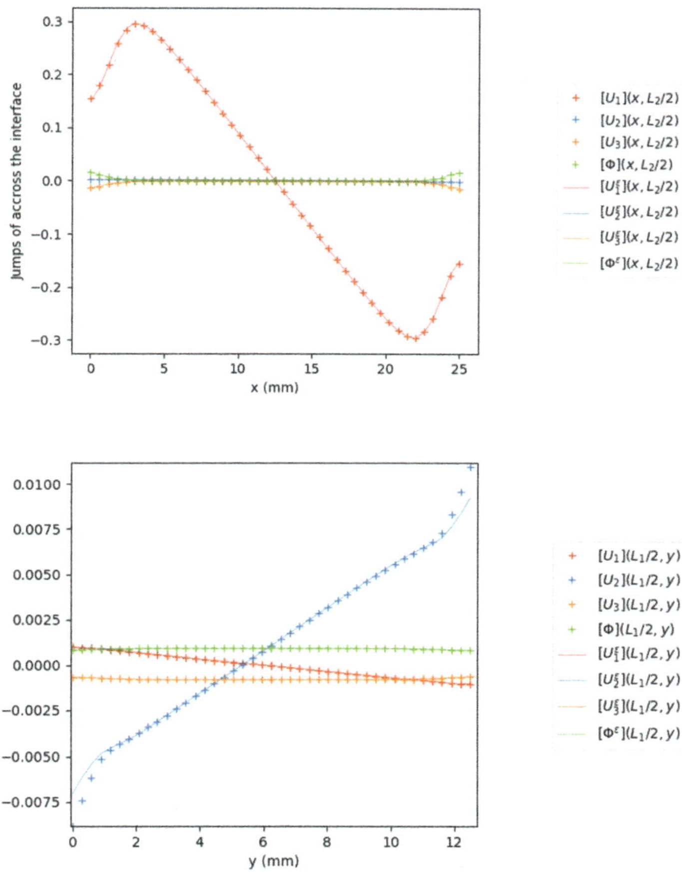

Fig. 18.7 Representation of the jumps in the displacements and the electric potential across the interface on a section along the x-axis and y-axis.

ditions link the generalized traction vector field and its jump to the multi-physic state vector field and its jump at the interface, which is the geometric limit of the interphase as its thickness parameter ε goes to zero. Three interface models (soft, hard and rigid) have been deduced by means of the asymptotic methods, by varying the rigidity ratios between the adhesive and adherents and considering the order 0 and order 1 corrector terms. Furthermore, these three different models have been condensed in one general imperfect interface model and its variational formulation has been presented. The weak formulation represents a key step towards the FEM simulation. A numerical example has been presented considering a piezoelectric composite plate, subject to an electric potential difference at the top and bottom

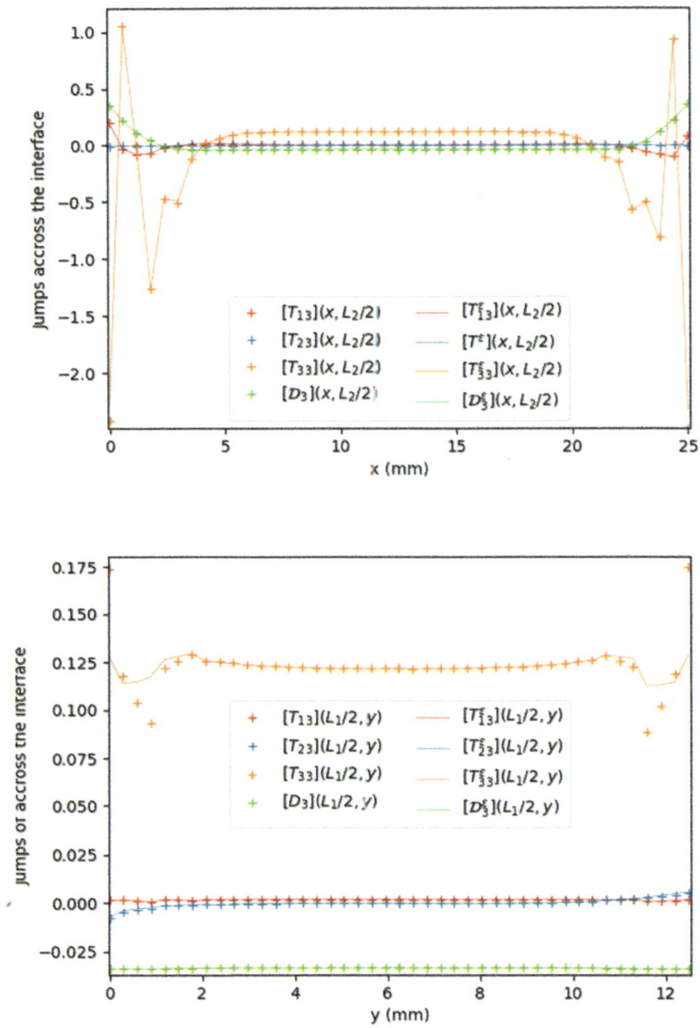

Fig. 18.8 Representation of the jumps of the stress vector and normal electric displacement across the interface on a section along the x-axis and y-axis.

faces. The numerical results show the convergence of the solution of the three-phases model towards the solution of the proposed model as ε tends to zero, highlighting the accuracy and "goodness" of the general interface conditions.

References

Alart P, Lebon F (1995) Solution of contact and friction problems using ILU and coarse-fine preconditioners. Computational Mechanics 16:98–105

Alart P, Barboteu M, Lebon F (1997) Solution of frictional contact problems by an EBE preconditioner. Computational Mechanics 20:370–378

Ascione F, Lamberti M, Razaqpur AG, Spadea S (2017) Strength and stiffness of adhesively bonded GFRP beam-column moment resisting connections. Composite Structures 160:1248–1257

Barbie L, Ramiere I, Lebon F (2015) An automatic multilevel refinement technique based on nested local meshes for nonlinear mechanics. Computers and Structures 147:14–25

Bessoud AL, Krasucki F, Michaille G (2009) Multi-materials with strong interface: variational modelings. Asymptotic Analysis 23:1–19

Bonaldi F, Geymonat G, Krasucki F, Serpilli M (2017) An asymptotic plate model for magneto-electro-thermo-elastic sensors and actuators. Mathematics and Mechanics of Solids 22:798–822

Breschi L, Mazzoni A A andRuggeri, Cadenaro M, Di Lenarda R, De Stefano Dorigo E (2008) Dental adhesion review: Aging and stability of the bonded interface. Dental Materials 24:90—-101

Caillerie D (1970) The effect of a thin inclusion of high rigidity in an elastic body. Mathematical Methods in the Applied Sciences 2:251—-270

Chatzigeorgiou G, Javili A, Steinmann P (2015) Multiscale modelling for composites with energetic interfaces at the micro-or nanoscale. Mathematics and Mechanics of Solids 20:1130–1145

Chen T (2008) Exact size-dependent connections between effective moduli of fibrous piezoelectric nanocomposites with interface effects. Acta Mechanica 196:205–217

Ciarlet PG (1997) Mathematical Elasticity, vol. II: Theory of Plates. North-Holland, Amsterdam

dell'Isola F, Romano A (1987) On the derivation of thermomechanical balance equations for continuous systems with a nonmaterial interface. International Journal of Engineering Science 25(11-12):1459–1468

Dumont S, Lebon F, Rizzoni R, Sacco S (2014) An asymptotic approach to the adhesion of thin films. Mechanics Research Communications 58:24–35

Dumont S, Lebon F, Rizzoni (2018) Soft and hard interface models for bonded elements. Composites Part B: Engineering 153:480–490

Eremeyev V (2019) Two-and three-dimensional elastic networks with rigid junctions: modeling within the theory of micropolar shells and solids. Acta Mechanica 230:3875–3887

Firooz S, Javili A (2019) Understanding the role of general interfaces in the overall behavior of composites and size effects. Composite Materials Science 162:245–254

Lebon F, Rizzoni R (2010) Asymptotic analysis of a thin interface: the case involving similar rigidity. International Journal of Engineering Science 48:473–486

Nairn JA (2007) Numerical implementation of imperfect interfaces. Computational Materials Science 40:1–19

Placidi L, Rosi G, Giorgio I, Madeo A (2014) Reflection and transmission of plane waves at surfaces carrying material properties and embedded in second-gradient materials. Mathematics and Mechanics of Solids 19(5):555–578

Rizzoni R, Dumont S, Lebon F, Sacco S (2014) Higher order model for soft and hard elastic interfaces. International Journal of Solids and Structures 51:4137–4148

Saeb S, Steinmann P, Javili A (2019) Designing tunable composites with general interfaces. International Journal of Solids and Structures 171:181–188

Serpilli M (2015) Mathematical modeling of weak and strong piezoelectric interfaces. Journal of Elasticity 121:235–254

Serpilli M (2017) Asymptotic interface models in magneto-electro-thermo-elastic composites. Meccanica 52:1407–1424

Serpilli M (2018) On modeling interfaces in linear micropolar composites. Mathematics and Mechanics of Solids 23:667–685

Serpilli M (2019) Classical and higher order interface conditions in poroelasticity. Annals of Solid and Structural Mechanics 11:1–22

Serpilli M, Lenci S (2016) An overview of different asymptotic models for anisotropic three-layer plates with soft adhesive. International Journal of Solids and Structures 81:130–140

Serpilli M, Krasucki F, Geymonat G (2013) An asymptotic strain gradient Reissner–Mindlin plate model. Meccanica 48:2007–2018

Wang B, Liu JT, Gu S (2017) An xfem/level set strategy for simulating the piezoelectric spring-type interfaces with apparent physical background. Finite Elements in Analysis and Design 133:62–75

Chapter 19
Modeling the Mechanical Response of Rubberised Concrete

Amedeo Gregori, Chiara Castoro, Micaela Mercuri, and Michele Angiolilli

Abstract The use of rubber particles from waste tyres in concrete as partial replacement of natural aggregates represents nowadays a significant recycling solution. In this paper, FEM models have been developed and analysed to investigate the effects on the compressive strength of the rubberised concrete. The natural fine and coarse aggregates have been substituted with rubber particles ranging from 3 mm to 30 mm in size, obtaining percentages from 0% to 80%. Results from the numerical analyses about the effects of the substitutions with rubber particles on the rubbercrete compressive strength confirmed the trend of literature experimental data. The effects of different values adopted for the reference concrete compressive strength are also confirmed to depend on the size of the rubber particles. Based on a large dataset, analytical relationships have been developed and proposed to analyse the Strength Reduction Factor (SRF) of rubberised concrete.

Keywords: Aggregates substitution · Rubberised concrete · Compressive strength · Finite element models

19.1 Introduction

Many authors have investigated in the last decades on the re-use of waste tyre rubber in concretes, as partial replacement of the natural aggregates, allowing to reduce the excessive consumption of the natural aggregates for concretes and to recycle waste tyres that are gathered in the world every year. The resulting composite material is called rubbercrete. Trucks or cars tyres are subjected to mechanical grinding in order to obtain rubber particles to be used in the concrete mixtures. Before their use in the

A. Gregori, C. Castoro, M. Mercuri, M. Angiolilli
University of L'Aquila DICEAA, Via G. Gronchi 18, 67100, L'Aquila, Italy
e-mail: amedeo.gregori@ing.univaq.it, chiara.castoro@graduate.univaq.it,
micaela.mercuri@graduate.univaq.it, michele.angiolilli@graduate.univaq.it

B. E. Abali and I. Giorgio (eds.), *Developments and Novel Approaches in Biomechanics and Metamaterials*, Advanced Structured Materials 132,
https://doi.org/10.1007/978-3-030-50464-9_19

mixture, rubber particles may be separated from their textile and metallic parts and may be subjected to pre-treatments, including also cryogenic processes. Rubbercrete shows several interesting properties that make it suitable for the construction industry (Wong and Ting, 2009; Gesoğlu and Güneyisi, 2011; Bignozzi and Sandrolini, 2006; Chung and Hong, 2009; Venslovas et al, 2011; Zhu et al, 2002; Ganjian et al, 2009).

In fact, it is demonstrated that rubbercrete usually exhibits better freeze-thaw resistance and better weathering properties compared to ordinary concrete (Chung and Hong, 2009). Moreover, the replacement of natural aggregates with rubber reduces the Young elastic modulus, allowing the material to absorb energy. The high vibration damping capacity, the impact resistance and the high thermal and acoustic insulation of rubbercrete (Venslovas et al, 2011) allow its use for specific applications including sub-foundations, road pavements, trench filling, Jersey barriers, etc. Rubbercrete also represents an interesting option for architectural applications thanks to its reduced specific weight.

On the other hand, the introduction of rubber aggregates in the mixture reduces the mechanical properties of the hardened concrete (Siddique and Naik, 2004; Skripkiūnas et al, 2007a,b), in particular of the compressive strength (Colom et al, 2006; Gesoğlu and Güneyisi, 2007; Güneyisi et al, 2004; Hernández-Olivares and Barluenga, 2004; Li et al, 2004; Papakonstantinou and Tobolski, 2006; Reda Taha et al, 2008; Ling, 2011). The compressive strength is observed to decrease when the amount of rubber increases and this reduction is observed to be greater when coarse aggregates are replaced with rubber, rather than fine ones (Eldin and Senouci, 1993, 1994; Lee et al, 1993; Topçu, 1995; Fattuhi and Clark, 1996; Fedroff et al, 1996; Topçu and Avcular, 1997; Khatib and Bayomy, 1999; Albano et al, 2005; Benazzouk et al, 2006).

There can be hypothesized different reasons for strength reduction in rubbercrete:

- Rubber particles have lower strength than concrete matrix around them.
- Moreover, rubber aggregates are characterized by a very low modulus of elasticity (in the order of 2-10 MPa) compared to that of concrete (in the order of 30000 MPa) and to that of natural aggregates. For this reason, rubber particles act like voids into concrete (Eldin and Senouci, 1993; Batayneh et al, 2008; Aiello and Leuzzi, 2010).
- In addition, there is poor adhesion between the rubber particles and the cement paste due to the hydrophobic nature of the rubber, and this further weaken the mechanical strength of the whole mixture (Güneyisi et al, 2004; Li et al, 2004; Siddique and Naik, 2004; Segre and Joekes, 2000).

On this purpose, many authors confirmed that rubber pre-treatments might increase the compressive strength of the mixture (Albano et al, 2005; Li et al, 2004). In particular, many authors proved that soaking waste tyre rubber in NaOH solution increases the rubber particles adhesion to the cementitious matrix (Güneyisi et al, 2004; Li et al, 2004; Papakonstantinou and Tobolski, 2006; Segre and Joekes, 2000; Siddique and Naik, 2004). Even using rubber particles washed in water was found to determine a 16% increased compressive strength compared to concrete mixtures containing untreated rubber (Benazzouk et al, 2006; Eldin and Senouci, 1994; Papakonstantinou and Tobolski, 2006).

In literature there is a large number of numerical models developed to estimate the properties of materials with defects (Christensen, 2012; Day et al, 1992; Huang et al, 1994; Isida and Igawa, 1991; Mura, 1982; Taya and Arsenault, 2016) or with additions in the mixture. In Misra (1998), the clay stabilization was developed adding Class C fly ash which behaves like a mixture of cementitious and pozzolanic materials, resulting in a stiffer and stronger material. In Scerrato et al (2014) a numerical model was developed for concretes modified by adding suitable inert additives, whose grains can fill crack voids and improve friction contact of crack lips. In that case, the micro-filler improved the damping performances without compromising the mechanical strength. In Giorgio and Scerrato (2017) a non-linear 3D model was developed to describe internal friction phenomena which occur in modified concrete under cyclic loading with different frequencies. Although a large number of numerical models have been proposed to study modified concretes, FEM simulations on rubbercrete are not found in the literature and most of the research only results from experimental investigations.

In this paper, several FE models have been developed and analysed using the software MIDAS FEA to describe the mechanical behaviour of rubbercrete. Effects on the compressive strength of the rubbercrete were investigated according to different characteristics of the rubber particles: their size, their elastic modulus and their amount in terms of percentage of substitution of the natural aggregates (with respect to the reference concrete volume).

19.2 Materials and Methods

The size, the amount and the distribution of the rubber particles in the rubbercrete matrix are very crucial with respect to the final mechanical properties of the resulting material. For this reason, these parameters have been taken into account as modelling preconditions. In this study, rubber particles were modeled with a cubic shape to overcome difficulties related with extremely expensive calculations. These cubic rubber particles were equally spaced along the three spatial directions and aligned parallel to each other. This regular spatial distribution corresponds to the idea of rubber particles well-dispersed in the concrete mixture, as attempted in real concrete preparation. Moreover, hypotheses about the rubber aggregates shape and spatial distribution allowed to use a much simpler and regular mesh, which also has a great influence on the quality of the results expected from the simulations.

The numerical analyses were carried out referring to a cubic specimen of concrete with a side length of 150 mm, in analogy with experimental tests performed for the qualification of concrete mixtures and in accordance with the European codes UNI EN 12390-1:2002 and UNI EN 12504-1:2002.

FE models were prepared considering the rubber particles and the concrete among them as two homogeneous different materials. Mesh length of 1.5 mm was set for all the analyses. The adopted mesh length allowed to model rubber particles with a minimum size of 3 mm, resulting to be suitable in case of rubber particles replacing both fine natural aggregates and coarse ones. Boundary conditions were introduced in the model to represent the concrete compressive test set up performed in real practice.

To overcome the high computational cost of the analyses, a reduced portion of the initial cubic specimen was actually modelled and tested, imposing the appropriate boundary conditions. This portion consists of a cube with a side length of 75 mm (Fig. 19.1), corresponding to one of the eight corners of the starting cubic specimen. As done in real concrete compressive tests, a gradually increasing vertical compressive stress was applied to the modelled sample of rubbercrete, in accordance with a displacement control procedure. More precisely, a total displacement of 0.45 mm was gradually imposed in z direction to the upper surface of the specimen, with the final displacement rationed in 30 equal increments of 0.015 mm each one.

Different material properties were assigned to the concrete and to the rubber particles.

Concerning concrete, mechanical properties were set in accordance to CEB-FIP (1993) and, in particular, a reference compressive strength f_c = 30 N/mm^2 was assumed together with a fracture energy G = 0.075 N/mm^2 and a Young elastic modulus E_c = 30000 N/mm^2. A tensile strength f_t = 3 N/mm^2 was assumed as result of an elastic, brittle stress-strain response. Mechanical parameters assumed for the reference concrete were kept constant in all the simulations while different amounts of rubber particles were modelled as result of different natural aggregates substitutions.

Concerning rubber particles, a linear elastic behavior was assumed (simple Cauchy model) and the elastic modulus E_r = 1.97 MPa has been assumed for rubber in accordance with literature recommendations (Aiello and Leuzzi, 2010)

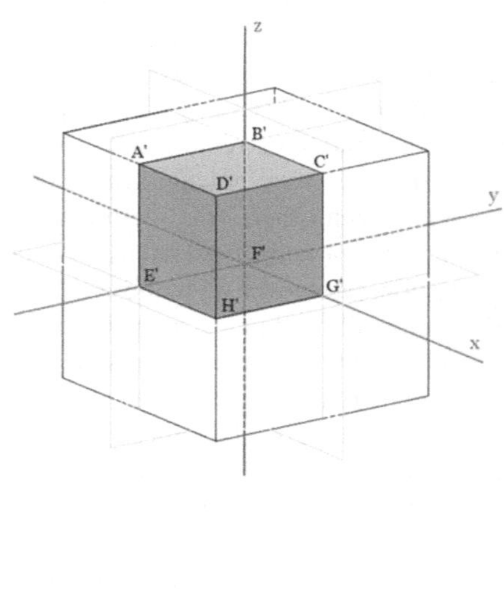

Fig. 19.1 Reduced model: corner cubic specimen with a side length of 75 mm.

19.2.1 The Numerical Simulation Cases

In this study, fourteen 3D FE models were developed and analysed using the MIDAS FEA software. A detailed list is given in Tab. 19.1. Case number 1 corresponds to the reference concrete with no rubber particles in the mixture and the remaining thirteen cases refer to concrete mixtures incorporating rubber particles of various dimensions (from 3 mm to 30 mm) and different amounts, corresponding to various natural aggregates substitutions. In particular, the original dosage of the natural aggregates in the reference concrete was assumed corresponding to 72% of the concrete volume, respectively divided in 46% coarse aggregates (9 mm - 30 mm) and 26% fine aggregates (3 mm - 6 mm). These proportions correspond to the reference concrete mixture also assumed by (Eldin and Senouci, 1993). The percentages of substitution of the natural aggregates with rubber particles were considered to vary in the range 0% - 83%, with respect to the original natural aggregates dosage. These percentages of substitution are reported in Table 1 and are also calculated with respect to the original concrete volume, so obtaining the volume of rubber per cubic meter of concrete. These percentages of defects resulted to be in the range of 0% - 37.32% with respect to the concrete volume. Their calculation allowed to conceive the spatial distribution of the defects to be modelled in the analyses.

Table 19.1 All the numerical simulation cases.

Case Number	Defects size mm	Number of rubber defects	% referred to the total concrete volume	% of substitution operated on coarse aggregates	% of substitution operated on fine aggregates
1 reference	0	0	0	0	0
2	3	3375	2.70	-	10.38
3	6	729	4.66	-	17.90
4	18	27	4.66	10.13	-
5	27	8	4.66	10.13	-
6	24	27	11.06	24.04	-
7	15	125	12.50	27.17	-
8	9	729	15.74	34.22	-
9	6	3375	21.60	-	83.00
10	18	125	21.60	46.95	-
11	30	27	21.60	46.95	-
12	21	125	34.30	74.56	-
13	12	729	37.32	81.00	-
14	18	216	37.32	81.00	-

The cases summarized in Tab. 19.1 were designed to investigate two different conditions: increasing the rubber particle sizes considering a fixed percentage of rubber (group of cases: 3, 4, 5-9, 10, 11-13, 14). An example is provided in Fig. 19.2. Increasing the amount of rubber considering a fixed size (groups of cases: 3, 9-4, 10, 14). An example is provided in Fig. 19.3.

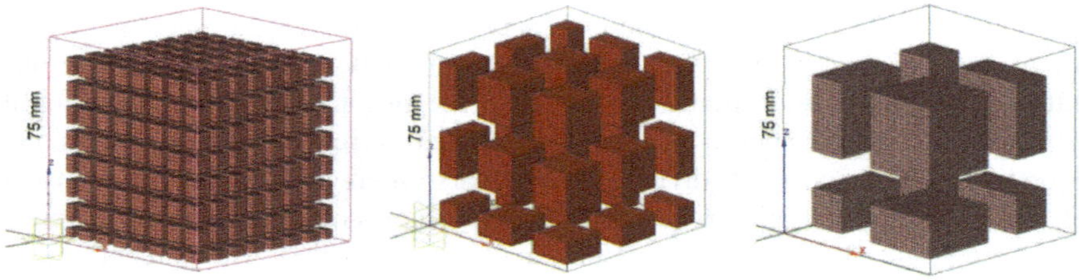

Fig. 19.2 Cases 9, 10, 11 of Tab. 19.1. The percentage of substitution of the natural aggregates is 21.6% and it was obtained with rubber particles measuring 6, 18 and 30 mm respectively.

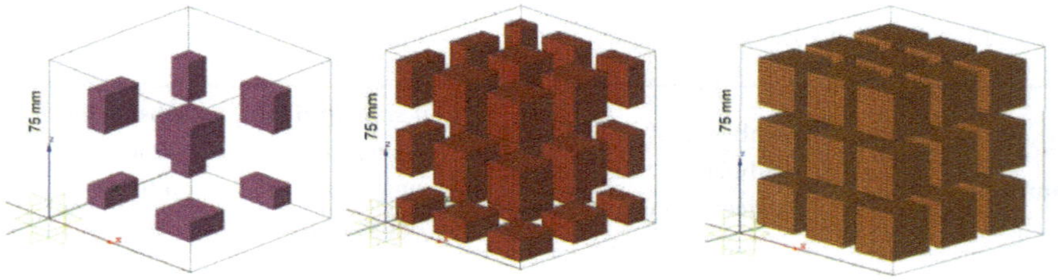

Fig. 19.3 Cases 4, 10, 14 of Tab. 19.1. The size of the rubber particles is fixed equal to 18 mm and different amount of defects are considered (different percentage of substitution of the natural aggregates equal to 4.66%, 21.6% and 37.32% respectively.

Since not any desired percentage of defects (rubber particles) could be modelled with a uniform spatial distribution, the selected percentages of substitution listed in Tab. 19.1 do not vary continuously in the range 0% - 37.32% and they were conceived in this way due to geometric construction requirements of the FE models.

Rubber particle size equal to 6 mm and 18 mm are those for which the larger number of defects percentages can be considered and compared each other.

19.3 Results

19.3.1 Numerical Prediction of the Compressive Strength

Predicted compressive strength values resulted from the numerical analyses are presented in Fig. 19.4. In this graph, the strength reduction factor (SRF), defined as the ratio among the rubbercrete compressive strength f_c and the reference concrete compressive strength f_{c0}, is plotted in relation to the percentage of the rubber particles with respect to the concrete volume. The SRF is unity at 0% rubber content, meaning the reference mix.

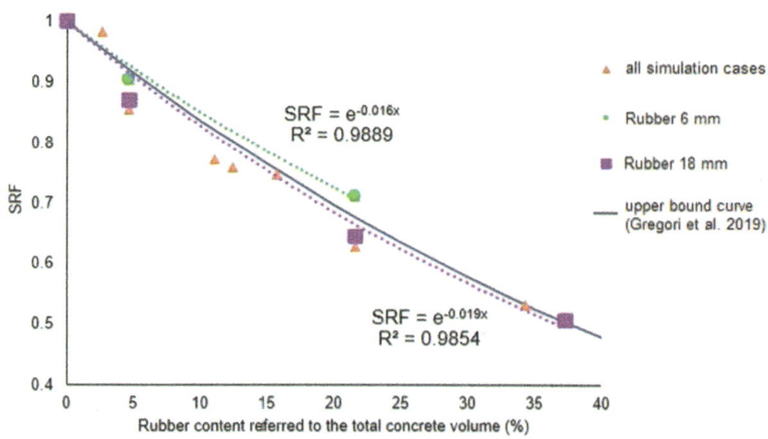

Fig. 19.4 Strength Reduction Factor (SRF) of rubbecretes with respect to the rubber content in the total volume .

As shown in Fig. 19.4, a reduction in the compressive strength is observed when the percentage of rubber, referred to the total volume of the concrete sample, increases from 0% to 37.32%. This percentages of rubber particles correspond to the range 0% - 83% of substitution of the original dosage of the aggregates. Mathematical regressions were performed to best fit the calculated data, in particular for cases with rubber particles measuring 6 mm in size (substitutions operated on fine aggregates) and 18 mm in size (substitutions operated on coarse aggregates).

The exponential form was assumed for the regressions, as done in (Gregori et al, 2019), and the equations obtained are:

- for rubber particles of 6mm:

$$SRF = e^{-0.016x} \tag{19.1}$$

- for rubber particles of 18mm:

$$SRF = e^{-0.019x} \tag{19.2}$$

The high values of R^2 calculated for the two equations confirm the good quality of the regressions.

Sub-ranges of variation can be identified in relation to a precise type of natural aggregate substitution operated with rubber: fine aggregates replacement or coarse aggregates replacement.

Figure 19.4 and the equivalent data summarised in Tab. 19.2 highlight that expectable SRF values tend to be lower in case of substitutions operated on bigger aggregates (18 mm rather than aggregates of 6 mm), this meaning that a larger loss in compressive strength may be expected when rubber aggregates are introduced in substitution of coarse natural aggregates compared to cases in which aggregate substitution is operated on fine ones.

Table 19.2 Comparing SRF values of rubbercrete with fine and coarse rubber particles.

Rubber content (%)	SRF size 6 mm	SRF size 18 mm
0	1	1
4.66	0.905	0.870
21.6	0.711	0.645
37.32		0.507

The upper bound curve illustrated in Fig. 19.4 is obtained from the large set of experimental data already analysed in Gregori et al (2019). In that work, SRF data resulting from several studies carried out by different authors and referring to partial substitutions of fine and coarse natural aggregates operated on various concrete mixes were reported.

Figure 19.4 shows that the results of the numerical simulations are in good agreement with the mentioned upper bound curve. According to literature, the use of NaOH for pre-treatment of the rubber particles improves the adhesion of these particles with the cement matrix, so resulting a lower decrease in the concrete compressive strength. The upper bound curve calculated in Gregori et al (2019) and plotted in Figure 19.4 refers to those cases.

In the FE models analysed in the present study, the hypothesis of perfect adhesion between the rubber particles and the cement matrix was made, and this explains why the results of the analyses are close to the upper bound curve of the experimental data. Since the ITZ between the cement paste and the rubber particles is intended to be weaker than ITZ in normal concretes, different and more complex hypothesis rather than perfect adhesion at the interfaces could be eventually introduced in the FE models, but this is currently out of the aim of this study.

19.3.2 Numerical Prediction of SRF at Varying the Reference Concrete Compressive Strength

The compressive strength of the reference concrete was assumed equal to 30 MPa and this value was maintained fixed in all the analyses.

In this section, variations of the predicted values of SRF are investigated with respect to the compressive strength value assumed for the reference concrete. In particular, the case of rubber particles content of 21.6% is represented considering different reference concrete compressive strength values, equal to 30, 40, 60 and 80 MPa respectively. Consequently, the elastic modulus assigned to the concrete was set in accordance to the formulation proposed by CEB-FIP (1993) as function of the assumed concrete compressive strength value. The analyses were performed for two different rubber particle sizes (6mm and 18 mm respectively) and results are summarised in Tab. 19.3.

Table 19.3 Predicted SRF values resulting from different values of the reference compressive strength assumed. SRF values refers to rubbercretes with an amount of 21.6% of rubber particles of 6mm and 18mm in size respectively.

Reference concrete compressive strength MPa	Reference concrete elastic modulus MPa	SRF rubber particles 6mm	SRF rubber particles 18mm
30	30000	0.7111	0.6450
40	35000	0.6900	0.6396
60	40000	0.6724	0.6242
80	45000	0.6619	0.6226

From data in Tab. 19.3, a slight variation in SRF values is noticed when different compressive strength values are assumed for the reference concrete. In particular, lower SRF values are found when a higher reference concrete compressive strength value is assumed. These reductions are larger in case of substitution operated on coarse aggregates (18 mm) rather than on fine ones (6 mm).

These conclusions are confirmed regardless of the percentage and of the size of the rubber particles in concrete. In this section, the rubber particles content of 21.6% (corresponding to the significant case of about 50% of substitution of the original dosage of the natural coarse aggregates or, in alternative, to about 80% of the fines) is representative because at this percentage modification in terms of compressive strength are relevant.

The range of variability of the predicted SRF values listed in Tab. 19.3 helps to partially explain the large difference sometimes also noticed among experimental SRF data reported in the literature by different authors. Actually, experimental studies from different authors certainly refer to concrete mixtures very different each other in terms of mix proportioning and, consequently, in terms of compressive strength.

Then, a wide range of experimental SRF values results from the literature, as shown in Gregori et al (2019).

19.4 Conclusion

In this study, several FE models were developed to investigate the mechanical response of rubbercrete, which is a concrete modified by partial replacement of the natural aggregates with rubber particles from recycled waste tyres.

The effects on the compressive strength were investigated according to the size, the elastic modulus, and the amount of the rubber particles. Results from the numerical analyses well reproduced the trends indicated by the experimental data in literature, confirming that numerical simulations may represent a promising, economic and complementary tool (in combination with usual experimental tests) for pre-qualification of rubbercrete mixtures.

Numerical simulations confirmed that the compressive strength of rubbercrete reduces when the amount of natural aggregates substituted with rubber particles increases, and this reduction is larger for coarse aggregates substitution, rather than fine one.

Moreover, the effects of different values adopted for the reference concrete compressive strength are also confirmed to depend on the size of the rubber particles, regardless the amount.

Predicting equations were provided for the estimation of the Strength Reduction Factor of the rubbercrete to be expected in case of substitutions of the natural fine and coarse aggregates with rubber particles, especially when rubber particles are pre-treated to enhance their adhesion with the cementitious matrix.

References

Aiello MA, Leuzzi F (2010) Waste tyre rubberized concrete: Properties at fresh and hardened state. Waste Management 30(8-9):1696–1704

Albano C, Camacho N, Reyes J, Feliu J, Hernández M (2005) Influence of scrap rubber addition to portland i concrete composites: destructive and non-destructive testing. Composite Structures 71(3-4):439–446

Batayneh MK, Marie I, Asi I (2008) Promoting the use of crumb rubber concrete in developing countries. Waste management 28(11):2171–2176

Benazzouk A, Douzane O, Mezreb K, Quéneudec M (2006) Physico-mechanical properties of aerated cement composites containing shredded rubber waste. Cement and Concrete Composites 28(7):650–657

Bignozzi M, Sandrolini F (2006) Tyre rubber waste recycling in self-compacting concrete. Cement and concrete research 36(4):735–739

CEB-FIP (1993) Model Code 1990 for Concrete Structures

Christensen RM (2012) Mechanics of composite materials. Courier Corporation

Chung KH, Hong YK (2009) Weathering properties of elastic rubber concrete comprising waste tire solution. Polymer Engineering & Science 49(4):794–798

Colom X, Cañavate J, Carrillo F, Velasco J, Pages P, Mujal R, Nogues F (2006) Structural and mechanical studies on modified reused tyres composites. European polymer journal 42(10):2369–2378

Day A, Snyder K, Garboczi E, Thorpe M (1992) The elastic moduli of a sheet containing circular holes. Journal of the Mechanics and Physics of Solids 40(5):1031–1051

Eldin NN, Senouci AB (1993) Rubber-tire particles as concrete aggregate. Journal of materials in civil engineering 5(4):478–496

Eldin NN, Senouci AB (1994) Measurement and prediction of the strength of rubberized concrete. Cement and Concrete Composites 16(4):287–298

Fattuhi N, Clark L (1996) Cement-based materials containing shredded scrap truck tyre rubber. Construction and building materials 10(4):229–236

Fedroff D, Ahmad S, Savas BZ (1996) Mechanical properties of concrete with ground waste tire rubber. Transportation Research Record 1532(1):66–72

Ganjian E, Khorami M, Maghsoudi AA (2009) Scrap-tyre-rubber replacement for aggregate and filler in concrete. Construction and building materials 23(5):1828–1836

Gesoğlu M, Güneyisi E (2007) Strength development and chloride penetration in rubberized concretes with and without silica fume. Materials and Structures 40(9):953–964

Gesoğlu M, Güneyisi E (2011) Permeability properties of self-compacting rubberized concretes. Construction and Building Materials 25(8):3319–3326

Giorgio I, Scerrato D (2017) Multi-scale concrete model with rate-dependent internal friction. European Journal of Environmental and Civil Engineering 21(7-8):821–839

Gregori A, Castoro C, Marano GC, Greco R (2019) Strength reduction factor of concrete with recycled rubber aggregates from tires. Journal of Materials in Civil Engineering 31(8):04019,146

Güneyisi E, Gesoğlu M, Özturan T (2004) Properties of rubberized concretes containing silica fume. Cement and concrete research 34(12):2309–2317

Hernández-Olivares F, Barluenga G (2004) Fire performance of recycled rubber-filled high-strength concrete. Cement and concrete research 34(1):109–117

Huang Y, Hu K, Wei X, Chandra A (1994) A generalized self-consistent mechanics method for composite materials with multiphase inclusions. Journal of the Mechanics and Physics of Solids 42(3):491–504

Isida M, Igawa H (1991) Analysis of a zig-zag array of circular holes in an infinite solid under uniaxial tension. International Journal of Solids and Structures 27(7):849–864

Khatib ZK, Bayomy FM (1999) Rubberized portland cement concrete. Journal of materials in civil engineering 11(3):206–213

Lee B, Burnett L, Miller T, Postage B, Cuneo J (1993) Tyre rubber/cement matrix composites. Journal of Materials Science Letters 12(13):967–968

Li G, Stubblefield MA, Garrick G, Eggers J, Abadie C, Huang B (2004) Development of waste tire modified concrete. Cement and Concrete Research 34(12):2283–2289

Ling TC (2011) Prediction of density and compressive strength for rubberized concrete blocks. Construction and Building Materials 25(11):4303–4306

Misra A (1998) Stabilization characteristics of clays using class c fly ash. Transportation Research Record 1611(1):46–54

Mura T (1982) Micromechanics of defects in solids. Martinus Nijhoff Pub lish ers, The Hague, Neth er lands

Papakonstantinou CG, Tobolski MJ (2006) Use of waste tire steel beads in portland cement concrete. Cement and Concrete Research 36(9):1686–1691

Reda Taha MM, El-Dieb A, Abd El-Wahab M, Abdel-Hameed M (2008) Mechanical, fracture, and microstructural investigations of rubber concrete. Journal of materials in civil engineering 20(10):640–649

Scerrato D, Giorgio I, Madeo A, Limam A, Darve F (2014) A simple non-linear model for internal friction in modified concrete. International Journal of Engineering Science 80:136–152

Segre N, Joekes I (2000) Use of tire rubber particles as addition to cement paste. Cement and concrete research 30(9):1421–1425

Siddique R, Naik TR (2004) Properties of concrete containing scrap-tire rubber–an overview. Waste management 24(6):563–569

Skripkiūnas G, Grinys A, Černius B (2007a) Deformation properties of concrete with rubber waste additives. Materials Science (Medžiagotyra) 13(3):219–222

Skripkiūnas G, Grinys A, Daukšys M (2007b) Using tires rubber waste for modification of concrete properties. In: Sustainable Construction Materials and Technologies International Conference, Coventry

Taya M, Arsenault RJ (2016) Metal matrix composites: thermomechanical behavior. Elsevier

Topçu IB (1995) The properties of rubberized concretes. Cement and Concrete Research 25(2):304–310

Topçu IB, Avcular N (1997) Collision behaviours of rubberized concrete. Cement and Concrete Research 27(12):1893–1898

Venslovas A, Aleksandravičiūte D, Idzelis RL (2011) Experimental investigation of scrap-tire crumb rubber application in noise-suppression structures. Baltic Journal of Road & Bridge Engineering (Baltic Journal of Road & Bridge Engineering) 6(2)

Wong SF, Ting SK (2009) Use of recycled rubber tires in normal-and high-strength concretes. ACI Materials Journal 106(4):325

Zhu H, Thong-On N, Zhang X (2002) Adding crumb rubber into exterior wall materials. Waste management & research 20(5):407–413

Chapter 20
Linear Dynamics of 2D Pantographic Metamaterials: Numerical and Experimental Study

Marco Laudato & Luca Manzari

Abstract In this paper, the results of numerical and experimental studies on the linear time-invariant dynamics of a 2D pantographic material are presented. The outcomes of a linear second gradient model enforcing the same symmetry of the microstructure is compared to experimental observations obtained via Digital Image Correlation (DIC).

Keywords: Pantographic material · Generalized continua · Dynamics · Mechanical metamaterials · Digital Image Correlation (DIC)

20.1 Introduction

Some of the most interesting features of mechanical metamaterials happen when they are subjected to dynamic loads. In this regime, the complex interactions among the elements of the microstructure give rise to unusual macroscopic mechanical behaviors—band gaps in the dispersion relations of the locally resonant metamaterials proposed by Liu et al (2000) are typical examples, see also di Cosmo et al (2018).

The mechanical metamaterials field has experienced a fast growth in the last few decades mainly due to the remarkable improvement of fast-prototyping techniques (such as 3D printing) that allow researchers to produce accurate specimens in a

---

Marco Laudato
Dipartimento di Ingegneria e Scienze dell'Informazione e Matematica, Università degli Studi dell'Aquila, Via Vetoio (Coppito 1), 67100 Coppito, L'Aquila, Italy
e-mail: marco.laudato@graduate.univaq.it

Luca Manzari
KTH Royal Institute of Technology, Marcus Wallenberg Laboratory for Sound and Vibration Research, SE-100 44, Stockholm, Sweden
e-mail: manzari@kth.se

B. E. Abali and I. Giorgio (eds.), *Developments and Novel Approaches in Biomechanics and Metamaterials*, Advanced Structured Materials 132,
https://doi.org/10.1007/978-3-030-50464-9_20

reasonable amount of time. However, the increasing complexity of the microstructure geometry requires elaborate mathematical models able to properly describe the resultant macroscopic behavior.

One of the possible modelling strategies consists in the definition of an analogous continuous mechanical system whose dynamics is obtained by means of a homogenization of the dynamics of the metamaterial underlying microstructure.

As expected, the more complex the geometry and the interactions of the microstructure, the more complex the homogenized continuous system will be. It is possible, indeed, that the resultant homogenized continuum model cannot be framed in the setting of the classical theory of elasticity. When this is the case, these analogous continuous systems are called *generalized continua*.

This kind of systems have been studied for the first time by Mindlin and Eshel (1968) (see for historical remarks dell'Isola et al, 2015, 2017a) and have been classified in two main classes, depending on the kinematical descriptors which appear in the definition of the mechanical energy of the system (see Rosi et al, 2018) for a complete description of their classification):

- Higher-order continua, in which the set of kinematical fields is extended w.r.t. the classical theory of elasticity. A typical example is the Cosserat model, in which local rotations are allowed in the kinematics of the system.
- Higher-grade continua, in which the set of kinematical fields is the same but higher gradients of the displacement field are involved in the definition of the energy of the system.

This work focuses on an exemplary system belonging to this last class called *pantographic material*. This mechanical metamaterial is characterised by a microstructure made of two or more parallel arrays of straight fibers which are oriented in mutual orthogonal directions. In the intersection points of the fibers, the two arrays are connected by a set of cylinders called pivots (see Fig. 20.1). It has been shown by mathematical (Alibert et al, 2003; Boutin et al, 2017; Placidi et al, 2020; Eremeyev et al, 2019) and experimental (dell'Isola et al, 2016; Golaszewski et al, 2019; Turco et al, 2019; Barchiesi et al, 2019b) methods, that in order to fully describe the macroscopic static behavior of this system a second gradient theory is needed.

However, the experimental investigation of the dynamical behavior of pantographic materials is very recent (Laudato et al, 2018). The main aim of this work is

Fig. 20.1 CAD visualization of the microstructure of a 2D pantographic material. Two arrays of straight parallel fibers are connected in the intersection points by the pivots.

to compare the outcomes of the experimental observations of a pantographic material undergoing a small periodic imposed displacement with the predictions of a linear second gradient model under the Linear Time Invariant (LTI) dynamics assumption.

The paper is organized as follows. In Sect. 20.2 the particular specimen under analysis is described. In Sect. 20.3 the details of the approach used to describe the LTI dynamics in both the experimental and numerical analysis are outlined. In Sect. 20.4 the linear model used for the numerical simulations is introduced, while in Sect. 20.5 the experimental setup is presented. Finally, in Sect. 20.6 the comparison between the experimental and numerical results is discussed.

20.2 2D Pantographic Metamaterials

Pantographic materials have been studied for long time (dell'Isola et al, 2019b,a; Placidi et al, 2016; Barchiesi and Placidi, 2017; Nejadsadeghi et al, 2019) and in several configurations, from 1D pantographic beams (see Barchiesi et al, 2019a) to 3D pantographic volumes (dell'Isola et al, 2017b). The microstructure of the so-called *pantographic sheet* studied in this work consists of two parallel planes connected in the intersection points of the fibers by a set of pivots as shown in Fig. 20.1. Some considerations are in order:

- Current state-of-the-art 3D printers are able to produce so-called "*perfect* pivots", meaning that there is almost no energy associated to their deformation. The specimens considered in this work have this feature and this must be accounted for during the modeling phase.
- Although the system is three-dimensional, its thickness is one order of magnitude smaller with respect to the other linear lengths and it has been experimentally observed (Laudato et al, 2019) that the out-of-plane oscillations can be considered negligible in such small deformations fashion. Therefore, for the sake of simplicity, the specimen is modelled as a two-dimensional system.

The physical specimen under investigation is a rectangular pantographic metamaterial sheet made of polyamide EOS PA2200 printed by a Formiga P100 3D printer. Its linear dimensions are $235\,\mathrm{mm} \times 78\,\mathrm{mm}$ with a thickness of $6.3\,\mathrm{mm}$ and its weight is $24.37\,\mathrm{g}$. The distance between two pivots is $7.00\,\mathrm{mm}$ and the fibers diameter is $1.8\,\mathrm{mm}$.

20.3 An Approach to the Analysis of LTI Systems

In this section the method adopted to analyse and compare the outcomes from experimental observations and numerical simulations of the dynamic behavior of the pantographic metamaterial is presented. The main assumption is that the dynamic behavior of the system in the observed configuration can be considered *linear time invariant* (LTI). This assumption implies that the relation between the input and the output signals of such a system is linear and that, for a given time T, the result of an input signal applied at time t and time $t+T$ will be the same except for a T delay. As a consequence, the output signal will be the result of the convolution of the input signal with the impulse response of the system in time domain or, equivalently, of the multiplication of the transfer function of the system by the Laplace transform of the input signal in the frequency domain. An interesting property is that if the input signal is a sinusoid at a given frequency, the output signal will still be in general a sinusoid with different amplitude and phase but same frequency.

Under this assumption, any mechanical system whose dynamic behavior can be modeled as LTI can be fully characterized by means of transfer functions relating the motion of its different parts. In this work, we will consider the outcomes from Digital Image Correlation (DIC) for the experimental side, and Finite Element Methods (FEM) for the numerical side. Indeed, both methods perform a discretisation of the system geometry and output the displacements in time for each point of the mesh.

Since the imposed displacement is sinusoidal, due to the LTI hypothesis, the output signals obtained by means of the aforementioned methods must be sinusoids as well. It can, as such, be fitted by the following general sinusoidal function, for each point of the discretised domain:

$$s_i(t) = A_i \cos\left(2\pi f_i t + \varphi_i\right) + B_i\,. \tag{20.1}$$

In this way, the information of the time behavior of each point of the system is fully represented by the set of parameters $(A_i, f_i, \varphi_i, B_i)$, where A_i is the amplitude of the output signal, f_i is the frequency of the oscillation, φ_i is the phase, B_i is a bias, and the index i runs on the independent directions in which is possible to decompose the resulting oscillations. It is now possible to plot the values of these parameters on the system domain, obtaining in this way an informative visualization of the time behavior of the system. The method applies in the same way for both numerical and experimental data, allowing for a meaningful comparison between the experimental observations and the outcomes from the model. For additional details about the data reduction procedure, along with the advantages it entails and other applications to experimental data from DIC (see Manzari et al, 2018).

20.4 Linear Model for 2D Pantographic Sheets

Mechanical metamaterials are inherently multi-scale systems (Barchiesi et al, 2019c; Del Vescovo and Giorgio, 2014). The different levels of descriptions are hierarchically connected and require different modelling approaches to grasp their particular features. Pantographic materials are no exception.

By assuming that the elements of the microstructure (pivots and fibers) are the elementary building blocks of our description[1], the three following scales of description of a pantographic sheet can be considered. At microscopic level, it is possible to model each element of the microstructure as a 3D continuum by means of the Cauchy theory of elasticity. However, although this approach would lead to a very accurate description of the microscopic behaviour of the system, it is extremely computationally expensive and, from a more pragmatic point of view, not useful since the interest of the current work lies in the observed overall macroscopic behaviour. Therefore it is convenient to introduce a mesoscopic scale of description, in which the elements of the microstructure are modeled in terms of mechanical elements (e.g. springs, bars, beams, etc.) arranged in a way able to mimic the behaviour of the microstructure. In case of the pantographic sheet, the main approximation is the passage from a three-dimensional model to a planar description. At this scale (for a more general description, see Giorgio et al, 2019), the distance between the two arrays of fibers (see Fig. 20.1) is neglected and the elements of the mesoscopic model are arranged in a planar lattice (see Fig. 20.2). In the literature, it is possible to find two different mesoscopic models for the pantographic sheet. One is the so-called Hencky-type model (Turco et al, 2016, 2018; Turco, 2018; Turco et al, 2017), in which the fibers are modeled as extensional springs and the pivots as rotational springs: an additional spring connects two orthogonal fibers.

In a different approach, instead, the fibers are modeled as Euler beams with finite extensional stiffness[2] and the pivots represent a hinge constraint in which four beams interact.

In addition to these two mesoscopic models, the pantographic sheet can be described as a continuous two-dimensional system. In order to embrace a variational approach, a continuous model will be defined by firstly specifying the kinematical fields describing the behaviour of the system and then by providing an appropriate form of the mechanical energy of such system as a functional of these fields. A

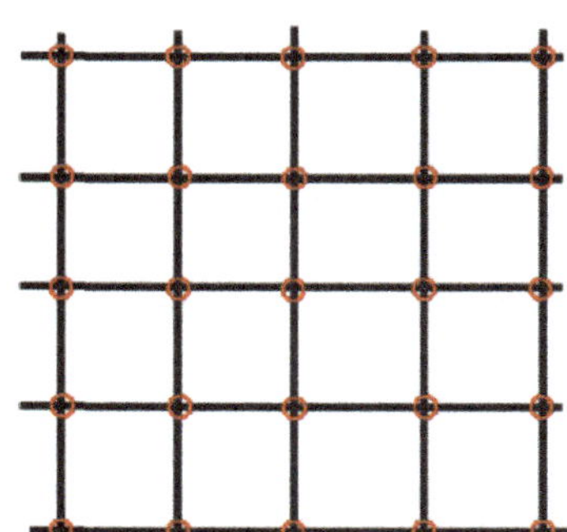

Fig. 20.2 Mesoscopic model of a pantographic sheet microstructure. The fibers (Euler extensible beams) form a 2D lattice in which the pivots (red circles) represent a hinge constraint.

[1] It means that the inner structure of the material which composes the elements of the microstructure is neglected (i.e. the granular nature of the polyamide).

[2] This assumption is driven by experimental observations showing that the fibers elongate up to 10 % when a pantographic sheet undergoes a bias extension test.

common approach in the metamaterial framework is to obtain the explicit form of such an energy by exploiting the multi-scale nature of the system. Indeed, there are several methods, usually called homogenization methods, which are able to yield a continuous model by means of some limit procedure of the semi-discrete description of the mesoscopic structure.

Several efforts have been devoted to obtain a homogenized continuous theory for the pantographic sheet starting from the mesoscopic description of its microstructure (dell'Isola et al, 2016; Barchiesi et al, 2018; Alibert and Della Corte, 2015; Alibert et al, 2003). Although these models can be slightly different, all of them entail a form of the mechanical energy which depends on the second gradient of the displacement field (Alibert and Della Corte, 2019; Giorgio et al, 2017). This is enough to consider homogenized models for pantographic sheets as higher-grade models, according to the aforementioned classification. A form of the mechanical energy is postulated in Sect. 20.4.1 and validated by comparing mathematical and numerical results with experimental observations. This will be an educated guess, driven by two main observations:

- As already anticipated, several mathematical and experimental results obtained during the investigation of the static behaviour of pantographic sheets suggest that it has to be considered as a higher-grade material to correctly forecast its macroscopic behavior. In particular, models in which the deformation energy depends upon the second gradient of the displacement field are able to perfectly forecast equilibrium configuration of pantographic sheet under quasi-static standard tests.
- The same symmetry properties which hold at the mesoscopic level must hold for the macroscopic description. Indeed, as it is evident from Fig. 20.2, the mesoscopic lattice shows a D4 symmetry, meaning that it is invariant under rotation of $\pi/2$ around the normal axis and under mirror transformations. The constitutive matrix of the macroscopic model must present the same symmetry properties.

The model will be formally defined in the rest of the section, but let us anticipate that the deformation energy has been introduced for the first time by Placidi et al (2017) and lies in the more general framework of the so-called strain gradient theory (Mindlin, 1965). The main reason behind this choice is that these kinds of models have been already successfully exploited to study the dynamical behaviour of higher grade mechanical metamaterials (Rosi and Auffray, 2019, 2016).

20.4.1 Kinematics and Dynamics

The reference configuration of the system is defined as a rectangular closed subset $\mathcal{B}$ of the Euclidean space $\mathbb{E}^2$. Consider a set of Cartesian coordinates such that the base directions (X_1, X_2) are parallel to the fibers of the pantographic sheet, as shown in Fig. 20.3.

A generic point of $\mathcal{B}$ will be identified by a vector $\boldsymbol{X} = (X_1, X_2)$. As per description in Sect. 20.5, the final output of the model will be rotated and described in terms of the (x, y) reference frame (in red in Fig. 20.3). Be the boundary $\partial\mathcal{B}$ the union of four subsets Σ_i of $\mathcal{B}$ $(i = 1, \ldots, 4)$ such that their intersection has zero 1D and 2D Lebesgue measure. In particular, Σ_1 is considered clamped and Σ_3 undergoes the imposed periodic oscillation along the y direction. The boundary of the boundary $\partial\partial\mathcal{B}$ is defined as the disjoint union of the four vertices V_i of $\mathcal{B}$ $(i = 1, \ldots, 4)$ (see Fig. 20.3). The position of a Lagrangian point $\boldsymbol{X} \in \mathcal{B}$ at time

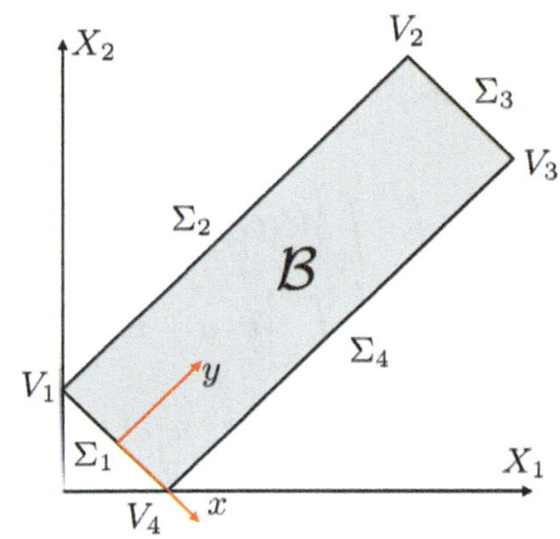

Fig. 20.3 Geometry of the model domain. To keep contact with the experimental setting the final output of the model will be rotated and described in terms of the (x, y) reference frame (in red).

t is defined by means of the so-called placement function $\chi : \mathcal{B} \times \mathbb{R} \longrightarrow \mathbb{E}^2$ such that $\boldsymbol{x} = \chi(\boldsymbol{X}, t)$. From the placement function it is possible to define the displacement function $\boldsymbol{u}(\boldsymbol{X}, t)$ as the difference w.r.t. the reference configuration, namely $\boldsymbol{u} = \chi(\boldsymbol{X}, t) - \boldsymbol{X}$.

In this model, the whole kinematics of the system is described by the placement field and its first and second derivatives. The gradient of the placement w.r.t. the Lagrangian coordinates is defined as $\boldsymbol{F}(\boldsymbol{X}, t) = \nabla_X \chi(\boldsymbol{X})$ and it is a second order tensor. This tensor can be used to define an objective deformation measure, i.e. a quantity able to take into account the amount of energy associated to a certain deformed configuration. In this work, the so-called Green–Saint Venant deformation measure is adopted,

$$\boldsymbol{G}(\boldsymbol{X}, t) = \frac{1}{2}\left[\boldsymbol{F}^\mathsf{T}\boldsymbol{F} - \boldsymbol{I}\right] , \tag{20.2}$$

where $(\cdot)^\mathsf{T}$ indicates the transposition operation and $\boldsymbol{I}$ is the identity matrix. If $\boldsymbol{F}$ is an orthogonal transformation, i.e. it preserves lengths and angles, the Green–Saint Venant tensor vanishes.

The dynamical behaviour of the pantographic sheet has to minimize the following action functional

$$\mathcal{A}(G, \nabla G) = \int_0^T \int_{\mathcal{B}} \frac{1}{2} \rho |\dot{\boldsymbol{u}}|^2 dX dt - \mathcal{E}_{\text{def}}(\boldsymbol{G}, \nabla \boldsymbol{G}) , \tag{20.3}$$

where ρ is the surface mass density of the sheet, $\dot{\boldsymbol{u}}$ indicates the time derivative of the displacement $\boldsymbol{u}$ and $\mathcal{E}_{\text{def}}$ is the deformation energy functional.

In the present formulation of the model the standard kinetic energy term has been used. It has been shown in (Battista et al, 2015) that the first modes in which the microintertia term (see Rosi et al, 2018, for a brief introduction) would play a crucial role are relative to a range of frequencies that are not considered in the present investigation. Therefore, this form of the kinetic energy term can be considered as a reliable first approximation for the present case. The effects of the microintertia will be investigated in a future work. The most general form of the deformation energy $\mathcal{E}_{\text{def}}$ is

$$\begin{aligned} \mathcal{E}_{\text{def}}(\boldsymbol{G}, \nabla \boldsymbol{G}) = & \int_0^T \int_{\mathcal{B}} \left[U(\boldsymbol{G}, \nabla \boldsymbol{G}) - \tilde{\boldsymbol{b}} \cdot \boldsymbol{u} \right] dX dt \\ & - \int_0^T \int_{\partial \mathcal{B}} \left[\tilde{\boldsymbol{t}} \cdot \boldsymbol{u} + \tilde{\boldsymbol{\tau}} \cdot [(\nabla \boldsymbol{u}) \hat{\boldsymbol{n}}] \right] ds dt \\ & - \int_0^T \int_{\partial \partial \mathcal{B}} \left[\tilde{\boldsymbol{f}} \cdot \boldsymbol{u} \right] dt . \end{aligned} \tag{20.4}$$

Let us analyse this expression term by term. $U(\boldsymbol{G}, \nabla \boldsymbol{G})$ is called strain gradient deformation energy, and it depends upon $\nabla \boldsymbol{G}$, i.e. upon the second gradient of the placement field, which is consistent with the discussion at the beginning of the section.

Once the explicit form of U is given, we will require for it to be invariant w.r.t. elements of the $D4$ symmetry group. All the elements labeled with a tilde represent external actions on the body. In particular (see dell'Isola et al, 2012; Auffray et al, 2015a), $\tilde{\boldsymbol{b}}$ is an external bulk force per unit area.

In the second term, $\tilde{\boldsymbol{t}}$ and $\tilde{\boldsymbol{\tau}}$ represent external force and double force per unit length. Finally, in the last term, $\tilde{\boldsymbol{f}}$ represents an external concentrated force[3] which acts on the vertices of the system (Placidi et al, 2017). In particular, the last two integrals are defined as:

$$\int_{\partial \mathcal{B}} g_1(\boldsymbol{X}) ds = \sum_{i=1}^{4} \int_{\Sigma_i} g_1(s) ds , \quad \int_{\partial \partial \mathcal{B}} g_2(\boldsymbol{X}) = \sum_{i=1}^{4} g_2(V_i) , \tag{20.5}$$

where s parameteris Σ_i and g_1, g_2 are generic function of $\boldsymbol{X}$ defined on $\partial \mathcal{B}$ and $\partial \partial \mathcal{B}$, respectively.

Before giving the explicit formula for the strain gradient energy U, in order to clarify the meaning of the following symbols, it is convenient to briefly introduce the

[3] When higher order derivatives of the displacement field are considered as independent variables, the set of admissible external actions on the system will include also terms which couple these with additional kinematic descriptors, such as double forces or forces acting on points.

constitutive equations of the linear strain-gradient elasticity theory. In this approach, formulated for the first time by Mindlin in the 1960s, the constitutive laws are given in terms of the usual infinitesimal strain tensor ε and its gradient, called strain gradient tensor $\boldsymbol{\eta} = \varepsilon \otimes \nabla$, by means of the following linear equations:

$$\begin{cases} \sigma_{ij} = C_{ijab}\varepsilon_{ab} + M_{ijabc}\eta_{abc} \\ s_{ijk} = M_{abijk}\varepsilon_{ab} + A_{ijkabc}\eta_{abc} \end{cases} . \tag{20.6}$$

In this expression, $\boldsymbol{\sigma}$ is the usual stress tensor, $\boldsymbol{s}$ is the so-called hyper-stress tensor, $\boldsymbol{C}$ is the usual elastic tensor, $\boldsymbol{M}$ is a tensor coupling ε and $\boldsymbol{\eta}$, and $\boldsymbol{A}$ is an additional sixth order elastic tensor.

In the case of a pantographic sheet, a bi-dimensional system invariant under rotation of $\pi/2$, the odd-order tensor $\boldsymbol{M}$ vanishes. Moreover, it is possible to show (Auffray et al, 2015b) that the two tensors $\boldsymbol{C}$ and $\boldsymbol{A}$ admit a matrix representation. The explicit formula for the strain gradient energy $U(\boldsymbol{G}, \nabla\boldsymbol{G})$ is therefore (Placidi et al, 2017)

$$U(\boldsymbol{G}, \nabla\boldsymbol{G}) = U(\varepsilon, \boldsymbol{\eta}) = \frac{1}{2}\left(\varepsilon^{\mathsf{T}}\boldsymbol{C}\varepsilon + \boldsymbol{\eta}^{\mathsf{T}}\boldsymbol{A}\boldsymbol{\eta}\right) \tag{20.7}$$

where ε and $\boldsymbol{\eta}$ are column vectors whose components are

$$\varepsilon = (G_{11}, G_{22}, \sqrt{2}G_{12}), \quad \boldsymbol{\eta} = (G_{11,1}, G_{22,2}, \sqrt{2}G_{12,2}, G_{11,2}, \sqrt{2}G_{12,1}), \tag{20.8}$$

and the comma represents derivation with respect to (X_1, X_2) directions. In order to take into account the symmetry w.r.t. the $D4$ dihedral group, we require the strain energy density to be invariant under the action of an element $Q \in D4$ on the Cartesian coordinate system, namely,

$$U(\boldsymbol{G}, \nabla\boldsymbol{G}) = U(Q_{ia}Q_{jb}G_{ab}, Q_{ia}Q_{jb}Q_{kc}G_{ab,c}). \tag{20.9}$$

This requirement implies that the tensors C and A admit the following matrix representations (for a detailed discussion about this result, refer to Auffray et al, 2009):

$$C^{D4}_{3\times 3} = \begin{pmatrix} c_{11} & c_{12} & 0 \\ c_{12} & c_{11} & 0 \\ 0 & 0 & c_{33} \end{pmatrix} \tag{20.10}$$

and

$$A^{D4}_{6\times 6} = \begin{pmatrix} A^{D4}_{3\times 3} & 0_{3\times 3} \\ 0_{3\times 3} & A^{D4}_{3\times 3} \end{pmatrix} \tag{20.11}$$

where

$$A^{D4}_{3\times 3} = \begin{pmatrix} a_{11} & a_{12} & a_{13} \\ a_{12} & a_{22} & a_{23} \\ a_{13} & a_{23} & a_{33} \end{pmatrix} \tag{20.12}$$

and c_{11}, c_{12}, and c_{33} replace the usual Lamé coefficients λ and μ and the tensor $\boldsymbol{A}$ depends on six parameters. In Rosi et al (2018) an identification procedure for these parameters is outlined and in our case they have been estimated by a comparison procedure w.r.t. static experimental observations.

Requiring the strain gradient energy U to be positive definite, further conditions on the matrices $\boldsymbol{C}$ and $\boldsymbol{A}$ can be obtained in terms of their eigenvalues. This remark is essential when the pantographic system under analysis comprises the so-called perfect pivots, i.e. pivots which can freely rotate without any resistance.

The main consequence of this assumption is that the action functional is not coercive anymore because there are several configurations in which the deformation consists only in a rotation of the fibers around the pivots, which have all the same deformation energy. From the modelling point of view, it translates into the condition $c_{33} = 0$, i.e. the strain gradient energy is semi-definite positive (Placidi et al, 2017).

However, since the best current manufacturing techniques are not able to produce completely perfect pivots, in our model the value of c_{33} has been estimated to be a positive number much smaller than the other matrix elements and therefore the resulting action functional in Eq. (20.3) is coercive.

In the case under investigation, one of the short sides of the system is clamped to the ground, while the other short side undergoes an imposed sinusoidal displacement. As a consequence, all the terms relative to external forces in Eq. (20.4) vanish in this case. Vanishing flux is imposed for the long sides of the system. The parameters of the model, i.e. the matrix elements in Eqs. (20.10) and (20.11) have been estimated by comparing the experimental and numerical results obtained by using the procedure outlined in Sect. 20.3 (see Table 20.1). By enforcing the least action principle on

Table 20.1 Values of the model parameters. Vanishing and symmetric parameters are not shown.

Parameter	Value
c_{11}	3.15×10^5 N/m
c_{22}	3.15×10^5 N/m
c_{33}	1 N/m
a_{11}	0.07134 J
a_{22}	0.07134 J
a_{23}	−0.10089 J
a_{33}	0.14268 J
a_{44}	0.07134 J
a_{55}	0.07134 J
a_{56}	−0.10089 J
a_{66}	0.14268 J

the functional (20.3), the dynamical behavior of the system is computed by means of a finite element scheme implemented in the software COMSOL Multiphysics (for the explicit form of the variation of the action functional we refer to Placidi and El Dhaba, 2017).

Like the majority of numerical integration algorithms, COMSOL has been optimized for the more used first gradient continua. Therefore, the Hellinger–Reissner variational principle has been applied by defining a new kinematical descriptor M and imposing the constraint $M = \nabla\chi$ by means of a Lagrange multiplier.

The required additional boundary condition has been fixed by imposing that the normal derivative of the displacement, written in terms of the new variable M, must vanish at the short sides of the system.

In this way, the integration algorithm can apply the finite element scheme on the sum of two first gradient energy functionals (see (Scerrato et al, 2016) for a detailed explanation of this method applied to second gradient continua). The discrete elements used are triangular cubic Hermite and quadratic Lagrange polynomials for the two terms of the energy functional, respectively, and the space of the test function is the H^1 Sobolev space.

20.4.2 Rayleigh Dissipation

For a meaningful physical representation, additional damping terms must be considered in the modeling phase. The focus of this work lies in the analysis of the steady state of the system, i.e. when the LTI condition holds. In the present model we have considered the following three Rayleigh dissipation terms which are relative to the velocity of the three deformation measures of a pantographic sheet:

$$\begin{aligned} \mathcal{D}_{\text{ext}} &= G_{\text{ext}}\left(\dot{u}_{1,1}\delta u_{1,1} + \dot{u}_{2,2}\delta u_{2,2}\right) , \\ \mathcal{D}_{\text{shear}} &= G_{\text{shear}}\left(\dot{u}_{1,2} + \dot{u}_{2,1}\right)\delta(u_{1,2} + u_{2,1}) , \\ \mathcal{D}_{\text{bend}} &= G_{\text{bend}}\left(\dot{u}_{2,11}\delta u_{2,11} + \dot{u}_{1,22}\delta u_{1,22}\right) , \end{aligned} \tag{20.13}$$

where $\mathcal{D}_{\text{ext}}$, $\mathcal{D}_{\text{shear}}$, and $\mathcal{D}_{\text{bend}}$ are the Rayleigh terms relative to the extension, shear, and bending deformation measures, respectively. The parameters G_{ext}, G_{shear}, and G_{bend}, whose values are listed in Table 20.2, have been estimated by comparison with the experimental outcomes. The approach adopted to describe the dissipation

Table 20.2 Values of the damping parameters.

Parameter	Value
G_{ext}	7×10^{-6} kg/s
G_{shear}	7×10^{-6} J s
G_{bend}	7×10^{-6} kg/s

behavior of a pantographic sheet leaves room for improvement. Indeed, the Rayleigh terms in Eq. (20.13) are describing a pure viscoelastic damping, whereas the physical mechanism behind the dissipation of a pantographic sheet should be found in the interaction between the elements of its microstructure.

Moreover, the values of the parameters in Table 20.2 should depend upon the imposed displacement frequency, due to their viscoelastic nature. The study of the dissipation mechanism at the level of the pantographic sheet microstructure, with particular attention to the behavior of the pivots, will be the object of a future work.

Currently, a simulated scenario in which the system undergoes a sinusoidal imposed displacement lasting one second is considered. Only the last 0.25 seconds of this time record, where the transient behavior is already negligible (see Fig. 20.4) due to the action of the dissipation terms presented in Eq. (20.13), is used for the subsequent data analysis.

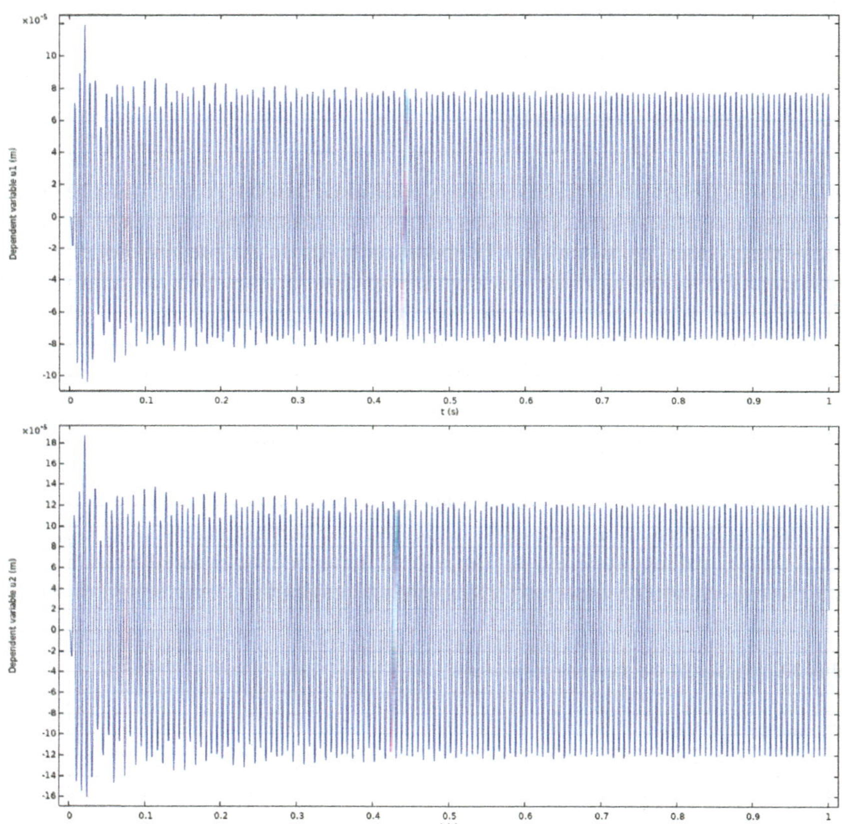

Fig. 20.4 Simulated oscillation along the X_1 (left panel) and X_2 (right panel) axes of a generic point at 140 Hz. In the last 0.25 s the system is considered at steady state due to the action of the damping terms.

20.5 Experimental Methods

In light of the methodology outlined in Sect. 20.3, an experimental rig is designed with the aim of measuring the displacement of a pantographic sheet excited by imposing a sinusoidal displacement at one of its short edges, while keeping the other edge fixed.

20.5.1 Mounting of the Specimen

In practice, the zero-displacement boundary condition is realized by clamping one of the short edges of the sample to the ground—which is assumed infinitely stiff given the extreme compliance of the sample—while the opposite edge is connected to an electrodynamic shaker.

The shaker, a Brüel & Kjær type 4809, is suspended from the ceiling by means of metal chains and soft springs, its height adjusted such that the pantographic sheet may be slightly prestressed in order to avoid the occurrance of buckling instabilities: as per description in Sect. 20.2 the relaxed unmounted sample is 235 mm long, while it measures 237 mm once mounted in between the shaker and the ground.

A laser distance meter and a laser level are used to carefully ensure consistent alignment and sample prestretch before each test. The excitation signal, a sinusoid at the frequency of interest, is generated using a NI-9263 digital-analog converter directly connected to a Brüel & Kjær type 2718 power amplifier.

The NI-9263 module is housed in a cDAQ-9178 chassis, controlled using the Matlab Data Acquisition toolbox. This mounting arrangement is shown in Figure 20.5, and a schematic representation of the whole rig is given in Figure 20.6.

20.5.2 Imaging Techniques, DIC

The actual measurement of the displacement field in time is performed employing Digital Image Correlation (DIC), an optical technique that relates correlation between subsets of pixel intensities to displacements. While there are many implementations of DIC and many different flavours of DIC-based analyses (e.g. global DIC (Besnard et al, 2012; Wittevrongel et al, 2014), Eulerian frequency analysis as in Venanzoni et al (2016)), the equipment available to the authors consists of a commercial white-light DIC platform that employs the time-proven Lucas–Kanade algorithm for optical flow estimation (LaVision GmbH, 2017; Fleet and Weiss, 2006; Bouguet, 2000).

To produce suitable stereo video footage, two Phantom v1612 high-speed cameras equipped with 200 mm macro telephoto lenses are mounted at a distance of ≈ 1.5 m from the sample. At such stand-off distance, this lens choice offers a field of view of 240 mm $\times$ 150 mm, with a scale factor of $\approx 5\,\mathrm{px\,mm^{-1}}$. This allows for a theoretical maximum resolving power of $\approx 10\,\mu\mathrm{m}$, given the use of subpixel-

Fig. 20.5 A detail of the experimental setup. The pantographic sheet is clamped to the ground via a metal adapter on one side, and connected to an electrodynamic shaker at the other end. The shaker is hanging from the ceiling suspended via soft springs, in turn connected to chains. The power amplifier used to drive the shaker is also shown, as well as the four spotlights that illuminate the sample. For the sake of clarity, cables to and from the shaker and the power amplifier have been removed before taking the picture.

interpolation with a 6th order spline (Research, 2017; Nikon, 2017). The system is calibrated using a pinhole camera model that accounts for radial and tangential distortion, taking 5 non-coplanar views of a two-level calibration plate and ultimately yielding a reprojection error of $0.13\,\text{px}$ RMS. The chosen interrogation window size is $19\,\text{px} \times 19\,\text{px}$, with a step of $4\,\text{px}$: this well agrees with the characteristic dimensions of the pantographic sheet, while keeping computational time reasonable. A bigger observation window would act as a spatial low-pass filter on the observed displacements, while a smaller observation window would increase computational costs without providing additional information. (and Rory Bigger et al, 2018)

20.5.3 Measurement Routine

For any excitation frequency of interest, a measurement proceeds as follows:

1. Four 250 W spotlights are turned on; this need arises from the relativelt short camera exposure time of 250 µs, necessary to avoid motion blur.
2. A sinusoid at the choosen frequency is fed to the power amplifier and in turn gets the sample moving.
3. After 2.5 s (a time deemed long enough for the sample to reach steady state) a trigger is sent to the cameras, that acquire 500 stereo frames at a rate of 2048 Hz. This frequency supersamples the maximum excitation frequency of interest (200 Hz) by a factor 4, and the record length is such that 9 whole periods at the minimum excitation frequency of interest (20 Hz) can be recorded. The spotlights and the signal generator are turned off, the video is transferred from the RAM of the cameras to persistent storage.

The assumption of steady state has been verified by looking at the envelope of the displacement in time, while the linearity assumption has been verified by running the measurement at three different excitation amplitudes.

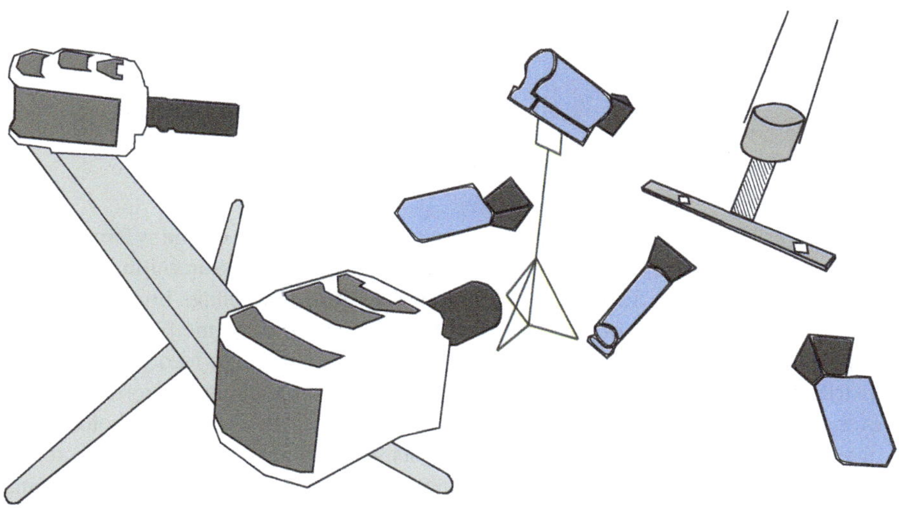

Fig. 20.6 A schematic representation of the core components constituting the measurement rig: the two high-speed cameras, four spotlights, the sample mounted in between the solid ground and the freely suspended shaker.

20.6 Results: Time-invariant Dynamics

In this section the results of the analysis of the numerical and experimental results obtained by means of the method discussed in Sect. 20.3 are presented. The goal of the analysis is to study the time-invariant dynamics of the specimen when one of its short sides undergoes a sinusoidal imposed displacement while the other short side is clamped to the ground. The direction of the imposed displacement is orthogonal to the ground. The observed physical quantity is the displacement vector field in time, namely the set of the deformations of the specimen points with respect to a given reference configuration. In the present investigation, the set of frequencies of the imposed sinusoidal displacement spans from 20 Hz to 220 Hz with a step of 20 Hz. A more comprehensive analysis of the spectrum, by means of an imposed signal with a frequency increasing in time, will be the object of a following work. Recall that, according to this scheme, the time behavior of all the points of the system has been fitted with the sinusoidal function in Eq. (20.1) under the assumption of LTI dynamics. In this way, the information on the dynamics of each point of the system has been compressed in the parameters $(A_i, f_i, \varphi_i, B_i)$, where A_i is the displacement amplitude, f_i the frequency of the oscillation, φ_i the phase, B_i a bias, and the index $i = x, y$, corresponding to the short side and long side directions of the specimen, respectively—in red in Fig. 20.3. A discussion follows, regarding the distribution on the reference configuration geometry of the values of the frequency parameter f_i and of the amplitude parameter A_i for the four relevant imposed oscillation frequencies of 40 Hz, 80 Hz, 160 Hz and 200 Hz.

20.6.1 Distribution of the Frequency Parameter

The interest in the distribution of the frequency parameter f on the reference configuration geometry of the system is relative to the linearity assumption. Indeed, as already anticipated in Sect. 20.3, the system is assumed to behave linearly, i.e. that all the non-nodal points of the system will oscillate with the same frequency of the imposed periodic signal. From Figs. 20.7-20.10 it is possible to observe that the distribution of the estimated frequencies lies in the immediate neighborhood of the excitation frequency. This is an indication that the linearity assumption holds in the investigated regime. In the experimental plots, some points are missing due to the application of a mask aimed at eliminating all non-convergent values of the fitting procedure. In particular, in all the experimental plots the missing regions are around the bottom part of the system, where the clamping inhibits the oscillations and the signal-to-noise ratio decreases. The same behavior happens in the upper region of the system when looking at oscillations in the x-direction: this is presumably due to the action of the imposed oscillation, which is directed along the y-direction. This behavior is encountered in the numerical simulations as well. The vertical central axis appears to be a set of nodal points for oscillations in the x direction. Furthermore, on the bottom part of the numerical plots relative to y-oscillations (fourth panels in Figs. 20.7-20.9) it is possible to observe the presence of a localized region of oscillation. A detailed investigation of this region, where the signal-to-noise ratio was too low to provide meaningful results with the current setup, will be the object of a future work.

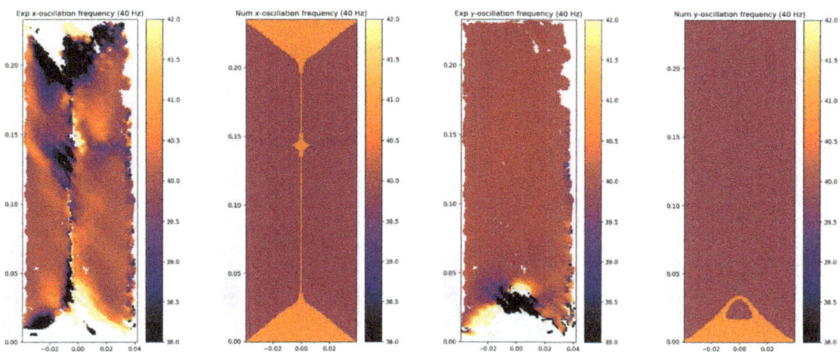

Fig. 20.7 Experimental vs numerical distribution of the frequency parameter f on the reference configuration geometry of the system. The first two and the second two panels are relative to oscillations along the x- and y-directions, respectively. The scales relative to the linear dimensions are in meters. The values of the color bar are in hertz. The imposed oscillation frequency is 40 Hz.

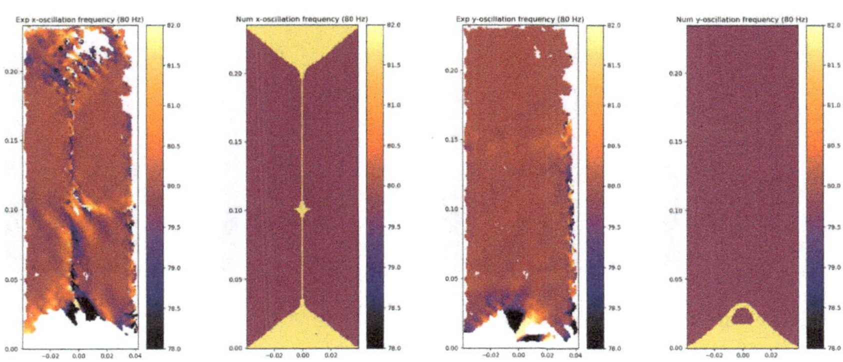

Fig. 20.8 Same plots with imposed oscillation frequency of 80 Hz.

20.6.2 Distribution of the Amplitude Parameter

In this section we present the plots relative to the distribution of the amplitude parameter A on the reference configuration geometry. The comparison between the experimental results and the outcomes of the numerical model yields a qualitative measure of the forecasting power of the proposed numerical model. Figures 20.11-20.14 show a remarkable correspondence between experimental and numerical results. In particular, they confirm that the second gradient model presented in Sect. 20.4 is able to forecast the linear time invariant dynamical behavior of a 2D pantographic material. In the same way of the frequency parameter plots presented in the previous section, the missing points in the experimental plots are due

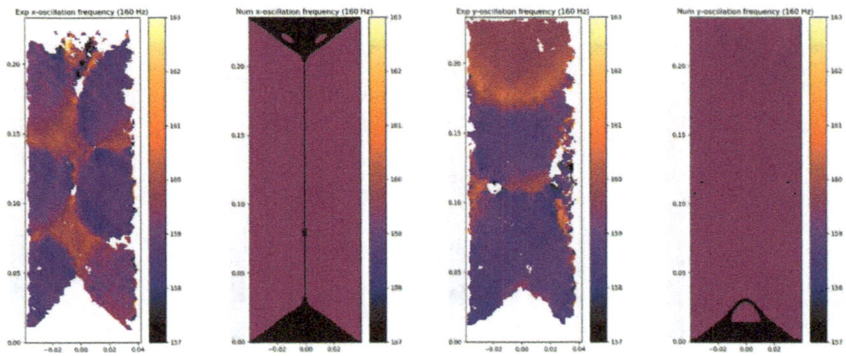

Fig. 20.9 Same plots with imposed oscillation frequency of 160 Hz.

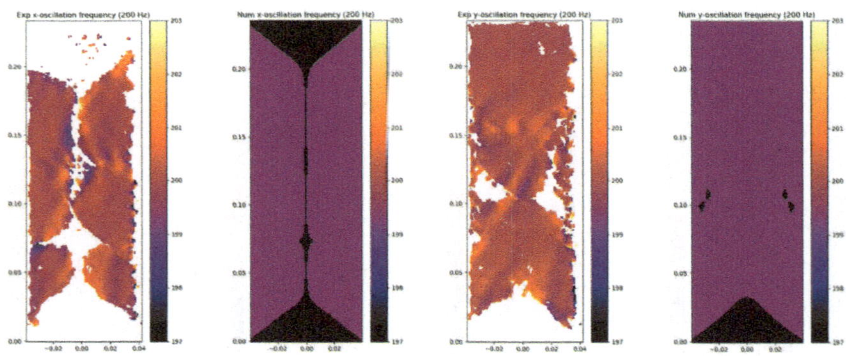

Fig. 20.10 Same plots with imposed oscillation frequency of 200 Hz.

to the action of the same masking procedure. Moreover, it is possible to observe in the experimental plots relative to oscillations in the y-direction (third plots in Figs. 20.11-20.14) that the central areas of oscillation are not perfectly circular as forecasted by the numerical model. This effect can be reasonably traced back to the microstructure of the pantographic material and its modeling will be investigated in a future work.

20.7 Conclusions and Perspectives

In this work, the results of the numerical and experimental study on the linear time-invariant dynamics of a 2D pantographic material have been presented. The experimental investigation has been carried out by means of high-speed stereo imag-

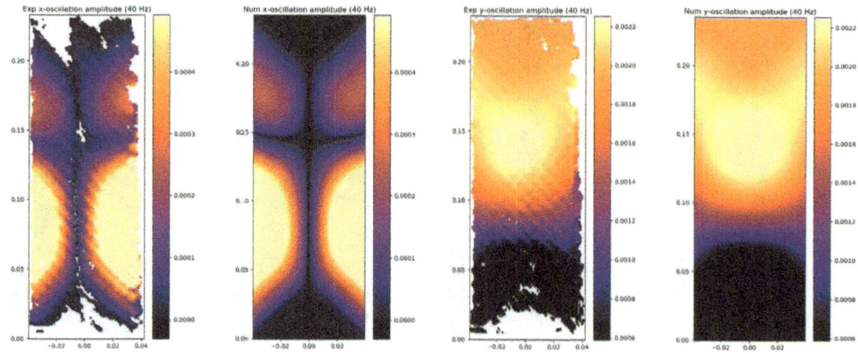

Fig. 20.11 Experimental vs numerical distribution of the amplitude parameter A on the reference configuration geometry of the system. The first two and the second two panels are relative to oscillations along the x- and y-directions, respectively. The scales relative to the linear dimensions and to the colorbar are in meters. The imposed oscillation frequency is 40 Hz.

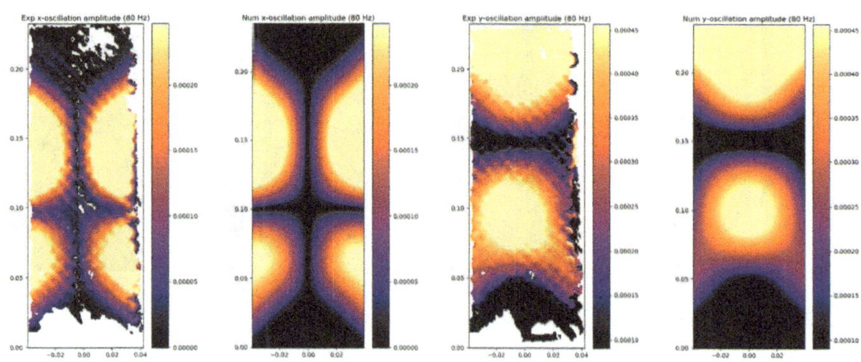

Fig. 20.12 Same plots with imposed oscillation frequency of 80 Hz.

ing and DIC. The numerical model has been developed starting from the results presented in Placidi et al (2017) and implemented to perform numerical simulations using the software COMSOL Multiphysics.

The data have been analysed by performing a fitting procedure of both numerical and experimental results with a sinusoidal function, under the hypothesis of linear time-invariant dynamics. The results have been presented in form of plots of the distribution on the reference configuration geometry of the values of the frequency and amplitude parameters of the fitting procedure. The comparison between experimental and numerical plots shows a remarkable correspondence, confirming the forecasting power of the presented second gradient mathematical model.

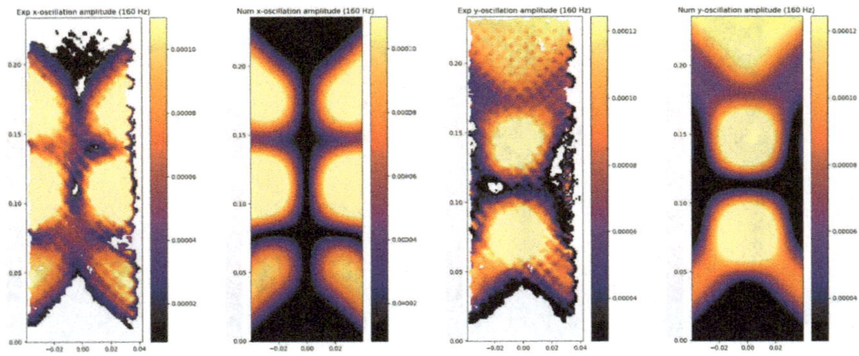

Fig. 20.13 Same plots with imposed oscillation frequency of 160 Hz.

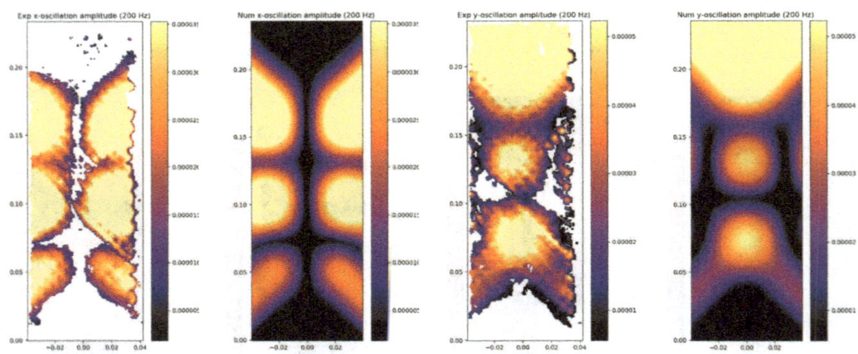

Fig. 20.14 Same plots with imposed oscillation frequency of 200 Hz.

The results presented in this paper have to be considered just as a promising starting point for future investigations. Indeed, as already remarked, the model can be enhanced in several ways: test introducing in the modeling scheme a microinertia term in the kinetic energy functional, as well as implementing a better dissipation mechanism which is closer to the physical behavior of the microstructure, will be the object of forthcoming works. From the experimental side, a dedicated investigation of the bottom-most region of the system is in the workings. A final goal is investigating the spectral properties of the system to check whether this second gradient model is able to forecast the eigenfrequencies and eigenmodes of a 2D pantographic structure.

The quality of the forecasting power of the model, despite all the limitations and the simplistic assumptions presented in the paper, is very encouraging and paves the way to the ambitious goal of the analysis of these kinds of systems in large deformation regime.

References

Alibert JJ, Della Corte A (2015) Second-gradient continua as homogenized limit of pantographic microstructured plates: a rigorous proof. Zeitschrift für angewandte Mathematik und Physik 66(5):2855–2870

Alibert JJ, Della Corte A (2019) Homogenization of nonlinear inextensible pantographic structures by γ-convergence. Mathematics and Mechanics of Complex Systems 7(1):1–24

Alibert JJ, Seppecher P, dell'Isola F (2003) Truss modular beams with deformation energy depending on higher displacement gradients. Mathematics and Mechanics of Solids 8(1):51–73

Auffray N, Bouchet R, Brechet Y (2009) Derivation of anisotropic matrix for bi-dimensional strain-gradient elasticity behavior. International Journal of Solids and Structures 46(2):440–454

Auffray N, dell'Isola F, Eremeyev VA, Madeo A, Rosi G (2015a) Analytical continuum mechanics à la Hamilton–Piola least action principle for second gradient continua and capillary fluids. Mathematics and Mechanics of Solids 20(4):375–417

Auffray N, Dirrenberger J, Rosi G (2015b) A complete description of bi-dimensional anisotropic strain-gradient elasticity. International Journal of Solids and Structures 69:195–206

Barchiesi E, Placidi L (2017) A review on models for the 3d statics and 2d dynamics of pantographic fabrics. In: Wave dynamics and composite mechanics for microstructured materials and metamaterials, Springer, pp 239–258

Barchiesi E, dell'Isola F, Laudato M, Placidi L, Seppecher P (2018) A 1d continuum model for beams with pantographic microstructure: Asymptotic micro-macro identification and numerical results. In: Advances in Mechanics of Microstructured Media and Structures, Springer, pp 43–74

Barchiesi E, Eugster SR, Placidi L, dell'Isola F (2019a) Pantographic beam: a complete second gradient 1d-continuum in plane. Zeitschrift für angewandte Mathematik und Physik 70(5):135

Barchiesi E, Ganzosch G, Liebold C, Placidi L, Grygoruk R, Müller WH (2019b) Out-of-plane buckling of pantographic fabrics in displacement-controlled shear tests: experimental results and model validation. Continuum Mechanics and Thermodynamics 31(1):33–45

Barchiesi E, Spagnuolo M, Placidi L (2019c) Mechanical metamaterials: a state of the art. Mathematics and Mechanics of Solids 24(1):212–234

Battista A, Cardillo C, Del Vescovo D, Rizzi NL, Turco E (2015) Frequency shifts induced by large deformations in planar pantographic continua. Nanoscience and Technology: An International Journal 6(2)

Besnard G, Leclerc H, Hild F, Roux S, Swiergiel N (2012) Analysis of image series through global digital image correlation. The Journal of Strain Analysis for Engineering Design 47(4):214–228

Bouguet JY (2000) Pyramidal implementation of the lucas kanade feature tracker. Intel Corporation, Microprocessor Research Labs

Boutin C, Giorgio I, Placidi L, et al (2017) Linear pantographic sheets: Asymptotic micro-macro models identification. Mathematics and Mechanics of Complex Systems 5(2):127–162

di Cosmo F, Laudato M, Spagnuolo M (2018) Acoustic metamaterials based on local resonances: homogenization, optimization and applications. In: Generalized Models and Non-classical Approaches in Complex Materials 1, Springer, pp 247–274

Del Vescovo D, Giorgio I (2014) Dynamic problems for metamaterials: review of existing models and ideas for further research. International Journal of Engineering Science 80:153–172

dell'Isola F, Seppecher P, Madeo A (2012) How contact interactions may depend on the shape of cauchy cuts in Nth gradient continua: approach "à la D'Alembert". Zeitschrift für angewandte Mathematik und Physik 63(6):1119–1141

dell'Isola F, Andreaus U, Placidi L (2015) At the origins and in the vanguard of peridynamics, non-local and higher-gradient continuum mechanics: an underestimated and still topical contribution of gabrio piola. Mathematics and Mechanics of Solids 20(8):887–928

dell'Isola F, Giorgio I, Pawlikowski M, Rizzi N (2016) Large deformations of planar extensible beams and pantographic lattices: heuristic homogenization, experimental and numerical examples of equilibrium. Proceedings of the Royal Society A: Mathematical, Physical and Engineering Sciences 472(2185):20150,790 dell'Isola F, Corte AD, Giorgio I (2017a) Higher-gradient continua: The legacy of Piola, Mindlin, Sedov and Toupin and some future research perspectives. Mathematics and Mechanics of Solids 22(4):852–872

dell'Isola F, Cuomo M, Greco L, Della Corte A (2017b) Bias extension test for pantographic sheets: numerical simulations based on second gradient shear energies. Journal of Engineering Mathematics 103(1):127–157

dell'Isola F, Seppecher P, Alibert JJ, Lekszycki T, Grygoruk R, Pawlikowski M, Steigmann D, Giorgio I, Andreaus U, Turco E, et al (2019a) Pantographic metamaterials: an example of mathematically driven design and of its technological challenges. Continuum Mechanics and Thermodynamics 31(4):851–884

dell'Isola F, Seppecher P, Spagnuolo M, Barchiesi E, Hild F, Lekszycki T, Giorgio I, Placidi L, Andreaus U, Cuomo M, et al (2019b) Advances in pantographic structures: design, manufacturing, models, experiments and image analyses. Continuum Mechanics and Thermodynamics pp 1–52

Eremeyev VA, Alzahrani FS, Cazzani A, dell'Isola F, Hayat T, Turco E, Konopińska-Zmysłowska V (2019) On existence and uniqueness of weak solutions for linear pantographic beam lattices models. Continuum Mechanics and Thermodynamics 31(6):1843–1861

Fleet D, Weiss Y (2006) Optical flow estimation. In: Handbook of Mathematical Models in Computer Vision, Springer-Verlag, pp 237–257

Giorgio I, Rizzi N, Turco E (2017) Continuum modelling of pantographic sheets for out-of-plane bifurcation and vibrational analysis. Proceedings of the Royal Society A: Mathematical, Physical and Engineering Sciences 473(2207):20170,636

Giorgio I, Rizzi NL, Andreaus U, Steigmann DJ (2019) A two-dimensional continuum model of pantographic sheets moving in a 3d space and accounting for the offset and relative rotations of the fibers. Mathematics and Mechanics of Complex Systems 7(4):311–325

Golaszewski M, Grygoruk R, Giorgio I, Laudato M, Di Cosmo F (2019) Metamaterials with relative displacements in their microstructure: technological challenges in 3d printing, experiments and numerical predictions. Continuum Mechanics and Thermodynamics 31(4):1015–1034

Laudato M, Manzari L, Barchiesi E, Di Cosmo F, Göransson P (2018) First experimental observation of the dynamical behavior of a pantographic metamaterial. Mechanics Research Communications 94:125–127

Laudato M, Di Cosmo F, Drobnicki R, Göransson P (2019) Dynamical vector fields on pantographic sheet: Experimental observations. In: New Achievements in Continuum Mechanics and Thermodynamics, Springer, pp 257–269

LaVision GmbH (2017) Product-Manual for DaVis 8.4: StrainMaster. LaVision GmbH

Liu Z, Zhang X, Mao Y, Zhu Y, Yang Z, Chan CT, Sheng P (2000) Locally resonant sonic materials. science 289(5485):1734–1736

Manzari L, Göransson P, Cuenca J, Lopez Arteaga I (2018) A fully automated high-speed optical rig for in vacuo, full field, non-contact vibration measurements for viscoelastic, anisotropic materials. In: NOVEM 2018 Proceedings – Noise and vibration emerging methods – The 6th conference, pp 91–102

Mindlin RD (1965) Second gradient of strain and surface-tension in linear elasticity. International Journal of Solids and Structures 1(4):417–438

Mindlin RD, Eshel N (1968) On first strain-gradient theories in linear elasticity. International Journal of Solids and Structures 4(1):109–124

Nejadsadeghi N, De Angelo M, Drobnicki R, Lekszycki T, dell'Isola F, Misra A (2019) Parametric experimentation on pantographic unit cells reveals local extremum configuration. Experimental Mechanics 59(6):927–939

Nikon (2017) AF Micro-Nikkor 200mm f/4d IF-ED from Nikon

Placidi L, El Dhaba AR (2017) Semi-inverse method à la saint-venant for two-dimensional linear isotropic homogeneous second-gradient elasticity. Mathematics and Mechanics of Solids 22(5):919–937

Placidi L, Barchiesi E, Turco E, Rizzi NL (2016) A review on 2d models for the description of pantographic fabrics. Zeitschrift für angewandte Mathematik und Physik 67(5):121

Placidi L, Andreaus U, Giorgio I (2017) Identification of two-dimensional pantographic structure via a linear d4 orthotropic second gradient elastic model. Journal of Engineering Mathematics 103(1):1–21

Placidi L, dell'Isola F, Barchiesi E (2020) Heuristic homogenization of euler and pantographic beams. In: Mechanics of Fibrous Materials and Applications, Springer, pp 123–155

Research V (2017) Phantom Ultrahigh-Speed Cameras UHS-12 Series Manual

and Rory Bigger, Blaysat B, Boo C, Grewer M, Hu J, Jones A, Klein M, Raghavan K, Reu P, Schmidt T, Siebert T, Simenson M, Turner D, Vieira A, Weikert T (2018) A good practices guide for digital image correlation. Tech. rep., International Digital Image Correlation Society

Rosi G, Auffray N (2016) Anisotropic and dispersive wave propagation within strain-gradient framework. Wave Motion 63:120–134

Rosi G, Auffray N (2019) Continuum modelling of frequency dependent acoustic beam focusing and steering in hexagonal lattices. European Journal of Mechanics-A/Solids p 103803

Rosi G, Placidi L, Auffray N (2018) On the validity range of strain-gradient elasticity: a mixed static-dynamic identification procedure. European Journal of Mechanics-A/Solids 69:179–191

Scerrato D, Zhurba Eremeeva IA, Lekszycki T, Rizzi NL (2016) On the effect of shear stiffness on the plane deformation of linear second gradient pantographic sheets. ZAMM-Journal of Applied Mathematics and Mechanics/Zeitschrift für Angewandte Mathematik und Mechanik 96(11):1268–1279

Turco E (2018) Discrete is it enough? The revival of Piola–Hencky keynotes to analyze three-dimensional Elastica. Continuum Mechanics and Thermodynamics 30(5):1039–1057

Turco E, dell'Isola F, Cazzani A, Rizzi NL (2016) Hencky-type discrete model for pantographic structures: numerical comparison with second gradient continuum models. Zeitschrift für angewandte Mathematik und Physik 67(4):85

Turco E, Golaszewski M, Giorgio I, Placidi L (2017) Can a Hencky-type model predict the mechanical behaviour of pantographic lattices? In: Mathematical Modelling in Solid Mechanics, Springer, pp 285–311

Turco E, Misra A, Pawlikowski M, dell'Isola F, Hild F (2018) Enhanced Piola–Hencky discrete models for pantographic sheets with pivots without deformation energy: numerics and experiments. International Journal of Solids and Structures 147:94–109

Turco E, Misra A, Sarikaya R, Lekszycki T (2019) Quantitative analysis of deformation mechanisms in pantographic substructures: experiments and modeling. Continuum Mechanics and Thermodynamics 31(1):209–223

Venanzoni A, Ryck LD, Cuenca J (2016) Eulerian frequency analysis of structural vibrations from high-speed video. Author(s)

Wittevrongel L, Lava P, Lomov SV, Debruyne D (2014) A self adaptive global digital image correlation algorithm. Experimental Mechanics 55(2):361–378

Chapter 21
Nonlinear Dispersion Properties of Acoustic Waveguides with Cubic Local Resonators

Marco Lepidi & Andrea Bacigalupo

Abstract Acoustic metamaterials are synthetic architected media featured by a periodic microstructured cell hosting one or more resonant oscillators. The cellular microstructure can be parametrically design to functionalize the dispersion properties of elastic waves. A one–dimensional crystal lattice, characterized by a diatomic periodic cell, is considered to prototypically simulate the essential undamped dynamics of weakly nonlinear acoustic waveguides. A cubic nonlinearity affects the intracellular elastic coupling between the primary atom and the secondary atom (resonator). In the small-amplitude oscillation range, the dispersion relations for the linear wavefrequencies $\omega(\beta)$ and linear waveforms $\boldsymbol{\phi}(\beta)$ are determined as analytical functions of the wavenumber β. A general asymptotic approach, based on the multiple scale method, is employed to determine the amplitude-dependent dispersion relations for the nonlinear wavefrequencies $\varpi(\beta)$ and nonlinear waveforms $\boldsymbol{\psi}(\beta)$. The actual existence of stable periodic oscillations orbits, confined on the invariant manifolds in the space of the two principal coordinates corresponding to the nonlinear waveforms, is successfully verified by numerical simulations.

Keywords: Crystal lattice · Acoustic metamaterials · Dispersion properties
Nonlinear waveforms · Cubic nonlinearity · Perturbation methods

21.1 Introduction

Acoustic or mechanical metamaterials are artificial architected media, microstructurally designed for functional applications oriented at the passive control of the

---

M. Lepidi
DICCA - Università di Genova, Via Montallegro 1, 16154 Genova, Italy
e-mail: marco.lepidi@unige.it

A. Bacigalupo
IMT School for Advanced Studies Lucca, Piazza S. Francesco 19, 55100 Lucca, Italy
e-mail: andrea.bacigalupo@imtlucca.it

B. E. Abali and I. Giorgio (eds.), *Developments and Novel Approaches in Biomechanics and Metamaterials*, Advanced Structured Materials 132,
https://doi.org/10.1007/978-3-030-50464-9_21

elastic wave propagation (Liu et al, 2000; Cummer et al, 2016). The metamaterial microstructure can be parametrically customized to obtain desired band structures in the dispersion spectrum, by leveraging the peculiar mechanism of *local resonances* (Deymier, 2013; Ma and Sheng, 2016; di Cosmo et al, 2018). Essentially, the physical realizations of locally resonant metamaterials is based on the internal coupling between the cellular microstructure and auxiliary tuned oscillators (resonators). The virtuous synergy between highly customizable periodic microstructures and fine-tunable local resonances paves the way for unconventional dynamic features, including—among the others—effective negative mass density, absolute stop bands, modal localizations, negative refraction indexes. Within this challenging scientific framework, acoustic metamaterials are presently attracting a spreading research attention, targeted at fascinating technological applications like phononic waveguiding, frequency filtering, flat superlensing, vibration shielding, acoustic cloaking, noise absorption, non-reciprocal propagation (Guenneau et al, 2007; Craster and Guenneau, 2012; Colquitt et al, 2014; Grinberg et al, 2018; El Sherbiny and Placidi, 2018; Bacigalupo and Gambarotta, 2019).

Since many emerging engineering trends are increasingly demanding ultra-light materials with super-low filtering properties, the microstructural optimization of acoustic metamaterials unavoidably tends to maximize the flexibility of local resonators, in order to conjugate the lowest resonant frequency with the minimal mass density (Manimala and Sun, 2014; Lepidi and Bacigalupo, 2018; Bacigalupo et al, 2019). Moreover, since enhanced dissipation properties are not strictly required to achieve superior performances of passive control in wave propagation, the local resonators tend to be slightly damped. This systematic tendency towards high flexibility combined with low damping motivates the renewed interest toward slender microstructures featured by geometric nonlinearities activated by high amplitude oscillations.

In the past decades, the standing and traveling waves charactering a monocoupled periodically-layered undamped system with cubic material nonlinearities have been studied by means of asymptotic methodologies (Vakakis and King, 1998). The chaotic and subharmonic oscillations of the traveling waves propagating in a periodic array of linear oscillators with local essentially nonlinear attachments have been analyzed (Rothos and Vakakis, 2009). The wave propagation properties in one–dimensional chains of mono- and bi-coupled nonlinear oscillators have been investigated by means of linearized and nonlinear maps (Romeo and Rega, 2006, 2008). In the nonlinear regime, the occurrence of quasi-periodic and chaotic solutions has been also recognized. The amplitude-dependent dispersion properties and transmission functions of one–dimensional chains with nonlinear local attachments modeled as damped cubic oscillators have been studied by using the harmonic balance method accompanied by numerical simulations (Lazarov and Jensen, 2007).

More recently, the amplitude-dependent band structure of one– and two–dimensional weakly-nonlinear lattices with monoatomic and diatomic periodic cells has been investigated through perturbation analyses and numerical simulations (Narisetti et al, 2010, 2011). Similar periodic systems, but featured by strong nonlinearities, have been analyzed by means of a semi-analytical series expansion method

(Fang et al, 2016). The nonlinear dynamic interactions between two internally resonant waves traveling through an undamped monoatomic chain of point masses interconnected by linear springs and cubic couplings have been analyzed (Manktelow et al, 2011). Different perturbation methods have been employed to determine the nonlinear dispersion functions and to describe the coupled motion in superharmonic and subharmonic resonance conditions between the wavefrequencies.

Starting from this scientific background, the primary objective of the paper is the analytical description of the amplitude-dependent dispersion properties of acoustic metamaterials equipped with highly flexible undamped resonators, characterized by weakly nonlinear coupling with the periodic microstructure. In this respect, the theoretical mechanical issue is twofold. First, the nonlinear dependence of both the frequencies and waveforms on the wave oscillation amplitude has to be assessed. Second, the existence and number of invariant manifolds of harmonic wave motion has to be determined. A further motivation is given by the crucial role played by the waveforms in the physical phenomena of wave polarization and energy transfers (Bacigalupo and Lepidi, 2018). Specifically, a one–dimensional periodic lattice characterized by a locally resonant diatomic cell with cubic interatomic coupling is considered. The nonlinear frequencies and waveforms are determined as amplitude-dependent analytical functions of the mechanical parameters. To this purpose, a general asymptotic strategy based on the perturbation method of multiple scales is adopted (Lacarbonara and Camillacci, 2004). The nonlinear dispersion properties are analytically determined in the absence of superharmonic internal resonance between the acoustic and optical branches of the frequency dispersion spectrum. Moreover, the invariant manifolds associated with the nonlinear waveforms are determined in the space of principal coordinates.

21.2 One–Dimensional Diatomic Lattice

A one–dimensional crystal lattice characterized by a diatomic periodic cell is assumed as minimal physical realization of a nonlinear acoustic metamaterial (Figure 21.1). The cellular microstructure can be synthesized by building up a waveguide made of an infinite chain of massive stiff rings (*principal atoms*). Each principal atom is assumed to exchange linear position-dependent forces (attraction or repulsion) with the adjacent elements of the chain. The local resonance mechanism can be realized by nesting heavy circular inclusions (*secondary atoms*) in the rings. Each inclusion is co-centered with the hosting ring and fully embedded in the soft matrix filling its internal volume. Each secondary atom is supposed to exchange nonlinear position-dependent forces with the corresponding principal atom. In the small amplitude range of oscillations, the secondary atoms play the role of *inertial resonators*, if their linear frequencies are properly tuned with certain wave frequencies of the principal atom chain.

From the mechanical viewpoint, the free dynamics of the acoustic metamaterial can be described by a low–dimensional Lagrangian model. All the atoms are

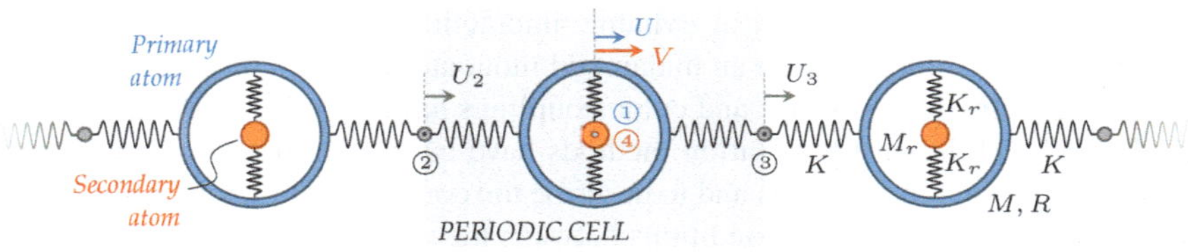

Fig. 21.1 Acoustic waveguide characterized by a diatomic periodic cell realizing a minimal nonlinear metamaterial with local resonances.

supposed to possess a single degree-of-freedom, aligned with the chain axis. The lattice is supposed to be perfectly non-dissipative. Considering the generic cell, the massive and stiff ring is modeled as a rigid annular body with mean radius R and translational mass M. Consequently, the unique degree-of-freedom of the principal atom is referred to the displacement U of the ring centroid, where the configurational node ① is located. The linear interactions with the principal atoms of the adjacent cells are simulated by two linear elastic springs (*principal springs*), with identical stiffness K. The springs connect the ring with the auxiliary massless nodes ② and ③, purposely located at the left and right sides of the cell boundary. Therefore, two extra degrees-of-freedom are related to the displacements U_2 and U_3 of the auxiliary nodes. The internal inclusion is modeled as a point body with mass M_r. Consequently, the unique degree-of-freedom of the secondary atom can be related to the displacement V of the configurational node ④. In the undeformed configuration, the positions of the nodes ① and ④ coincide with each other. Considering the ring filler embedding the inclusion as a hardening elastic medium, the nonlinear coupling between the principal and secondary atoms can be simulated by a pair of chain-orthogonal pre-stressed springs (*secondary springs*), which realize the simplest retaining system exhibiting odd nonlinearities. The secondary springs have stiffness K_r and are identically prestressed with pretension H, for the sake of static equilibrium.

Considering known a certain dimensional length L of the periodic cell (namely $L = R$ in the following), it is convenient to introduce the following nondimensional independent and dependent variables

$$\tau = \Omega t, \quad u = \frac{U}{L}, \quad v = \frac{V}{L}, \quad u_2 = \frac{U_2}{L}, \quad u_3 = \frac{U_3}{L} \tag{21.1}$$

where $\Omega^2 = K/M$ is the reference (squared) frequency of the simple undamped oscillator with the ring mass and the principal spring stiffness. The displacement difference $w = v - u$ is an auxiliary nondimensional variable that conveniently describes the inter-atomic drift.

A minimal set of independent nondimensional parameters, sufficient to fully describe the geometric, elastic and inertial properties of the model, is

$$\varrho^2 = \frac{M_r}{M}, \quad \mu = \frac{H}{KR}, \quad \eta = \frac{K_r}{K} \tag{21.2}$$

where ϱ^2 is the *mass ratio* between the secondary and principal atoms, while μ and η account for the *geometric* and *elastic stiffnesses* of the secondary springs. All the parameters are strictly positive by construction.

Adopting the undeformed prestressed configuration as initial reference for the dynamic equilibrium, a second-order kinematic formulation can be adopted to determine the elastic energy of the diatomic lattice (Lepidi and Bacigalupo, 2019). Therefore, the Hamilton's Principle can be applied to obtain the geometrically nonlinear equations of motion governing the free dynamics of the periodic cell

$$(1+\varrho^2)\ddot{u} + \varrho^2\ddot{w} + 2u - u_2 - u_3 = 0 \tag{21.3}$$

$$\varrho^2\ddot{u} + \varrho^2\ddot{w} + 2\mu\, w + \eta\, w^3 = 0 \tag{21.4}$$

where dot indicates derivative with respect to time τ. The dynamically *active* displacements u and w can be distinguished from the *passive* displacements u_2 and u_3. The passive displacements are related to the massless boundary nodes, which do not develop inertial forces. The displacements u_2 and u_3 are quasi-statically coupled with the active displacement u by the linear laws

$$u_2 - u = f_2, \qquad u_3 - u = f_3 \tag{21.5}$$

where $f_2 = F_2/(KL)$ and $f_3 = F_3/(KL)$ are the nondimensional variables accounting for the inter-cellular forces F_2 and F_3 exerted at the two sides of the cell boundary by the adjacent cells.

21.3 Asymptotic Strategy

The free wave propagation through crystal lattices is classically analyzed by means of the Floquet–Bloch theory, valid for linear periodic structures (Brillouin, 1946). Therefore, the multiple scale method can represent the proper mathematical tool to approach the wave dispersion problem for the nonlinear acoustic metamaterial. Indeed, according to this asymptotic method the nonlinear equations of motion are expanded into an ordered hierarchy of linear equations, so that the Floquet–Bloch conditions of quasi-periodicity can be imposed at each order (Georgiou and Vakakis, 1996; Narisetti et al, 2011).

Introducing the *active* displacement vector $\mathbf{u}_a = (u, w)$ and the *passive* displacement vector $\mathbf{u}_p = (u_2, u_3)$, the nonlinear equations (21.3)-(21.5) can be expressed in the partitioned matrix form

$$\begin{bmatrix} \mathbf{M} & \mathbf{O} \\ \mathbf{O} & \mathbf{O} \end{bmatrix} \begin{pmatrix} \ddot{\mathbf{u}}_a \\ \ddot{\mathbf{u}}_p \end{pmatrix} + \begin{bmatrix} \mathbf{K}_{aa}^n & \mathbf{K}_{ap} \\ \mathbf{K}_{pa} & \mathbf{K}_{pp} \end{bmatrix} \begin{pmatrix} \mathbf{u}_a \\ \mathbf{u}_p \end{pmatrix} = \begin{pmatrix} \mathbf{0} \\ \mathbf{f}_p \end{pmatrix} \tag{21.6}$$

where the passive force vector is $\mathbf{f}_p = (f_2, f_3)$ and the governing matrices are

$$\mathbf{M} = \begin{bmatrix} 1+\varrho^2 & \varrho^2 \\ \varrho^2 & \varrho^2 \end{bmatrix}, \quad \mathbf{K}^n_{aa} = \begin{bmatrix} 2 & 0 \\ 0 & 2\mu + \eta w^2 \end{bmatrix}, \quad \mathbf{K}_{ap} = \begin{bmatrix} -1 & -1 \\ 0 & 0 \end{bmatrix} \tag{21.7}$$

with also $\mathbf{K}_{pa} = \mathbf{K}^{\top}_{ap}$ and $\mathbf{K}_{pp} = \mathbf{I}$. The nonlinear stiffness submatrix can be written $\mathbf{K}^n_{aa} = \mathbf{K}_{aa} + \mathbf{N}(\mathbf{u}_a)$, where $\mathbf{N}(\mathbf{u}_a)$ accounts for the nonlinearities.

Following a general asymptotic strategy for determining the nonlinear normal modes through the multiple scale method (Rosenberg, 1966; Lacarbonara and Camillacci, 2004), a small dimensionless parameter $\epsilon \ll 1$ can be introduced and the solution of the nonlinear equation (21.6) can be expanded in ϵ-power series as

$$\begin{aligned} \mathbf{u}_a &= \epsilon\,\mathbf{u}_{a1}(T_0,T_1,T_2) + \epsilon^2\mathbf{u}_{a2}(T_0,T_1,T_2) + \epsilon^3\mathbf{u}_{a3}(T_0,T_1,T_2) + \mathcal{O}(\epsilon^4) \\ \mathbf{u}_p &= \epsilon\,\mathbf{u}_{p1}(T_0,T_1,T_2) + \epsilon^2\mathbf{u}_{p2}(T_0,T_1,T_2) + \epsilon^3\mathbf{u}_{p3}(T_0,T_1,T_2) + \mathcal{O}(\epsilon^4) \\ \mathbf{f}_p &= \epsilon\,\mathbf{f}_{p1}(T_0,T_1,T_2) + \epsilon^2\mathbf{f}_{p2}(T_0,T_1,T_2) + \epsilon^3\mathbf{f}_{p3}(T_0,T_1,T_2) + \mathcal{O}(\epsilon^4) \end{aligned} \tag{21.8}$$

where $T_0 = \tau$ is the fast time-scale characterizing the harmonic wave motions at the linear frequencies $\omega^{\pm}$ and $T_j = \epsilon^j \tau$ are slow time-scales ($j = 1, 2$). Accordingly, the ordinary time-derivative $d/d\tau$ is expressed through the partial derivatives $D_j = \partial/\partial T_j$ as $D_0 + \epsilon D_1 + \epsilon^2 D_2 + \ldots$. According to the variable expansion (21.8), the nonlinear part of the stiffness matrix can be expanded in ϵ-power series as $\mathbf{N}(\mathbf{u}_a) = \epsilon^2\mathbf{N}_2(\mathbf{u}_{a1}) + \epsilon^3\mathbf{N}_3(\mathbf{u}_{a1}, \mathbf{u}_{a2}) + \mathcal{O}(\epsilon^4)$.

Introducing the variable expansion into the nonlinear differential equation (21.6) and equating all terms of like ϵ-powers up to the third order, an ordered hierarchy of linear differential equations (perturbation equations) is stated

- Order ϵ :
$$\begin{bmatrix} \mathbf{M} & \mathbf{O} \\ \mathbf{O} & \mathbf{O} \end{bmatrix} \begin{pmatrix} D_0^2\mathbf{u}_{a1} \\ D_0^2\mathbf{u}_{p1} \end{pmatrix} + \begin{bmatrix} \mathbf{K}_{aa} & \mathbf{K}_{ap} \\ \mathbf{K}_{pa} & \mathbf{K}_{pp} \end{bmatrix} \begin{pmatrix} \mathbf{u}_{a1} \\ \mathbf{u}_{p1} \end{pmatrix} = \begin{pmatrix} \mathbf{0} \\ \mathbf{f}_{p1} \end{pmatrix} \tag{21.9}$$
- Order ϵ^2 :
$$\begin{bmatrix} \mathbf{M} & \mathbf{O} \\ \mathbf{O} & \mathbf{O} \end{bmatrix} \begin{pmatrix} D_0^2\mathbf{u}_{a2} \\ D_0^2\mathbf{u}_{p2} \end{pmatrix} + \begin{bmatrix} \mathbf{K}_{aa} & \mathbf{K}_{ap} \\ \mathbf{K}_{pa} & \mathbf{K}_{pp} \end{bmatrix} \begin{pmatrix} \mathbf{u}_{a2} \\ \mathbf{u}_{p2} \end{pmatrix} = \begin{pmatrix} \mathbf{f}_{a2} \\ \mathbf{f}_{p2} \end{pmatrix} \tag{21.10}$$
- Order ϵ^3 :
$$\begin{bmatrix} \mathbf{M} & \mathbf{O} \\ \mathbf{O} & \mathbf{O} \end{bmatrix} \begin{pmatrix} D_0^2\mathbf{u}_{a3} \\ D_0^2\mathbf{u}_{p3} \end{pmatrix} + \begin{bmatrix} \mathbf{K}_{aa} & \mathbf{K}_{ap} \\ \mathbf{K}_{pa} & \mathbf{K}_{pp} \end{bmatrix} \begin{pmatrix} \mathbf{u}_{a3} \\ \mathbf{u}_{p3} \end{pmatrix} = \begin{pmatrix} \mathbf{f}_{a3} \\ \mathbf{f}_{p3} \end{pmatrix} \tag{21.11}$$

where the active forcing vectors at the high orders are $\mathbf{f}_{a2} = -2\mathbf{M}D_0D_1\mathbf{u}_{a1}$ and $\mathbf{f}_{a3} = -2\mathbf{M}D_0D_1\mathbf{u}_{a2} - 2\mathbf{M}D_0D_2\mathbf{u}_{a1} - \mathbf{M}D_1^2\mathbf{u}_{a1} + \mathbf{N}_2\mathbf{u}_{a1}$. It may be worth noting that the ordinary differential problems (21.9)-(21.11) are governed by the same mass and stiffness matrices at all the ϵ-orders.

The Floquet–Bloch theory for the free wave propagation can be applied to the linear perturbation equations by separately imposing independent quasi-periodicity conditions for the passive displacements $u_{3j} = u_{2j}\exp(-\imath\beta)$ and the passive forces $f_{3j} = -f_{2j}\exp(-\imath\beta)$ at each order ($j = 1, 2, 3$).

Focusing on the first order of the perturbation equations, the linear differential problem (21.9) can be tackled by separating the dynamic (upper) part from the quasi-static (lower) part, yielding

- Order ϵ :
$$\begin{aligned} \mathbf{M}D_0^2\mathbf{u}_{a1} + \mathbf{K}_{aa}\mathbf{u}_{a1} + \mathbf{K}_{ap}\mathbf{u}_{p1} &= \mathbf{0} \\ \mathbf{K}_{pa}\mathbf{u}_{a1} + \mathbf{K}_{pp}\mathbf{u}_{p1} &= \mathbf{f}_{p1} \end{aligned} \tag{21.12}$$

Imposing the quasi-periodicity conditions on the passive displacements $\mathbf{u}_{p1}$ and forces $\mathbf{f}_{p1}$ and subsequently solving the quasi-static part of the equation (21.12), the condensation rules $\mathbf{u}_{p1} = \mathbf{L}_{pa}\mathbf{u}_{a1}$ and $\mathbf{f}_{p1} = (\mathbf{K}_{pa} + \mathbf{K}_{pp}\mathbf{L}_{pa})\,\mathbf{u}_{a1}$ are obtained. Consequently, the homogeneous dynamic equation (21.12) can be condensed in the active displacement vector only, reading

- Order ϵ :
$$\mathbf{M}D_0^2\mathbf{u}_{a1} + \mathbf{K}(\beta)\mathbf{u}_{a1} = \mathbf{0} \tag{21.13}$$

where the generalized stiffness matrix $\mathbf{K}(\beta) = (\mathbf{K}_{aa} + \mathbf{K}_{ap}\mathbf{L}_{pa})$ is Hermitian and depends on the wavenumber β through the auxiliary matrix

$$\mathbf{L}_{pa} = B^{-1}\begin{pmatrix} \mathbf{K}_{pa}^{(2)} + \mathbf{K}_{pa}^{(1)}\mathrm{e}^{-\imath\beta} \\ \mathbf{K}_{pa}^{(2)}\mathrm{e}^{-\imath\beta} + \mathbf{K}_{pa}^{(1)}\mathrm{e}^{-2\imath\beta} \end{pmatrix} \tag{21.14}$$

where $\mathbf{K}_{pa}^{(k)}$ is the k-th row of the matrix $\mathbf{K}_{pa}$ ($k = 1, 2$) and the quantity

$$B = -\left[K_{pp}^{(21)} + \left(K_{pp}^{(22)} + K_{pp}^{(11)}\right)\mathrm{e}^{-\imath\beta} + K_{pp}^{(12)}\mathrm{e}^{-2\imath\beta}\right] \tag{21.15}$$

where $K_{pp}^{(hk)}$ is the generic component of the matrix $\mathbf{K}_{pp}$ ($h, k = 1, 2$).

Repeating the procedure of imposing the periodicity conditions and condensing the passive variables for the higher orders of the perturbation equations, the linear dynamic problems (21.10) and (21.11) can be condensed in the active displacement vector only, reading

- Order ϵ^2 :
$$\mathbf{M}D_0^2\mathbf{u}_{a2} + \mathbf{K}(\beta)\mathbf{u}_{a2} = \mathbf{f}_{a2} \tag{21.16}$$
- Order ϵ^3 :
$$\mathbf{M}D_0^2\mathbf{u}_{a3} + \mathbf{K}(\beta)\mathbf{u}_{a2} = \mathbf{f}_{a3} \tag{21.17}$$

where it can be noted that the ordinary differential problems (21.13), (21.16) and (21.17) are governed by the same mass and condensed stiffness matrices and are also formally identical, apart for the right-hand term.

21.3.1 Linear Dispersion Properties

The linear dispersion properties characterizing the small-amplitude waves propagating through the acoustic metamaterial can be determined by solving the equation (21.13). To this purpose, the harmonic mono-frequent solution $\mathbf{u}_{a1} = \boldsymbol{\phi}\exp(\imath\omega\tau)$ can be imposed, where ω is the unknown nondimensional frequency. Eliminating the dependence on time gives the linear eigenproblem

$$(\mathbf{K}(\beta) - \lambda \mathbf{M})\,\boldsymbol{\phi} = \mathbf{0} \tag{21.18}$$

The eigensolution includes two non-defective eigenpairs $(\lambda, \boldsymbol{\phi})$, where the eigenvalue λ is related to the frequency by the relation $\lambda = \omega^2$ and the eigenvector $\boldsymbol{\phi} = (\phi_1, \phi_2)$ represents the corresponding waveform.

The eigenproblem (21.18) can be solved analytically. Therefore, the two linear frequencies of the harmonic waves $\mathbf{u}_{a1} = \boldsymbol{\phi} \exp(\imath\omega\tau - \imath\beta)$ that freely propagate with a certain wavenumber β and unitary-amplitude through the diatomic lattice characterized by the mechanical parameters (ϱ^2, μ) are

$$\omega^{\mp} = \left[\frac{1}{2}(1 - \cos\beta) - \mu\left(1 + \frac{1}{\varrho^2}\right) \mp \frac{S(\beta)}{2\varrho^2}\right]^{\frac{1}{2}} \tag{21.19}$$

while the corresponding waveforms are 2-by-1 vectors

$$\boldsymbol{\phi}^{\mp} = \left(1, \frac{2\varrho^2(1 - \cos\beta)}{2\mu\,(1 + \varrho^2) - \varrho^2(1 - \cos\beta) \pm S(\beta)}\right) \tag{21.20}$$

whose amplitude remains undetermined. Since the β-dependent quantity

$$S(\beta) = \left[(4\mu)^2\varrho^2 + \left(\varrho^2(1 - \cos\beta) - 2\mu\,(1 - \varrho^2)\right)^2\right]^{\frac{1}{2}} \tag{21.21}$$

is real-valued, the waveforms are real-valued vectors. Considering positive valued frequencies ($\omega^{\mp} \geq 0$), the *forward* propagating waves characterized by positive wavenumbers β ranging in the $\mathcal{B}$-subdomain $\mathcal{B}_1 = [0, \pi]$ can be studied. According to the assumptions made for the physical parameters, the quantity $S(\beta)$ is strictly positive. Consequently, the strict inequality $\omega^- < \omega^+$ holds in the entire $\mathcal{B}_1$-domain. Therefore, the dispersion functions $\omega^-(\beta)$ and $\omega^+(\beta)$ describe the lower-frequency curve (*acoustic branch*) and the higher-frequency curve (*optical branch*) of the dispersion spectrum, respectively.

The analytical solution of the eigenproblem allows to demonstrate—first—that the acoustic and optical branches do not cross each other, so that a double frequency (perfect $1:1$ internal resonance) cannot occur and—second—that a total band gap exists between the acoustic and optical frequency bands. Specifically, if the secondary springs are weakly pretensioned (namely $2\mu < \varrho^2/(1 + \varrho^2)$), the band gap is caused by a *veering* phenomenon between the dispersion curves. This spectral phenomenon implies strong linear interactions between the corresponding waves (Perkins and Mote Jr, 1986; Pierre, 1988; Lacarbonara et al, 2005; Vidoli and Vestroni, 2005; Lepidi, 2013). As first consequence, the dispersion curves can assume high curvature values in the veering region. From the physical viewpoint, high dispersion curvatures may entail strong changes in the group velocity of the propagating waves (Bacigalupo and Lepidi, 2016). As second consequence, the waveforms associated with the veering dispersion curves can undergo a *hybridization* phenomenon, merging their respective forms in a rapid but continuous way. From the physical viewpoint, the waveform hybridization can determine strong variations in the flux of the mechanical energy transported by the propagating waves (Bacigalupo and Lepidi, 2018). A complete analysis of the linear dispersion properties is reported in Lepidi and Bacigalupo (2019).

21.3.2 Nonlinear Dispersion Properties

Collecting column-wise the linear mass-normalized waveforms $\varphi^{\mp}$ in the 2-by-2 matrix $\mathbf{\Phi} = \left[\varphi^{-}, \varphi^{+}\right]$, the change of variable $\mathbf{u}_a = \mathbf{\Phi}\mathbf{q}$ can be introduced, with $\mathbf{q} = (q, p)$ working as vector of *principal* coordinates, that is consistently ordered according to the ϵ-power series

$$\mathbf{q} = \epsilon\, \mathbf{q}_1(T_0, T_1, T_2) + \epsilon^2 \mathbf{q}_2(T_0, T_1, T_2) + \epsilon^3 \mathbf{q}_3(T_0, T_1, T_2) + \mathcal{O}(\epsilon^4) \qquad (21.22)$$

Therefore, applying the change of variable in the equations (21.13), (21.16) and (21.17), the ordered hierarchy of perturbation equations read

- Order ϵ: $\quad D_0^2\, \mathbf{q}_1 + \mathbf{\Lambda}\mathbf{q}_1 = \mathbf{0} \qquad (21.23)$
- Order ϵ^2: $\quad D_0^2\, \mathbf{q}_2 + \mathbf{\Lambda}\mathbf{q}_2 = -2D_0 D_1\, \mathbf{q}_1 \qquad (21.24)$
- Order ϵ^3: $\quad D_0^2\, \mathbf{q}_3 + \mathbf{\Lambda}\mathbf{q}_3 = -2D_0 D_1\, \mathbf{q}_2 - 2D_0 D_2\, \mathbf{q}_1 - D_1^2 \mathbf{q}_1 + \mathbf{c}(\mathbf{q}_1) \qquad (21.25)$

where $\mathbf{\Lambda} = \text{diag}\,(\lambda^{-}, \lambda^{+})$ collects the eigenvalues corresponding to the wave frequencies. The nonlinearity vector $\mathbf{c}(\mathbf{q}_1) = -\eta\, \varphi_2(\varphi_2^{\top} \mathbf{q}_1)(\varphi_2^{\top} \mathbf{q}_1)(\varphi_2^{\top} \mathbf{q}_1)$ depends on the auxiliary vector $\varphi_2 = (\varphi_2^{-}, \varphi_2^{+})$, collecting column-wise the second components of the *mass-normalized* linear waveforms $\varphi^{\mp}$.

The differential problem stated by the coupled equations (21.23)-(21.25) in the unknowns $\mathbf{q}_1, \mathbf{q}_2, \mathbf{q}_3$ is solved in the following for the two fundamental cases occurring in the absence of superharmonic internal resonance between the wave frequencies (*non-resonant* lattices satisfying the spectral condition $\omega^{+} - 3\omega^{-} = \mathcal{O}(1)$). The complementary fundamental case, occurring in the presence of superharmonic 3:1 internal resonance or nearly-resonance between the optical and acoustic frequencies (*internally resonant* lattices satisfying the spectral condition $\omega^{+} - 3\omega^{-} = \mathcal{O}(\epsilon)$), is solved and analyzed in Lepidi and Bacigalupo (2019), where the parametric conditions for the occurrence of the superharmonic resonance are also determined analytically.

The general solution of the homogeneous equation (21.23) at the lowest ϵ-order is the superposition of two real-valued, T_0-periodic and mutually orthogonal solutions $\mathbf{q}_1^{-}$ and $\mathbf{q}_1^{+}$ (*generating solutions*), reading

$$\mathbf{q}_1^{\mp} = A^{\mp}(T_1, T_2)\mathbf{a}^{\mp} \mathrm{e}^{\imath \omega^{\mp} T_0} + cc \qquad (21.26)$$

where $\mathbf{a}_0^{-} = (1, 0)$ and $\mathbf{a}_0^{+} = (0, 1)$ are the canonical base vectors of the two-dimensional $\mathbf{q}$-space, $A^{-}(T_1, T_2)$ and $A^{+}(T_1, T_2)$ are complex-valued amplitudes and cc stands for the complex conjugate of the preceding terms.

Therefore, substituting one or the other of the generating solutions (21.26) in the ϵ^2-order equations (21.24) and imposing solvability conditions to remove the secular terms allows to determine that the oscillation amplitudes $A^{\mp}$ are T_1-independent. As major consequence, the inhomogeneous term of the equation (21.24) vanishes. Hence, the second-order solution $\mathbf{q}_2^{\mp}$ can be neglected, because its particular part is null and its complementary part (*parallel* to the first-order solution $\mathbf{q}_1^{\mp}$) can be normalized to zero without loss of generality.

Substituting again one or the other of the generating solutions (21.26) in the ϵ^3-order equations (21.25), the solvability condition required to remove the secular terms returns the modulation equation

$$D_2 A^{\mp} = \frac{3\imath\eta(\phi_2^{\mp})^4}{2\omega^{\mp}}(A^{\mp})^2\bar{A}^{\mp} \tag{21.27}$$

Introducing the convenient polar form $A^{\mp}(T_2) = \frac{1}{2}a^{\mp}(T_2)\exp(\imath\gamma^{\mp}(T_2))$, the equation (21.27) allows to conclude that the amplitude $a^{\mp}$ is actually T_2-independent, while the phase $\gamma^{\mp}$ linearly depends the slow T_2-time scale

$$\gamma^{\mp} = \gamma_\circ^{\mp} + \frac{3}{8}\frac{\eta(\phi_2^{\mp})^4}{\omega^{\mp}}(a^{\mp})^2\, T_2 \tag{21.28}$$

where $\gamma_\circ^{\mp}$ is a constant depending on the initial conditions. Coming back to the real time scale (by recalling that $T_0 = \tau$ and $T_2 = \epsilon^2\tau$), the equation (21.28) can be reformulated as $\gamma^{\mp} = \gamma_\circ^{\mp} + \epsilon^2\varpi_2^{\mp}\tau$. Therefore, the first-order solution (21.26) can be expressed in the equivalent exponential or trigonometric forms

$$\mathbf{q}_1^{\mp} = \frac{1}{2}a^{\mp}\mathbf{a}_0^{\mp}\mathrm{e}^{\imath\theta^{\mp}} + cc = a^{\mp}\mathbf{a}_0^{\mp}\cos\theta^{\mp} \tag{21.29}$$

The solution is featured by mono-harmonic oscillations, governed by the time-dependent auxiliary variable $\theta^{\mp}(\tau) = \gamma_\circ^{\mp} + (\omega^{\mp} + \epsilon^2\varpi_2^{\mp})\tau$, where the τ-multiplier can be interpreted as the *nonlinear frequencies* of the harmonically oscillating waves, quadratically depending on the oscillation amplitude

$$\varpi^{\mp} = \omega^{\mp} + \epsilon^2\varpi_2^{\mp} = \omega^{\mp} + \frac{3}{8}\epsilon^2\frac{\eta(\varphi_2^{\mp})^4}{\omega^{\mp}}(a^{\mp})^2 \tag{21.30}$$

where $\kappa_2^{\mp} = \frac{3}{8}\eta(\varphi_2^{\mp})^4/\omega^{\mp}$ is the (positive) *effective nonlinearity coefficient*.

Once the solvability condition (21.27) is satisfied, the third-order equation (21.25) admits the multi-harmonic solution in trigonometric form

$$\mathbf{q}_3^{\mp} = \frac{1}{4}(a^{\mp})^3\mathbf{a}_2^{\mp}\cos\theta^{\mp} + \frac{1}{4}(a^{\mp})^3\mathbf{c}_2^{\mp}\cos(3\theta^{\mp})$$

which includes nonlinear contributions from both the principal coordinates. Indeed, $\mathbf{a}_2^{\mp}$ and $\mathbf{c}_2^{\mp}$ are linear combinations of the canonical base vectors

$$\mathbf{a}_2^{\mp} = \mathcal{P}_2^{\pm}\mathbf{a}_0^{\mp} + \mathcal{C}_2^{\mp}\mathbf{a}_0^{\pm}, \qquad \mathbf{c}_2^{\mp} = \mathcal{D}_2^{\mp}\mathbf{a}_0^{\mp} + \mathcal{G}_2^{\pm}\mathbf{a}_0^{\pm} \tag{21.31}$$

where $\mathcal{P}_2^{\pm} = 0$, since the complementary part of the solution $\mathbf{q}_3^{\mp}$ (*parallel* to $\mathbf{q}_1^{\mp}$) can be normalized to zero. The coefficient $\mathcal{C}_2^{\pm}$ of the $\omega^{\mp}$-harmonic term (*orthogonal* to $\mathbf{q}_1^{\mp}$) and the coefficients $\mathcal{D}_2^{\mp}$ and $\mathcal{G}_2^{\pm}$ of the two superharmonic $3\omega^{\mp}$-harmonic terms (respectively *parallel* and *orthogonal* to $\mathbf{q}_1^{\mp}$) contributing to the particular part of the solution $\mathbf{q}_3^{\mp}$ are

$$\mathcal{C}_2^{\pm} = \frac{3\eta(\varphi_2^{\mp})^3\varphi_2^{\pm}}{(\omega^{\mp})^2 - (\omega^{\pm})^2}, \quad \mathcal{D}_2^{\mp} = \frac{\eta(\varphi_2^{\mp})^4}{2(2\omega^{\mp})^2}, \quad \mathcal{G}_2^{\pm} = \frac{\eta(\varphi_2^{\mp})^3\varphi_2^{\pm}}{(3\omega^{\mp})^2 - (\omega^{\pm})^2} \tag{21.32}$$

Substituting the trigonometric form of the solutions $\mathbf{q}_1^{\mp}$ and $\mathbf{q}_3^{\mp}$ (and recalling that $\mathbf{q}_2^{\mp} = \mathbf{0}$) in the equation (21.22), the reconstructed solution reads

$$\mathbf{q}^{\mp} = \epsilon\, a^{\mp}\mathbf{a}_0^{\mp}\cos\theta^{\mp} + \frac{1}{4}\epsilon^3(a^{\mp})^3\mathbf{a}_2^{\mp}\cos\theta^{\mp} + \frac{1}{4}\epsilon^3(a^{\mp})^3\mathbf{c}_2^{\mp}\cos(3\theta^{\mp}) \tag{21.33}$$

where terms up to $\mathcal{O}(\epsilon^3)$ are retained. The reconstructed solution allows to determine the third-order approximation of the two–dimensional invariant manifolds $\mathcal{M}^-$ and $\mathcal{M}^+$, associated with the nonlinear waveform ψ^- and ψ^+, respectively (Shaw and Pierre, 1991). The manifold $\mathcal{M}^-$ (or $\mathcal{M}^+$) is fully determined in the space of the principal coordinates $\mathbf{q} = (q, p)$ once the second coordinate p (first coordinate q) is analytically related to the first coordinate q (second coordinate p) in the reconstructed solution $\mathbf{q}^-$ (or $\mathbf{q}^+$). The manifolds are approximated by the coordinate relations

$$\mathcal{M}^-:\ p = \mathcal{H}_2^+ q^3 + \frac{\mathcal{K}_2^+}{(\omega^-)^2} q(\mathring{q})^2, \qquad \mathcal{M}^+:\ q = \mathcal{H}_2^- p^3 + \frac{\mathcal{K}_2^-}{(\omega^+)^2} p(\mathring{p})^2 \tag{21.34}$$

where $\mathring{q}$ and $\mathring{p}$ stand for the derivatives $\partial q/\partial T_0|_{T_0=\tau}$ and $\partial p/\partial T_0|_{T_0=\tau}$ and the parameters $\mathcal{H}_2^{\pm} = \frac{1}{4}\left(\mathcal{C}_2^{\pm} - \mathcal{G}_2^{\pm}\right)$ and $\mathcal{K}_2^{\pm} = \frac{1}{4}\left(\mathcal{C}_2^{\pm} - 3\mathcal{G}_2^{\pm}\right)$ have been used.

The manifolds $\mathcal{M}^-$ and $\mathcal{M}^+$ can also be expressed in the space of the configuration vector $\mathbf{u}$ by recalling the change of variables $\mathbf{u} = \mathbf{\Phi}\mathbf{q}$

$$\mathcal{M}^-:\ \mathbf{u} = q\boldsymbol{\varphi}^- + q\mathcal{Q}^+(q, \mathring{q})\boldsymbol{\varphi}^+ \qquad \mathcal{M}^+:\ \mathbf{u} = p\boldsymbol{\varphi}^+ + p\mathcal{Q}^-(p, \mathring{p})\boldsymbol{\varphi}^- \tag{21.35}$$

where the quadratic functions $\mathcal{Q}^+(q, \mathring{q}) = \mathcal{H}_2^+ q^2 + \mathcal{K}_2^+(\mathring{q}/\omega^-)^2$ and $\mathcal{Q}^-(p, \mathring{p}) = \mathcal{H}_2^- p^2 + \mathcal{K}_2^-(\mathring{p}/\omega^+)^2$. The initial conditions $q(\tau_\circ) = a$ and $\mathring{q}(\tau_\circ) = 0$ can be assigned into $\mathbf{u}$ for $\mathcal{M}^-$, and analogous initial conditions $p(\tau_\circ) = a$ and $\mathring{p}(\tau_\circ) = 0$ can be assigned into $\mathbf{u}$ for $\mathcal{M}^+$. Therefore, dividing by a, the third-order approximation of the nonlinear waveforms is

$$\boldsymbol{\psi}^{\mp} = \boldsymbol{\varphi}^{\mp} + a^2\mathcal{H}_2^{\pm}\boldsymbol{\varphi}^{\pm} \tag{21.36}$$

or, making explicit the dependence on the linear dispersion properties

$$\boldsymbol{\psi}^{\mp} = \boldsymbol{\varphi}^{\mp} + \frac{a^2}{4}\left[\frac{3\eta(\varphi_2^{\mp})^3\varphi_2^{\pm}}{(\omega^{\mp})^2 - (\omega^{\pm})^2} + \frac{\eta(\varphi_2^{\mp})^3\varphi_2^{\pm}}{(3\omega^{\mp})^2 - (\omega^{\pm})^2}\right]\boldsymbol{\varphi}^{\pm} \tag{21.37}$$

where the nonlinear correction of each linear waveform is directly proportional to the square oscillation amplitude multiplying the other (orthogonal) waveform. The asymptotic consistency of the nonlinear correction can be recognized to break down for the resonance conditions $\omega^+ : \omega^- \approx 1$ (internal resonance) or $\omega^+ : \omega^- \approx 3$ (superharmonic internal resonance). The first possibility is excluded by the systematic presence of the band gap, the second possibility is analyzed in Lepidi and Bacigalupo (2019).

21.3.3 Parametric Analyses and Numerical Validation

The amplitude-dependence of the nonlinear frequencies $\varpi^\pm$ is parametrically analyzed. The effective nonlinearity coefficient $\kappa_2^\mp$ is positive and determines the curvature of the frequency-amplitude curve (*backbone curve*). Therefore, the frequency systematically increases for growing oscillation amplitudes (*hardening backbone*), as expected for cubic nonlinearities. From the mechanical viewpoint, the curvature mainly depends on the work done by the nonlinear waveforce $\mathbf{N}_2(\varphi^\mp)\varphi^\mp$ in the *same* waveform $\varphi^\mp$. Figure 21.2 shows the nonlinear frequencies $\varpi^\mp$ versus the mass ratio ϱ^2 for increasing amplitudes $a^\mp$. The parameter range ($\mu = 1/2$, $1/2 < \varrho^2 < 3/2$) is conveniently selected to avoid superharmonic resonances in the limit of short wavelengths ($\beta = \pi$). From the quantitative viewpoint, the results highlight that the nonlinear increment of the acoustic frequency ϖ^-/ω^- (backbone curves marked by continuous mesh lines in Figure 21.2a) is much lower than the increment of the optical frequency ϖ^+/ω^+ (backbone curves in Figure 21.2b). From the qualitative viewpoint, a certain mass ratio maximizes the nonlinear increment of the acoustic frequency ϖ^-/ω^- at large amplitudes (iso-amplitude curves marked by dashed mesh lines in Figure 21.2a). On the contrary, the nonlinear increment of the optical frequency ϖ^+/ω^+ monotonically increases for decreasing mass ratios (iso-amplitude dashed curves in Figure 21.2b).

The invariant manifolds $\mathcal{M}^-$ and $\mathcal{M}^+$ are portrayed in Figure 21.3 in the space $(p, q, \mathring{q})$ and $(q, p, \mathring{p})$, respectively. The lattice parameters ($\mu = 1/2$, $\eta = 1/2$, $\beta = \pi$) are selected in the range explored for the nonlinear frequencies in Figure 21.2, while the mass ratio ($\varrho^2 = 4/5$) is fixed to maximize the effective nonlinearity coefficients. The origin is the tangency points between the manifolds and the invariant

Table 21.1 Initial conditions for the nonlinear equations governing the wave motion. *Master* and *slave* coordinates indicate independent and dependent variables in equations (21.34)

Initial conditions for $\mathcal{M}^-$		Initial conditions for $\mathcal{M}^+$	
Master $(q_0, \dot{q}_0)$	Slave $(p_0, \dot{p}_0)$	Master $(p_0, \dot{p}_0)$	Slave $(q_0, \dot{q}_0)$
$(0.10, 0)$	$(-0.22151, 0) \times 10^{-4}$	$(0.10, 0)$	$(2.06646, 0) \times 10^{-4}$
$(0.15, 0)$	$(-0.74761, 0) \times 10^{-4}$	$(0.15, 0)$	$(6.97429, 0) \times 10^{-4}$
$(0.20, 0)$	$(-1.77211, 0) \times 10^{-4}$	$(0.20, 0)$	$(16.5317, 0) \times 10^{-4}$

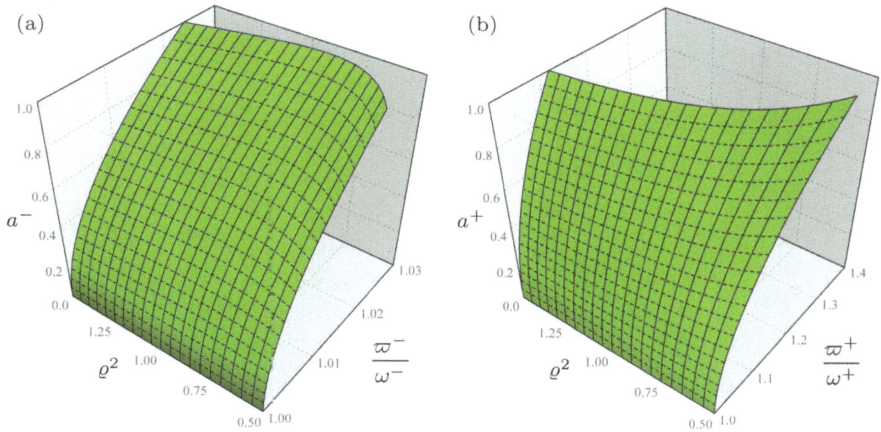

Fig. 21.2 Nonlinear frequencies of the lattice with mechanical parameters $\mu = 1/2$, $\eta = 1/2$ at the limit of short wavelengths ($\beta = \pi$).

planes $(q, \mathring{q})$ and $(p, \mathring{p})$ associated with the corresponding linear waveforms. From the mechanical viewpoint, the principal curvatures of the manifolds mainly depends on the work done by the nonlinear waveforce $\mathbf{N}_2(\varphi^{\mp})\varphi^{\mp}$ in the *other* waveform $\varphi^{\pm}$. As major qualitative remark, the principal curvatures of the optical manifold $\mathcal{M}^+$ (normalized with respect to p) tend to be systematically higher than those of the acoustic manifold $\mathcal{M}^-$ (normalized with respect to q). In order to verify the asymptotic results, numerical results are finally carried out by directly integrating the nonlinear equations of wave motion in the space of the principal coordinate vector $\mathbf{q} = (q, p)$. The free oscillation responses are determined for three different sets of initial conditions belonging to the asymptotic approximation of the invariant manifolds (displacement $\mathbf{q}_0$ and velocity $\dot{\mathbf{q}}_0$ at time $\tau = \tau_0$ reported in Table 21.1). The numerical time-histories of motion $\mathbf{q}(\tau)$ are successfully verified to describe periodic stable orbits (red curves in Figure 21.3), sitting on the manifold surfaces with fine approximation.

21.4 Conclusions

A one–dimensional crystal lattice, characterized by a diatomic periodic cell, is considered as minimal physical realization of an acoustic metamaterial with weakly nonlinear microstructure. The free undamped dynamics of the microstructured cell is described by a Lagrangian two-degrees-of-freedom model. Cubic nonlinearities are introduced to enrich the linear elastic coupling between the principal atom and the secondary atom—serving as tunable local resonator. According to the Floquet–Bloch theory, the linearized eigenproblem governing the free wave propagation in the small-

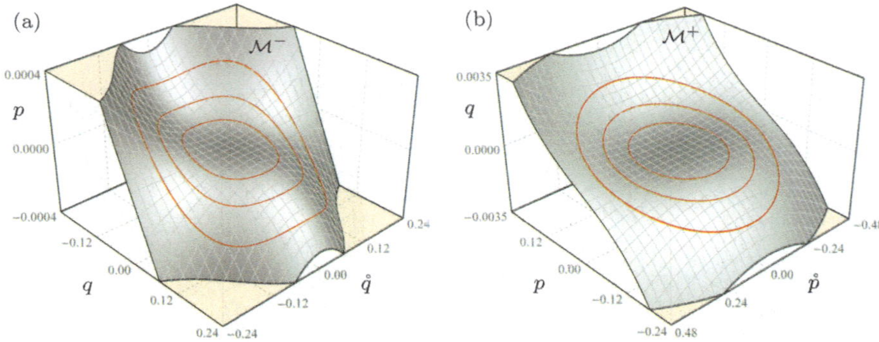

Fig. 21.3 Comparison between numerical solutions (red curves) and analytical invariant manifolds associated to the nonlinear waveforms: (a) manifold $\mathcal{M}^-$, (b) manifold $\mathcal{M}^+$.

amplitude oscillation regime is solved analytically to determine the linear dispersion properties (frequencies and waveforms). The analytical eigensolution shows that the linear waveforms are real-valued. In the frequency dispersion spectrum, a stop band systematically separates the pass bands corresponding to the low-frequency acoustic branch and the high-frequency optical branch.

Starting from the linear solution, a general asymptotic approach is employed to determine the nonlinear dispersion properties. Specifically, the multiple scale method is adopted to achieve a perturbation-based approximation of the free wave motion up to the third order. The nonlinear frequencies and waveforms are obtained as analytical—although asymptotically approximate—functions of the mechanical parameters and the oscillation amplitudes. The nonlinear acoustic and optical frequencies are found to quadratically depend on the wave amplitudes, with backbone curves exhibiting the characteristic hardening behavior. Each nonlinear waveform is analytically obtained as amplitude-dependent combination of the two linear waveforms. Furthermore, the invariant manifolds associated with the nonlinear waveforms are parametrically determined in the space of the principal coordinates. Finally, the analytical solutions are successfully validated by running out numerical simulations of the amplitude-dependent free wave oscillations.

Acknowledgements The authors acknowledge financial support of the (MIUR) Italian Ministry of Education, University and Research in the framework of the research MIUR Prin15 project 2015LYYXA8, "Multi-scale mechanical models for the design and optimization of micro-structured smart materials and metamaterials", coordinated by prof. A. Corigliano. The authors also acknowledge financial support by National Group of Mathematical Physics (GNFM-INdAM).

References

Bacigalupo A, Gambarotta L (2019) Generalized micropolar continualization of 1D beam lattices. International Journal of Mechanical Sciences 155:554–570

Bacigalupo A, Lepidi M (2016) High-frequency parametric approximation of the Floquet-Bloch spectrum for anti-tetrachiral materials. International Journal of Solids and Structures 97-98:575–592

Bacigalupo A, Lepidi M (2018) Acoustic wave polarization and energy flow in periodic beam lattice materials. International Journal of Solids and Structures 147:183–203

Bacigalupo A, Lepidi M, Gnecco G, Vadalà F, Gambarotta L (2019) Optimal design of the band structure for beam lattice metamaterials. Frontiers in Materials 6:2

Brillouin L (1946) Wave propagation in periodic structures: electric filters and crystal lattices. Dover Publications, New York

Colquitt DJ, Brun M, Gei M, Movchan AB, Movchan NV, Jones IS (2014) Transformation elastodynamics and cloaking for flexural waves. Journal of the Mechanics and Physics of Solids 72:131–143

di Cosmo F, Laudato M, Spagnuolo M (2018) Acoustic metamaterials based on local resonances: homogenization, optimization and applications. In: Generalized Models and Non-classical Approaches in Complex Materials 1, Springer, pp 247–274

Craster RV, Guenneau S (2012) Acoustic metamaterials: Negative refraction, imaging, lensing and cloaking, vol 166. Springer Science & Business Media, Heidelberg, Germany

Cummer SA, Christensen J, Alù A (2016) Controlling sound with acoustic metamaterials. Nature Reviews Materials 1(3):16,001

Deymier PA (2013) Acoustic metamaterials and phononic crystals, vol 173. Springer Science & Business Media

El Sherbiny MG, Placidi L (2018) Discrete and continuous aspects of some metamaterial elastic structures with band gaps. Archive of Applied Mechanics 88(10):1725–1742

Fang X, Wen J, Yin J, Yu D, Xiao Y (2016) Broadband and tunable one-dimensional strongly nonlinear acoustic metamaterials: Theoretical study. Physical Review E 94(5):052,206

Georgiou IT, Vakakis AF (1996) An invariant manifold approach for studying waves in a one-dimensional array of non-linear oscillators. International journal of non-linear mechanics 31(6):871–886

Grinberg I, Vakakis AF, Gendelman OV (2018) Acoustic diode: Wave non-reciprocity in nonlinearly coupled waveguides. Wave Motion 83:49–66

Guenneau S, Movchan A, Pétursson G, Ramakrishna SA (2007) Acoustic metamaterials for sound focusing and confinement. New Journal of physics 9(11):399

Lacarbonara W, Camillacci R (2004) Nonlinear normal modes of structural systems via asymptotic approach. International Journal of Solids and Structures 41(20):5565–5594

Lacarbonara W, Arafat HN, Nayfeh AH (2005) Non-linear interactions in imperfect beams at veering. International Journal of Non-Linear Mechanics 40(7):987–1003

Lazarov BS, Jensen JS (2007) Low-frequency band gaps in chains with attached non-linear oscillators. International Journal of Non-Linear Mechanics 42(10):1186–1193

Lepidi M (2013) Multi-parameter perturbation methods for the eigensolution sensitivity analysis of nearly-resonant non-defective multi-degree-of-freedom systems. Journal of Sound and Vibration 332(4):1011–1032

Lepidi M, Bacigalupo A (2018) Multi-parametric sensitivity analysis of the band structure for tetrachiral acoustic metamaterials. International Journal of Solids and Structures 136-137:186–202

Lepidi M, Bacigalupo A (2019) Wave propagation properties of one-dimensional acoustic metamaterials with nonlinear diatomic microstructure. Nonlinear Dynamics 98(4):2711–2735

Liu Z, Zhang X, Mao Y, Zhu Y, Yang Z, Chan CT, Sheng P (2000) Locally resonant sonic materials. Science 289(5485):1734–1736

Ma G, Sheng P (2016) Acoustic metamaterials: From local resonances to broad horizons. Science advances 2(2):e1501,595

Manimala JM, Sun C (2014) Microstructural design studies for locally dissipative acoustic metamaterials. Journal of Applied Physics 115(2):023,518

Manktelow K, Leamy MJ, Ruzzene M (2011) Multiple scales analysis of wave–wave interactions in a cubically nonlinear monoatomic chain. Nonlinear Dynamics 63(1-2):193–203

Narisetti RK, Leamy MJ, Ruzzene M (2010) A perturbation approach for predicting wave propagation in one-dimensional nonlinear periodic structures. Journal of Vibration and Acoustics 132(3):031,001

Narisetti RK, Ruzzene M, Leamy MJ (2011) A perturbation approach for analyzing dispersion and group velocities in two-dimensional nonlinear periodic lattices. Journal of Vibration and Acoustics 133(6):061,020

Perkins N, Mote Jr C (1986) Comments on curve veering in eigenvalue problems. Journal of Sound and Vibration 106(3):451–463

Pierre C (1988) Mode localization and eigenvalue loci veering phenomena in disordered structures. Journal of Sound and Vibration 126(3):485–502

Romeo F, Rega G (2006) Wave propagation properties in oscillatory chains with cubic nonlinearities via nonlinear map approach. Chaos, Solitons & Fractals 27(3):606–617

Romeo F, Rega G (2008) Propagation properties of bi-coupled nonlinear oscillatory chains: analytical prediction and numerical validation. International Journal of Bifurcation and Chaos 18(07):1983–1998

Rosenberg R (1966) On nonlinear vibrations of systems with many degrees of freedom. In: Advances in applied mechanics, vol 9, Elsevier, pp 155–242

Rothos V, Vakakis A (2009) Dynamic interactions of traveling waves propagating in a linear chain with an local essentially nonlinear attachment. Wave Motion 46(3):174–188

Shaw S, Pierre C (1991) Non-linear normal modes and invariant manifolds. Journal of Sound and Vibration 150(1):170–173

Vakakis AF, King ME (1998) Resonant oscillations of a weakly coupled, nonlinear layered system. Acta mechanica 128(1-2):59–80

Vidoli S, Vestroni F (2005) Veering phenomena in systems with gyroscopic coupling. Journal of Applied Mechanics 72(5):641–647

Chapter 22
Deformation and Failure Onset of Random Elastic Beam Networks Generated From the Same Type of Random Graph

Uwe Mühlich

Abstract Deformation and failure onset of random elastic beam networks is investigated numerically. Different types of planar beam networks are generated from a random graph for which geometrical and topological characteristics can be defined easily. Sampling according to these characteristics is performed by means of the Ising-model.
The networks are exposed to an overall strain and the deformation response is discussed in terms of the network's strain energy density W. The results show, that W correlates well with the characteristics of the underlying graph. Intensity of failure onset, on the other hand, is examined using an energy based edge centrality measure. The results suggest, that randomness causes in most cases a decrease in stiffness, but it reduces as well severity of failure onset.
In addition, the results show, that for the beam networks considered here, correlations can be discussed as well in terms of mean cell sidedness and mean product moment of cell area, i.e., measures which do not rely on the existence of a specific underlying graph.

Keywords: Random beam networks · Random graphs · Ising model · Strain energy

22.1 Introduction

Insight into complex physical processes such as deformation and fracture of random cellular materials or flow through porous media is at the heart of intelligent and targeted materials design and engineering. Network models have proven to be convenient idealizations for studying such complex phenomena efficiently.

U. Mühlich
Universidad Austral de Chile, General Lagos 2086, Campus Miraflores, Valdivia, Chile
e-mail: Uwe.Muehlich@yahoo.com

B. E. Abali and I. Giorgio (eds.), *Developments and Novel Approaches in Biomechanics and Metamaterials*, Advanced Structured Materials 132,
https://doi.org/10.1007/978-3-030-50464-9_22

Following Estrada (2015), we use the term network to express the synthesis of a graph's geometric representation and a mathematical model describing some process, for instance, by virtue of differential equations defined on the graph's edges.

For that reason, the response of a network is determined by three sets of characteristics. On the one hand, it obviously depends on parameters characterizing the process, such as order of the differential equation(s), magnitude of coefficients like stiffness, conductivity, etc., and of course, boundary conditions. On the other hand, the response is strongly influenced by the graph's node-edge system as it determines the topology of the network. Last but not least, geometrical properties of a network are a direct consequence of the graph's geometric realization. In short, the response of a network is controlled by topological, geometrical and process related characteristics. In addition, the response of random networks will be affected significantly by the random variations of these characteristics throughout the network.

Here, we focus on deformation and failure onset of random beam networks in the context of materials modelling. Investigations regarding elastic deformation of random networks usually aim to derive eventually an effective continuum description within the framework of classical continua. Such investigations were performed for elastic truss networks in the spirit of solid state physics by Garboczi (1987) and in the context of strength of materials and structural mechanics, respectively, by Hansen et al (1996); Ostoja-Starzewski (2002); Zeman et al (2011). Similar investigations for random beam networks are reported by Jang et al (2008); Zhu et al (2001); Roberts and Garboczi (2002). In the context of generalized continua, such networks are studied by Diebels and Steeb (2002); Tekoglu and Onck (2008).

Extensive literature exists as well on simulation of fracture in networks. Fracture in regular central force networks with deterministic and randomly distributed material properties is investigated, e.g., by Beex et al (2014); Dimas et al (2014); Wilbrink et al (2013). Regular beam networks with randomly distributed failure strength are considered by Herrmann et al (1989); Savija et al (2013), whereas random beam networks are subject of investigation in Kadashevich and Stoyan (2010). Considerable differences in the results are observed for using different failure criteria by Schlangen and Garboczi (1997) in a study including regular as well as random beam networks. Failure in pantographic networks is studied for instance by Turco et al (2016). A general survey regarding the most important aspects of simulating fracture in networks, including rigidity loss and percolation concepts, is given by Alava et al (2006) in the context of statistical fracture mechanics. Most recently, defect sensitivity of random brittle foams is discussed by Chen and Isaksson (2019).

It is worth noting that none of the previously mentioned contributions in the context of materials research intends to investigate rigorously the influence of network's topology on its physical response. A possible reason may lie in the fact that topological features can hardly be controlled independently in more complex random networks. However, their importance is reported implicitly by Roberts and Garboczi (2002) and pointed out explicitly in the general context of materials science by Gupta and Saxena (2014). Topological effects are addressed explicitly by Vogel (2002) regarding permeability of porous media. The effect of topology on elastic properties of foams and bone scaffolds is studied by Nachtrab et al (2012). Both contributions focus, however, exclusively on variations in the Euler characteristic.

Topological effects for truss networks with fixed Euler characteristic are systematically studied by Mühlich et al (2015) using a particular planar graph which allows for controlling directly topological features. The results show, that also second order topological properties can affect significantly the overall response. In this paper, we extend this study towards deformation and failure onset of random beam networks. Therefore, the graph proposed by Mühlich et al (2015) is briefly revisited and graphs derived from it for generating random beam networks are introduced. The methodology employed here in terms of load cases and definitions used to evaluate results is presented afterwards before discussing the numerical results of this study in detail.

22.2 Graph Construction and Characterization

We consider a regular square lattice, assigning exactly one diagonal with random direction (/ or \) to each of the $M \times M$ squares, as shown in Fig. 22.1. This type of graph is called "0 graph" in the sequel. There are 2^{M^2} possible realizations of this "0 graph". For every realization X_k $(k = 1...2^{M^2})$ there exists a corresponding discrete field $\{\xi_{i,j}\}$ with $i, j = 1, .., M$ and

$$\xi_{i,j} = \begin{cases} 0 & \text{if diagonal } i, j \; / \\ 1 & \phantom{\text{if diagonal } i, j} \; \backslash \end{cases} . \tag{22.1}$$

Therefore, a "0"graph can be described by corresponding random field characteristics. The first-order characteristic is the empirical mean of $\{\xi_{i,j}\}$

$$\rho = \langle \xi_{i,j} \rangle = \frac{1}{M^2} \sum_{i,j} \xi_{i,j} \, . \tag{22.2}$$

Geometrically, ρ is a directional characteristic, the ratio between the numbers of "/" and "\" diagonals. All problems considered here are invariant under the operation of flipping all diagonals, formally expressed by $\xi_{i,j} = 1 - \xi_{i,j}$, $\xi_{i,j} \in \{0, 1\}$. This is tantamount to symmetry with respect to $\rho = 0.5$. Therefore, according to Mühlich et al (2015),

$$\bar{\rho} = \min(\rho, 1 - \rho) \tag{22.3}$$

is used instead of ρ. A simple second-order characteristic is

$$\mu = \langle \mu_{i,j} \rangle \tag{22.4}$$

with

$$\mu_{i,j} = \sum_{\substack{k,l\\d=1}} (\xi_{i,j} - \xi_{k,l})^2 \tag{22.5}$$

where $d = |i - k| + |j - l|$, i.e., the summation goes over the nearest neighbour cells. Hence, $\mu_{i,j}$ is a nearest neighbour correlation measure. Its empirical mean, μ, captures to some extent the irregularity of a configuration.

Certain values for μ and $\bar{\rho}$ indicate perfectly regular configurations. If all diagonals are of the same type, either / or \, then μ evaluates to zero. The configuration with perfectly alternating diagonals corresponds to $\mu = 4$ and $\bar{\rho} = 0.5$.

This type of graph is used here, because it is easy to generate and to characterize. Furthermore, varying diagonal directions affects directly the topology of the graph, i.e., topological characteristics can be varied systematically. In addition, the construction principle guarantees a constant total number of edges. In the context of elastic deformation, this is tantamount to ensuring the same amount of material for all possible realizations of a particular network.

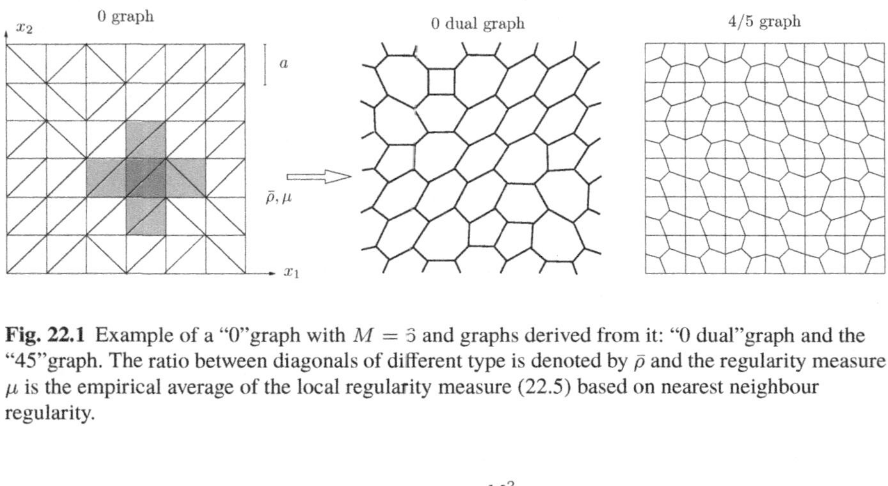

Fig. 22.1 Example of a "0"graph with $M = 5$ and graphs derived from it: "0 dual"graph and the "45"graph. The ratio between diagonals of different type is denoted by $\bar{\rho}$ and the regularity measure μ is the empirical average of the local regularity measure (22.5) based on nearest neighbour regularity.

As already mentioned above, there are 2^{M^2} possible configurations, which is intractable even for moderate values of M. Therefore, samples of "0 graphs" are generated, as reported in Mühlich et al (2015) by making use of the $M \times M$ Ising model with fixed magnetization as defined in Gulminelli et al (2003), Carmona et al (1998). The spin variables in this model are interpreted as the directions of diagonals. According to Honerkamp (2012), we write the probability distribution of the model as

$$P(X = X_k) = \frac{1}{Z} \exp(-\beta\mu(X_k)) \tag{22.6}$$

with partition function Z. As above, X_k denotes a configuration, $\mu(X_k)$ is μ as in (22.4) for X_k, whereas β is the control parameter of the model. By means of β, the degree of regularity/irregularity of the configurations can be controlled. Here, "regularity" refers to regularity in terms of the directions of the diagonal edges.

Samples are generated by simulating according to (22.6) employing the Metropolis algorithm constraint by $\bar{\rho}$ as described in Carmona et al (1998). The number of iterations N_{MR} is set according to $N_{\text{MR}} = 10000(1 + 100|\beta|)M$. While $\bar{\rho}$ can be controlled directly, μ is only controlled indirectly via β in (22.6). Thus, we arrive at a two-parameter model, with the parameters β and $\bar{\rho}$.

As illustrated in Mühlich et al (2015), bijective relations exist between β and μ for given $\bar{\rho} \neq 0$. The most irregular configurations are obtained for $\beta = 0$, whereas high degree of regularity is enforced in the limits $\beta \rightarrow \pm\infty$. Given $\bar{\rho} = 0.5$, for instance, $\beta \rightarrow \infty$ generates configurations consisting of exactly two clusters, one cluster of "/" diagonals and a second cluster of "\" diagonals. On the other hand, $\beta \rightarrow -\infty$ samples perfectly regular configurations with alternating diagonals which corresponds to $\mu = 4$ and $\bar{\rho} = 0.5$, see above.

Different graphs can be constructed from a "0"graph as shown in Fig.22.1. A natural choice is the corresponding dual graph. In addition, a so-called "45"graph is obtained from the initial graph by shorten the diagonals and connecting their endpoints with the center points of those edges which define the squares.

22.3 Methodology

22.3.1 Deformation

Here, networks are interpreted as periodic arrangements of a basic pattern derived from a $M \times M$ "0"graph with $a = 10\text{mm}$ as shown in Fig.22.1. $M = 20$ and $M = 15$ are used for "0 dual"graphs and "45"graphs, respectively. Nodes are interpreted as rigid connections between compressible Euler-Bernoulli beams, i.e., classical beams with additional axial force. Beams with quadratic cross section of side length $h = 1\text{mm}$ and Young's modulus $E = 1000\text{N}/\text{mm}^2$ are considered.

Representative network patterns are exposed to an effective strain state by prescribing the nodal displacements

$$\mathbf{u} = \boldsymbol{E} \cdot \mathbf{x} \tag{22.7}$$

at the corner nodes of the pattern. $\boldsymbol{E}$ is an effective strain tensor and the position vector $\mathbf{x}$ refers to the Cartesian coordinate system shown in Fig. 22.1. For "0 dual"based networks, corner nodes are the first and last nodes on a line defining a segment of the boundary $\partial\mathcal{B}$. The rotation at the corner nodes is prescribed to zero. For all remaining nodes, periodicity is enforced for displacements, rotations and corresponding reaction forces by extending straightforwardly the approach described in Mühlich et al (2015).

Three load cases are considered: uni-axial tension in x_1 direction ($E_{11} = \varepsilon, E_{22} = E_{12} = 0$), shear ($E_{12} = \varepsilon, E_{11} = E_{22} = 0$) and combined loading ($E_{11} = E_{22} = E_{12} = \varepsilon$). In all simulations, $\varepsilon = 0.01$ is used.

In order to compute the response of the networks, a finite-element scheme has been developed specifically for this purpose, following the standard procedure as given e.g. in Zienkiewicz and Taylor (2005). It is worth noting that deviations from the exact solution are only due to floating point operations because the loading is exclusively applied at the nodes.

The response of the networks is discussed in terms of total strain energy per M^2, denoted by W, for it allows to classify the networks' mechanical response by a single scalar measure regardless of the degree of anisotropy. Note that for the displacement controlled tests, a lower W implies lower stiffness. Since all networks are eventually constructed out of a "0" graph with characteristics $\bar{\rho}$ and μ, the hypothesis

$$W = f_W(\bar{\rho}, \mu, \boldsymbol{E}, \alpha_k) \tag{22.8}$$

is proposed and examined, where the α_k are parameters like Young's modulus, cross section area. The latter are not varied in this study.

Table 22.1 Possible descriptors for characterizing geometric realizations of planar graphs. The first three measures are purely topological whereas the remaining ones are purely geometrical. Given values indicate, that the corresponding measure evaluates always to this value for the considered type of graph, which renders it useless for distinguishing between different configurations.

		0 graph		0 dual graph		4/5 graph	
measure	symbol	mean	var	mean	var	mean	var
node degree	n_d	6	✓	3	0	3.5	0.25
cell degree	c_d	3	0	6		4.5	0.25
sidedness (nearest neighbours)	s_R	3	0				
cell area	A_C	$A_\triangle$	0*	100†		$\frac{A_4 + A_5}{2}$	$\frac{(A_4 - A_5)^2}{2}$
area moments	I_{11}, I_{22}, I_{12}						
edge angle	$\tan\gamma = \frac{\sum l_e \sin\alpha_e}{\sum l_e \cos\alpha_e}$	✓					

*) without varying node coordinates †) for dimensions used here

22.3.2 Failure Onset

Fracture simulations are rather expensive for a variety of reasons. However, considerable insight regarding the onset of failure, and even beyond, can be gained already from inspecting the deformation results accordingly. We assume brittle failure based on the hypothesis, that an edge fails if its strain energy reaches a given critical value. In this case, the importance of an edge for failure onset is due to its strain energy in comparison with the highest individual strain energy in the network. The more edges of high importance exist in a network, the more sever the onset of failure will be, because more edges will fail at once. Hence, we define an importance measure for the edge i by

$$C_i^{\mathrm{W}} = \frac{W_i}{W_{\mathrm{e,max}}}, \tag{22.9}$$

where the strain energy of edge i is denoted by W_i and $W_{\mathrm{e,max}}$ is the maximum edge strain energy in the network. In the language of graph theory, C_i^{W} could be interpreted as a kind of energy based centrality measure.

22.3.3 Testing Selected General Descriptors

Since for networks commonly used in materials modelling a "0" graph does not exist, the measures $\bar{\rho}$ and μ are not available in such cases. Therefore, correlations between physical response and graph characteristics like $\bar{\rho}$ and μ are of rather limited value. Deducing useful indications from these studies requires the transition to descriptors applicable for general planar graphs such as planar Voronoi or Laguerre tessellations.

The fundamental entities of a planar graph are nodes and edges also called vertexes and bonds, respectively. Although the cell structure is a direct consequence of the relations between nodes and edges, sometimes it is rather the cell structure which provides insight. This suggests some simple and obvious descriptors, such as node degree n_d, i.e., the number of edges connected by a node, and cell degree c_d, i.e., the number of edges which form a cell.

For random networks, these descriptors are not represented by a single number but by a discrete probability distribution. Simple global measures can be obtained by computing empirical mean, empirical variance etc. of these descriptors for a given configuration.

A more elaborated measure is the so-called sidedness ratio s_R, see, e.g., Hilhorst (2006), which is defined as the average cell degree of the neighbouring cells divided by the degree of the cell under consideration. The sidedness-ratio depends whether or not next nearest neighbours are included. Next nearest neighbours share only a node whereas nearest neighbours share an edge. Here, only nearest neighbours are taken into account. Other possible descriptors exist, such as shortest path length, centrality measures of different kind, etc., which are not considered here. The interested reader is referred, for instance, to Chiu et al (2013), Ranjan and Zhang (2013) or Costa et al (2007).

The descriptors discussed so far are purely topological. Geometrical properties can be captured, for instance, by means of edge angle, variation in cell size and shape, etc.

The descriptors considered here are summarized in Table 22.1. Given values indicate, that the corresponding measure evaluates always to this value for the considered type of graph, which renders it useless for distinguishing between different configurations.

22.4 Results and Discussion

22.4.1 Deformation

Results for W are plotted against $\bar{\rho}$ and β. The graphs are shown for "0 dual"and "45"networks in Fig. 22.2 and Fig. 22.4, respectively. Every point in a graph corresponds to the strain energy per M^2 for a configuration with given $\bar{\rho}$ and β. In addition polynomial approximations obtained from the data points are visualized by means of grids.

Fairly good correlations between $\bar{\rho}$, β and W can be observed for all cases. Due to the existence of a bijective relation between β and μ, the results confirm hypothesis (22.8). Loading conditions affect the correlation functions qualitatively and quantitatively.

For the "0 dual"networks the lowest stiffness is obtained for $\bar{\rho} = 0.5$ and $\beta = -1$ in shear but for $\bar{\rho} = 0.5$ and $\beta = 1$ in uni-axial tension. Please note, that both configurations are regular or of high degree of regularity, respectively. In all cases, the perfectly regular configurations with $\bar{\rho} = 0$ are the stiffest.

"45"networks behave in certain aspects considerably different. For instance, the variation in magnitude is considerably smaller for uni-axial tension ($\approx 25\%$) compared to shear ($\approx 75\%$). Contrary to the "0 dual"networks, perfectly regular "45"networks with $\bar{\rho} = 0$ exhibit the lowest stiffness under combined loading.

Testing more general descriptors revealed, that the results for W correlate reasonably well with mean sidedness ration $\langle s_R \rangle$ and mean product moment $\langle I_{12} \rangle$ as shown in Fig. 22.4. The results suggest, that certain combinations between $\langle s_R \rangle$ and $\langle I_{12} \rangle$ are not possible. The limit curve which separates the $\langle s_R \rangle - \langle I_{12} \rangle$ plane accordingly appears to be linear for "0 dual"networks and at least quadratic for "45"networks, respectively. The reasons for this are unclear at the moment and further theoretical investigation regarding this question is required. In addition, the correlation functions, especially for "0 dual"networks, are significantly more complex compared to all other correlation functions observed in this research. Based on the results, we propose the hypothesis

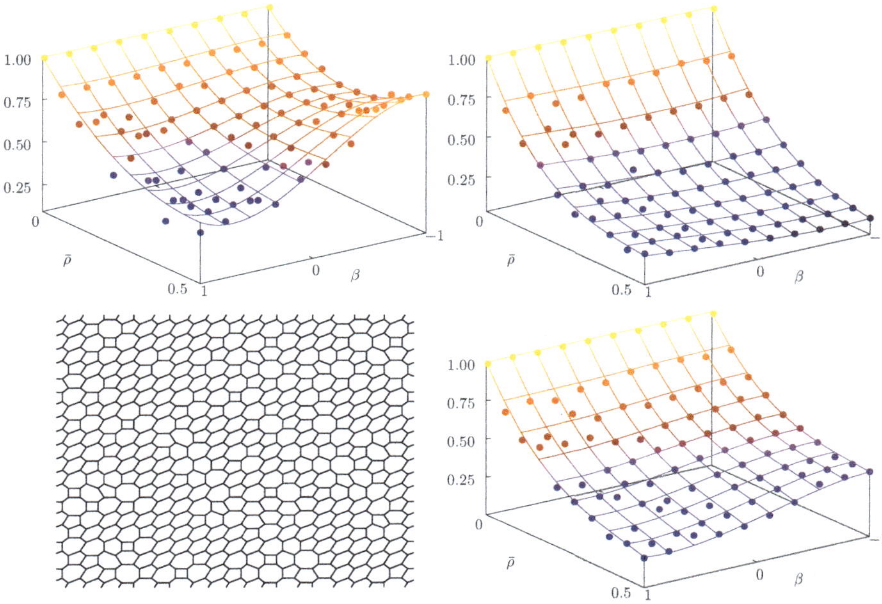

Fig. 22.2 Strain energy density W of the networks based on "0"graphs as functions of $\bar{\rho}$ and β for uni-axial tension ($E_{11} = 0.01$) (upper left), shear ($E_{12} = 0.01$) (upper right) and combined loading ($E_{11} = E_{22} = E_{12} = 0.01$) (lower right). Dots represent data points. Corresponding fit functions are visualized by their surface grid representations. The data are normalized with respect to $W(\bar{\rho} = 0, \beta)$, i.e., the strain energy density of the perfectly regular networks. In addition a configuration for $\bar{\rho} = 0.227$ and $\beta = -0.111$ is shown exemplary.

$$W = \bar{f}_W(\langle s_R \rangle, \langle I_{12} \rangle, \boldsymbol{E}, \alpha_k) \tag{22.10}$$

for planar beam networks.

22.4.2 Failure Onset

Onset of failure is discussed by means of the importance measure C_i^W defined by (22.9). Results are shown exemplary for "0 dual"networks with $\bar{\rho} = 0.5$ under shear loading in Fig. 22.5. The results indicate a transition from a bi-modal or even multi-modal distribution to a uni-modal distribution. The mean of C_i^W correlates well only within certain limits but not for the whole range of $\bar{\rho}$ and μ considered here. A possible reason is seen in the fact, that the empirical average underestimates severity of failure onset for cases with bi-modal distributions. This subject requires further investigation.

The effect of a small random variation of the node coordinates is shown in Fig. 22.5. The positions of the inner nodes of the network are randomly displaced

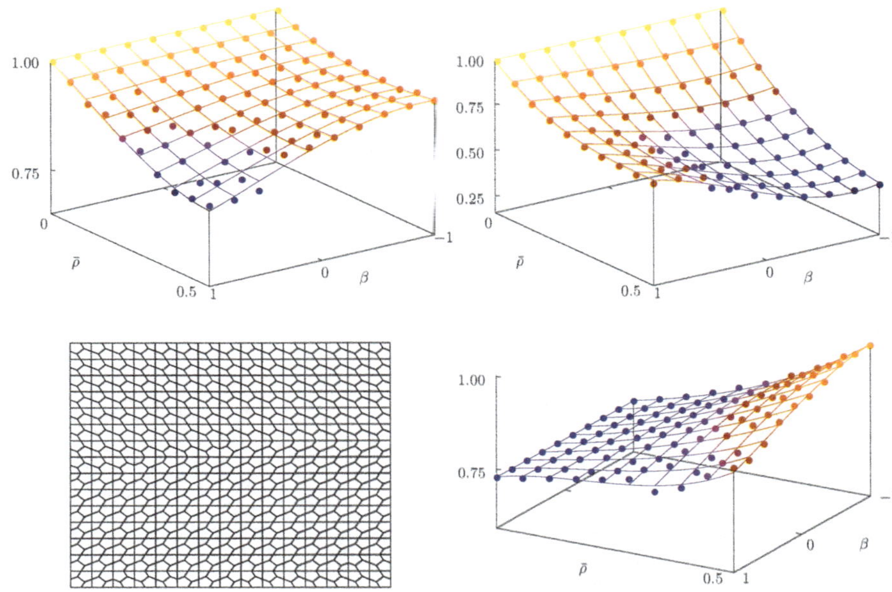

Fig. 22.3 Strain energy density W of the networks based on "45" graphs as functions of $\bar{\rho}$ and β for uni-axial tension ($E_{11} = 0.01$) (upper left), shear ($E_{12} = 0.01$) (upper right) and combined loading ($E_{11} = E_{22} = E_{12} = 0.01$) (lower right). Dots represent data points. Corresponding fit functions are visualized by their surface grid representations. The data for uniaxial tension and combined loading are normalized with respect to $W(\bar{\rho} = 0, \beta)$, i.e., the strain energy density of the perfectly regular networks. The data for shear loading are normalized by $W(\bar{\rho} = 0.5, \beta = -1)$. In addition a configuration for $\bar{\rho} = 0.5$ and $\beta = 1$ is shown exemplary.

by $a \cdot \Delta/10$ with Δ uniformly distributed in the interval $[-1, 1]$. Especially for configurations with $\beta < -0.75$, a significant smoothening in the C_i^W distribution can be observed, which will decrease the severity of the failure process at least at the beginning.

One of the most remarkable networks in the context of failure onset is the "45"network with $\bar{\rho} = 0.5$ and $\beta = -1$. The uni-axial deformation of a typical representative is shown in Fig. 22.7. Although, there is a considerable amount of edges with C_i^W equal or close to one, respectively, these edges are all separated by small substructures. In the event, that all edges with C_i^W equal or close to one fail, the remaining structure is still connected and consists of edges with reasonably low C_i^W values, i.e., the structure can still bear load. The implications are twofold. First, provided, the structure is monitored, onset of failure will cause a significant signal. Second, although the structure is significantly damaged after onset of failure, it still has certain structural integrity to avoid complete collapse.

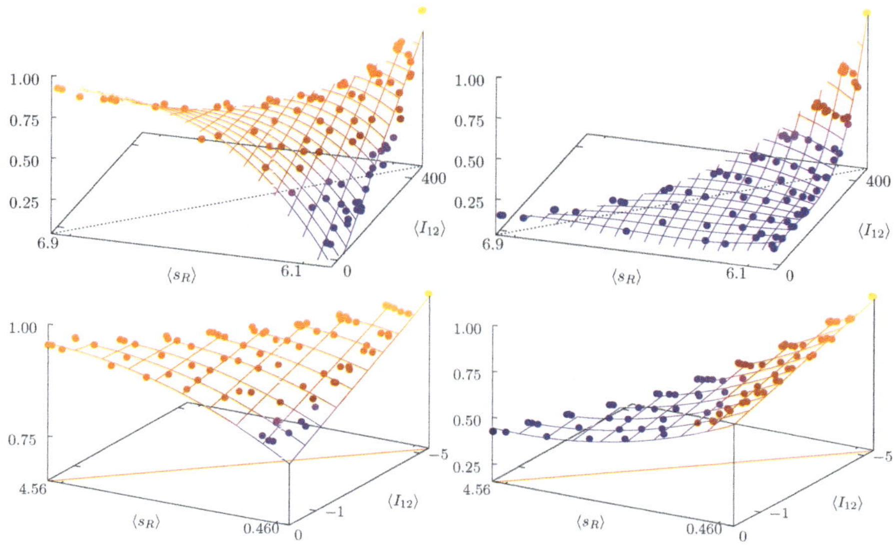

Fig. 22.4 Strain energy density W as functions of $\langle S_R \rangle$ and $\langle I_{12} \rangle$ for uni-axial tension ($E_{11} = 0.01$) (left) and shear ($E_{12} = 0.01$) (right). Dots represent data points. Corresponding fit functions are visualized by their surface grid representations. The data are normalized with respect to $W(\bar{\rho} = 0, \beta)$, i.e., the strain energy density of the perfectly regular networks. The results for the "0"graph based networks and the "45"graph based networks are shown in the top and in the second row, respectively.

22.5 Summary and Outlook

The investigation strategy proposed in Mühlich et al (2015) for truss networks is equally applicable for planar random beam networks. Exactly one topological and one geometrical descriptor is necessary to correlate the deformation response with corresponding network properties for the cases considered here.

For the truss networks investigated in Mühlich et al (2015), irregularity implies always a decrease in stiffness compared to perfectly regular configurations. Although this observation can be confirmed here for most cases, there exist exceptions for which regular configurations exhibit the lowest stiffness.

Randomness in topology is beneficial for moderating onset of failure. Similarly, randomizing node positions leads to less severe initial failure behaviour.

A hypothesis for more general planar beam networks, i.e., networks derived from Voronoi tessellations, is proposed. Verification of this hypothesis, however, requires to sample planar Voronoi tessellations according to sidedness ratio and product moment.

Some useful information emerged from this study regarding damage tolerant design of regular beam networks.

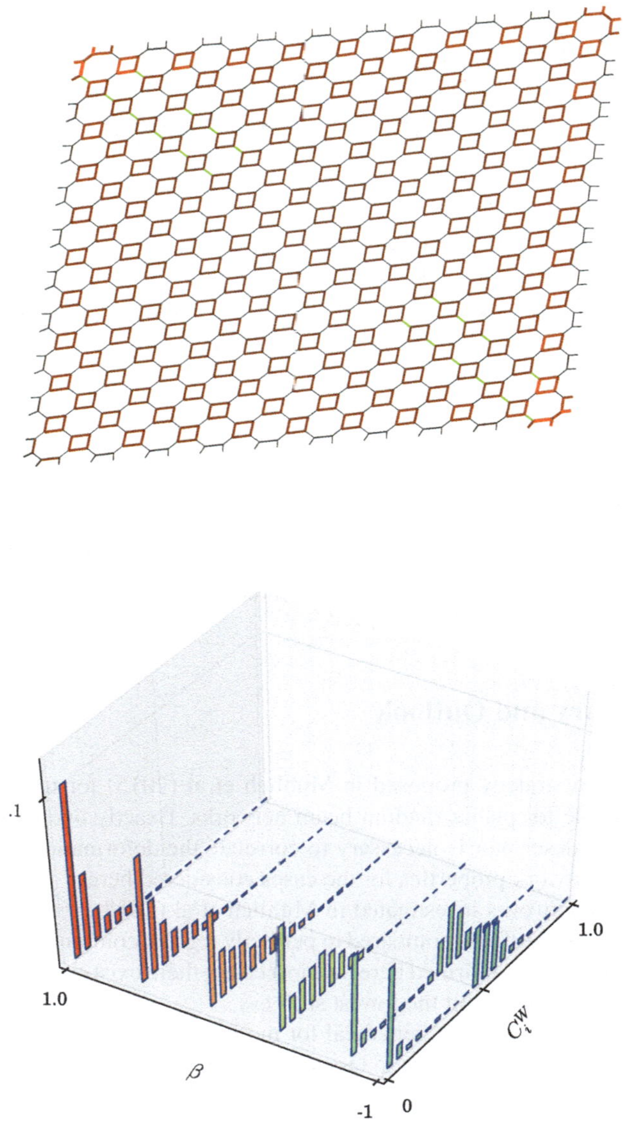

Fig. 22.5 Histogram of the edge importance measure C_i^W for "0 dual"networks with $\bar{\rho} = 0.5$ and different values of β together with a corresponding example configurations for $\bar{\rho} = 0.5$ and $\beta = -1$. The magnitude of C_i^W is indicated in the example configuration by edge thickness and color, i.e., highest values of C_i^W correspond to thick red lines.

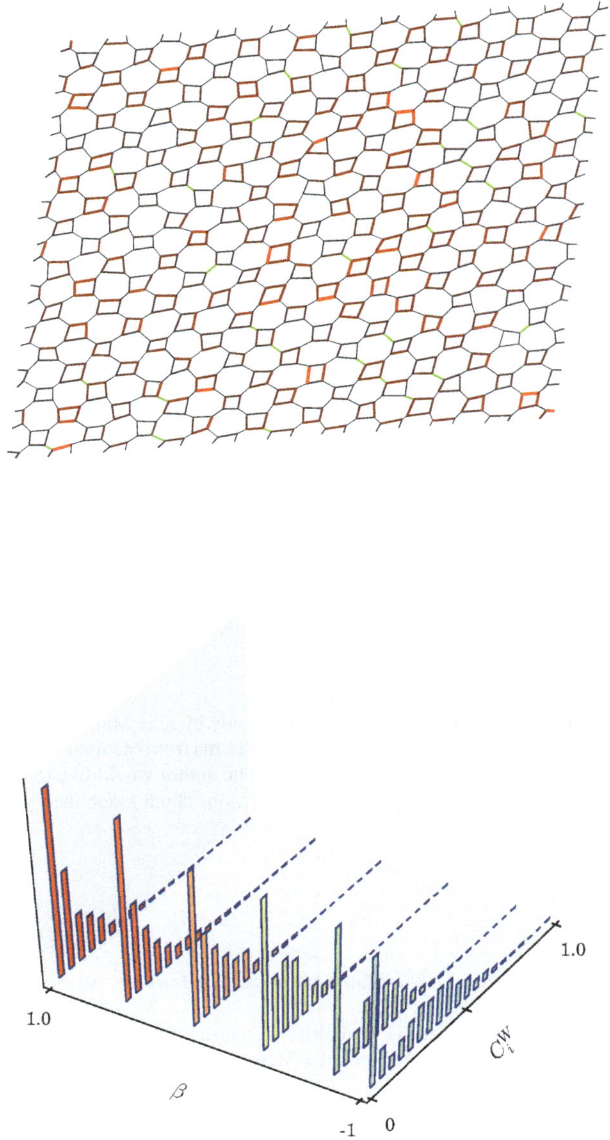

Fig. 22.6 Histogram of the edge importance measure C_i^W for "0 dual"networks with $\bar{\rho} = 0.5$ and different values of β together with a corresponding example configurations for $\bar{\rho} = 0.5$ and $\beta = -1$. The magnitude of C_i^W is indicated in the example configuration by edge thickness and color, i.e., highest values of C_i^W correspond to thick red lines. The results are obtained for a small random variation of the node coordinates.

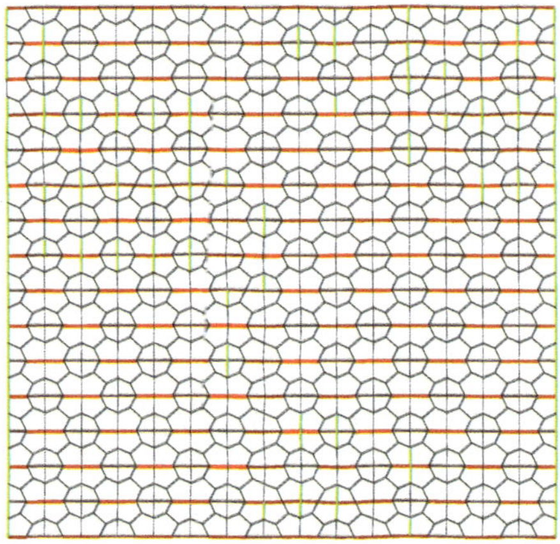

Fig. 22.7 C_i^W distribution in a "45"network with $\bar{\rho} = 0.5$ and $\beta = -1$ in uni-axial tension. The magnitude of C_i^W is indicated by edge thickness and color, i.e., highest values of C_i^W correspond to thick red lines. Edges with $C_i^W = 0$ are displayed in green.

Acknowledgements This research was supported partially by Jezy Mazik, Jolanta Budzejko from Vilnius Gediminas Technical University, and Smritika Sharma from Manipal Institute of Technology during their Erasmus stays at University of Antwerp. The author would like to thank as well Rudi Penne from University of Antwerp for insightful discussions about graph theory.

References

Alava M, Nukala P, Zapperi S (2006) Statistical models of fracture. Advances in Physics 55(3-4):349–476

Beex L, Peerlings R, Geers M (2014) A multiscale quasicontinuum method for lattice models with bond failure and fiber sliding. Computer Methods in Applied Mechanics and Engineering 269:108–122

Carmona J, Richert J, Tarancon A (1998) A model for nuclear matter fragmentation: phase diagram and cluster distributions. Nuclear Physics A 643:115–134

Chen S, Isaksson P (2019) A note on the defect sensitivity of brittle solid foams. Engineering Fracture Mechanics 206:541 – 550

Chiu S, Stoyan D, Kendall W, Mecke J (2013) Stochastic Geometry and its Applications, 3rd edn. Wiley

Costa LDF, Rodrigues FA, Travieso G, Villas Boas PR (2007) Characterization of complex networks: A survey of measurements. Advances in Physics 56(1):167–242

Diebels S, Steeb H (2002) The size effect in foams and its theoretical and numerical investigation. Proceedings of the Royal Society of London A: Mathematical, Physical and Engineering Sciences 458(2028):2869–2883

Dimas L, Giesa T, Buehler M (2014) Coupled continuum and discrete analysis of random heterogeneous materials: Elasticity and fracture. Journal of the Mechanics and Physics of Solids 63(1):481–490

Estrada E (2015) Graph and network theory. In: Grinfeld M (ed) Mathematical Tools for Physicists, WILEY, chap 4, pp 111–157

Garboczi E (1987) Cauchy relations for central-force random networks. Physical Review B 36(4):2115–2120

Gulminelli F, Carmona JM, Chomaz P, Richert J, Jiménez S, Regnard V (2003) Transient backbending behavior in the ising model with fixed magnetization. Phys Rev E 68:026,119

Gupta S, Saxena A (2014) A topological twist on materials science. MRS Bulletin 39:265–279

Hansen J, Skalak R, Chien S, Hoger A (1996) An Elastic Network Model Based on the Structure of the Red Blood Cell Membrane Skeleton. Biophysical Journal 70:1146–166

Herrmann H, Hansen A, Roux S (1989) Fracture of disordered, elastic lattices in two dimensions. Physical Review B 39(1):637–648

Hilhorst HJ (2006) Planar Voronoi cells: the violation of Aboav's law explained. Journal of Physics A Mathematical General 39(23):7227–7243

Honerkamp J (2012) Statistical Physics, 3rd edn. Graduate Texts in Physics, Springer

Jang WY, Kraynik A, Kyriakides S (2008) On the microstructure of open-cell foams and its effect on elastic properties. International Journal of Solids and Structures 45:1845–1875

Kadashevich I, Stoyan D (2010) Simulation of Brittle Fracture of Autoclaved Aerated Concrete. Computers and Concrete 7:39–51

Mühlich U, Ballani F, Stoyan D (2015) Influence of randomness in topology, geometry and material properties on the mechanical response of elastic central-force networks. Probabilistic Engineering Mechanics 40:36–41

Nachtrab S, Kapfer S, Rietzel D, Drummer D, Madadi M, Arns C, Kraynik A, Schröder-Turk G, Mecke K (2012) Tuning elasticity of open-cell solid foams and bone scaffolds via randomized vertex connectivity. Advanced Engineering Materials 14(1-2):120–124

Ostoja-Starzewski M (2002) Lattice models in micromechanics. Applied Mechanics Reviews 55(1):35–60

Ranjan G, Zhang ZL (2013) Geometry of complex networks and topological centrality. Physica A: Statistical Mechanics and its Applications 392(17):3833 – 3845

Roberts A, Garboczi E (2002) Elastic properties of model random three-dimensional open-cell solids. Journal of the Mechanics and Physics of Solids 50:33–55

Savija B, Lukovic M, Pacheco J, Schlangen E (2013) Cracking of the concrete cover due to reinforcement corrosion: A two-dimensional lattice model study. Construction and Building Materials 44:626 – 638

Schlangen E, Garboczi E (1997) Fracture simulations of concrete using lattice models: computational aspects. Engineering Fracture Mechanics 57(2-3):319–332

Tekoglu C, Onck PR (2008) Size effects in two-dimensional voronoi foams: A comparison between generalized continua and discrete models. Journal of the Mechanics and Physics of Solids 56(12):3541 – 3564

Turco E, dell'Isola F, Rizzi NL, Grygoruk R, Müller WH, Liebold C (2016) Fiber rupture in sheared planar pantographic sheets: Numerical and experimental evidence. Mechanics Research Communications 76:86 – 90

Vogel HJ (2002) Topological characterization of porous media. In: Mecke K, Stoyan D (eds) Morphology of Condensed Matter, Lecture Notes in Physics, vol 600, Springer Berlin Heidelberg, pp 75–92

Wilbrink D, Beex L, Peerlings R (2013) A discrete network model for bond failure and frictional sliding in fibrous materials. International Journal of Solids and Structures 50(9):1354–1363

Zeman J, Peerlings R, Geers M (2011) Non-local energetics of random heterogeneous lattices. Journal of the Mechanics and Physics of Solids 59(6):1214–1230

Zhu H, Hobdell J, Windle A (2001) Effects of cell irregularity on the elastic properties of 2d voronoi honeycombs. Journal of the Mechanics and Physics of Solids 49(4):857 – 870
Zienkiewicz O, Taylor R (2005) The Finite Element Method Set. Elsevier Science

Chapter 23
Mechanical Behavior Investigation of 3D Printed Pantographic Unit Cells via Tension and Compression Tests

Nima Nejadsadeghi, Marco Laudato, Michele De Angelo, and Anil Misra

Abstract In this work we present the results of an experimental parametric investigation of the mechanical behavior of the pantographic unit cells. The main goal is to determine extreme configurations of the geometric parameters of the microstructure. This work extends the set of measurements presented in a previous paper by analyzing a larger set of samples with different geometrical features. The considered set of 3D printed pantographic unit cells reveals local extremum configurations that are different from the ones observed in the previous paper. A discussion regarding these differences and concluding remarks are provided.

Keywords: Metamaterials · Pantographic unit cells · 3D printed samples

---

N. Nejadsadeghi
Mechanical Engineering Department, University of Kansas, 1530 W. 15th Street, Learned Hall, Lawrence, KS 66045-7609, USA
e-mail: nima.nejadsadeghi@ku.edu

M. Laudato
Dipartimento di Ingegneria e Scienze dell'Informazione e Matematica, Università degli Studi dell'Aquila. Via Vetoio, Coppito, 67100 L'Aquila, Italy
e-mail: marco.laudato@graduate.univaq.it

M. De Angelo, A. Misra
Civil, Environmental and Architectural Engineering Department, The University of Kansas, 1530 W. 15th Street, Lawrence, KS 66045-7609, USA
e-mail: michele.deangelo.aq@gmail.com, amisra@ku.edu

M. De Angelo
Dipartimento di Ingegneria Civile, Edile-Architettura e Ambientale, Università degli Studi dellAquila, Via Giovanni Gronchi 18 - Zona industriale di Pile, 67100 LAquila, Italy
e-mail: michele.deangelo.aq@gmail.com

B. E. Abali and I. Giorgio (eds.), *Developments and Novel Approaches in Biomechanics and Metamaterials*, Advanced Structured Materials 132,
https://doi.org/10.1007/978-3-030-50464-9_23

23.1 Introduction

Mechanical metamaterials are a class of dynamical systems whose macroscopic behavior is determined by the features of their underlying microstructure (Barchiesi et al, 2019b; Del Vescovo and Giorgio, 2014). Following the metamaterial paradigm, indeed, it is possible to obtain a desired macroscopic behavior by tailoring the geometric properties of a precise microstructure made of mechanical elements. The range of application for this class of mechanical system is therefore determined by the amount of control which is possible to exercise during the fabrication procedure and on the forecasting power that the available mathematical models can provide (Di Cosmo et al, 2018; dell'Isola et al, 2019a; Andreaus et al, 2018).

During the last few decades, the framework of fast-prototyping methods (such as 3D printing) has experienced an exceptional growth and it is nowadays possible to produce fine-tuned specimens in a reasonable amount of time, providing researchers with a powerful and reliable trial and error mechanism. On the other hand, the efforts to obtain solid mathematical and numerical tools to investigate the behavior of such complex systems have brought a new understanding on their multi-scale nature (e.g., homogenization procedures as done in Boutin et al (2017); Placidi et al (2020); Abdoul-Anziz and Seppecher (2018); Abdoul-Anziz et al (2019)). However, the challenging problems that current research on mechanical metamaterials is tackling in the last years seem to require a shift from classical approaches to more general ones. An example are the so-called generalized continua (dell'Isola et al, 2015a; Seppecher et al, 2011; Nejadsadeghi and Misra, 2019b,a; Nejadsadeghi et al, 2019b; Poorsolhjouy and Misra, 2019; Misra and Nejadsadeghi, 2019; Misra and Poorsolhjouy, 2016; Chróścielewski et al, 2020; Auffray et al, 2015). This class of mechanical metamaterials shows a macroscopic behavior that cannot be framed in the setting of classical elastic theory (Alibert et al, 2003). The main reason is the complex geometric features of their microstructure (Turco et al, 2017a,b; Giorgio et al, 2018; Scerrato and Giorgio, 2019).

A paradigmatic example of generalized continua is the pantographic material (dell'Isola et al, 2019b). The microstructure of this metamaterial is made of two arrays of straight fibers oriented with a relative angle of 90° (dell'Isola et al, 2015b, 2016). The fibers are arranged on two parallel planes and in the intersection points they are connected by means of cylinders called pivots. It has been shown by several theoretical (Eremeyev et al, 2019, 2018; Steigmann and dell'Isola, 2015; Abali, 2019) and experimental results that the macroscopic behavior of pantographic structure cannot be effectively described by classical theory of elasticity (dell'Isola et al, 2017, 2016; Giorgio et al, 2017; Misra et al, 2018; Abali et al, 2017). Indeed, to produce correct forecasts, a generalization of classical elasticity theory in which also the second gradient of the displacement field is considered must be used (Barchiesi et al, 2019a). This mechanical metamaterial shows some interesting features that could be exploited for technological applications. For example, it is able to stretch in a large deformation regime keeping an elastic behavior and from the first to the ultimate rupture it can store the same amount of energy needed for the first rupture

(Golaszewski et al, 2019; Placidi et al, 2018; dell'Isola et al, 2019c; Turco et al, 2016b; Spagnuolo et al, 2017).

In this work, we analyze the mechanical behavior of the pantographic unit cells via tensile and compression tests. We aim at extending the results on the behavior of this system obtained in Nejadsadeghi et al (2019a); Yang et al (2018); Abali et al (2016); Placidi et al (2015) by analyzing a larger set of specimens showing different geometric features. The main goal is to determine extreme configurations in the space of the microstructure's geometric parameters and, at the same time, to exclude the role of the particular material used to produce the specimens. For this reason, a different polyamide powder has been used during the printing phase. The paper is organized as follows. In section 2, the preparation procedure of the samples and the experimental setup is described in detail. In section 3, the results of the experimental campaign are presented and discussed. Finally, in the conclusion section, we discuss some possible applications of the results and future perspectives.

23.2 Fabrication and Experimental Setup

The specimens considered in this paper characterize the unit cells of a larger pantographic metamaterial sheet. A scheme of a pantographic unit cell and its geometric parameters are presented in Fig. 23.1. Experimental testing results on pantographic unit cells with specific geometrical parameter values have been previously presented in Nejadsadeghi et al (2019a). In this paper, we consider a broader range for the variation of geometrical parameters involved in describing such pantographs. The considered range for such geometrical parameters encompasses the ones reported in Nejadsadeghi et al (2019a). However, as the properties of the powder and the 3D printer parameters used to fabricate the samples alters the mechanical properties of the unit cells, the measured values of the forces for these samples may differ from the ones presented before. The fabrication procedure of the samples is as follows. Firstly, the 3D geometry of the pantographic unit cells was modelled in the commercial CAD software SolidWorks (Dassault System SolidWorks Corporation, Waltham, MA, USA). A 3D printed Formiga P 100 (EOS GmbH, Munich, Germany) was then utilized to fabricate the pantographic unit cells from Polyamide powder (PA2200) with an average grain size of 0.06 mm, using a selective laser sintering (SLS) technology. The considered pantographic unit cells share several constant geometrical parameters provided in Table 23.1. The differences between the current set of pantographic unit cells lie in their spacing, p, and pivot heights, h (see Fig. 23.1). The naming prescription used in Nejadsadeghi et al (2019a) is adopted here for conciseness in referring to the samples with different geometrical parameters. We refer to the samples using a compound name made up of a letter in the beginning and a number at the end. We label the specimens as A, B, C, D, and E, depending on their geometrical parameter p value being 5 mm, 8 mm, 12 mm, 15 mm, and 18 mm, respectively. Numbers of 1, 2, 3, and 4, corresponding to the values of 1 mm, 1.5 mm, 2 mm, and 3 mm, respectively, follow the alphabetic labels in order to differentiate samples with different values for their geometrical parameter h. As an example, the sample B3 has its p and h values, respectively, as 8 mm and 2 mm. The tensile and compression experiments on the specimens were conducted using a Bose ElectroForce 3200 testing device. The software WinTest Material Testing System was used to control the testing device. The reaction force was measured with a load cell with a measurement range of ± 22 N, a measurement uncertainty of 0.1% and precision of 0.001 N. The displacement was measured with the built-in transducer with a range of ± 6.5 mm, a measurement uncertainty of 0.1% and precision of 0.001 mm. Using a Mojo 3D printer (FDM Fused Deposition Modeling, by Stratasys), two clamps were fabricated to facilitate gripping of the specimens. The clamps were made of acrylonitrile butadiene styrene (ABS) thermoplastic and are shown in Fig. 23.2. All the experiments were done in displacement control, meaning that one end of the pantographic unit cell is fixed, and the other end undergoes a predefined displacement. The reaction force then is read from the load cell attached to the fixed end of the specimen. The studied pantographic unit cells display viscoelastic behavior, therefore, the measured force-displacement curves are specific to a loading rate of 0.1 mm/s.

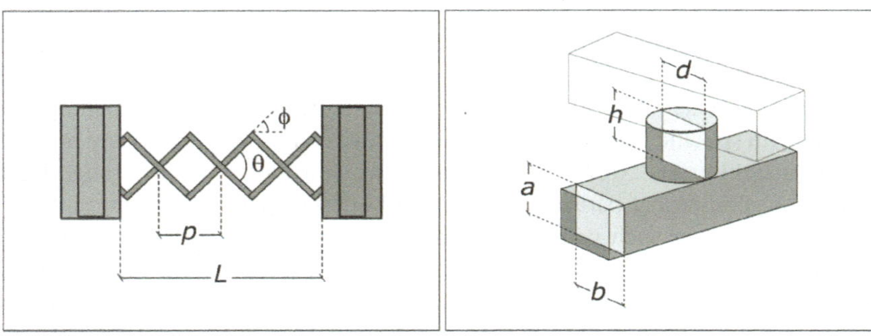

Fig. 23.1 Geometric parameters characterizing the pantographic structure.

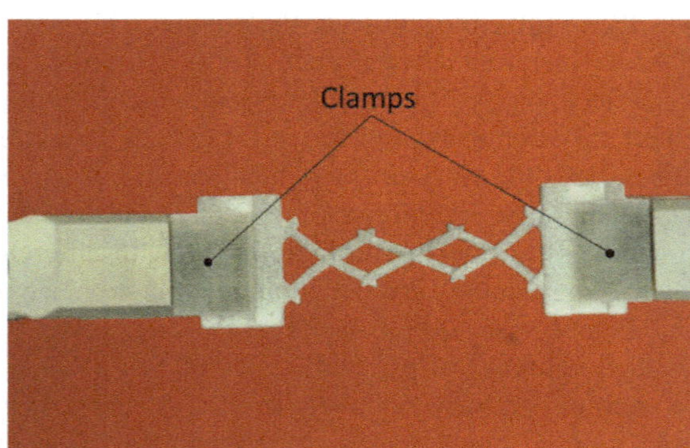

Fig. 23.2 Experimental setup of the tests.

Table 23.1 Constant geometric parameters for all the specimens.

a	b	d	L	θ	ϕ
1.2 mm	1 mm	1 mm	5 mm	$\pi/2$	$\pi/4$

23.3 Results and Discussion

The assigned deformation values for the pantographic unit cells in both the tension and compression tests were chosen such that the unit cells do not experience rupture (in tension) or major change in deformation mode Nejadsadeghi et al (2019a) (in compression) during the experiment, while also ensuring that the limits of the testing machine are not exceeded. Fig. 23.3 shows the results for the tensile tests performed on the pantographic unit cells, grouped for different spacing, p, values. It can be observed that force vs displacement plots of sample A2, A3, A4 exhibit a linear elastic behavior while sample A1 has a non-linear trend (Fig 23.3(a)). Similar results appear by looking at plots of group B-D (Fig 23.3(b-d) respectively) where samples B1, C1, D1 show non-linear behavior with hardening starting at higher displacement values, while samples B(2-4), C(2-4) and D(2-4) have linear trend. Differently, the mechanical behavior of group E (Fig. 23.3(e)) indicate a non-linear behavior for all the samples (Fig. 23.3(e)). Overall, the result depicted in Fig. 23.3 reveal an indirect proportionality between the slope of the curves and the pivot length, but an aberration can be noted for group B and group D. In order to analyze how each geometrical parameter alters the stiffness of the pantographic unit cells, one can plot for particular displacements and constant p values, the reaction force values associated with unit cells with different h values. The plots corresponding to this analysis are presented in Fig. 23.4, which shows that for a given displacement the measured reaction force is decreasing with the length of the pivot. A slight difference emerges from the plot in Fig. 23.4(b) and Fig. 23.4(d) for group B and D respectively, where samples B2 and D2 represent a local minimum. However, their values are very similar to what has been calculated for the samples B3 and D3. Alternatively, one can plot for particular displacements and constant pivot heights, h, the reaction force values associated with unit cells with different spacing, p. The latter analysis is reported in Fig. 23.5. It can be observed that in each plot of Fig. 23.5 - where the abscissa reports the specimens of each group having the same height of the pivot while ordinate reports the force level - there is an indirect proportionality between the measured tension force and the spacing, p. In this case, there is no local extremum. Fig. 23.6 shows the results for the compression tests performed on the pantographic unit cells, grouped for different spacing, p. Overall, the samples exhibit a linear behavior for the very initial values of imposed displacement, then a softening branch can be observed. Additionally, for groups A and B a hardening part occurs after a compression displacement of 6 mm and 10 mm respectively, which is related to the contact of the beams. The rapid increase of the force level is not present in the plots for groups C, D, E (Fig. 23.6 (c-e)) because the instrumentation used did not allow the attainment of a deformation capable of bringing the beams into contact. As for tension test, the samples of group A show an indirect proportionality between compression force and pivot length (Fig. 23.6(a)). Similar considerations can be made for group D and E, whose results under compression are reported in Fig. 23.6(d) and Fig. 23.6(e). A substantial deviation emerges from the result plotted in Fig. 23.6(b) and Fig. 23.6(c), for groups B and C respectively. Indeed, sample B2 does not follow the mentioned inverse proportionality, and it exhibits the minimum reaction force. Similarly, in Fig 23.6(c) the curve of sample C1 has a reaction force comparable to sample C4, while the curve of sample C2 and C3 present higher values. Fig. 23.7 shows the reaction force values associated with unit cells for selected displacement values and constant spacing, p, for different pivot heights, h. It is remarkable that for group A, D and E the force measured for a fixed displacement decreases for higher values of pivot length while groups B and C reveal extremum configurations. Indeed, sample B2 shows minimum reaction force while samples C2 and C3 exert the highest reaction force values. Further, one can plot for particular displacements and constant pivot heights, h, the reaction force values associated with unit cells with different spacing, p. These plots are shown in Fig. 23.8. It can be observed that in each plot of Fig. 23.8 - where the abscissa reports the specimens of each group having the same height of the pivot while ordinate reports the force level- there is an indirect proportionality between the measured compression force and the parameter p. In this case, we do not observe any local extremum in the behavior of specimens.

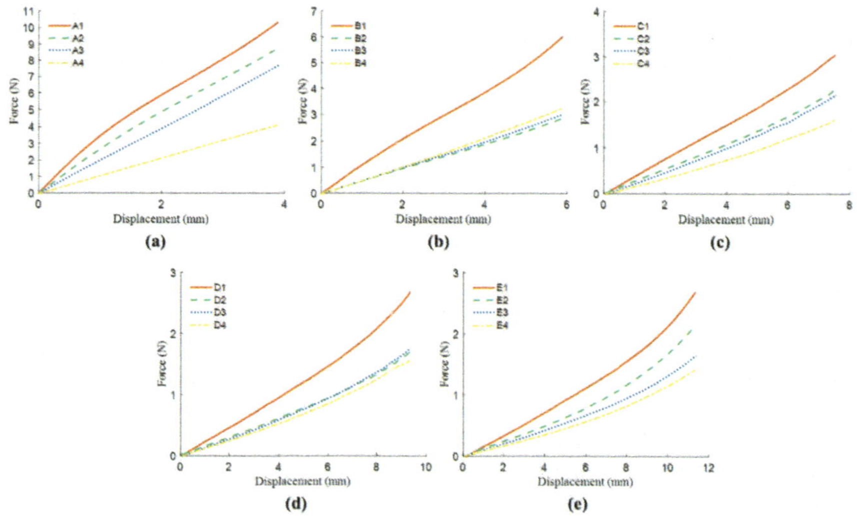

Fig. 23.3 Plots of the data from experimental tensile tests for (a) group A (b) group B (c) group C (d) group D and (e) group E.

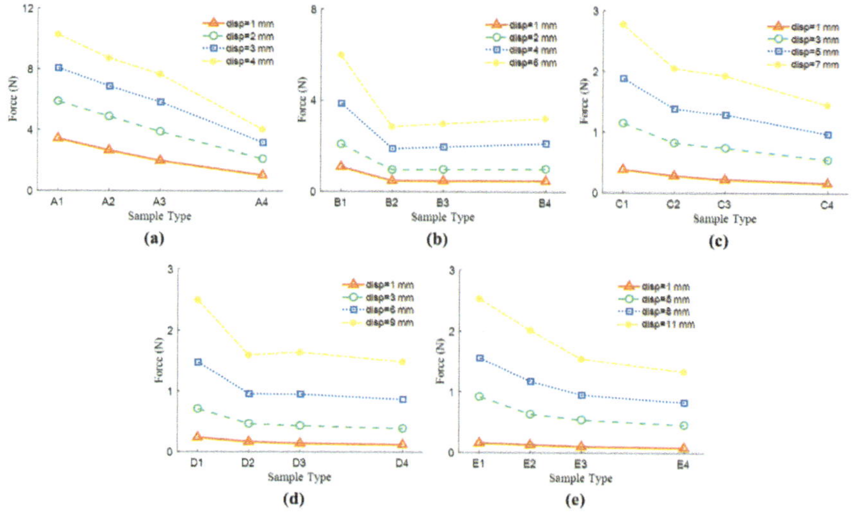

Fig. 23.4 Interpolation plots of the force values with respect to the length of the pivots for tensile tests. Each line refers to a specific value of the imposed displacement ("disp" in the legend) for (a) group A (b) group B (c) group C (d) group D and (e) group E.

23.4 Conclusions and Future Perspectives

In this work, we have analyzed the mechanical behavior of pantographic unit cells by means of tensile and compressing tests. The unit cells under investigation are characterized by different values of the microstructure's geometrical parameters p, the spacing between two pivots, and h, the height of the pivots (see Fig. 23.1). Based on figures, when we fix the h value and compare p values for specimens (as in Fig. 23.5 and Fig. 23.8), there is no aberration in behavior and for specimens with larger spacing, p, we observe a decreased value of reaction force. However, for the case where we fix the p value and compare specimens with different pivot heights, h (as in Fig. 23.4 and Fig. 23.7), there is local extremum observed for some specimens. Interestingly, the observed aberration for samples in group B is different from the one reported in Nejadsadeghi et al (2019a). The sample B1 is reported to be more compliant than the sample B2 in both tension and compression in Nejadsadeghi et al (2019a), making specimen B2 to be a local maximum in its group. On the other hand, we observe here the sample B2 as a local minimum in both tension and compression tests in its group. These differences have been consistent in each 3D printed set of unit cells, as the results of multiple tests on different group B samples revealed (not shown here), but as noted, are not consistent in different sets of prints. Further, we remark the difference in the mechanical behavior of geometrically

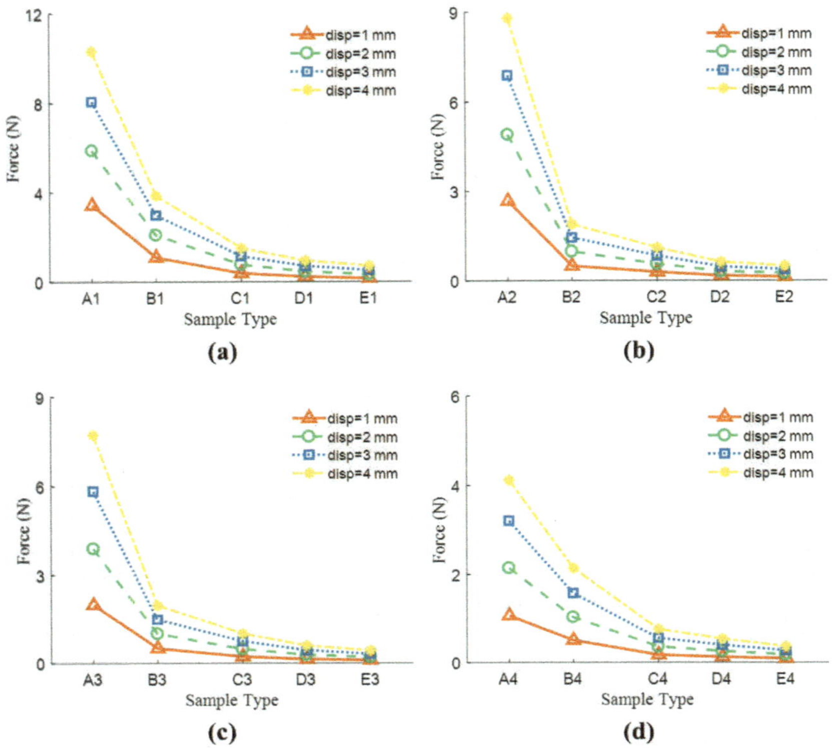

Fig. 23.5 Interpolation plots of the force values with respect to the parameter p for tensile tests. Each line refers to a specific value of the imposed displacement ("disp" in the legend). Specimens are grouped for equal values of the parameter h, being the length of the pivot equal to (a) 1 mm (b) 1.5 mm (c) 2 mm and (d) 3 mm.

identical specimens in each 3D printed set of samples. Fig. 23.9 shows the tensile test results for 10 geometrically identical samples with particular chosen geometric parameters. Based on the results shown in Fig. 23.9, the average reaction force read for a total of 4 mm displacement is 3.39 N while the minimum and maximum forces are reported as 2.75 N and 3.82 N, respectively, which are lower and higher than the average force value by 19% and 13%. This range of value may be enough for some of the local extremum observed here to disappear, and may be pertinent to the inhomogeneity of the constituent powder used to fabricate the specimens, experimental setup, the feature size of the 3D printer, and the effect of relaxation of the samples when being mounted on the testing machine. However, it is unlikely that the differences between the group with spacing, $p = 8$ mm, in the current paper and the previous paper be a result of this dispersion. One convincing reason for the change in mechanical behavior in the mentioned group of specimens is that the mechanical behavior of the pantographic unit cells is not changing linearly with the change of

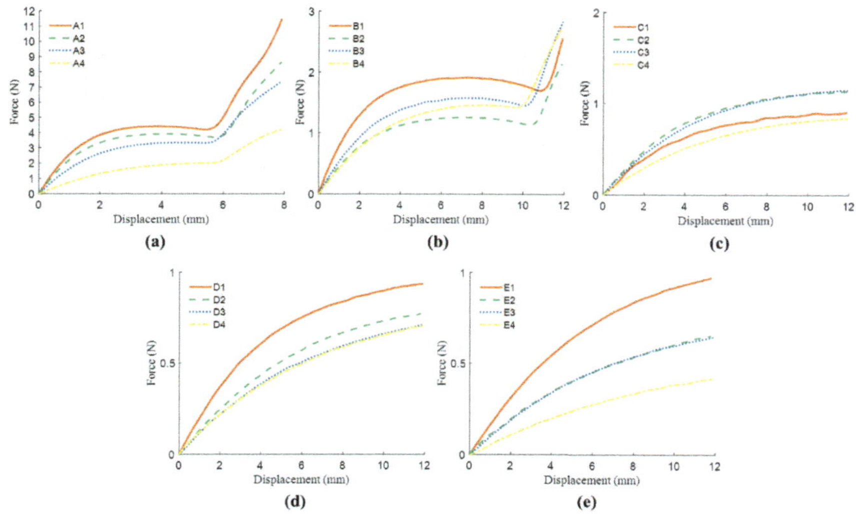

Fig. 23.6 Plots of the data from experimental compression tests for (a) group A (b) group B (c) group C (d) group D and (e) group E.

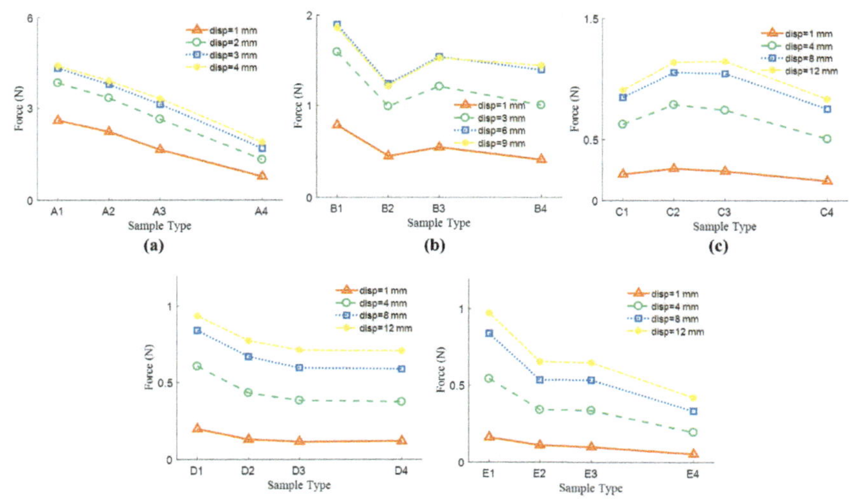

Fig. 23.7 Interpolation plots of the force values with respect to the length of the pivots for compression tests. Each line refers to a specific value of the imposed displacement ("disp" in the legend) for (a) group A (b) group B (c) group C (d) group D and (e) group E.

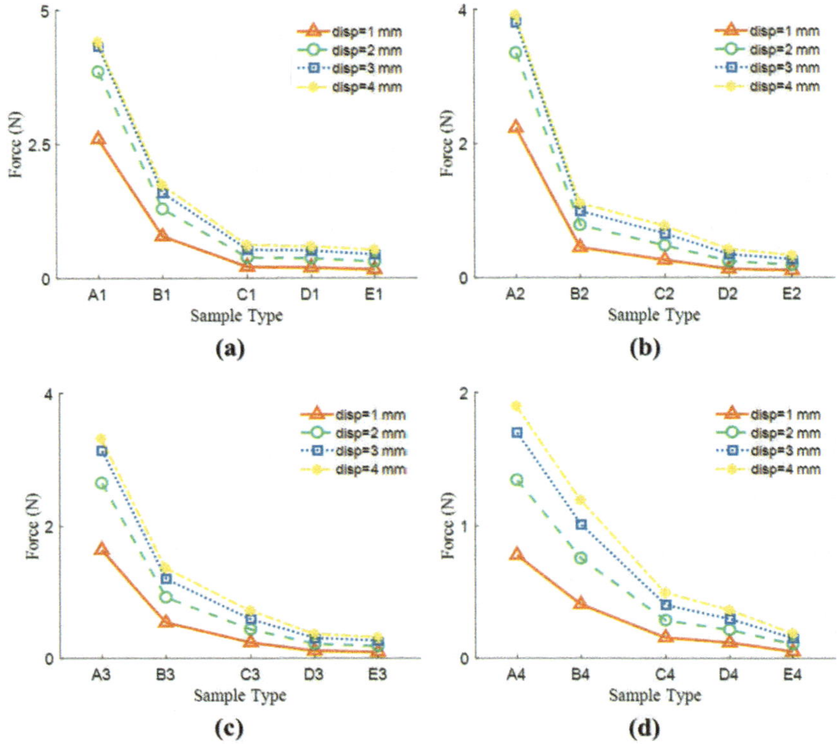

Fig. 23.8 Interpolation plots of the force values with respect to the parameter p for compression tests. Each line refers to a specific value of the imposed displacement ("disp" in the legend). Specimens are grouped for equal values of the parameter h, being the length of the pivot equal to (a) 1 mm (b) 1.5 mm (c) 2 mm and (d) 3 mm.

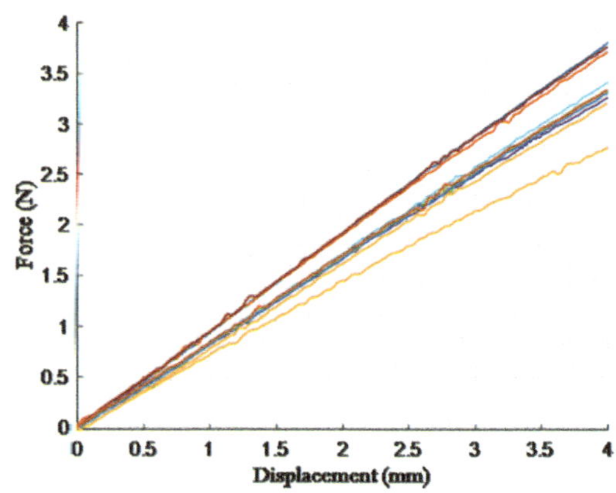

Fig. 23.9 *Tensile test results for 10 geometrically identical samples with particular chosen geometric parameters.*

their constituent powder. In other words, the deformation mechanisms involved in the mechanical behavior of the pantographic unit cells may change in a nonlinear fashion if the constituent material is changed. The results of this analysis could be usefully exploited to better determine and characterize computationally efficient mesoscopic models such as the so-called Hencky-type models Turco et al (2016a, 2017c, 2018). This kind of approach is very promising from the computational point of view since it allows to determine equilibrium configurations of pantographic structure by studying mesoscopic discrete systems directly inspired by the discrete nature of the microstructure without introducing a continuous description (see also Turco, 2018; Baroudi et al, 2019; Giorgio and Del Vescovo, 2019; Giorgio, 2020; Wang et al, 2015). Of course, the outcomes obtained from the analysis presented in this work can be exploited to optimally calibrate this kind of models (Turco, 2019). Starting from the results discussed in this paper, in a future work we aim at analyzing the viscoelastic properties of the pantographic unit cell by imposing a dynamical load on the system (the first experimental observation of the dynamical behavior of pantographic material is discussed in Laudato et al (2018, 2019)). This analysis would provide a clearer understanding on the microscopic dynamical behavior of this kind of system and to yield new indications on how to build effective mesoscopic models for dynamical simulations which encompass also the viscoelastic behavior.

References

Abali BE (2019) Revealing the physical insight of a length-scale parameter in metamaterials by exploiting the variational formulation. Continuum Mechanics and Thermodynamics 31(4):885–894

Abali BE, Wu CC, Müller WH (2016) An energy-based method to determine material constants in nonlinear rheology with applications. Continuum Mechanics and Thermodynamics 28(5):1221–1246

Abali BE, Müller WH, dell'Isola F (2017) Theory and computation of higher gradient elasticity theories based on action principles. Archive of Applied Mechanics 87(9):1495–1510

Abdoul-Anziz H, Seppecher P (2018) Strain gradient and generalized continua obtained by homogenizing frame lattices. Mathematics and mechanics of complex systems 6(3):213–250

Abdoul-Anziz H, Seppecher P, Bellis C (2019) Homogenization of frame lattices leading to second gradient models coupling classical strain and strain-gradient terms. Mathematics and Mechanics of Solids 24(12):3976–3999

Alibert JJ, Seppecher P, dell'Isola F (2003) Truss modular beams with deformation energy depending on higher displacement gradients. Mathematics and Mechanics of Solids 8(1):51–73

Andreaus U, Spagnuolo M, Lekszycki T, Eugster SR (2018) A Ritz approach for the static analysis of planar pantographic structures modeled with nonlinear Euler–Bernoulli beams. Continuum Mechanics and Thermodynamics 30(5):1103–1123

Auffray N, dell'Isola F, Eremeyev VA, Madeo A, Rosi G (2015) Analytical continuum mechanics à la Hamilton–Piola least action principle for second gradient continua and capillary fluids. Mathematics and Mechanics of Solids 20(4):375–417

Barchiesi E, Eugster SR, Placidi L, dell'Isola F (2019a) Pantographic beam: a complete second gradient 1D-continuum in plane. Zeitschrift für angewandte Mathematik und Physik 70(5):135

Barchiesi E, Spagnuolo M, Placidi L (2019b) Mechanical metamaterials: a state of the art. Mathematics and Mechanics of Solids 24(1):212–234

Baroudi D, Giorgio I, Battista A, Turco E, Igumnov LA (2019) Nonlinear dynamics of uniformly loaded elastica: Experimental and numerical evidence of motion around curled stable equilibrium configurations. ZAMM-Zeitschrift für Angewandte Mathematik und Mechanik 99(7):e201800,121

Boutin C, Giorgio I, Placidi L, et al (2017) Linear pantographic sheets: Asymptotic micro-macro models identification. Mathematics and Mechanics of Complex Systems 5(2):127–162

Chróścielewski J, dell'Isola F, Eremeyev VA, Sabik A (2020) On rotational instability within the nonlinear six-parameter shell theory. International Journal of Solids and Structures doi: 10.1016/j.ijsolstr.2020.04.030

Del Vescovo D, Giorgio I (2014) Dynamic problems for metamaterials: review of existing models and ideas for further research. International Journal of Engineering Science 80:153–172

dell'Isola F, Andreaus U, Placidi L (2015a) At the origins and in the vanguard of peridynamics, non-local and higher-gradient continuum mechanics: An underestimated and still topical contribution of Gabrio Piola. Mathematics and Mechanics of Solids 20(8):887–928

dell'Isola F, Lekszycki T, Pawlikowski M, Grygoruk R, Greco L (2015b) Designing a light fabric metamaterial being highly macroscopically tough under directional extension: first experimental evidence. Zeitschrift für angewandte Mathematik und Physik 66(6):3473–3498

dell'Isola F, Giorgio I, Pawlikowski M, Rizzi N (2016) Large deformations of planar extensible beams and pantographic lattices: heuristic homogenization, experimental and numerical examples of equilibrium. Proceedings of the Royal Society A: Mathematical, Physical and Engineering Sciences 472(2185):20150,790

dell'Isola F, Della Corte A, Giorgio I (2017) Higher-gradient continua: The legacy of Piola, Mindlin, Sedov and Toupin and some future research perspectives. Mathematics and Mechanics of Solids 22(4):852–872

dell'Isola F, Seppecher P, Alibert JJ, et al (2019a) Pantographic metamaterials: an example of mathematically driven design and of its technological challenges. Continuum Mechanics and Thermodynamics 31(4):851–884

dell'Isola F, Seppecher P, Spagnuolo M, et al (2019b) Advances in pantographic structures: design, manufacturing, models, experiments and image analyses. Continuum Mechanics and Thermodynamics pp 1–52

dell'Isola F, Turco E, Misra A, Vangelatos Z, Grigoropoulos C, Melissinaki V, Farsari M (2019c) Force–displacement relationship in micro-metric pantographs: Experiments and numerical simulations. Comptes Rendus Mécanique 347(5):397–405

Di Cosmo F, Laudato M, Spagnuolo M (2018) Acoustic metamaterials based on local resonances: homogenization, optimization and applications. In: Generalized Models and Non-classical Approaches in Complex Materials 1, Springer, pp 247–274

Eremeyev VA, dell'Isola F, Boutin C, Steigmann D (2018) Linear pantographic sheets: existence and uniqueness of weak solutions. Journal of Elasticity 132(2):175–196

Eremeyev VA, Alzahrani FS, Cazzani A, dell'Isola F, Hayat T, Turco E, Konopińska-Zmysłowska V (2019) On existence and uniqueness of weak solutions for linear pantographic beam lattices models. Continuum Mechanics and Thermodynamics pp 1–19

Giorgio I (2020) A discrete formulation of Kirchhoff rods in large-motion dynamics. Mathematics and Mechanics of Solids 25(5):1081–1100

Giorgio I, Del Vescovo D (2019) Energy-based trajectory tracking and vibration control for multilink highly flexible manipulators. Mathematics and Mechanics of Complex Systems 7(2):159–174

Giorgio I, Rizzi N, Turco E (2017) Continuum modelling of pantographic sheets for out-of-plane bifurcation and vibrational analysis. Proceedings of the Royal Society A: Mathematical, Physical and Engineering Sciences 473(2207):20170,636

Giorgio I, dell'Isola F, Steigmann D (2018) Axisymmetric deformations of a 2nd grade elastic cylinder. Mechanics Research Communications 94:45–48

Golaszewski M, Grygoruk R, Giorgio I, Laudato M, Di Cosmo F (2019) Metamaterials with relative displacements in their microstructure: technological challenges in 3D printing, experiments and numerical predictions. Continuum Mechanics and Thermodynamics 31(4):1015–1034

Laudato M, Manzari L, Barchiesi E, Di Cosmo F, Göransson P (2018) First experimental observation of the dynamical behavior of a pantographic metamaterial. Mechanics Research Communications 94:125–127

Laudato M, Di Cosmo F, Drobnicki R, Göransson P (2019) Dynamical vector fields on pantographic sheet: Experimental observations. In: New Achievements in Continuum Mechanics and Thermodynamics, Springer, pp 257–269

Misra A, Nejadsadeghi N (2019) Longitudinal and transverse elastic waves in 1D granular materials modeled as micromorphic continua. Wave Motion 90:175–195

Misra A, Poorsolhjouy P (2016) Granular micromechanics based micromorphic model predicts frequency band gaps. Continuum Mechanics and Thermodynamics 28(1-2):215–234

Misra A, Lekszycki T, Giorgio I, Ganzosch G, Müller WH, dell'Isola F (2018) Pantographic metamaterials show atypical Poynting effect reversal. Mechanics Research Communications 89:6–10

Nejadsadeghi N, Misra A (2019a) Axially moving materials with granular microstructure. International Journal of Mechanical Sciences 161:105,042

Nejadsadeghi N, Misra A (2019b) Extended granular micromechanics approach: a micromorphic theory of degree n. Mathematics and Mechanics of Solids 25(2):407–429

Nejadsadeghi N, De Angelo M, Drobnicki R, Lekszycki T, dell'Isola F, Misra A (2019a) Parametric experimentation on pantographic unit cells reveals local extremum configuration. Experimental Mechanics 59:927–939

Nejadsadeghi N, Placidi L, Romeo M, Misra A (2019b) Frequency band gaps in dielectric granular metamaterials modulated by electric field. Mechanics Research Communications 95:96–103

Placidi L, Andreaus U, Della Corte A, Lekszycki T (2015) Gedanken experiments for the determination of two-dimensional linear second gradient elasticity coefficients. Zeitschrift für angewandte Mathematik und Physik 66(6):3699–3725

Placidi L, Barchiesi E, Misra A (2018) A strain gradient variational approach to damage: a comparison with damage gradient models and numerical results. Mathematics and Mechanics of Complex Systems 6(2):77–100

Placidi L, dell'Isola F, Barchiesi E (2020) Heuristic homogenization of Euler and pantographic beams. In: Mechanics of Fibrous Materials and Applications, Springer, pp 123–155

Poorsolhjouy P, Misra A (2019) Granular micromechanics based continuum model for grain rotations and grain rotation waves. Journal of the Mechanics and Physics of Solids 129:244–260

Scerrato D, Giorgio I (2019) Equilibrium of two-dimensional cycloidal pantographic metamaterials in three-dimensional deformations. Symmetry 11(12):1523

Seppecher P, Alibert JJ, dell'Isola F (2011) Linear elastic trusses leading to continua with exotic mechanical interactions. In: Journal of Physics: Conference Series, IOP Publishing, vol 319, p 012018

Spagnuolo M, Barcz K, Pfaff A, dell'Isola F, Franciosi P (2017) Qualitative pivot damage analysis in aluminum printed pantographic sheets: numerics and experiments. Mechanics Research Communications 83:47–52

Steigmann DJ, dell'Isola F (2015) Mechanical response of fabric sheets to three-dimensional bending, twisting, and stretching. Acta Mechanica Sinica 31(3):373–382

Turco E (2018) Discrete is it enough? the revival of piola–hencky keynotes to analyze three-dimensional elastica. Continuum Mechanics and Thermodynamics 30(5):1039–1057

Turco E (2019) How the properties of pantographic elementary lattices determine the properties of pantographic metamaterials. In: New Achievements in Continuum Mechanics and Thermodynamics, Springer, pp 489–506

Turco E, dell'Isola F, Cazzani A, Rizzi NL (2016a) Hencky-type discrete model for pantographic structures: numerical comparison with second gradient continuum models. Zeitschrift für angewandte Mathematik und Physik 67(4):85

Turco E, dell'Isola F, Rizzi NL, Grygoruk R, Müller WH, Liebold C (2016b) Fiber rupture in sheared planar pantographic sheets: numerical and experimental evidence. Mechanics Research Communications 76:86–90

Turco E, Giorgio I, Misra A, dell'Isola F (2017a) King post truss as a motif for internal structure of (meta) material with controlled elastic properties. Royal Society open science 4(10):171,153

Turco E, Golaszewski M, Giorgio I, D'Annibale F (2017b) Pantographic lattices with non-orthogonal fibres: experiments and their numerical simulations. Composites Part B: Engineering 118:1–14

Turco E, Golaszewski M, Giorgio I, Placidi L (2017c) Can a Hencky-type model predict the mechanical behaviour of pantographic lattices? In: Mathematical Modelling in Solid Mechanics, Springer, pp 285–311

Turco E, Misra A, Pawlikowski M, dell'Isola F, Hild F (2018) Enhanced Piola–Hencky discrete models for pantographic sheets with pivots without deformation energy: numerics and experiments. International Journal of Solids and Structures 147:94–109

Wang CM, Zhang H, Gao RP, Duan WH, Challamel N (2015) Hencky bar-chain model for buckling and vibration of beams with elastic end restraints. International Journal of Structural Stability and Dynamics 15(07):1540,007

Yang H, Ganzosch G, Giorgio I, Abali BE (2018) Material characterization and computations of a polymeric metamaterial with a pantographic substructure. Zeitschrift für angewandte Mathematik und Physik 69(4):105

Chapter 24
DIC Measurements on Single Struts of Ni/PU Hybrid Foams—Damage Behaviour During Three-Point Bending

Martin Reis, Stefan Diebels, and Anne Jung

Abstract Hybrid foams are an evolution of classic metal foams. They consist of an open-cell foam template coated with a second material. Thereby, the coating material dominates the material behaviour of the hybrid foams. Ni/PU hybrid foams are a suitable example of hybrid foams. An inexpensive polyurethane (PU) foam is coated with nanocrystalline nickel (Ni). The coating strongly increases the stiffness of the material while the price is lower than for an aluminium alloy open-cell foam. The relatively low costs and the high stiffness of the structure are promising to be actually used in real-world applications. But first, a mechanical characterisation of the macroscopic hybrid structure is needed. The macroscopic deformation results in pore deformation on the mesoscale. Pore deformation consists of bending and buckling of struts, therefore, the damage behaviour of single struts during three-point bending needs more investigation. In this contribution, these experiments are presented and the influence of the individual geometry of each single strut on the measured force is shown. For a better comparability of the experiments, a simplification using the Euler–Bernoulli beam theory and the assumption of a constant triangular strut cross section is applied. The resulting stress-strain-curves show matching slopes in the linear elastic part of the curves, which proves the used strategy.

Keywords: Experimental mechanics · DIC evaluation · Open-cell foams · Ni/PU hybrid foams · Three-point bending · Individual struts

M. Reis, S. Diebels, A. Jung
Applied Mechanics, Saarland University, Campus A4.2, Saarbrücken, Germany
e-mail: martin.reis@uni-saarland.de, s.diebels@mx.uni-saarland.de, anne.jung@mx.uni-saarland.de

B. E. Abali and I. Giorgio (eds.), *Developments and Novel Approaches in Biomechanics and Metamaterials*, Advanced Structured Materials 132,
https://doi.org/10.1007/978-3-030-50464-9_24

24.1 Introduction

Open-cell foams can find their use in a wide field of applications. They can be used as filters or crash absorbers. For hybrid foams, the field of application is determined by the coating material. A Ni/PU hybrid foam is a polyurethane (PU) foam coated with nanocrystalline nickel (Ni). The inexpensive PU foam is the template material with almost no influence on the material behaviour of the hybrid foam. Due to the Ni coating, the stiffness of the material strongly increases, without a rise in weight in the same rate. The low weight and high stiffness allows the use of these structures in crash absorbers for high performance applications. For a mechanical valid design and the use in real applications, a full knowledge of the material behaviour is needed. Macroscopic foams are hierarchically structured, consisting of pores on the mesoscale and struts on the microscale. For a detailed understanding of the characteristic material behaviour mechanical tests in all scales are required. Thereby, the main deformation modes for individual struts are bending and buckling. Nevertheless, in literature only one three-point bending experiment on a specimen extracted from a cell wall of a closed-cell foam can be found in Fíla et al (2014). Due to the complex structure of the struts and the scattering in grain orientation, the load case has a strong influence on the material behaviour. Hence, bending experiments are needed to allow material characterisation. For this reason, the contribution presents three-point bending tests on single struts. The small strut dimensions and rough surface of the samples only allow digital image correlation (DIC) for strain evaluation. DIC is a common tool to measure inhomogeneous deformation fields on macroscopic samples as used by Jung and Diebels (2017) and dell'Isola et al (2019). The complex geometry and small size of the struts are thereby extremely challenging for the evaluation and require a high precession in the measurement system (Reis et al, 2019). Although this is very complex, successful full-field strain measurements on individual struts are presented in this work. For the evaluation a simplification to linear bending theory shows good potential for further investigations.

24.2 Experiments and Discussion

The electrochemical coating of an electrically non conductive material such as PU is a two step process. The PU foam has a pore size of 20 pores per inch (ppi) and is purchased from Jumpinoo Schaumstoff Direkt, Enger, Germany. In the first step of the coating process a layer of conductive material is needed. For the presented foams graphite lacquer (Graphit 33, CRC INDUSTRIES Deutschland GmbH, Iffezheim, Germany) is applied in a dip coating process. All surplus lacquer was removed with compressed air. The method produces a fine layer of graphite all over the PU foam sample. In the second step, the conductive foam is connected to a power supply and immersed in an electrolyte. The current flow between the foam and a nickel cathode initiates an electrodeposition process. For a detailed description of the process see Jung et al (2010), Jung and Diebels (2016). A custom-build micro mechanical testing

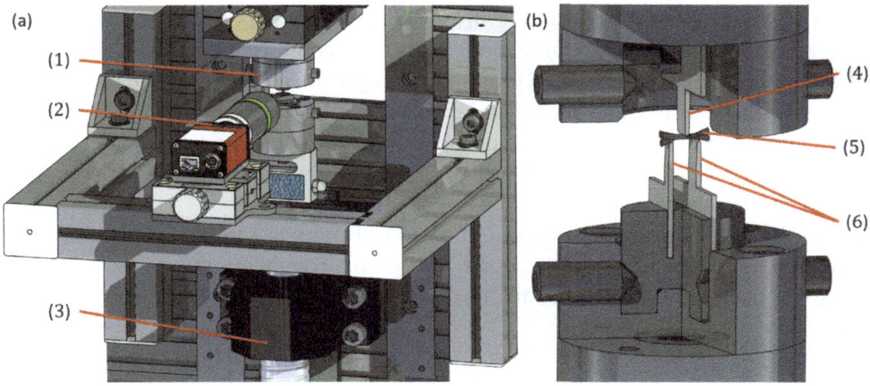

Fig. 24.1 Experimental setup: (a) total view with (1) sample holder, (2) camera with telecentric lens, (3) precision screw drive; (b) detail view of three-point bending setup with (4) bending mandrel, (5) idealised strut, (6) lower bearings

device, constructed for the special requirements of pore and strut testing, is used. The translational movement is realised by a stepper motor (mDrive 23, Schneider Electric GmbH, Rating, Germany) in combination with a high precision screw drive. The stroke accuracy of the device is 0.1 μm with a reproducibility of 0.02 μm. The setup is controlled by a custom-build software based on LabVIEW™ (National Instruments Germany GmbH, Munich, Germany). The software controls the motion, records the forces and captures the images during the experiment. Using different bearings, the device allows to perform tensile and compression tests on pores and single struts as well as three-point bending tests on single struts (Jung et al, 2015). For the three-point bending tests two blades are fixed on the lower moving part of the setup (see Fig. 24.1). The blades are fixed with a distance of 2.2 mm from mid-to-mid. The specimen is placed on the blades without any fixation. A third blade is used as bending mandrel. The blades have a bottom width of 1.05 mm and a double facet top with an angle of 108 degrees.

The commercial software Istra4D™ (Dantec Dynamics, Skovlunde, Denmark) is used for the DIC evaluation. To achieve high accuracy, the ideal parameters (lighting, subset size, etc.) were found using a three-step routine by Reis et al (2019). Before each experiment 20 images without any movement of camera or specimen are captured and the displacement is measured. The resulting displacement can be taken as quality indicator for this experiment. A 9MPx CCD camera (Manta 917B, Allied Vision Technologies GmbH, Stadtroda, Germany) in combination with a telecentric lens (DCTM16.6-110, VICO Technology CO., Guangdong, China) having a pixel size of $3.69 \times 3.69\ \mu m^2$ takes the images. Due to small dimensions of the strut ($0.4 \times 2\ mm^2$) small pixels are needed to get a valid number of pixels over the strut's surface. Resulting from that, a full-field strain and displacement measurement can be achieved.

For the evaluation of the bending stress, the real geometry of the strut is needed. Therefore, a photogrammetric method is used. The method requires a special pattern below the specimens to calculate the rotation angle in each image. The shadows on the specimen's surface and the edges of the specimen combined with the rotation information from the pattern enable the geometric reconstruction of the sample. Thereby, the resolution of the method is comparable to the resolution of computed tomography models.

The analysis of photogrammetric geometrical models shows a triangular cross section area of all struts (see Fig. 24.2). Hence, in a rough approximation struts can be seen as triangular shaped beams. Using the Euler–Bernoulli beam theory allows

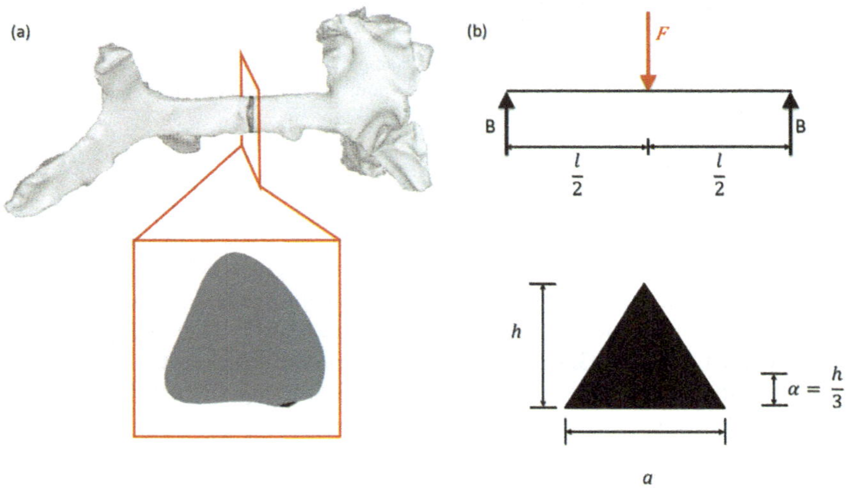

Fig. 24.2 (a) real strut geometry (top) with real cross section(bottom); (b) idealised beam with force of bending mandrel F and lower bearings B (top), simplified triangular cross section with height h, width a and α as horizontal distance to the neutral phase (bottom)

to compare the first bending tests performed on individual struts. The theory for three-point bending starts with the bending moment M_b by

$$M_b = \frac{F\,l}{4} \quad , \tag{24.1}$$

using the measured force F and the distance between the lower bearings l. In addition, the section modulus W_Δ is defined as

$$W_\Delta = \frac{\mathrm{I}}{\alpha} \quad , \tag{24.2}$$

with the area moment of inertia I and the distance between the neutral phase and the searched section modulus α. The bending stress σ_b can be calculated as quotient of the bending moment (Eq. 24.1) and the section modulus (Eq. 24.2)

$$\sigma_b = \frac{M_b}{W_\Delta} \quad . \tag{24.3}$$

With the assumption that the height h of the triangle is similar to the width a, the bending stress σ_b results in

$$\sigma_b = \frac{18\,F\,l}{A\,\sqrt{A}} \quad , \tag{24.4}$$

with $A = h\,a\,/\,2$ as cross sectional area of the individual strut. This assumption is only valid during the linear elastic deformation of the strut due to the simplification to Euler–Bernoulli beam theory, allowing the estimation of homogeneous material parameters from experimental data. Furthermore, it can only be used for a rough estimation and comparison of the results of the different experiments of individual struts. Besides the stress, the strain is needed to generate a stress-strain-curve. For this purpose a DIC measurement is used. The strain is exported as gauge line strain in the area with the largest positive strain close to the lower bearings. Different deformation steps of an exemplary strut can be found in Fig. 24.3.

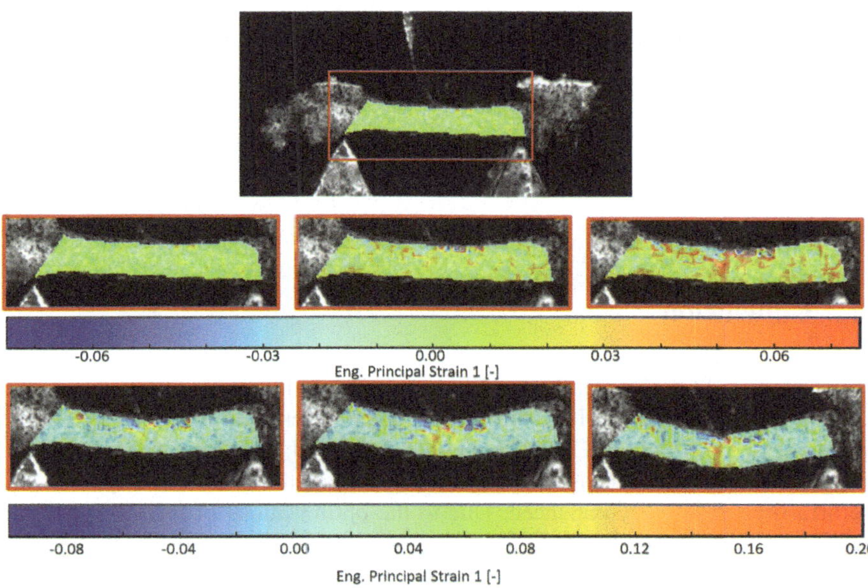

Fig. 24.3 Full-field strain measurement on a single strut in six steps of the deformation during three-point bending test: starting steps with small strains from -0.07 to 0.07 strain (top), steps with bigger deformation with strains from -0.1 to 0.2 strain (bottom)

During the deformation, the main strain appears near the bending mandrel with only small deformations in the remaining strut. This is affected by the brittle material characteristic of the nanocrystalline Ni coating. A similar strut behaviour can be found during the deformation of a macroscopic foam sample and seems to be realistic. The experimental results of four struts are shown in Fig. 24.4. Regarding the force-strain-curves, a huge scattering of the slope in the linear elastic region is recognisable. The struts vary in their geometry which makes the comparison of slope and maximums in force-strain-curves unreliable. The stress-strain-curves are similar for all experiments. The scattering of the stiffness's, which are equal to the slope in the linear elastic area of the curves, is smaller as for the force-strain-curves. The stiffness' can be found in Table 24.1, with a mean of 830.75 GPa $\pm$ 193.27 GPa, the scattering is at around 23%. The use of an elastic theory, only allows the investigation of this elastic part of the curves. The improvement and the comparability of the results confirm the use of the simplified geometrical models. Nevertheless, in future, the evaluation should be changed to the real geometry of the struts taking the correct area moment of inertia into account.

Table 24.1 Stiffness of specimen 1-4 from linear part of stress-strain diagrams in Fig. 24.4 (bottom)

specimen	stiffness [GPa]
1	702
2	1089
3	867
4	665

24.3 Conclusion

The custom-build device can be used for large range of experiments on pores and struts of open-cell foams. Here, three-point bending tests on single struts and the post processed full-field DIC measurement is described. To reach an accuracy high enough to perform DIC measurements on single struts the setup was improved and the measurement quality enhanced. Due to the optimisations of the setup and the measurement quality full-field strain measurements on individual struts during a three-point bending test were possible for the first time. The individual geometry of each strut is a particular challenge during the stress calculation. A first simplification using Euler–Bernoulli beam theory and the assumption of a constant triangular cross section over the whole strut show an agreement for the stiffness of the different struts. Nevertheless, the evaluation should be changed to the real geometry of the strut and the real area moment of inertia.

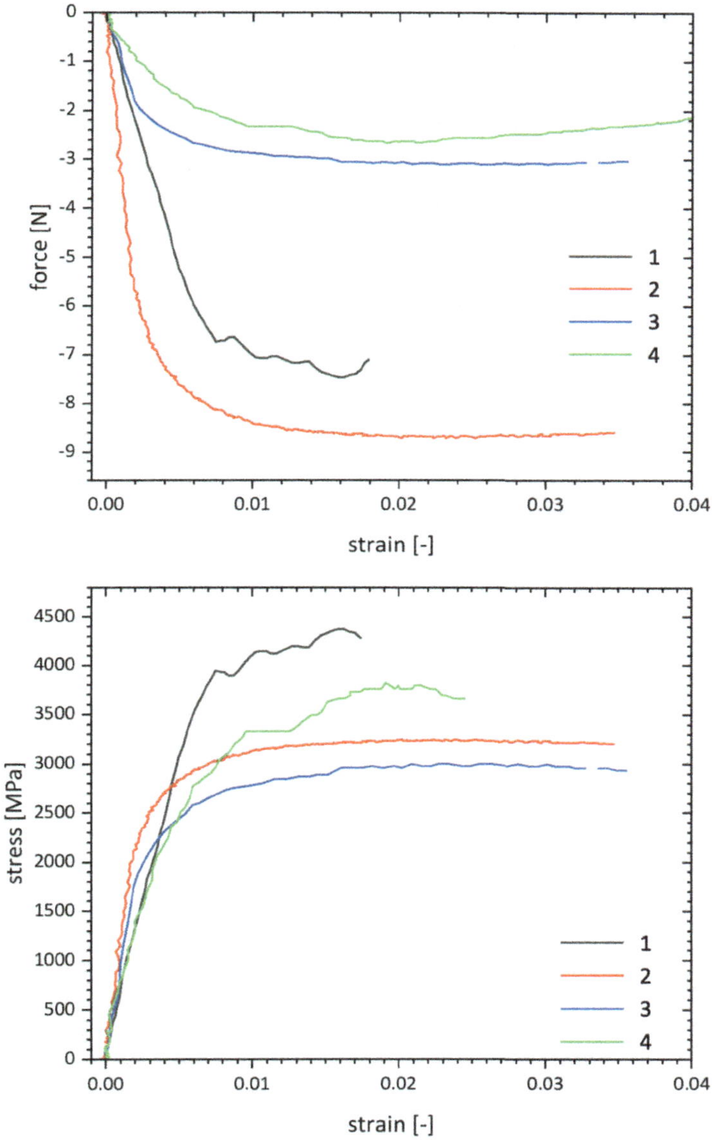

Fig. 24.4 Force-strain-diagrams of four individual struts recorded during three-point bending (top), stress-strain-diagrams of four individual struts recorded during three-point bending (bottom)

References

dell'Isola F, Seppecher P, Spagnuolo M, Barchiesi E, Hild F, Lekszycki T, Giorgio I, Placidi L, Andreaus U, Cuomo M, et al (2019) Advances in pantographic structures: design, manufacturing, models, experiments and image analyses. Continuum Mechanics and Thermodynamics pp 1–52

Fíla T, Kytýř D, Koudelka P, Doktor T, Zlámal P, Jiroušek O (2014) Micro-Mechanical Testing of Metal Foam Cell Walls Using Miniature Three-Point Bending. In: Key Engineering Materials, Trans Tech Publ, vol 586, pp 120–124

Jung A, Diebels S (2016) Synthesis and mechanical properties of novel Ni/PU hybrid foams: a new economic composite material for energy absorbers. Advanced Engineering Materials 18(4):532–541

Jung A, Diebels S (2017) Microstructural characterisation and experimental determination of a multiaxial yield surface for open-cell aluminium foams. Materials & Design 131:252–264

Jung A, Natter H, Hempelmann R, Diebels S, Koblischka MR, Hartmann U, Lach E (2010) Electrodeposition of nanocrystalline metals on open cell metal foams: improved mechanical properties. ECS Transactions 25(41):165–172

Jung A, Wocker M, Chen Z, Seibert H (2015) Microtensile testing of open-cell metal foams—experimental setup, micromechanical properties. Materials & Design 88:1021–1030

Reis M, Adorna M, Jiroušek O, Jung A (2019) Improving DIC accuracy in experimental setups. Advanced Engineering Materials p 1900092

Chapter 25
Apparent Bending and Tensile Stiffness of Lattice Beams with Triangular and Diamond Structure

Yury Solyaev, Sergey Lurie, and Anastasia Ustenko

Abstract In this study we examine apparent elastic properties of slender lattice structures consist of triangular and diamond unit cells. We reveal that mechanical response of such structures can be described by using non-classical Euler–Bernoulli beam theories developed in the framework of strain gradient elasticity. At first, we evaluate the size-dependent behavior of long and narrow lattice sheets under pure bending, cantilever bending and uniaxial tension tests by using finite element modeling and classical Cauchy continuum model. Apparent elastic properties are introduced based on the direct approach and estimated for the lattices of different length and thickness. Secondly, from the comparison of obtained numerical results and corresponding analytical solutions for the gradient Euler–Bernoulli beams we get an assessment for the standard and gradient materials constant of the last ones. As result, we show that the non-classical size-dependent elastic response of considered structures with diamond unit cells can be described by using so-called gradient beam theories with "uniaxial stress state."

Keywords: Lattice beams · Second gradient elasticity · Gradient beam theories · Apparent elastic properties · Size effects

25.1 Introduction

Development of new structural metamaterials with non-trivial mechanical behavior is a challenging task related to a lot of important industrial applications, such as damage resistant and lightweight structures in aerospace and civil engineering (Barchiesi

Y. Solyaev, S. Lurie, A. Ustenko
Institute of Applied Mechanics of Russian Academy of Sciences, Leningradskiy ave., 7, Moscow, Russia
Moscow Aviation Institute, Moscow, Russia
e-mail: yos@iam.ras.ru, salurie@mail.ru,audlord@ya.ru

B. E. Abali and I. Giorgio (eds.), *Developments and Novel Approaches in Biomechanics and Metamaterials*, Advanced Structured Materials 132,
https://doi.org/10.1007/978-3-030-50464-9_25

et al, 2019), intelligent materials for robotics (Mirzaali et al, 2018), damping and sound absorption materials in acoustics (Del Vescovo and Giorgio, 2014) etc.

Considering special metamaterials with incorporated long-range interactions provided by the internal structural crosslinks, one have to apply a non-classical models different from the Cauchy continuum which does not allow to study, e.g., size effects or chirality effects, and can not provide a precise description of such structures in all range of material parameters and configurations of their microstructure. Examples of such structures have been widely studied recently. Stochastic fiber networks is one of non-classical materials which exist in nature and exhibit strong size effects (Shahsavari and Picu, 2013). Mechanical behavior of such materials was described by using second gradient elasticity and couple stress theories (Berkache et al, 2019, 2017). Another example are auxetic foams and granular systems, which can be described by using micropolar continuum (Rueger and Lakes, 2016; De Angelo et al, 2020). Three-dimensional printing was used recently to obtain a non-isotropic micropolar structures made of small metallic unit cells with chirality effects (Frenzel et al, 2017). Lattice metamaterials with micro-dilatation effects were studied in Lurie et al (2018); Solyaev et al (2019b). Pantographic metamaterials is an important class of artificial structures which exhibit non-classical behavior, such as large elastic deformations and high damage resistance, and can be described based on the strain gradient elasticity theory (Della Corte et al, 2016; Eremeyev et al, 2018; Yang et al, 2018; dell'Isola et al, 2019a,b; Boutin et al, 2017).

In the present work we determine the apparent elastic properties of long and narrow sheets of lattice materials consisting of triangular and diamond unit cells. From a macroscopic point of view, such structures can be modeled by using beam-type theories, e.g., the most simple Euler–Bernoulli beam model can be involved to predict deformations of such structures under the actions of concentrated and distributed loads and moments. Note, that similar problems have been considered recently in a number of works. Different models of the generalized continuum mechanics have been used for the refined analysis of sandwich structures, lattice beams, truss beams etc (see Abdoul-Anziz and Seppecher, 2018; Romanoff and Reddy, 2014; Khakalo et al, 2018; Karttunen et al, 2019). Modified couple stress, micropolar, strain gradient theories have been used in these works.

In the present study we use the two variants of the strain gradient Euler–Bernoulli beam theory. The first one is the so-called model with "uniaxial stress state" of the gradient beam, which was proposed in Papargyri-Beskou et al (2003). Another formulation with “general stress state” of gradient beams was developed later in Lazopoulos and Lazopoulos (2010). Recently, the correctness of these theories and their comparison was widely studied in literature (Lurie and Solyaev, 2019, 2018; Niiranen et al, 2019; Jafari et al, 2016). Main differences between these theories appear in bending problems, where the size-dependent bending stiffness arise in the formulation of the theory with “general stress state.” However, in Lurie and Solyaev (2018), it was shown that this effect does not arise in the semi-inverse three-dimensional solutions for the pure bending problems. Refined variational formulation for the gradient beam theory with “uniaxial stress state” has been proposed later in Lurie and Solyaev (2019). In the present study we show that the lattices with diamond unit cells behave in bending in accordance with gradient beam theory proposed by Papargyri-Beskou et al (2003), in opposite to the triangular unit cells that corresponds to the gradient beams proposed by Lazopoulos and Lazopoulos (2010) (this was also shown earlier in Khakalo et al (2018)). Additionally, we consider the uniaxial tension problem to show that the length-related size effect, which arises under conditions of non-uniform uniaxial deformations, can be described by the strain gradient beam theory.

25.2 Finite Element Models of the Lattice Beams

We consider long and narrow sheets of lattice materials made of repeated unit cells of diamond and triangular structure (Fig. 25.1). Mechanical behavior of considered lattices was studied at first by using finite element (FE) simulations realized in Femap/Nastran. In this numerical analysis we used the classical Euler–Bernoulli model, i.e. the one-dimensional beam-type model derived based on the first order Cauchy continuum. Thus, in numerical simulations we consider the lattice geometry and take into account the behavior of each individual structural element. In the following estimations we consider the structures made of unit cells, which size is $d =$ 40 mm. Beams inside the unit cells are made of polyamide PA6 (Young's modulus 2.5 GPa, Poisson's ratio 0.35) and have circular cross section, which diameter is 1 mm. Beams are connected to each other with a clamp constraints. Angle between beams in the diamond unit cells is chosen to be 90 deg. and they are rotated 45 deg., such that the realized structure has orthotropic symmetry with principal directions along and perpendicular to the beam axis. The difference between diamond and triangular unit cells is the presence of horizontal cross links in the last ones (see Fig. 25.1b). As result, boundaries of the sheets with diamond unit cells are formed by the vertices, which are the intersections of the nearest beams, while in the sheets with triangular unit cells upper and lower boundaries are formed with the added horizontal beams. This difference in the boundaries structure is the main reason that results in the different mechanical response and size effects in these structures that will be shown below.

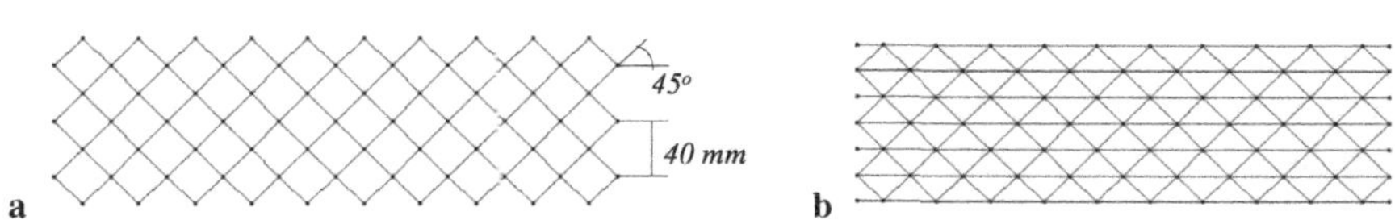

Fig. 25.1 Lattice structures with diamond (**a**) and triangular (**b**) unit cells

Apparent elastic properties of the lattice beams is found by using direct approach. From the FE simulations we determine the displacements of the specimen under prescribed loading and use the classical mechanics of materials analytical solutions to evaluate its apparent elastic properties. In such approach we compare only the tip displacements of the considered lattice beam and corresponding classical beam with the same dimensions and under the same magnitude of loading. From this comparison we find the value of the Young's modulus of classical beam, which will have the same displacements as the lattice beam. Generally say, the determined apparent characteristic depends on the boundary conditions, on the type of loading and on the dimensions of the specimen. That's why the found value of the Young's modulus is called "apparent" for the lattice beam as it is not only a material characteristic, but also a structural. Note, that such approach is convenient and can be used for the analysis of experimental data with non-classical materials (see, e.g.Rueger and Lakes (2016)) and it is also used in the mechanics of composites, e.g. when the apparent interlaminar shear strength is determined from the short beam shear test according to ASTM D2344. Thus, under the pure bending, cantilever bending and uniaxial tension we can find the corresponding apparent elastic modulus of the lattice beam by using following well known solutions:

$$E_{PB} = \frac{ML^2}{2Iu_3(L)}, \qquad E_{CB} = \frac{PL^3}{3Iu_3(L)}, \qquad E_T = \frac{FL}{Au_1(L)}, \tag{25.1}$$

where M is the total bending moment applied at the beam end in the pure bending test (Fig. 25.2a); P is the total bending force applied at the cantilever beam end (Fig. 25.2b); F is the total tensile force applied at the beam end under tension (Fig. 25.2c); L is the length of the lattice beam, $I = \frac{bh^3}{12}$ and $A = bh$ is the apparent moment of inertia and cross section area of the lattice beam, which thickness is h and width is b; u_1 and u_3 are the displacements of the beam end, which are found by using numerical FE simulations in Nastran.

Boundary conditions which were used in the FE simulations are shown in Fig. 25.2a-c. Illustrations are given for the beams with diamond unit cells and for the triangular lattices they are the same. Presented conditions are standard and usually used in the FE modeling of similar problems. However, in Fig. 25.2d we show the non-classical type of constraints, which provide zero displacements of the first several segments of the lattice beams. Length of this constrained area of the beam is denoted as x_0. Such type of boundary conditions results in a non-uniform strain state of the lattice beam even under uniaxial tension. In this case apparent elastic modulus determined by using Eq. (25.1) becomes size-dependent. However, as it is shown below, this type of constraint (Fig. 25.2d) is related with additional high-order boundary conditions, which arise in the strain gradient theories for the normal gradient of displacements.

25.3 Strain Gradient Beam Theories

Let's consider a straight beam of length L with rectangular cross section $h \times b$ made of centrosymmetric linear elastic second gradient material. Equilibrium equations

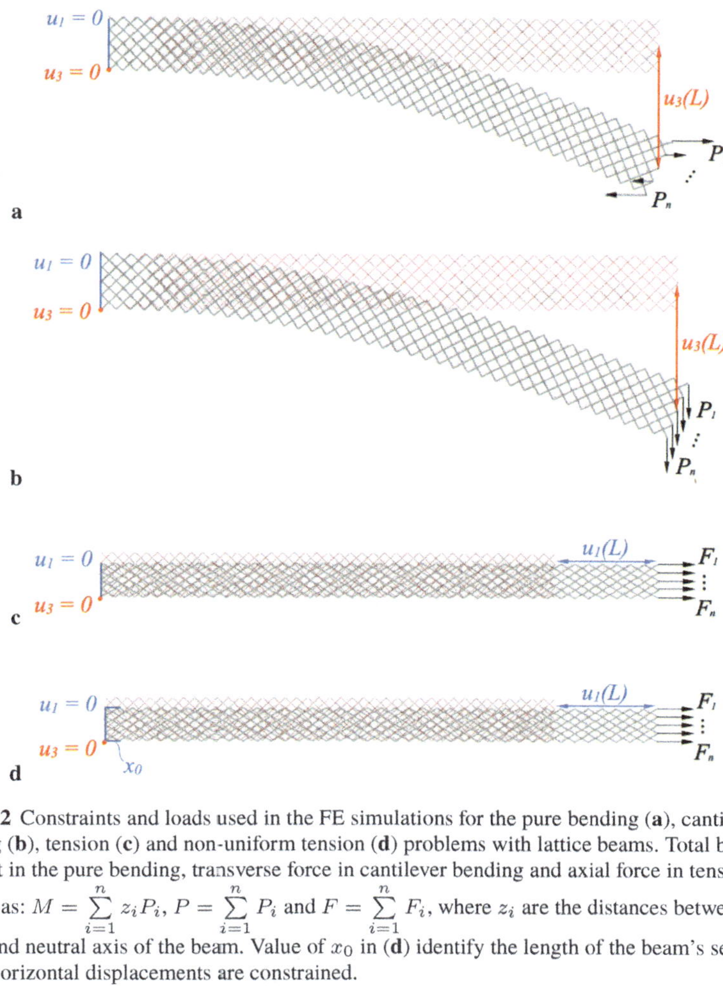

Fig. 25.2 Constraints and loads used in the FE simulations for the pure bending (**a**), cantilever bending (**b**), tension (**c**) and non-uniform tension (**d**) problems with lattice beams. Total bending moment in the pure bending, transverse force in cantilever bending and axial force in tension are defined as: $M = \sum_{i=1}^{n} z_i P_i$, $P = \sum_{i=1}^{n} P_i$ and $F = \sum_{i=1}^{n} F_i$, where z_i are the distances between loaded nodes and neutral axis of the beam. Value of x_0 in (**d**) identify the length of the beam's segments, which horizontal displacements are constrained.

of gradient model can be presented in the following form (Papargyri-Beskou et al (2003); Lurie and Solyaev (2019); Alibert et al (2003)):

$$\begin{aligned} &M'' - M_h''' + q = 0, \qquad x \in (0, L) \\ &N' - N_h'' + f = 0, \qquad x \in (0, L) \end{aligned} \tag{25.2}$$

where $M = -EIw''$ is bending moment, $M_h = -\ell^2 EIw'''$ is high order bending moment, $N = EAu'$ is the resultant axial force, and $N_h = l^2 EAu''$ is the high order axial force, respectively; E is Young's modulus of the beam; l is the length scale parameter; $u = u_1(x)$ and $w = u_3(x)$ are the axial displacements and deflections of the beam, respectively; q and f are the distributed transverse and axial loading.

Note, that above in (25.2) we use the definition of bending moment M according to the so-called model with "uniaxial stress state" developed in Papargyri-Beskou et al (2003), which correctness was validated recently in Lurie and Solyaev (2018, 2019). In the following we will denote this model as USS. Another formulation of Euler–Bernoulli gradient beam model with "general stress state" was proposed in Lazopoulos and Lazopoulos (2010), which formulation results in the another definition of bending moment $M = -E(I + l^2A)w''$. In the following we will call this model GSS. Definitions of other resultants are the same in both models, as well as the form of boundary conditions in terms of resultants:

$$\begin{array}{l} \underline{x = 0,\, L:} \\ N - N_h' = N_0 \quad or \quad u = u_0, \\ N_h = N_{h0} \quad or \quad u' = \varepsilon_0, \\ M - M_h' = M_0 \quad or \quad w' = \theta_0 \\ M' - M_h'' = Q_0 \quad or \quad w = w_0 \\ M_h = M_{h0} \quad or \quad w'' = \kappa_0 \end{array} \tag{25.3}$$

where $N_0, u_0, N_{h0}, \varepsilon_0, M_0, \theta_0, Q_0, w_0, M_{h0}$ and κ_0 are the axial load, axial displacement, axial high order force, axial strain, bending moment, rotation angle, transverse force, deflection, high order moment and curvature, which can be prescribed on the beam ends in the strain gradient theory.

Uniaxial tension and bending of lattices, considered in previous section, can be described by using one-dimensional beam type model (25.2)-(25.3). Solution for the pure bending problem, obviously, has the form:

$$\begin{cases} x \in (0, L): & M'' - M_h''' = 0. \\ x = 0, L: & M - M_h' = M_0, \\ & M' - M_h'' = 0, \\ & M_h = 0 \end{cases} \implies \quad M \equiv M_0, \quad M_h \equiv 0 \tag{25.4}$$

Thus, for the USS model we obtain that the apparent elastic modulus of the beam under pure bending loading should be constant and equals to the Young's modulus that can be determined from the uniaxial tension $E_{PB} = E_T$. In the GSS model we have a size dependent behavior of the beam even under the pure bending, such that $E_{PB} = E_T(1 + l^2A/I) = E_T(1 + 12\frac{l^2}{h^2})$.

Let us consider then the cantilever beam bending problem with the following statement:

$$\begin{cases} x \in (0, L): & M'' - M_h''' = 0, \\ x = 0: & w = 0,\ w' = 0,\ w'' = 0 \\ x = L: & M - M_h' = 0,\ M' - M_h'' = P,\ M_h = 0 \end{cases} \tag{25.5}$$

Solution of this problem can be found analytically and provides us the following results for the apparent elastic modulus of the cantilever beam in USS model (Lurie and Solyaev (2018)):

$$E_{CB} = E_T \frac{\bar{L}^3 \cosh \bar{L}}{\bar{L}(3+\bar{L}^2)\cosh \bar{L} + 3(1-\bar{L}^2)\sinh \bar{L} - 6\bar{L}} \tag{25.6}$$

and in GSS model:

$$E_{CB} = E_T \frac{\widehat{L}^3 \sqrt{1+12\frac{l^2}{h^2}} \cosh \widehat{L}}{\widehat{L}(3+\widehat{L}^2)\cosh \widehat{L} + 3(1-\widehat{L}^2)\sinh \widehat{L} - 6\widehat{L}} \tag{25.7}$$

where $\bar{L} = L/l$ is the normalized length of the beam and $\widehat{L} = \bar{L}\sqrt{1+12\frac{l^2}{h^2}}$.

From the obtained results one can see that in GSS model the non-classical bending behavior of gradient beam significantly depends on its thickness (more precisely, on the ratio between thickness and length scale parameter h/l). In opposite, in USS model all non-classical effects defined by the normalized length of the beam L/l, that can be treated as an influence of the boundary conditions.

It is easy to find the solutions for the uniaxial tension problems, which coincide in both considered gradient models. Homogeneous uniaxial tension can be used to determine the Young's modulus E_T of the lattice according to the solution of the following problem:

$$\begin{cases} x \in (0,L): & N' - N_h'' = 0, \\ x = 0: & u = 0,\ N_h = 0 \\ x = L: & N - N_h' = F,\ N_h = 0 \end{cases} \quad \Longrightarrow \quad N \equiv N, \quad N_h \equiv 0 \tag{25.8}$$

that means the absence of size-effects under homogeneous uniaxial deformations.

Non-homogeneous uniaxial deformations of the beam are realized in the case of prescribed restriction for strain at the fixed end of the beam:

$$\begin{cases} x \in (0,L): & N' - N_h'' = 0, \\ x = 0: & u = 0,\ u' = 0 \\ x = L: & N - N_h' = F,\ N_h = 0 \end{cases} \tag{25.9}$$
$$\Longrightarrow \quad N = F\left(1 - \tfrac{\cosh(\bar{L}-\bar{x})}{\cosh \bar{L}}\right), \quad N_h = l\, F \tfrac{\sinh(\bar{L}-\bar{x})}{\cosh \bar{L}}$$

In this case one can find the following solution for the apparent elastic modulus of the beam (similar solution was used, e.g. in Solyaev et al (2019a)):

$$E_T = E \frac{\bar{L}}{\bar{L} - \tanh \bar{L}} \tag{25.10}$$

25.4 Size-Dependent Mechanical Behavior of the Lattice Beams

Now we can compare the size-dependent behavior of the lattices (Section 25.2) and corresponding continuum models (Section 25.3). From the comparison of the apparent elastic properties determined from the numerical solutions for the beams with lattice structures (25.1) and analytical solutions for the gradient Euler–Bernoulli beams (25.6), (25.7), (25.10) we can identify the material parameters of the last one. In such a way, we can realize a homogenization of the lattice structures for the effective continuum media described by the second gradient elasticity.

At first, let us consider the behavior of lattice beams under cantilever bending. In Figure 25.3 we present the dependences of the effective bending modulus E_{CB} on the dimensions of the beams with different type of the unit cells. In this figure points indicate the FE solution for the lattice structure and lines correspond to the gradient beam solutions, which were fitted by choosing the length scale parameter l. Presented values of the apparent moduli are normalized to the Young's modulus E_T of the lattices, which was found from the uniaxial tension test with homogeneous deformations (Fig. 25.2c). For the diamond-celled lattices we have $E_T = 8$ MPa and for the triangular unit cells $E_T = 62$ MPa.

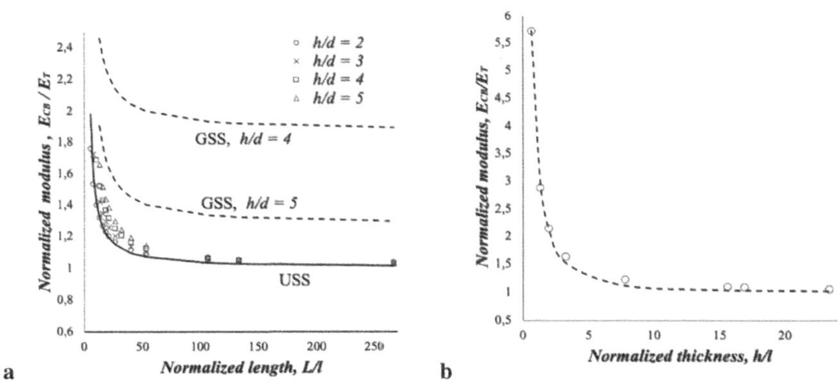

Fig. 25.3 Apparent elastic modulus of lattice beams with diamond (a) and triangular (b) structure evaluated based on the cantilever bending test. Points – FE solution, lines – analytical solution of the gradient beam theories (solid line – USS model, dashed lines - GSS model)

In the Fig. 25.3a we present the results for the beams with diamond unit cells. It is seen, that the effective properties of these structures correspond to the USS model: apparent modulus in the cantilever beam test E_{CB} significantly depends only on the length of the beam and almost independent on its thickness (see (25.6)). Length scale parameter identified for this structure equals to $l = 15$ mm. This value was found for the thin beams with two unit cells across thickness. Some deviations for the thicker beams arise between numerical and fitted analytical USS solutions for very short beams with length-to-thickness ratio smaller than 10, which is inappropriate for the Euler–Bernoulli theory. Nevertheless, one can see that the GSS model can not be applied for the diamond-celled lattices, because it predicts the significant dependence of the apparent modulus on its thickness even for the slender beams (see dashed lines in Fig. 25.3a), which does not arise in FE simulations with these structures.

For the beams with triangular unit cells we have the results similar to those one presented in Khakalo et al (2018). These type of lattices obey the behavior of the GSS model and their effective moduli significantly depends on the beam thickness. This dependence is shown in Fig. 25.3b. Determined value of the length scale parameter for these structures is $l = 31$ mm.

For the pure bending loading (Fig. 25.2a) we found that for the diamond lattices E_{PB} is independent on the sheet dimensions and equals to E_T, while in the triangular lattices E_{PB} becomes thickness-dependent and equals to E_{CB} for these structures. This result one more time confirms our conclusions about correspondence between unit cells geometry and USS/GSS gradient beams models.

Next, let us consider the uniaxial tension test with non-uniform deformations across beams length (Fig. 25.2d). Found apparent elastic moduli E'_T of the lattices for this type of loading are presented in Fig. 25.4. One can see that that size effects are similar in both type of lattices – apparent modulus E'_T increases with decrease of the length of the beam. In diamond-celled lattices (Fig. 25.4a) there exist some deviations between numerical solutions for the thick beams and fitted analytical solution (25.10). As it was mentioned above, it is the consequence of the considered simplified beam theory, which can not be applied to the short beams. For the triangular structures coincidence between numerical and analytical solutions are exact (Fig. 25.4b). These structures are much stiffer then the diamond-celled lattices and deviations between 1D solutions and 3D FE simulations are not so large.

Note, that the length scale parameters used in the analytical solutions for uniaxial tension (25.10) were the same to those one found previously in the bending tests. Interesting result here is the found values of the length x_0 which was constrained in the FE simulations to provide the non-uniform state of the lattices (see Fig. 25.4d). It is found that obtained consistency between FE simulations and beam-type theories is realized when the constrained length x_0 equals to the length scale parameter l, which was identified in previous bending tests. In other words, the physical meaning of additional boundary conditions for the strain of gradient beam (in our case it was $x = 0 : u' = 0$ in (25.9)) is that in real structure the length of the fixed segment at the boundary of the beam is of the order of its length scale parameter.

25.5 Conclusion

Two main contributions of the present study are the following:

1. We present an example of the lattice structure with diamond unit cells which exhibit the mechanical behavior related to the gradient elastic beam theory with "uniaxial stress state." Previously such examples were given only for GSS model.

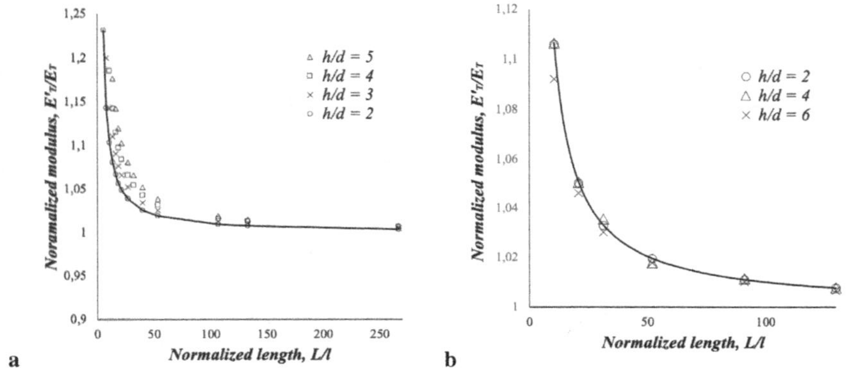

Fig. 25.4 Apparent elastic modulus of lattice beams with diamond (a) and triangular (b) structure evaluated based on the uniaxial tension test. Points – FE solution, lines – analytical solution of the gradient beam theories.

2. We show that additional boundary conditions of gradient beam theory for strains imply that there exist a constrained area in the corresponding three-dimensional structure, which has zero-value displacements and which length is equal to the length scale parameter of the gradient model.

In the present study, found values of the length scale parameters for considered lattices are rather small (of the order of unit cells size). Use of the similar pantographic-type lattices (Della Corte et al, 2016) will provide more extensive gradient effects and non-classical behavior of these materials, thought remain the same nature of the thickness-related and length-related size effects.

Acknowledgements This work was supported by the RFBR grants 18-08-00643, 18-01-00553 and President Grant MK-2856.2019.8.

References

Abdoul-Anziz H, Seppecher P (2018) Strain gradient and generalized continua obtained by homogenizing frame lattices. Mathematics and mechanics of complex systems 6(3):213–250

Alibert JJ, Seppecher P, dell'Isola F (2003) Truss modular beams with deformation energy depending on higher displacement gradients. Mathematics and Mechanics of Solids 8(1):51–73

Barchiesi E, Spagnuolo M, Placidi L (2019) Mechanical metamaterials: a state of the art. Mathematics and Mechanics of Solids 24(1):212–234

Berkache K, Deogekar S, Goda I, Picu R, Ganghoffer JF (2017) Construction of second gradient continuum models for random fibrous networks and analysis of size effects. Composite Structures 181:347–357

Berkache K, Deogekar S, Goda I, Picu RC, Ganghoffer JF (2019) Identification of equivalent couple-stress continuum models for planar random fibrous media. Continuum Mechanics and Thermodynamics 31(4):1035–1050

Boutin C, Giorgio I, Placidi L, et al (2017) Linear pantographic sheets: asymptotic micro-macro models identification. Mathematics and Mechanics of Complex Systems 5(2):127–162

De Angelo M, Placidi L, Nejadsadeghi N, Misra A (2020) Non-standard timoshenko beam model for chiral metamaterial: Identification of stiffness parameters. Mechanics Research Communications 103:103,462

Del Vescovo D, Giorgio I (2014) Dynamic problems for metamaterials: review of existing models and ideas for further research. International Journal of Engineering Science 80:153–172

Della Corte A, Giorgio I, Scerrato D, et al (2016) Pantographic 2d sheets: Discussion of some numerical investigations and potential applications. International Journal of Non-Linear Mechanics 80:200–208

dell'Isola F, Seppecher P, Alibert JJ, Lekszycki T, Grygoruk R, Pawlikowski M, Steigmann D, Giorgio I, Andreaus U, Turco E, et al (2019a) Pantographic metamaterials: an example of mathematically driven design and of its technological challenges. Continuum Mechanics and Thermodynamics 31(4):851–884

dell'Isola F, Seppecher P, Spagnuolo M, Barchiesi E, Hild F, Lekszycki T, Giorgio I, Placidi L, Andreaus U, Cuomo M, et al (2019b) Advances in pantographic structures: design, manufacturing, models, experiments and image analyses. Continuum Mechanics and Thermodynamics 31(4):1231–1282

Eremeyev VA, dell'Isola F, Boutin C, Steigmann D (2018) Linear pantographic sheets: existence and uniqueness of weak solutions. Journal of Elasticity 132(2):175–196

Frenzel T, Kadic M, Wegener M (2017) Three-dimensional mechanical metamaterials with a twist. Science 358(6366):1072–1074

Jafari A, Shah-enayati SS, Atai AA (2016) Size dependency in vibration analysis of nano plates; one problem, different answers. European Journal of Mechanics-A/Solids 59:124–139

Karttunen AT, Reddy J, Romanoff J (2019) Two-scale constitutive modeling of a lattice core sandwich beam. Composites Part B: Engineering 160:66–75

Khakalo S, Balobanov V, Niiranen J (2018) Modelling size-dependent bending, buckling and vibrations of 2d triangular lattices by strain gradient elasticity models: applications to sandwich beams and auxetics. International Journal of Engineering Science 127:33–52

Lazopoulos K, Lazopoulos A (2010) Bending and buckling of thin strain gradient elastic beams. European Journal of Mechanics-A/Solids 29(5):837–843

Lurie S, Solyaev Y (2018) Revisiting bending theories of elastic gradient beams. International Journal of Engineering Science 126:1–21

Lurie S, Solyaev Y (2019) On the formulation of elastic and electroelastic gradient beam theories. Continuum Mechanics and Thermodynamics pp 1–13

Lurie SA, Kalamkarov AL, Solyaev YO, Ustenko AD, Volkov AV (2018) Continuum micro-dilatation modeling of auxetic metamaterials. International Journal of Solids and Structures 132:188–200

Mirzaali M, Janbaz S, Strano M, Vergani L, Zadpoor AA (2018) Shape-matching soft mechanical metamaterials. Scientific reports 8(1):965

Niiranen J, Balobanov V, Kiendl J, Hosseini S (2019) Variational formulations, model comparisons and numerical methods for euler–bernoulli micro-and nano-beam models. Mathematics and Mechanics of Solids 24(1):312–335

Papargyri-Beskou S, Tsepoura K, Polyzos D, Beskos D (2003) Bending and stability analysis of gradient elastic beams. International Journal of Solids and Structures 40(2):385–400

Romanoff J, Reddy J (2014) Experimental validation of the modified couple stress timoshenko beam theory for web-core sandwich panels. Composite Structures 111:130–137

Rueger Z, Lakes RS (2016) Cosserat elasticity of negative poisson's ratio foam: experiment. Smart Materials and Structures 25(5):054,004

Shahsavari A, Picu R (2013) Size effect on mechanical behavior of random fiber networks. International Journal of Solids and Structures 50(20-21):3332–3338

Solyaev Y, Lurie S, Barchiesi E, Placidi L (2019a) On the dependence of standard and gradient elastic material constants on a field of defects. Mathematics and Mechanics of Solids p 1081286519861827

Solyaev Y, Lurie S, Ustenko A (2019b) Numerical modeling of a composite auxetic metamaterials using micro-dilatation theory. Continuum Mechanics and Thermodynamics 31(4):1099–1107
Yang H, Ganzosch G, Giorgio I, Abali BE (2018) Material characterization and computations of a polymeric metamaterial with a pantographic substructure. Zeitschrift für angewandte Mathematik und Physik 69(4):105

Chapter 26
On the Relations between Direct and Energy Based Homogenization Approaches in Second Gradient Elasticity

Yury Solyaev, Sergey Lurie, and Anastasia Ustenko

Abstract In this chapter we consider Eshelby integral formulas in the framework of second gradient elasticity theory (SGET). Based on these formulas we provide a comparison between direct and energy based homogenization approaches for the second gradient elastic media containing inhomogeneities. These approaches are widely used in micromechanics for the estimation of the effective properties of composites assuming the equivalence of averaged field variables (direct approach) or the equivalence of strain energy (energy based approach) in heterogeneous and equivalent homogeneous media. Equivalence between these approaches in classical elasticity can be proven by using Eshelby formulas. In the present work we use these formulas generalized for SGET to show, that the direct and energy based homogenization approaches are equivalent in SGET only in the case of homogeneous boundary conditions, while for the more general type of loading these approaches may provide different predictions, resulting, e.g., in different values of the effective high-order material constants estimated for the composite materials and mechanical metamaterials.

26.1 Introduction

Integral formulas derived by Eshelby in his celebrated work Eshelby (1956) is an important result of classical elasticity that allow to estimate the difference between strain energy of heterogeneous material (consisting of matrix and inclusions) and corresponding pure matrix under the same loading conditions. In classical micromechanics Eshelby formula is used to prove the equivalence between so-called *direct* and *energy* based homogenization approaches (Aboudi, 2013; Christensen, 2012).

Y. Solyaev, S. Lurie, A. Ustenko
Institute of Applied Mechanics of the Russian Academy of Sciences, Moscow, Russia
Moscow Aviation Institute, Moscow, Russia
e-mail: yos@iam.ras.ru, salurie@mail.ru, audlord@ya.ru

B. E. Abali and I. Giorgio (eds.), *Developments and Novel Approaches in Biomechanics and Metamaterials*, Advanced Structured Materials 132,
https://doi.org/10.1007/978-3-030-50464-9_26

In *direct* approach one should use the relations between averaged stress and strain fields to find the effective properties of heterogeneous media. In *energy* based approach effective properties are determined based on the assumption of equivalence between strain energies stored in the composite and equivalent homogeneous media. Considering basic relations of the direct approach one can obtain Eshelby integral formulas by using simple algebraic manipulations and in such a way the equivalence between direct and energy based approaches can be proven (Aboudi, 2013; Christensen, 2012).

Direct approach is widely used in classical micromechanics, e.g. in the framework of Mori–Tanaka method, self-consistent method, differential scheme, dilute approximation etc. (Aboudi, 2013; Benveniste, 1987; Kachanov and Sevostianov, 2018) because it provides a simple evaluation of the effective properties of composites, that can be found in some cases analytically in a closed form. Direct approach is also widely used in homogenization problems in more general framework of elastoplacticity, thermoelasticity, electroelasticity (Lagoudas et al, 1991; Dunn and Taya, 1993; Siboni and Benveniste, 1991) etc.

Energy approach may be more complicated than the direct one as it involves a volumetric integration of the field variables. However, energy approach is also widely used in classical micromechanics, e.g. in generalized self consistent method, in which Eshelby formula is applied to reduce the volumetric integration of strain energy to the simple surface integral of special type (Christensen, 2012). Such energy based approach is also used in homogenization problems with viscoelastic materials (Gusev and Lurie, 2009; Lurie et al, 2014, 2019), in piezoelectricity (Jiang et al, 2001), etc.

In classical micromechanics it is well known that different homogenization methods (Mori–Tanaka, self-consistent, Maxwell etc.) may provide significantly different predictions for the effective properties of composites (Christensen, 1990; Dunn and Taya, 1993; Aboudi, 2013; Lurie et al, 2018a). Thus, the choice of homogenization method (scheme) is important and should be validated by using, e.g., experimental data. However, the choice between direct and energy approach for estimation of the effective properties is not significant as their equivalence is proven. In other words, in classical models considering certain homogenization scheme and representative element of composite material one can estimate its effective properties as the ratio between corresponding averaged field variables or by estimating its total potential energy and results will be the same.

In the present work we derive Eshelby integral formulas in the framework of second gradient elasticity theory (SGET). This theory attract high attention during last decades as it allows to obtain size-dependent solutions in micromechanics (Fleck and Hutchinson, 1993; Ma and Gao, 2014; Lurie et al, 2006), non-singular solutions in linear elastic fracture mechanics (Lurie and Belov, 2008; Askes and Aifantis, 2011; Lurie and Belov, 2014; Mousavi and Aifantis, 2016), in contact problems (Vasil'ev and Lurie, 2017) and in dislocation problems (Lazar and Maugin, 2006a; Lazar and Po, 2018), it provides a continuous description of mechanical metamaterials (Alibert et al, 2003; dell'Isola et al, 2019a,b; Boutin et al, 2017; dell'Isola et al, 2017; Eremeyev et al, 2018; Scerrato and Giorgio, 2019), refinement analysis of thin structures (Papargyri-Beskou et al, 2003; Askes and Aifantis, 2009; Lurie and Solyaev, 2019, 2018) and mesh independent FE solutions for the problems with non-smooth geometries and concentrated loads (Lazar and Maugin, 2006b; Vasil'ev and Lurie, 2017; Andreaus et al, 2016; Reiher et al, 2017).

In the previous work of authors (Lurie et al, 2018b), it was found that different homogenization methods of micromechanics may provide significantly different predictions for the effective properties of elastic composites, which constituents exhibit strain gradient effects. The reason of this difference is not obvious, namely, it may be the consequence of considered different homogenization methods (Mori–Tanaka/generalized self consistent) or the consequence of different homogenization approaches (direct/energy)[1]. In the present work we show, that the equivalence between direct and energy based homogenization approaches in SGET is realized only in the case of so-called homogeneous boundary conditions (Hashin, 1983). This type of boundary conditions provides a uniform strain and stress fields in the corresponding equivalent homogeneous media, which elastic moduli can be found then from the solution of the problem. However, it will be shown that in the case of more general boundary conditions (non-homogeneous Forest, 1998, 2002; Forest and Trinh, 2011) the equivalence between direct and energy homogenization is not valid in SGET. This result may be important in the case of evaluation of the effective high-order material constants and apparent length scale parameters of composite materials and mechanical metamaterials that are widely studied in recent years (Ma and Gao, 2014; Delfani and Latifi Shahandashti, 2017; Bacca et al, 2013; Forest, 1998; Goda and Ganghoffer, 2016; Auffray et al, 2010; Trinh et al, 2012; Turco et al, 2018; Giorgio et al, 2018; dell'Isola et al, 2019a; Abali et al, 2017).

26.2 Second Gradient Elasticity

Let us consider an isotropic linear elastic body occupying the region Ω with smooth boundary $\partial\Omega$ without any edge. The strain energy density of isotropic second gradient material is given by Mindlin (1964); Mindlin and Eshel (1968),

$$w(\boldsymbol{\varepsilon}, \nabla\boldsymbol{\varepsilon}) = \frac{1}{2}\boldsymbol{\varepsilon} : \mathbf{C} : \boldsymbol{\varepsilon} + \frac{1}{2}\boldsymbol{\eta} \,\vdots\, \mathbf{A} \,\vdots\, \boldsymbol{\eta} \tag{26.1}$$

where $\mathbf{C}$ and $\mathbf{A}$ are the fourth- and sixth-order tensors of the elastic moduli; $\boldsymbol{\varepsilon} = \frac{1}{2}(\nabla\boldsymbol{u} + (\nabla\boldsymbol{u})^T)$ is an infinitesimal strain tensor, $\boldsymbol{\eta} = \nabla\boldsymbol{\varepsilon}$ is the strain gradient tensor, $\boldsymbol{u}(\mathbf{r})$ is the vector of mechanical displacements at a point $\mathbf{r} = \{x_1, x_2, x_3\}$; ∇ is the 3D nabla operator.

Constitutive equations for the Cauchy stress tensor $\boldsymbol{\tau}$, and third-order double stress tensor $\boldsymbol{\mu}$:

[1] Note, that Mori–Tanaka method is related to the direct homogenization approach and it involves the averaged filed quantities in estimations of the effective properties, while generalized self consistent method is realized following energy based approach

$$\boldsymbol{\tau} = \frac{\partial w}{\partial \boldsymbol{\varepsilon}} = \mathbf{C} : \boldsymbol{\varepsilon}, \qquad \boldsymbol{\mu} = \frac{\partial w}{\partial \boldsymbol{\eta}} = \mathbf{A} \,\vdots\, \boldsymbol{\eta} \tag{26.2}$$

Boundary value problem statement of gradient theory can be obtained based on the variational approach and can be presented as follows (Mindlin and Eshel, 1968; Gao and Park, 2007; Polizzotto, 2015):

$$\begin{aligned} \nabla \cdot \boldsymbol{\sigma} + \bar{\boldsymbol{b}} = 0, \qquad & \mathbf{x} \in \Omega \\ \boldsymbol{t} = \bar{\boldsymbol{t}}, \quad or \quad \boldsymbol{u} = \bar{\boldsymbol{u}}, \qquad & \mathbf{x} \in \partial\Omega \\ \boldsymbol{m} = \bar{\boldsymbol{m}} \quad or \quad \partial_n \boldsymbol{u} = \bar{\boldsymbol{g}}, \qquad & \mathbf{x} \in \partial\Omega \end{aligned} \tag{26.3}$$

where the edge boundary conditions and corresponding line loads along surface edges are omitted as we assume that the entire surface $\partial\Omega$ is smooth (Gao and Park, 2007); $\bar{\boldsymbol{b}}$ is the body force within Ω; $\bar{\boldsymbol{t}}$, $\bar{\boldsymbol{m}}$, $\bar{\boldsymbol{u}}$ and $\bar{\boldsymbol{g}}$ are the traction, double traction, displacement vector and normal gradient of displacement, respectively, prescribed on $\partial\Omega$; total stress tensor $\boldsymbol{\sigma}$, traction vector $\boldsymbol{t}$ and double traction vector $\boldsymbol{m}$ that persist in (26.3) are given by

$$\begin{aligned} \boldsymbol{\sigma} &= \boldsymbol{\tau} - \nabla \cdot \boldsymbol{\mu}, \\ \boldsymbol{t} &= \mathbf{n} \cdot \boldsymbol{\sigma} - \nabla_S \cdot (\mathbf{n} \cdot \boldsymbol{\mu}) - H \, \mathbf{nn} : \boldsymbol{\mu}, \\ \boldsymbol{m} &= \mathbf{nn} : \boldsymbol{\mu} \end{aligned} \tag{26.4}$$

where $\mathbf{n}$ is the vector of the external unit normal to $\partial\Omega$; $H = -\nabla_S \cdot \mathbf{n}$ is the mean curvature of $\partial\Omega$; $\nabla_S = \nabla - \mathbf{n}\partial_n$ is the surface gradient operator.

26.3 Eshelby Formulas

In classical elasticity Eshelby formulas allow to estimate the strain energy of the media containing inhomogeneities by using particular type of surface integration (Eshelby, 1956). In this section we derive the generalized variant of Eshelby formula suitable for SGET. We follow Christensen (1979), where this formula was involved in the classical micromechanics problems, and Lurie et al (2011), where this formula was derived in the framework of gradient theory of interphase layer that is the special case of SGET.

Let us consider a body containing single inclusion and subjected to the externally applied surface traction $\bar{\boldsymbol{t}}$ and double traction $\bar{\boldsymbol{m}}$ (Fig. 26.1, a). Body occupies region Ω which can be divided into inclusion phase Ω_1 and matrix phase Ω_2, such that $\Omega = \Omega_1 \cup \Omega_2$. External surface $\partial\Omega$ and interface $\partial\Omega_1$ are smooth and do not contain edges. Total strain energy of the body Ω in SGET is given by

$$W = \frac{1}{2} \int\limits_{\Omega} (\boldsymbol{\tau} : \boldsymbol{\varepsilon} + \boldsymbol{\mu} \,\vdots\, \boldsymbol{\eta}) dv \tag{26.5}$$

For the purpose of the following derivations let us consider three auxiliary problems:

1. Body Ω consisted of entirely matrix material subjected to the same external loading $\bar{\boldsymbol{t}}$, $\bar{\boldsymbol{m}}$ (Fig. 26.1b). Field variables of this problem will be denoted with subscript "0" as $\boldsymbol{u}_0$, $\boldsymbol{\tau}_0$ etc. Strain energy of this body is given by:

$$W_0 = \frac{1}{2} \int_{\Omega} (\boldsymbol{\tau}_0 : \boldsymbol{\varepsilon}_0 + \boldsymbol{\mu}_0 \vdots \boldsymbol{\eta}_0) dv \tag{26.6}$$

2. Body Ω consisted of entirely matrix material subjected to the same external loading $\bar{\boldsymbol{t}}$, $\bar{\boldsymbol{m}}$ with additional set of body forces $\bar{\boldsymbol{b}}$ acted inside the inclusion region Ω_1 and produced exactly the same state of the field variables in the region outside the inclusion (Fig. 26.1c). In this problem, field variables, which will be denoted with subscript "1", satisfy following conditions:

$$\boldsymbol{u}_1 = \boldsymbol{u}, \quad \boldsymbol{\varepsilon}_1 = \boldsymbol{\varepsilon}, \quad \boldsymbol{\eta}_1 = \boldsymbol{\eta}, \quad \boldsymbol{\tau}_1 = \boldsymbol{\tau}, \quad \boldsymbol{\mu}_1 = \boldsymbol{\mu}, \qquad \mathbf{x} \in \Omega_2 \tag{26.7}$$

3. Body Ω consisted of entirely matrix material subjected only to the set of body forces $\bar{\boldsymbol{b}}$ acted inside the inclusion region Ω_1 (Fig. 26.1d). Field variables in this problem will be denoted with subscript "2" as $\boldsymbol{u}_2$, $\boldsymbol{\tau}_2$ etc. In this problem, stress free boundary conditions are prescribed on the external surface of the body:

$$\boldsymbol{t}_2 = 0, \quad \boldsymbol{m}_2 = 0, \qquad \mathbf{x} \in \partial\Omega \tag{26.8}$$

Using (26.6), strain energy of the inhomogeneous body (26.5) can be evaluated as

$$W = W_0 + \frac{1}{2} \int_{\Omega} (\boldsymbol{\tau} : \boldsymbol{\varepsilon} - \boldsymbol{\tau}_0 : \boldsymbol{\varepsilon}_0 + \boldsymbol{\mu} \vdots \boldsymbol{\eta} - \boldsymbol{\mu}_0 \vdots \boldsymbol{\eta}_0) dv \tag{26.9}$$

Taking into account symmetry of stress tensors and by using chain rule, from (26.9) we obtain:

$$\begin{aligned} W =& W_0 + \frac{1}{2} \int_{\Omega} (\boldsymbol{\tau} : \nabla \boldsymbol{u} - \boldsymbol{\tau}_0 : \nabla \boldsymbol{u}_0 + \boldsymbol{\mu} \vdots \nabla\nabla \boldsymbol{u} - \boldsymbol{\mu}_0 \vdots \nabla\nabla \boldsymbol{u}_0 dv \\ =& W_0 + \frac{1}{2} \int_{\Omega} (\nabla \cdot (\boldsymbol{\tau} \cdot \boldsymbol{u}) - (\nabla \cdot \boldsymbol{\tau}) \cdot \boldsymbol{u} - \nabla \cdot (\boldsymbol{\tau}_0 \cdot \boldsymbol{u}_0) + (\nabla \cdot \boldsymbol{\tau}_0) \cdot \boldsymbol{u}_0 \\ &+ \nabla \cdot (\boldsymbol{\mu} : \nabla \boldsymbol{u}) - (\nabla \cdot \boldsymbol{\mu}) : \nabla \boldsymbol{u} - \nabla \cdot (\boldsymbol{\mu}_0 : \nabla \boldsymbol{u}_0) + (\nabla \cdot \boldsymbol{\mu}_0) : \nabla \boldsymbol{u}_0) dv \end{aligned} \tag{26.10}$$

Accounting for the definitions of total stress tensor $\boldsymbol{\sigma}$, $\boldsymbol{\sigma}_0$ (26.4.1) and equilibrium equations (26.3.1), we can simplify (26.10) as follows:

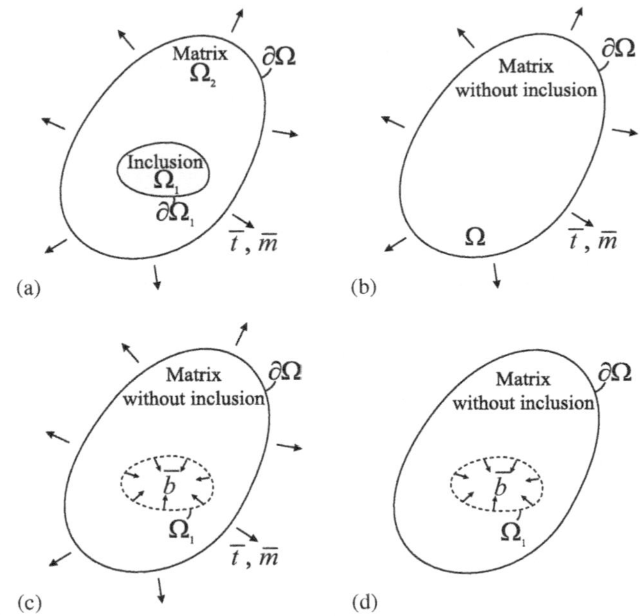

Fig. 26.1 To the derivation of Eshelby formula in second gradient elasticity

$$W = W_0 + \frac{1}{2}\int_{\Omega}(\nabla\cdot(\boldsymbol{\sigma}\cdot\boldsymbol{u}) - \nabla\cdot(\boldsymbol{\sigma}_0\cdot\boldsymbol{u}_0) + \nabla\cdot(\boldsymbol{\mu}:\nabla\boldsymbol{u}) - \nabla\cdot(\boldsymbol{\mu}_0:\nabla\boldsymbol{u}_0))dv \tag{26.11}$$

By using divergence theorem, (26.11) becomes

$$W = W_0 + \frac{1}{2}\int_{\partial\Omega}(\mathbf{n}\cdot\boldsymbol{\sigma}\cdot\boldsymbol{u} - \mathbf{n}\cdot\boldsymbol{\sigma}_0\cdot\boldsymbol{u}_0 + \mathbf{n}\cdot\boldsymbol{\mu}:\nabla\boldsymbol{u} - \mathbf{n}\cdot\boldsymbol{\mu}_0:\nabla\boldsymbol{u}_0)ds \tag{26.12}$$

Last two terms under the integral in (26.12) can be simplified by using identities, similar to those one used by Mindlin in Mindlin (1964) and proven by Gao and Park (2007),

$$\begin{aligned}\mathbf{n}\cdot\boldsymbol{\mu}:\nabla\boldsymbol{u} &= \mathbf{n}\cdot\boldsymbol{\mu}:\nabla_S\boldsymbol{u} + \mathbf{nn}:\boldsymbol{\mu}\cdot\partial_n\boldsymbol{u}\\ &= \nabla_S\cdot(\mathbf{n}\cdot\boldsymbol{\mu}\cdot\boldsymbol{u}) - \nabla_S\cdot(\mathbf{n}\cdot\boldsymbol{\mu})\cdot\boldsymbol{u} + \mathbf{nn}:\boldsymbol{\mu}\cdot\partial_n\boldsymbol{u} \quad (26.13)\\ &= -H\mathbf{nn}:\boldsymbol{\mu}\cdot\boldsymbol{u} + \mathbf{n}\cdot\nabla\times(\mathbf{n}\times(\mathbf{n}\cdot\boldsymbol{\mu}\cdot\boldsymbol{u}))\\ &- \nabla_S\cdot(\mathbf{n}\cdot\boldsymbol{\mu})\cdot\boldsymbol{u} + \mathbf{nn}:\boldsymbol{\mu}\cdot\partial_n\boldsymbol{u}\end{aligned}$$

and the same one for the field variables with "0" subscript.

By using Stokes's theorem, surface integration of terms with cross products that arise in (26.13) can be replaced by the line integration, and as far as the considered surfaces are smooth, these terms are vanished. As result, based on (26.12), (26.13), we obtain

$$W = W_0 + \frac{1}{2} \int\limits_{\partial\Omega} (\boldsymbol{t} \cdot \boldsymbol{u} - \boldsymbol{t}_0 \cdot \boldsymbol{u}_0 + \boldsymbol{m} \cdot \partial_n \boldsymbol{u} - \boldsymbol{m}_0 \cdot \partial_n \boldsymbol{u}_0) ds, \qquad (26.14)$$

in which the definitions of tractions $\boldsymbol{t}$, $\boldsymbol{t}_0$ and double tractions $\boldsymbol{m}$, $\boldsymbol{m}_0$ are used according to (26.4).

Finally, taking into account that the boundary conditions for the problems presented in Fig. 26.1a and Fig. 26.1b are the same, i.e. that $\boldsymbol{t} = \boldsymbol{t}_0 = \bar{\boldsymbol{t}}$ and $\boldsymbol{m} = \boldsymbol{m}_0 = \bar{\boldsymbol{m}}$ if $\mathbf{x} \in \partial\Omega$, from (26.14) we found the variant of Eshelby formula generalized for SGET in the case of body with smooth surface

$$W = W_0 + \frac{1}{2} \int\limits_{\partial\Omega} (\bar{\boldsymbol{t}} \cdot (\mathbf{u} - \mathbf{u}_0) + \bar{\boldsymbol{m}} \cdot (\partial_n \mathbf{u} - \partial_n \mathbf{u}_0)) ds. \qquad (26.15)$$

Following Christensen (2012), we will derive the variant of Eshelby formula that is more suitable for the homogenization problems. To do this let us consider the second and the third auxiliary problems presented in Fig. 26.1c, d. At first we note, that the field variables in the problem presented in Fig. 26.1c can be decomposed into the sum of those ones of the problems presented in Fig. 26.1b and in Fig. 26.1d. Therefore, we can write

$$\begin{aligned} \boldsymbol{u}_1 &= \boldsymbol{u}_0 + \boldsymbol{u}_2, \\ \boldsymbol{\varepsilon}_1 &= \boldsymbol{\varepsilon}_0 + \boldsymbol{\varepsilon}_2, \\ \boldsymbol{\eta}_1 &= \boldsymbol{\eta}_0 + \boldsymbol{\eta}_2, \\ \boldsymbol{\tau}_1 &= \boldsymbol{\tau}_0 + \boldsymbol{\tau}_2, \\ \boldsymbol{\mu}_1 &= \boldsymbol{\mu}_0 + \boldsymbol{\mu}_2 \end{aligned} \qquad (26.16)$$

Energy of the homogeneous media with equivalent state in matrix phase (Fig. 26.1c) can be presented then in the following form:

$$\begin{aligned} W_1 &= \frac{1}{2} \int\limits_{\Omega} \Big((\boldsymbol{\tau}_0 + \boldsymbol{\tau}_2) : (\boldsymbol{\varepsilon}_0 + \boldsymbol{\varepsilon}_2) + (\boldsymbol{\mu}_0 + \boldsymbol{\mu}_2) \,\vdots\, (\boldsymbol{\eta}_0 + \boldsymbol{\eta}_2) \Big) dv \\ &= W_0 + W_2 + W_{int} \end{aligned} \qquad (26.17)$$

where the total strain energy of the third auxiliary problem W_2 (Fig. 26.1d) and the interaction energy of the two stress states W_{int} are defined by

$$W_2 = \frac{1}{2}\int\limits_{\Omega}(\boldsymbol{\tau}_2 : \boldsymbol{\varepsilon}_2 + \boldsymbol{\mu}_2 \vdots \boldsymbol{\eta}_2)dv$$

$$W_{int} = \frac{1}{2}\int\limits_{\Omega}(\boldsymbol{\tau}_0 : \boldsymbol{\varepsilon}_2 + \boldsymbol{\tau}_2 : \boldsymbol{\varepsilon}_0 + \boldsymbol{\mu}_0 \vdots \boldsymbol{\eta}_2 + \boldsymbol{\mu}_2 \vdots \boldsymbol{\eta}_0)dv \tag{26.18}$$

Taking into account symmetry conditions

$$\boldsymbol{\tau}_2 : \boldsymbol{\varepsilon}_0 = \boldsymbol{\varepsilon}_2 : \mathbf{C} : \boldsymbol{\varepsilon}_0 = \boldsymbol{\varepsilon}_0 : \mathbf{C} : \boldsymbol{\varepsilon}_2 = \boldsymbol{\tau}_0 : \boldsymbol{\varepsilon}_2$$
$$\boldsymbol{\mu}_2 \vdots \boldsymbol{\eta}_0 = \boldsymbol{\eta}_2 \vdots \mathbf{A} \vdots \boldsymbol{\eta}_0 = \boldsymbol{\eta}_0 \vdots \mathbf{A} \vdots \boldsymbol{\eta}_2 = \boldsymbol{\mu}_0 \vdots \boldsymbol{\eta}_2 \tag{26.19}$$

and providing the similar derivations as it was done in (26.9)-(26.15), from (26.18) we obtain

$$W_{int} = \int\limits_{\Omega}(\boldsymbol{\tau}_0 : \boldsymbol{\varepsilon}_2 + \boldsymbol{\mu}_0 \vdots \boldsymbol{\eta}_2)dv = \int\limits_{\partial\Omega}(\bar{\boldsymbol{t}} \cdot \boldsymbol{u}_2 + \bar{\boldsymbol{m}} \cdot \partial_n \mathbf{u}_2)ds \tag{26.20}$$

Next, we notice that on the external boundary $\partial\Omega$ field variables of the initial problem with inhomogeneous body (Fig. 26.1a) are coincides with those one of the second auxiliary problem (Fig. 26.1c), such that decomposition (26.16) gives us $\boldsymbol{u}_2 = \boldsymbol{u}_1 - \boldsymbol{u}_0 = \boldsymbol{u} - \boldsymbol{u}_0$ and $\partial_n \boldsymbol{u}_2 = \partial_n \boldsymbol{u}_1 - \partial_n \boldsymbol{u}_0 = \partial_n \boldsymbol{u} - \partial_n \boldsymbol{u}_0$, if $\mathbf{x} \in \partial\Omega$. Therefore, integrals in (26.20) and in (26.15) coincide, and we can state that

$$W = W_0 + \frac{1}{2}W_{int}, \tag{26.21}$$

Then we consider a closed surface Σ outside of the inclusion that separate the domain Ω into internal part Ω_I containing inclusion ($\Omega_1 \subset \Omega_I$) and external part Ω_{II}. As result, representation of W_{int} (26.20) can be rewritten as follows:

$$\begin{aligned} W_{int} &= \int\limits_{\Omega_I}(\boldsymbol{\tau}_0 : \boldsymbol{\varepsilon}_2 + \boldsymbol{\mu}_0 \vdots \boldsymbol{\eta}_2)dv + \int\limits_{\Omega_{II}}(\boldsymbol{\tau}_0 : \boldsymbol{\varepsilon}_2 + \boldsymbol{\mu}_0 \vdots \boldsymbol{\eta}_2)dv \\ &= \int\limits_{\Omega_I}(\boldsymbol{\tau}_0 : \boldsymbol{\varepsilon}_2 + \boldsymbol{\mu}_0 \vdots \boldsymbol{\eta}_2)dv + \int\limits_{\Omega_{II}}(\boldsymbol{\tau}_2 : \boldsymbol{\varepsilon}_0 + \boldsymbol{\mu}_2 \vdots \boldsymbol{\eta}_0)dv \\ &= \int\limits_{\Sigma}(\boldsymbol{t}_0 \cdot \boldsymbol{u}_2 + \boldsymbol{m}_0 \cdot \partial_n \boldsymbol{u}_2)ds - \int\limits_{\Sigma}(\boldsymbol{t}_2 \cdot \boldsymbol{u}_0 + \boldsymbol{m}_2 \cdot \partial_n \boldsymbol{u}_0)ds \\ &+ \int\limits_{\partial\Omega}(\bar{\boldsymbol{t}} \cdot \boldsymbol{u}_0 + \bar{\boldsymbol{m}} \cdot \partial_n \boldsymbol{u}_0)ds \end{aligned} \tag{26.22}$$

where we use the symmetry conditions (26.19) and take into account different signs of the unit normal, which it has on Σ in domains Ω_I and Ω_{II}.

Stress free boundary conditions for the third auxiliary problem (26.8) provide us that the last integral in (26.22) vanishes. Then, taking into account relations (26.4),

(26.7) and (26.16), from (26.22) we obtain

$$W_{int} = \int_{\Sigma}(\boldsymbol{t}_0 \cdot \boldsymbol{u} - \boldsymbol{t} \cdot \boldsymbol{u}_0)ds + \int_{\Sigma}(\boldsymbol{m}_0 \cdot \partial_n \boldsymbol{u} - \boldsymbol{m} \cdot \partial_n \boldsymbol{u}_0)ds \tag{26.23}$$

Finally, substituting (26.23) into (26.21) we find another variant of Eshelby formula generalized for the second gradient elasticity:

$$W = W_0 + \frac{1}{2}\int_{\Sigma}(\boldsymbol{t}_0 \cdot \boldsymbol{u} - \boldsymbol{t} \cdot \boldsymbol{u}_0)ds + \frac{1}{2}\int_{\Sigma}(\boldsymbol{m}_0 \cdot \partial_n \boldsymbol{u} - \boldsymbol{m} \cdot \partial_n \boldsymbol{u}_0)ds \tag{26.24}$$

Note, that considering other sets of boundary conditions on the external surface $\partial\Omega$, one can obtain, generally say, four different variants of Eshelby formula in SGET. Formula (26.24) is given for the prescribed traction and double traction $\boldsymbol{t} = \bar{\boldsymbol{t}}$, $\boldsymbol{m} = \bar{\boldsymbol{m}}$ on $\partial\Omega$. For example, if instead of the boundary conditions for traction, we prescribe the displacements, i.e. $\boldsymbol{u} = \bar{\boldsymbol{u}}$, $\boldsymbol{m} = \bar{\boldsymbol{m}}$ on $\partial\Omega$, we will have

$$W = W_0 + \frac{1}{2}\int_{\Sigma}(\boldsymbol{t} \cdot \boldsymbol{u}_0 - \boldsymbol{t}_0 \cdot \boldsymbol{u})ds + \frac{1}{2}\int_{\Sigma}(\boldsymbol{m}_0 \cdot \partial_n \boldsymbol{u} - \boldsymbol{m} \cdot \partial_n \boldsymbol{u}_0)ds \tag{26.25}$$

(the differences arise in the placement of subscripts "0" in the first integral).

Classical variants of Eshelby formulas follows from (26.24), (26.25) if we assume the absence of gradient effects in the media. In this case gradient moduli equals to zero in (26.2), double stresses do not arise $\boldsymbol{m} = \boldsymbol{m}_0 = 0$ and we have classical definitions of tractions $\boldsymbol{t} = \boldsymbol{t}_{clas} = \mathbf{n} \cdot \boldsymbol{\tau}$ and $\boldsymbol{t}_0 = (\boldsymbol{t}_0)_{clas} = \mathbf{n} \cdot \boldsymbol{\tau}_0$.

26.4 Relations between Direct and Energy Based Homogenization in SGET

Based on derived Eshelby formulas we can compare the direct and energy based homogenization approaches that can be applied for the estimation of the effective properties of non-homogeneous materials in SGET.

Let us consider a body Ω, consisting of inclusion Ω_1 and matrix Ω_2, which volumes are V_1 and V_2, respectively. Volume fraction of the inclusion is defined by $c = V_1/V$, where $V = V_1 + V_2$. For this two-phase composite the average Cauchy stress $\langle\boldsymbol{\tau}\rangle$ and strain $\langle\boldsymbol{\varepsilon}\rangle$ can be defined by

$$\begin{aligned} \langle\boldsymbol{\tau}\rangle &= c\langle\boldsymbol{\tau}^{(1)}\rangle + (1-c)\langle\boldsymbol{\tau}^{(2)}\rangle, \\ \langle\boldsymbol{\varepsilon}\rangle &= c\langle\boldsymbol{\varepsilon}^{(1)}\rangle + (1-c)\langle\boldsymbol{\varepsilon}^{(2)}\rangle \end{aligned} \tag{26.26}$$

where the averaged quantities are defined by relations

$$\langle \boldsymbol{\tau} \rangle = \frac{1}{V} \int\limits_{\Omega} \boldsymbol{\tau} dv, \qquad \langle \boldsymbol{\tau}^{(1)} \rangle = \frac{1}{V_1} \int\limits_{\Omega} \boldsymbol{\tau}^{(1)} dv, \qquad \ldots$$

By using constitutive relations of phases $\boldsymbol{\varepsilon}^{(1,2)} = \boldsymbol{S}^{(1,2)} : \boldsymbol{\tau}^{(1,2)}$, in which $\boldsymbol{S}^{(1,2)}$ are the compliance tensors, from (26.26.2) we obtain

$$\langle \boldsymbol{\varepsilon} \rangle = c\, \boldsymbol{S}^{(1)} : \langle \boldsymbol{\tau}^{(1)} \rangle + (1 - c) \boldsymbol{S}^{(2)} : \langle \boldsymbol{\tau}^{(2)} \rangle \qquad (26.27)$$

Then we introduce the stress concentration tensors, that define the relations between averaged Cauchy stress in phases and averaged Cauchy stress in a whole composite:

$$\begin{aligned} \langle \boldsymbol{\tau}^{(1)} \rangle &= \boldsymbol{T}^{(1)} : \langle \boldsymbol{\tau} \rangle \\ \langle \boldsymbol{\tau}^{(2)} \rangle &= \boldsymbol{T}^{(2)} : \langle \boldsymbol{\tau} \rangle \end{aligned} \qquad (26.28)$$

Additionally we use the general definition of the effective properties $\boldsymbol{S}^*$ (compliances) of the composite that are defined in the framework of direct approach as follows:

$$\langle \boldsymbol{\varepsilon} \rangle = \boldsymbol{S}^* : \langle \boldsymbol{\tau} \rangle \qquad (26.29)$$

Substituting (26.28) and (26.29) into (26.27) we finally find the relations between the effective properties of composite material and properties of its phases

$$\boldsymbol{S}^* = \boldsymbol{S}^{(2)} + c\, (\boldsymbol{S}^{(1)} - \boldsymbol{S}^{(2)}) : \boldsymbol{T}^{(1)} \qquad (26.30)$$

where we take into account the relation between stress concentration tensors $c\, \boldsymbol{T}^{(1)} + (1 - c) \boldsymbol{T}^{(2)} = \boldsymbol{I}$, that follows from (26.26.1), (26.28); and $\boldsymbol{I}$ is an identity tensor.

Note, that the main difference between derived relation (26.30) and similar one in classical micromechanics (see, e.g., Hashin, 1983; Christensen, 2012; Aboudi, 2013) lies in the definition of the stress concentration tensor $\boldsymbol{T}^{(1)}$. In classical elasticity it is well known that the stress field inside the inclusion is uniform under prescribed homogeneous boundary conditions, such that components of the stress concentration tensor can be defined as $\boldsymbol{T}^{(1)} : \langle \boldsymbol{\tau}^{(1)} \rangle = \boldsymbol{\tau}^{(1)}$ (without averaging of stress field inside the inclusion).

However in SGET, boundary layer effects leads to the non-uniform strain and stress fields inside the inclusions even under prescribed homogeneous boundary conditions (Lurie et al, 2005; Gao and Ma, 2010; Ma and Gao, 2014; Lurie et al, 2018b), such that $\langle \boldsymbol{\tau}^{(1)} \rangle \neq \boldsymbol{\tau}^{(1)}$, and concentration tensors becomes position dependent. Thus, in opposite to classical elasticity, in SGET, averaging procedure of the field variables inside the inclusion becomes important. Considering the direct homogenization approach in the framework of SGET one should find the averaged values of concentration tensor $\boldsymbol{T}^{(1)} = \langle \boldsymbol{T}^{(1)}(\mathbf{r}) \rangle$ and use it to find the effective elastic constants based in relation similar to (26.30). Such an approach for the strain concentration tensor was used, e.g., in Lurie et al (2018b), and by using its representation through the components of averaged Eshelby tensor in Ma and Gao (2014).

Now, let us compare a direct and energy based homogenization approaches in SGET. In classical elasticity Eshelby formula can be derived from (26.30) and this proves the equivalence between these homogenezation approaches (Aboudi, 2013; Christensen, 2012; Eremeyev and Konopińska-Zmysłowska, 2019). Thus, let us try to derive the Eshelby formula from the relation (26.30) in the case of SGET. To do this, let us assume that the following static boundary conditions are prescribed on the body surface (Fig. 26.1a),

$$\boldsymbol{t} = \bar{\boldsymbol{t}}, \qquad \bar{\boldsymbol{m}} = \boldsymbol{0}, \qquad \boldsymbol{r} \in \partial\Omega \tag{26.31}$$

Such type of boundary conditions can be defined as "homogeneous" and they should be used to estimate the classical effective elastic constants of the composite because they produce the uniform stress field $\langle\boldsymbol{\tau}\rangle = \boldsymbol{\tau}_0$, $\boldsymbol{\mu} \equiv 0$ inside the homogeneous media Lurie et al (2018b); Ma and Gao (2014). Contracting (26.30) twice with $\boldsymbol{\tau}_0$ and taking into account (26.28.1), we obtain

$$\boldsymbol{\tau}_0 : \boldsymbol{S}^* : \boldsymbol{\tau}_0 = \boldsymbol{\tau}_0 : \boldsymbol{S}^{(2)} : \boldsymbol{\tau}_0 + c\,\boldsymbol{\tau}_0 : (\boldsymbol{S}^{(1)} - \boldsymbol{S}^{(2)}) : \langle\boldsymbol{\tau}^{(1)}\rangle \tag{26.32}$$

This equation contains the strain energy density of inhomogeneous body $2w = \boldsymbol{\tau}_0 : \boldsymbol{S}^* : \boldsymbol{\tau}_0$ (Fig. 26.1a) and homogeneous body $2w_0 = \boldsymbol{\tau}_0 : \boldsymbol{S}^{(2)} : \boldsymbol{\tau}_0$ (Fig. 26.1b) evaluated through the effective compliance constants and matrix compliance constants, respectively. Double stress do not arise in these definitions of the strain energy density due to their vanishing for the considered type of boundary conditions (26.31). Integrating (26.32) over the body volume V and using the definition of averaged stress field $\langle\boldsymbol{\tau}^{(1)}\rangle$, we obtain

$$W = W_0 + \frac{1}{2}\int\limits_{\Omega_1} \boldsymbol{\tau}_0 : (\boldsymbol{S}^{(1)} - \boldsymbol{S}^{(2)}) : \boldsymbol{\tau}^{(1)}\, dv \tag{26.33}$$

Substituting constitutive equations for the matrix and inclusion materials in (26.33) yields the relation:

$$W = W_0 + \frac{1}{2}\int\limits_{\Omega_1} (\boldsymbol{\tau}_0 : \boldsymbol{\varepsilon} - \boldsymbol{\tau} : \boldsymbol{\varepsilon}_0) dv \tag{26.34}$$

where the field variables without indexes are related to the problem with inhomogeneous body (Fig. 26.1a), because inside the inclusion domain Ω_1 it is valid that $\boldsymbol{\tau}^{(1)} = \boldsymbol{\tau}$ and $\boldsymbol{\varepsilon}^{(1)} = \boldsymbol{S}^{(1)} : \boldsymbol{\tau}^{(1)} = \boldsymbol{\varepsilon}$; however, stress $\boldsymbol{\tau}_0$ and strain $\boldsymbol{\varepsilon}_0 = \boldsymbol{S}^{(2)} : \boldsymbol{\tau}_0$ are related to the problem with pure matrix (Fig. 26.1b).

Relation (26.34) can be rewritten by using the similar procedure as we used during derivation of Eshelby formula in Section 26.3. After using the chain rule, divergence theorem and accounting for the symmetry conditions (26.19), equilibrium equations (26.3.1), identities (26.13) and definitions (26.4) we can arrive in (26.34) to the following relation:

$$W = W_0 + \frac{1}{2} W_{int} + \frac{1}{2} W'_{int} \tag{26.35}$$

where the interaction energy W_{int} is defined by (26.23) in which we should assume $\Sigma = \partial\Omega_1$ and additional interaction energy W'_{int} is related with gradient effects as follows:

$$W'_{int} = \int_{\Omega_1} (\boldsymbol{\mu} \vdots \boldsymbol{\eta}_0 - \boldsymbol{\mu}_0 \vdots \boldsymbol{\eta}) dv \tag{26.36}$$

As it was mentioned above, for the prescribed type of boundary conditions (26.31), solution for the homogeneous media will contain zero values of double stresses and strain gradients in whole region Ω, i.e. $\boldsymbol{\mu}_0 \equiv 0$, $\boldsymbol{\eta}_0 \equiv 0$. Therefore, in considered problem we have $W'_{int} = 0$ and the direct approach implied by the relation (26.30) will provide us the equivalence of strain energies stored in composite media and in the equivalent homogeneous media which can be presented in the form of Eshelby formula (26.24).

Thus, we prove the equivalence between energy and direct homogenization approaches in SGET for the case of classical elastic constants estimations, that can be done by using homogeneous boundary conditions (26.31). However, in the case of more general boundary conditions (with prescribed $\bar{\boldsymbol{m}} \neq 0$ or $\partial_n \boldsymbol{u}$ on $\partial\Omega$) the differences between approaches may arise due to non-zero interaction term W'_{int} that persist in (26.35). For example, such type of non-uniform boundary conditions should be used to find the effective gradient moduli of composite materials or the apparent length scale parameters of mechanical metamaterials (Bacca et al, 2013; Forest and Trinh, 2011).

26.5 Conclusions

In this work we derived generalized variant of Eshelby integral formulas (26.15), (26.24), (26.25) that can be used in micromechanics applications, namely, it can be used in the generalized self consistent method realized for the inclusion problems in SGET (Lurie et al, 2018b, 2016).

Based on the derived Eshelby formulas we compare a direct ("straightforward" Auffray et al (2010)) and energy based homogenization approaches in SGET. We found that, similarly to classical elasticity, these approaches are equivalent in SGET when one consider a homogeneous boundary conditions and estimate classical elastic properties of composite material. However, in the case of more general loading and, e.g. non-homogeneous boundary conditions, that one should apply to estimate the effective high-order material constants of composite materials, direct and energy based approaches may provide different results. Deviations between these approaches may be treated as errors of direct approach as it can not take into account a portion of strain energy related with high order interactions persisting in the media.

Derived Eshelby formulas can be additionally generalized for the second gradient media with multiple inclusions and for the bodies with non-smooth geometry. In the last case additional edge-type boundary conditions and line loads will be important and should be taken into consideration (dell'Isola et al, 2016, 2015).

Acknowledgements This work was supported by the RFBR grant 18-31-20043.

References

Abali BE, Müller WH, dell'Isola F (2017) Theory and computation of higher gradient elasticity theories based on action principles. Archive of Applied Mechanics 87(9):1495–1510

Aboudi J (2013) Mechanics of composite materials: a unified micromechanical approach, vol 29. Elsevier

Alibert JJ, Seppecher P, dell'Isola F (2003) Truss modular beams with deformation energy depending on higher displacement gradients. Mathematics and Mechanics of Solids 8(1):51–73

Andreaus U, dell'Isola F, Giorgio I, Placidi L, Lekszycki T, Rizzi NL (2016) Numerical simulations of classical problems in two-dimensional (non)linear second gradient elasticity. International Journal of Engineering Science 108:34–50

Askes H, Aifantis EC (2009) Gradient elasticity and flexural wave dispersion in carbon nanotubes. Physical Review B 80(19):195,412

Askes H, Aifantis EC (2011) Gradient elasticity in statics and dynamics: An overview of formulations, length scale identification procedures, finite element implementations and new results. International Journal of Solids and Structures 48(13):1962–1990

Auffray N, Bouchet R, Brechet Y (2010) Strain gradient elastic homogenization of bidimensional cellular media. International Journal of Solids and Structures 47(13):1698–1710

Bacca M, Bigoni D, Dal Corso F, Veber D (2013) Mindlin second-gradient elastic properties from dilute two-phase cauchy-elastic composites. part i: Closed form expression for the effective higher-order constitutive tensor. International Journal of Solids and Structures 50(24):4010–4019

Benveniste Y (1987) A new approach to the application of mori-tanaka's theory in composite materials. Mechanics of materials 6(2):147–157

Boutin C, dell'Isola F, Giorgio I, Placidi L (2017) Linear pantographic sheets: Asymptotic micro-macro models identification. Mathematics and Mechanics of Complex Systems 5(2):127–162

Christensen RM (1979) Mechanics of composite materials. John Wiley and Sons

Christensen RM (1990) A critical evaluation for a class of micro-mechanics models. Journal of the Mechanics and Physics of Solids 38(3):379–404

Christensen RM (2012) Mechanics of composite materials. Courier Corporation

Delfani M, Latifi Shahandashti M (2017) Elastic field of a spherical inclusion with non-uniform eigenfields in second strain gradient elasticity. Proceedings of the Royal Society A: Mathematical, Physical and Engineering Sciences 473(2205):20170,254

dell'Isola F, Andreaus U, Placidi L (2015) At the origins and in the vanguard of peridynamics, non-local and higher-gradient continuum mechanics: An underestimated and still topical contribution of Gabrio Piola. Mathematics and Mechanics of Solids 20(8):887–928

dell'Isola F, Madeo A, Seppecher P (2016) Cauchy tetrahedron argument applied to higher contact interactions. Archive for Rational Mechanics and Analysis 219(3):1305–1341

dell'Isola F, Cuomo M, Greco L, Della Corte A (2017) Bias extension test for pantographic sheets: numerical simulations based on second gradient shear energies. Journal of Engineering Mathematics 103(1):127–157

dell'Isola F, Seppecher P, Alibert JJ, et al (2019a) Pantographic metamaterials: an example of mathematically driven design and of its technological challenges. Continuum Mechanics and Thermodynamics 31(4):851–884

dell'Isola F, Seppecher P, Spagnuolo M, et al (2019b) Advances in pantographic structures: design, manufacturing, models, experiments and image analyses. Continuum Mechanics and Thermodynamics 31(4):1231–1282

Dunn ML, Taya M (1993) Micromechanics predictions of the effective electroelastic moduli of piezoelectric composites. International Journal of Solids and Structures 30(2):161–175

Eremeyev VA, Konopińska-Zmysłowska V (2019) On the correspondence between two-and three-dimensional Eshelby tensors. Continuum Mechanics and Thermodynamics 31(6):1615–1625

Eremeyev VA, dell'Isola F, Boutin C, Steigmann D (2018) Linear pantographic sheets: existence and uniqueness of weak solutions. Journal of Elasticity 132(2):175–196

Eshelby J (1956) The continuum theory of lattice defects. In: Solid state physics, vol 3, Elsevier, pp 79–144

Fleck NA, Hutchinson JW (1993) A phenomenological theory for strain gradient effects in plasticity. Journal of the Mechanics and Physics of Solids 41(12):1825–1857

Forest S (1998) Mechanics of generalized continua: construction by homogenizaton. Le Journal de Physique IV 8(PR4):Pr4–39

Forest S (2002) Homogenization methods and the mechanics of generalized continua. Theoretical and applied mechanics 28(29):113–144

Forest S, Trinh DK (2011) Generalized continua and non-homogeneous boundary conditions in homogenisation methods. ZAMM-Journal of Applied Mathematics and Mechanics/Zeitschrift für Angewandte Mathematik und Mechanik 91(2):90–109

Gao XL, Ma HM (2010) Strain gradient solution for Eshelby's ellipsoidal inclusion problem. Proceedings of the Royal Society A: Mathematical, Physical and Engineering Sciences 466(2120):2425–2446

Gao XL, Park SK (2007) Variational formulation of a simplified strain gradient elasticity theory and its application to a pressurized thick-walled cylinder problem. International Journal of Solids and Structures 44(22-23):7486–7499

Giorgio I, Harrison P, dell'Isola F, Alsayednoor J, Turco E (2018) Wrinkling in engineering fabrics: a comparison between two different comprehensive modelling approaches. Proceedings of the Royal Society A: Mathematical, Physical and Engineering Sciences 474(2216):20180,063

Goda I, Ganghoffer JF (2016) Construction of first and second order grade anisotropic continuum media for 3d porous and textile composite structures. Composite Structures 141:292–327

Gusev AA, Lurie SA (2009) Loss amplification effect in multiphase materials with viscoelastic interfaces. Macromolecules 42(14):5372–5377

Hashin Z (1983) Analysis of composite materials - a survey. Journal of Applied Mechanics 50(3):481–505

Jiang C, Tong Z, Cheung Y (2001) A generalized self-consistent method for piezoelectric fiber reinforced composites under antiplane shear. Mechanics of Materials 33(5):295–308

Kachanov M, Sevostianov I (2018) Micromechanics of materials, with applications, vol 249. Springer

Lagoudas D, Gavazzi A, Nigam H (1991) Elastoplastic behavior of metal matrix composites based on incremental plasticity and the mori-tanaka averaging scheme. Computational Mechanics 8(3):193–203

Lazar M, Maugin GA (2006a) Dislocations in gradient elasticity revisited. Proceedings of the Royal Society A: Mathematical, Physical and Engineering Sciences 462(2075):3465–3480

Lazar M, Maugin GA (2006b) A note on line forces in gradient elasticity. Mechanics Research Communications 33(5):674–680

Lazar M, Po G (2018) Singularity-free dislocation continuum theory for anisotropic crystals. PAMM 18(1):e201800,095

Lurie S, Belov P (2008) Cohesion field: Barenblatt's hypothesis as formal corollary of theory of continuous media with conserved dislocations. International Journal of Fracture 150(1-2):181–194

Lurie S, Belov P (2014) Gradient effects in fracture mechanics for nano-structured materials. Engineering Fracture Mechanics 130:3–11

Lurie S, Solyaev Y (2018) Revisiting bending theories of elastic gradient beams. International Journal of Engineering Science 126:1–21

Lurie S, Solyaev Y (2019) On the formulation of elastic and electroelastic gradient beam theories. Continuum Mechanics and Thermodynamics pp 1–13

Lurie S, Belov P, Tuchkova N (2005) The application of the multiscale models for description of the dispersed composites. Composites Part A: Applied Science and Manufacturing 36(2):145–152

Lurie S, Belov P, Volkov-Bogorodsky D, Tuchkova N (2006) Interphase layer theory and application in the mechanics of composite materials. Journal of materials science 41(20):6693–6707
Lurie S, Volkov-Bogorodsky D, Leontiev A, Aifantis E (2011) Eshelby's inclusion problem in the gradient theory of elasticity: applications to composite materials. International Journal of Engineering Science 49(12):1517–1525
Lurie S, Minhat M, Tuchkova N, Soliaev J (2014) On remarkable loss amplification mechanism in fiber reinforced laminated composite materials. Applied Composite Materials 21(1):179–196
Lurie S, Volkov-Bogorodskii D, Tuchkova N (2016) Exact solution of eshelby–christensen problem in gradient elasticity for composites with spherical inclusions. Acta Mechanica 227(1):127–138
Lurie S, Solyaev Y, Rabinskiy L, Polyakov P, Sevostianov I (2018a) Mechanical behavior of porous si3n4 ceramics manufactured with 3d printing technology. Journal of Materials Science 53(7):4796–4805
Lurie S, Solyaev Y, Shramko K (2018b) Comparison between the mori-tanaka and generalized self-consistent methods in the framework of anti-plane strain inclusion problem in strain gradient elasticity. Mechanics of Materials 122:133–144
Lurie S, Solyaev Y, Ustenko A (2019) Optimal damping behavior of a composite sandwich beam reinforced with coated fibers. Applied Composite Materials 26(1):389–408
Ma HM, Gao XL (2014) A new homogenization method based on a simplified strain gradient elasticity theory. Acta Mechanica 225(4-5):1075–1091
Mindlin RD (1964) Micro-structure in linear elasticity. Archive for Rational Mechanics and Analysis 16(1):51–78
Mindlin RD, Eshel N (1968) On first strain-gradient theories in linear elasticity. International Journal of Solids and Structures 4(1):109–124
Mousavi SM, Aifantis EC (2016) Dislocation-based gradient elastic fracture mechanics for in-plane analysis of cracks. International Journal of Fracture 202(1):93–110
Papargyri-Beskou S, Tsepoura K, Polyzos D, Beskos D (2003) Bending and stability analysis of gradient elastic beams. International Journal of Solids and Structures 40(2):385–400
Polizzotto C (2015) A unifying variational framework for stress gradient and strain gradient elasticity theories. European Journal of Mechanics, A/Solids 49:430–440
Reiher JC, Giorgio I, Bertram A (2017) Finite-Element Analysis of Polyhedra under Point and Line Forces in Second-Strain Gradient Elasticity. Journal of Engineering Mechanics 143(2):04016,112
Scerrato D, Giorgio I (2019) Equilibrium of two-dimensional cycloidal pantographic metamaterials in three-dimensional deformations. Symmetry 11(12):1523
Siboni G, Benveniste Y (1991) A micromechanics model for the effective thermomechanical behaviour of multiphase composite media. Mechanics of Materials 11(2):107–122
Trinh DK, Janicke R, Auffray N, Diebels S, Forest S (2012) Evaluation of generalized continuum substitution models for heterogeneous materials. International Journal for Multiscale Computational Engineering 10(6)
Turco E, Misra A, Pawlikowski M, dell'Isola F, Hild F (2018) Enhanced Piola–Hencky discrete models for pantographic sheets with pivots without deformation energy: numerics and experiments. International Journal of Solids and Structures 147:94–109
Vasil'ev V, Lurie S (2017) New solution of axisymmetric contact problem of elasticity. Mechanics of Solids 52(5):479–487

Chapter 27
The Mechanical Diode: On the Tracks of James Maxwell Employing Mechanical–Electrical Analogies in the Design of Metamaterials

Mario Spagnuolo & Daria Scerrato

Abstract Following the pioneering idea of James Clerk Maxwell, who first introduced mechanical-electrical analogies in the 19th century, we want to exploit his innovative vision in the design of new metamaterials. Indeed, metamaterials can be defined as engineered materials designed to implement one or more specific functionalities. In particular, we have in mind to realize a mechanical system whose behavior is characterized by a useful response, which is typical of an electrical system. In other words, we want to design a mechanical system whose constitutive behavior is the same as an electrical system, namely a diode. To this end, a pantographic material with perfect pivots can be seen as a mechanical diode. Indeed, it is constituted by two families of straight fibers, with remarkable deformation due to the presence of the hinges and a threshold behavior when the fibers tend to be aligned.

27.1 Introduction

Nowadays, the 3D printing process offers an unparalleled possibility to realize highly complex mechanical systems characterized by internal structures that provide "exotic" (not present in nature) behaviors (dell'Isola et al, 2015b; Milton et al, 2017). From this point of view, the technological advancement we are witnessing is particular fruitfully for the development of new materials so-called metamaterials (Del Vescovo and Giorgio, 2014; Barchiesi et al, 2019b). Indeed, metamaterials are artificially conceived materials that inherit the macroscopic behavior by their microstructure rather than the material of which they are made. Among the panoply of mechanical metamaterials, we want to focus on pantographic sheets, namely mechanical systems, which can experience a very high deformation in the elastic range.

M. Spagnuolo & D. Scerrato
International Research Center for the Matematics and Mechanics of Complex Systems, M&MoCS, Università degli Studi dell'Aquila, L'Aquila, Italy
e-mail: mario.spagnuolo.memocs@gmail.com, daria.scerrato@gmail.com

B. E. Abali and I. Giorgio (eds.), *Developments and Novel Approaches in Biomechanics and Metamaterials*, Advanced Structured Materials 132,
https://doi.org/10.1007/978-3-030-50464-9_27

The microstructure of a pantographic sheet is at a microscopic level of observation, a mechanism—namely, a pantograph—periodically repeated that allows a local motion spreading above the majority of the unit cells of which the system is made of and accordingly results in a considerable deformation on a macroscopic scale (Barchiesi and Placidi, 2017; dell'Isola et al, 2019a,b; Turco et al, 2017a; Misra et al, 2018). To be more precise, the architecture of the system at hand consists of two parallel fiber families that are joined each other in correspondence to the intersection points of the fibers. In particular, these connections can be realized by pin devices, which are equivalent to hinges (characterized by free rotations). Details of this research are given in (dell'Isola et al, 2015b; Scerrato et al, 2017; Turco et al, 2017b; Placidi et al, 2016; Giorgio et al, 2017; Turco et al, 2016; Boutin et al, 2017; Steigmann and dell'Isola, 2015; Barchiesi et al, 2019a; Giorgio et al, 2018; Turco et al, 2020).

In this paper, we resort to mechanical-electrical analogies to support in developing mechanical systems that can provide behaviors typical of electrical networks. Since the first proposal from James C. Maxwell in the 19th century, as it has been already done in the past, this analogy is very useful in the design of a system of one of these two domains using the achievements of the other one. Hence, for example, one can design mechanical systems that are realized and developed by means of electric networks (Bloch, 1945; Kron, 1945). A further example can be found in Bloch (1944), where a purely mechanical coupler, i.e. the gyroscopic, is discussed and a better understanding of its behavior is attained by a translation into the electrical analogy.

Pursue this line of thought, the results obtained for pantographic sheets from experimental tests and numerical simulations shown an analogy between the force-displacement diagram curve and the constitutive law voltage-current characteristic of electric diodes. We observed that this is particularly valid when the connections are perfect hinges. Indeed, if the connections are made with deformable elastic cylinders this specific behavior is less evident (Turco et al, 2018).

27.2 Experimental Observations: A Mechanical Diode

A diode is an electric component characterized by an asymmetric conductance. The current that flows in it is almost null in the case that the applied voltage is less or equal than a certain threshold. On the other hand, when the external voltage exceeds that value, the diode behaves like a voltage generator. Therefore, the current will be a consequence of the electric network where the diode is placed (provided that the maximum allowed current is not exceeded).

The same behavior exhibited by an electric diode can be reproduced in the mechanical field by a pantographic sheet with connections, also called pivots, realized by "perfect" hinges (Fig. 27.1).

Indeed, in the case of a bias extension test, the force-displacement curve of a pantographic sheet can be summarized into two principal steps: firstly, the reaction force is caused by a locally spread bending deformation of fibers, which has a very

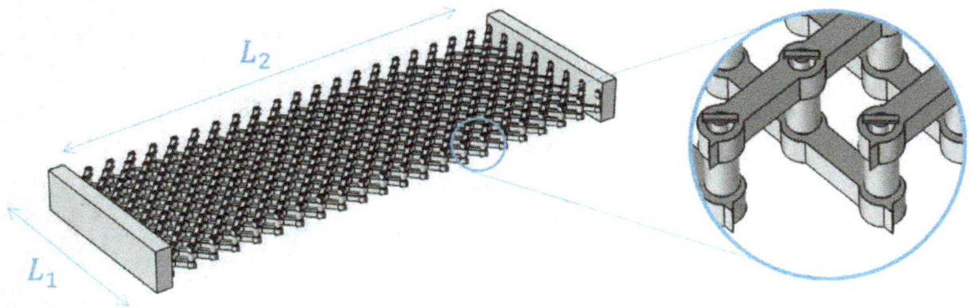

Fig. 27.1 A three-dimensional scheme of a pantographic structure with perfect pivots and a zoom of the connections.

low value (almost negligible) compared to the maximum value of the exerted force; secondly, the deformation turns to an elongation of the fibers and the slope of the force increases drastically, especially after the fibers come into contact each other.

In this mechanical case, therefore, the response of the system resembles that of an electric diode. Here, the macroscopic deformation is a consequence of the deformation of the fundamental constituents at the micro-level, namely fibers and pivots. Specifically, it is worth noticing that when the pivots are almost perfect (see Fig. 27.5), and the energy associated with their deformation is negligible, the macroscopic behavior is mainly driven by the deformation of the fibers which can only bend and stretch (see Fig. 27.2).

To be more precise, in the bias extension test the presence of perfect pivots allows: (i) the relative rotation of the fibers belonging to the different families connected by the pivot itself with very small deformation energy due to the pantographic architecture; (ii) initially only a flexion of the fibers is triggered in narrow localized strips, which is related to almost negligible storage of elastic energy, (iii) then, when the fibers become increasingly aligned, in the central portion of the structure, the elongation of them becomes predominant, and the increase of reaction force needed to deform the sample rises considerably (see Fig. 27.3).

To illustrate what has been explained before, some experimental tests were performed together with some numerical simulations that corroborate the theoretical formulations already developed in the literature about pantographic sheets with perfect pivot. The experiments were carried out on a specimen made of polyamide. The results of the bias extension test are reported as a force-displacement plot shown in Fig. 27.4.

The plot of Fig. 27.4 shows that the reaction force is negligible until the imposed displacement reached a value of about 35 mm, namely ca. 16.7% of the specimen length which is 210 mm. Indeed in this first part of the experiment, the level of the

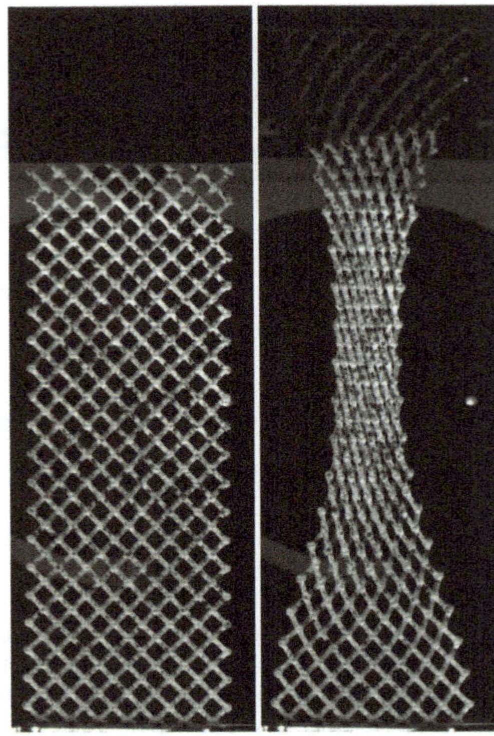

Fig. 27.2 Pantographic sheet with perfect pivots under a bias extension test. Reference configuration (left), deformed configuration with the characteristic hourglass shape (right).

force is of the same order of magnitude of the experimental noise. After this phase, the force increases as in the case of an electric diode.

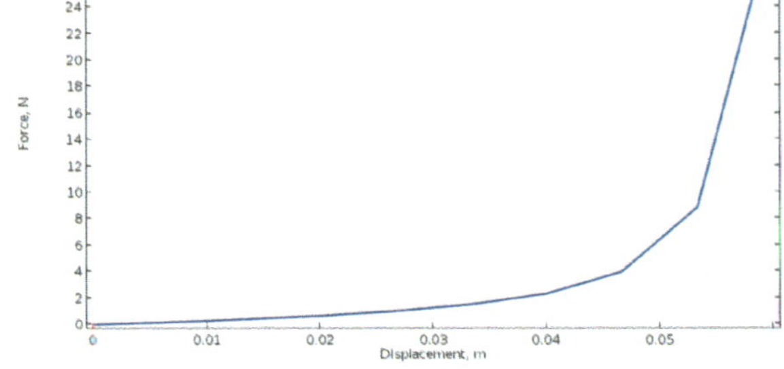

Fig. 27.3 Force-displacement plot of a pantographic sheet obtained by FEM numerical simulations.

27.3 Modeling Metamaterials: Deformation Energy of Pantographic Sheets

To design a new metamaterial, the first step to be done is to establish the performance required in the ordinary functioning of it. Subsequently, the microstructure that characterizes the macroscopic behavior should be synthesized to get the prescribed requirements, namely the desired mechanical response. In the case of the pantographic sheet, the desired behavior consists of large deformation of the material with a zero expense of energy, or negligible. In Seppecher et al (2011); Alibert et al (2003) it is proven that such a material can be obtained by utilizing a pantographic microstructure (see Fig. 27.1).

According to the innovative ideas of Germain, Tupin, Sedov, and Piola (dell'Isola et al, 2017), and following the seminal work of Seppecher et al (2011), an assembly of fibers as shown in Fig. 27.1 can be proved to have a homogenized macroscopic model with a term of energy that involves the presence of the second gradient of the displacement (Auffray et al, 2015; dell'Isola et al, 2015a, 2012; Eugster and dell'Isola, 2017, 2018a,b). The model resulting from the homogenization has the following properties:

1. it consists of two infinite distributions of mutually orthogonal fibers, that are linked in their intersecting points with the so-called pivots (see Fig.27.1);
2. the pivots behave as perfect hinges, i.e no energy is involved in their deformations;
3. when clamping the short sides of a rectangular sample, its deformation is due to elongation and flexion of the fibers; in particular, the latter is modeled by a second gradient energetic term.

From a mathematical point of view, the macroscopic model can be expressed as a continuum elastic surface characterized by some second-order derivatives of the displacement field. It is worth noticing that the considered plate is actually a second gradient continuum however not all the second derivatives of displacement are present, some of them miss. For this reason, the solution to the problem under

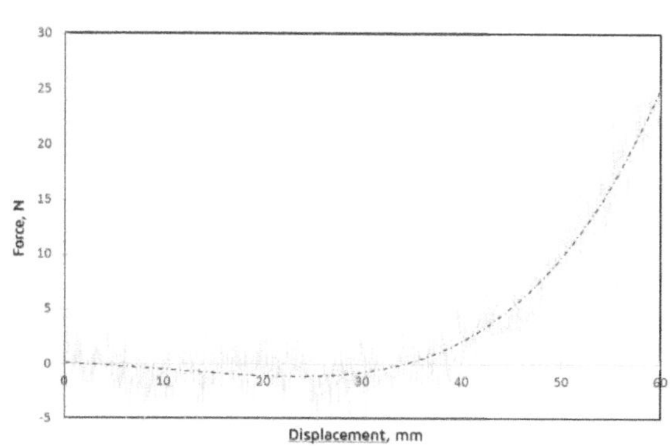

Fig. 27.4 Force-displacement plot of a pantographic sheet obtained by experiments (grey solid line) and interpolation (red dash-dotted line).

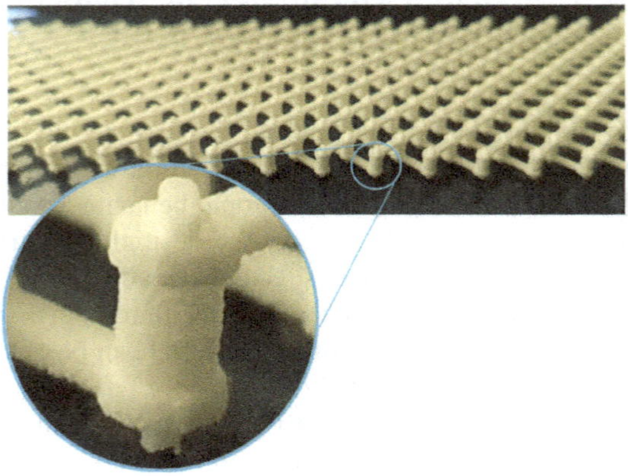

Fig. 27.5 Real sample of a pantographic sheet with perfect pivots printed with Polyamide and zoom of the pivot.

study can be difficult to handle some times from a numerical viewpoint if the proper precautions are not taken (Eremeyev et al, 2018; Pideri and Seppecher, 1997; Camar-Eddine and Seppecher, 2003; Bouchitté et al, 2019; dell'Isola et al, 2016a; Eugster et al, 2014, 2019).

Specifically, let us consider a rectangular sample in the reference configuration given by the domain $\Omega = [0, L_1] \times [0, L_2] \subset R^2$ (see Fig. 27.1) where L_1 and L_2 are the lengths of the sides of the rectangle which contains the pantographic specimen) and assume that the motion is planar, thus the current configuration of Ω is described by the macro-placement

$$\chi : \Omega \to R^2 \tag{27.1}$$

According to dell'Isola et al (2016b), the continuum deformation energy $\mathfrak{U}$ stored in a pantographic fabric can be postulated as

$$\begin{aligned}\mathfrak{U}(\chi(\cdot)) = &\int_\Omega \sum_\alpha \frac{\mathrm{K}_e}{2}(\|\boldsymbol{F}\mathrm{D}_\alpha\| - 1)^2 \,\mathrm{d}\Omega + \\ &+ \int_\Omega \sum_\alpha \frac{\mathrm{K}_b}{2}\left[\frac{\nabla\boldsymbol{F}|\mathrm{D}_\alpha \otimes \mathrm{D}_\alpha \cdot \nabla\boldsymbol{F}|\mathrm{D}_\alpha \otimes \mathrm{D}_\alpha}{\|\boldsymbol{F}\mathrm{D}_\alpha\|^2} - \right. \\ &\left. - \left(\frac{\boldsymbol{F}\mathrm{D}_\alpha}{\|\boldsymbol{F}\mathrm{D}_\alpha\|} \cdot \frac{\nabla\boldsymbol{F}|\mathrm{D}_\alpha \otimes \mathrm{D}_\alpha}{\|\boldsymbol{F}\mathrm{D}_\alpha\|}\right)^2\right] \mathrm{d}\Omega\end{aligned} \tag{27.2}$$

where $\boldsymbol{F} = \nabla\chi$ is the classical deformation gradient tensor, while D_α are the unit vectors along the fibers in the reference configuration, $\alpha = \{1, 2\}$ being the index representing the particular family of fibers. The material parameters K_e and K_b are the elongation and bending stiffnesses, respectively. It should be noticed that in the expression of the energy (27.2) are identifiable the elongation measure of the fibers

$$\varepsilon_\alpha = ||\boldsymbol{F}\mathrm{D}_\alpha|| - 1 \tag{27.3}$$

in the two principal directions of fibers, and the curvature of them

$$\kappa_\alpha = \sqrt{\frac{\nabla\boldsymbol{F}|\mathrm{D}_\alpha \otimes \mathrm{D}_\alpha \cdot \nabla\boldsymbol{F}|\mathrm{D}_\alpha \otimes \mathrm{D}_\alpha}{||\boldsymbol{F}\mathrm{D}_\alpha||^2} - \left(\frac{\boldsymbol{F}\mathrm{D}_\alpha}{||\boldsymbol{F}\mathrm{D}_\alpha||} \cdot \frac{\nabla\boldsymbol{F}|\mathrm{D}_\alpha \otimes \mathrm{D}_\alpha}{||\boldsymbol{F}\mathrm{D}_\alpha||}\right)^2} \tag{27.4}$$

In the case that the interconnecting pivots are not perfect hinges, a further term must be added into the energy. Deformable pivots realized as small cylinders, indeed, can be considered alternately to the perfect hinges, and therefore they introduce a torsion stored energy at micro-level which corresponds to a shear deformation at macro-level of observation. Moreover, it could be also considered another term in the energy that models the relative sliding between the two fiber layers in correspondence of the interconnections. In this paper, the sliding effect is neglected, while it has been experimentally observed and theoretically investigated in previous works (Spagnuolo et al, 2017; Andreaus et al, 2018).

27.4 Conclusions and Future Perspectives

In this paper, we show that a particular mechanical system, namely the pantographic sheet, is characterized by a constitutive behavior force-displacement that resembles very closely to that of an electric diode concerning the voltage-current curve. From this point of view, we can interpret the pantographic sheet, when the internal connections are made of perfect pivots, as a mechanical diode. The presence of this kind of connection, indeed, implies vanishing shear energy, at a macroscopic scale, in the homogenized model responsible for an almost zero reaction force below a particular threshold. Above that threshold, the terms of energy associated with the elongation of the fibers become predominant and therefore the reaction force arises significantly.

This kind of new material can be used in applications that need to provide a reaction force above a certain threshold. This behavior can be useful in the field of automotive, aerospace as a sort of suspension, bumpers, i.e. shock absorbers, as well as in civil engineering for structural elements in earthquake-proof constructions.

The particular behavior of such a system can be employed in connection with other materials to form composites with enhanced performance, for instance using granular materials (Jia et al, 2017; Misra and Poorsolhjouy, 2017; Turco et al, 2019; Eremeyev, 2018), laminate plates (Cazzani et al, 2018; Altenbach et al, 2015; Altenbach and Eremeyev, 2008), micropolar materials (Pietraszkiewicz and Eremeyev, 2009; Altenbach and Eremeyev, 2009; Eremeyev and Pietraszkiewicz, 2016), multi-physics materials (Abali and Zohdi, 2020; Abali and Queiruga, 2019; Abali and Reich, 2018, 2017).

As a future development, the behavior of the proposed pantographic structure can be also investigated in the plastic regime and in case of damage (see, e.g., Placidi and Barchiesi, 2018; Placidi et al, 2018b,a).

The pantographic sheet can be considered as a particular example of a second gradient material, thus, like many metamaterials of new conception, it needs proper numerical tools to handle in an efficient way the issues related to its formulation (see for more details Cazzani et al, 2016; Luongo et al, 2008; Giorgio, 2016; Placidi et al, 2017; Greco et al, 2019b,a; Niiranen et al, 2019; Balobanov and Niiranen, 2018; De Angelo et al, 2019).

References

Abali BE, Queiruga AF (2019) Theory and computation of electromagnetic fields and thermomechanical structure interaction for systems undergoing large deformations. Journal of Computational Physics 394:200–231

Abali BE, Reich FA (2017) Thermodynamically consistent derivation and computation of electro-thermo-mechanical systems for solid bodies. Computer Methods in Applied Mechanics and Engineering 319:567–595

Abali BE, Reich FA (2018) Verification of deforming polarized structure computation by using a closed-form solution. Continuum Mechanics and Thermodynamics pp 1–16

Abali BE, Zohdi TI (2020) Multiphysics computation of thermal tissue damage as a consequence of electric power absorption. Computational Mechanics 65:149–158

Alibert JJ, Seppecher P, dell'Isola F (2003) Truss modular beams with deformation energy depending on higher displacement gradients. Mathematics and Mechanics of Solids 8(1):51–73

Altenbach H, Eremeyev V (2009) On the linear theory of micropolar plates. ZAMM-Journal of Applied Mathematics and Mechanics/Zeitschrift für Angewandte Mathematik und Mechanik 89(4):242–256

Altenbach H, Eremeyev VA (2008) Direct approach-based analysis of plates composed of functionally graded materials. Archive of Applied Mechanics 78(10):775–794

Altenbach H, Eremeyev VA, Naumenko K (2015) On the use of the first order shear deformation plate theory for the analysis of three-layer plates with thin soft core layer. ZAMM-Journal of Applied Mathematics and Mechanics/Zeitschrift für Angewandte Mathematik und Mechanik 95(10):1004–1011

Andreaus U, Spagnuolo M, Lekszycki T, Eugster SR (2018) A Ritz approach for the static analysis of planar pantographic structures modeled with nonlinear Euler–Bernoulli beams. Continuum Mechanics and Thermodynamics 30:1103–1123

Auffray N, dell'Isola F, Eremeyev V, Madeo A, Rosi G (2015) Analytical continuum mechanics à la Hamilton–Piola least action principle for second gradient continua and capillary fluids. Mathematics and Mechanics of Solids 20(4):375–417

Balobanov V, Niiranen J (2018) Locking-free variational formulations and isogeometric analysis for the timoshenko beam models of strain gradient and classical elasticity. Computer Methods in Applied Mechanics and Engineering 339:137–159

Barchiesi E, Placidi L (2017) A review on models for the 3D statics and 2D dynamics of pantographic fabrics. In: Wave Dynamics and Composite Mechanics for Microstructured Materials and Metamaterials, Springer, pp 239–258

Barchiesi E, Ganzosch G, Liebold C, Placidi L, Grygoruk R, Müller WH (2019a) Out-of-plane buckling of pantographic fabrics in displacement-controlled shear tests: experimental results and model validation. Continuum Mechanics and Thermodynamics 31(1):33–45

Barchiesi E, Spagnuolo M, Placidi L (2019b) Mechanical metamaterials: a state of the art. Mathematics and Mechanics of Solids 24(1):212–234

Bloch A (1944) XXXVIII. A new approach to the dynamics of systems with gyroscopic coupling terms. The London, Edinburgh, and Dublin Philosophical Magazine and Journal of Science 35(244):315–334

Bloch A (1945) Electromechanical analogies and their use for the analysis of mechanical and electromechanical systems. Journal of the Institution of Electrical Engineers-Part I: General 92(52):157–169

Bouchitté G, Mattei O, Milton GW, Seppecher P (2019) On the forces that cable webs under tension can support and how to design cable webs to channel stresses. Proceedings of the Royal Society A 475(2223):20180,781

Boutin C, Giorgio I, Placidi L, et al (2017) Linear pantographic sheets: Asymptotic micro-macro models identification. Mathematics and Mechanics of Complex Systems 5(2):127–162

Camar-Eddine M, Seppecher P (2003) Determination of the closure of the set of elasticity functionals. Archive for rational mechanics and analysis 170(3):211–245

Cazzani A, Malagù M, Turco E (2016) Isogeometric analysis: a powerful numerical tool for the elastic analysis of historical masonry arches. Continuum Mechanics and thermodynamics 28(1-2):139–156

Cazzani A, Serra M, Stochino F, Turco E (2018) A refined assumed strain finite element model for statics and dynamics of laminated plates. Continuum Mechanics and Thermodynamics pp 1–28

De Angelo M, Barchiesi E, Giorgio I, Abali BE (2019) Numerical identification of constitutive parameters in reduced-order bi-dimensional models for pantographic structures: application to out-of-plane buckling. Archive of Applied Mechanics 89(7):1333–1358

Del Vescovo D, Giorgio I (2014) Dynamic problems for metamaterials: review of existing models and ideas for further research. International Journal of Engineering Science 80:153–172

dell'Isola F, Seppecher P, Madeo A (2012) How contact interactions may depend on the shape of Cauchy cuts in nth gradient continua: approach "à la D'Alembert". Zeitschrift für angewandte Mathematik und Physik 63(6):1119–1141

dell'Isola F, Andreaus U, Placidi L (2015a) At the origins and in the vanguard of peridynamics, non-local and higher-gradient continuum mechanics: An underestimated and still topical contribution of Gabrio Piola. Mathematics and Mechanics of Solids 20(8):887–928

dell'Isola F, Steigmann D, Della Corte A (2015b) Synthesis of fibrous complex structures: Designing microstructure to deliver targeted macroscale response. Applied Mechanics Reviews 67(6):060,804

dell'Isola F, Della Corte A, Esposito R, Russo L (2016a) Some cases of unrecognized transmission of scientific knowledge: from antiquity to Gabrio Piola's peridynamics and generalized continuum theories. In: Generalized continua as models for classical and advanced materials, Springer, pp 77–128

dell'Isola F, Giorgio I, Pawlikowski M, Rizzi N (2016b) Large deformations of planar extensible beams and pantographic lattices: heuristic homogenization, experimental and numerical examples of equilibrium. Proc R Soc A 472(2185):20150,790

dell'Isola F, Della Corte A, Giorgio I (2017) Higher-gradient continua: The legacy of Piola, Mindlin, Sedov and Toupin and some future research perspectives. Mathematics and Mechanics of Solids 22(4):852–872

dell'Isola F, Seppecher P, Alibert JJ, et al (2019a) Pantographic metamaterials: an example of mathematically driven design and of its technological challenges. Continuum Mechanics and Thermodynamics 31(4):851–884

dell'Isola F, Seppecher P, Spagnuolo M, et al (2019b) Advances in pantographic structures: design, manufacturing, models, experiments and image analyses. Continuum Mechanics and Thermodynamics 31:1231–1282

Eremeyev VA (2018) On the material symmetry group for micromorphic media with applications to granular materials. Mechanics Research Communications 94:8–12

Eremeyev VA, Pietraszkiewicz W (2016) Material symmetry group and constitutive equations of micropolar anisotropic elastic solids. Mathematics and Mechanics of Solids 21(2):210–221

Eremeyev VA, dell'Isola F, Boutin C, Steigmann D (2018) Linear pantographic sheets: existence and uniqueness of weak solutions. Journal of Elasticity 132(2):175–196

Eugster S, Hesch C, Betsch P, Glocker C (2014) Director-based beam finite elements relying on the geometrically exact beam theory formulated in skew coordinates. International Journal for Numerical Methods in Engineering 97(2):111–129

Eugster S, Steigmann D, et al (2019) Continuum theory for mechanical metamaterials with a cubic lattice substructure. Mathematics and Mechanics of Complex Systems 7(1):75–98

Eugster SR, dell'Isola F (2017) Exegesis of the Introduction and Sect. I from "Fundamentals of the Mechanics of Continua"** by E. Hellinger. ZAMM-Journal of Applied Mathematics and Mechanics/Zeitschrift für Angewandte Mathematik und Mechanik 97(4):477–506

Eugster SR, dell'Isola F (2018a) Exegesis of sect. II and III. A from "Fundamentals of the mechanics of continua" by E. Hellinger. ZAMM-Journal of Applied Mathematics and Mechanics/Zeitschrift für Angewandte Mathematik und Mechanik 98(1):31–68

Eugster SR, dell'Isola F (2018b) Exegesis of Sect. III. B from "Fundamentals of the Mechanics of Continua" by E. Hellinger. ZAMM-Journal of Applied Mathematics and Mechanics/Zeitschrift für Angewandte Mathematik und Mechanik 98(1):69–105

Giorgio I (2016) Numerical identification procedure between a micro-Cauchy model and a macro-second gradient model for planar pantographic structures. Zeitschrift für angewandte Mathematik und Physik 67(4)(95)

Giorgio I, Rizzi N, Turco E (2017) Continuum modelling of pantographic sheets for out-of-plane bifurcation and vibrational analysis. Proc R Soc A 473(2207):20170,636

Giorgio I, Harrison P, dell'Isola F, Alsayednoor J, Turco E (2018) Wrinkling in engineering fabrics: a comparison between two different comprehensive modelling approaches. Proceedings of the Royal Society A: Mathematical, Physical and Engineering Sciences 474(2216):20180,063

Greco L, Cuomo M, Contrafatto L (2019a) A quadrilateral G1-conforming finite element for the Kirchhoff plate model. Computer Methods in Applied Mechanics and Engineering 346:913–951

Greco L, Cuomo M, Contrafatto L (2019b) Two new triangular G1-conforming finite elements with cubic edge rotation for the analysis of Kirchhoff plates. Computer Methods in Applied Mechanics and Engineering 356:354–386

Jia H, Misra A, Poorsolhjouy P, Liu C (2017) Optimal structural topology of materials with micro-scale tension-compression asymmetry simulated using granular micromechanics. Materials & Design 115:422–432

Kron G (1945) Numerical solution of ordinary and partial differential equations by means of equivalent circuits. Journal of Applied Physics 16(3):172–186

Luongo A, Zulli D, Piccardo G (2008) Analytical and numerical approaches to nonlinear galloping of internally resonant suspended cables. Journal of Sound and Vibration 315(3):375–393

Milton G, Briane M, Harutyunyan D (2017) On the possible effective elasticity tensors of 2-dimensional and 3-dimensional printed materials. Mathematics and Mechanics of Complex Systems 5(1):41–94

Misra A, Poorsolhjouy P (2017) Grain-and macro-scale kinematics for granular micromechanics based small deformation micromorphic continuum model. Mechanics Research Communications 81:1–6

Misra A, Lekszycki T, Giorgio I, Ganzosch G, Müller WH, dell'Isola F (2018) Pantographic metamaterials show atypical poynting effect reversal. Mechanics Research Communications 89:6–10

Niiranen J, Balobanov V, Kiendl J, Hosseini S (2019) Variational formulations, model comparisons and numerical methods for Euler–Bernoulli micro-and nano-beam models. Mathematics and Mechanics of Solids 24(1):312–335

Pideri C, Seppecher P (1997) A second gradient material resulting from the homogenization of an heterogeneous linear elastic medium. Continuum Mechanics and Thermodynamics 9(5):241–257

Pietraszkiewicz W, Eremeyev V (2009) On natural strain measures of the non-linear micropolar continuum. International Journal of Solids and Structures 46(3):774–787

Placidi L, Barchiesi E (2018) Energy approach to brittle fracture in strain-gradient modelling. Proc R Soc A 474(2210):20170,878

Placidi L, Barchiesi E, Turco E, Rizzi NL (2016) A review on 2D models for the description of pantographic fabrics. Zeitschrift für angewandte Mathematik und Physik 67(5):121

Placidi L, Barchiesi E, Battista A (2017) An inverse method to get further analytical solutions for a class of metamaterials aimed to validate numerical integrations. In: Mathematical Modelling in Solid Mechanics, Springer, pp 193–210

Placidi L, Misra A, Barchiesi E (2018a) Simulation results for damage with evolving microstructure and growing strain gradient moduli. Continuum Mechanics and Thermodynamics pp 1–21

Placidi L, Misra A, Barchiesi E (2018b) Two-dimensional strain gradient damage modeling: a variational approach. Zeitschrift für angewandte Mathematik und Physik 69(3):56

Scerrato D, Eremeeva IZ, Lekszycki T, Rizzi N (2017) On the shear stiffness influence for modelling of deformations of pantographic sheets. In: Shell Structures: Theory and Applications Volume 4: Proceedings of the 11th International Conference" Shell Structures: Theory and Applications,(SSTA 2017), October 11-13, 2017, Gdansk, Poland, CRC Press, p 161

Seppecher P, Alibert JJ, dell'Isola F (2011) Linear elastic trusses leading to continua with exotic mechanical interactions. In: Journal of Physics: Conference Series, IOP Publishing, vol 319, p 012018

Spagnuolo M, Barcz K, Pfaff A, dell'Isola F, Franciosi P (2017) Qualitative pivot damage analysis in aluminum printed pantographic sheets: Numerics and experiments. Mechanics Research Communications 83:47–52

Steigmann D, dell'Isola F (2015) Mechanical response of fabric sheets to three-dimensional bending, twisting, and stretching. Acta Mechanica Sinica 31(3):373–382

Turco E, dell'Isola F, Cazzani A, Rizzi N (2016) Hencky-type discrete model for pantographic structures: numerical comparison with second gradient continuum models. Zeitschrift für angewandte Mathematik und Physik 67

Turco E, Giorgio I, Misra A, dell'Isola F (2017a) King post truss as a motif for internal structure of (meta) material with controlled elastic properties. Royal Society open science 4(10):171,153

Turco E, Golaszewski M, Giorgio I, D'Annibale F (2017b) Pantographic lattices with non-orthogonal fibres: Experiments and their numerical simulations. Composites Part B: Engineering 118:1–14

Turco E, Misra A, Pawlikowski M, dell'Isola F, Hild F (2018) Enhanced Piola–Hencky discrete models for pantographic sheets with pivots without deformation energy: Numerics and experiments. International Journal of Solids and Structures 147:94–109

Turco E, dell'Isola F, Misra A (2019) A nonlinear Lagrangian particle model for grains assemblies including grain relative rotations. International Journal for Numerical and Analytical Methods in Geomechanics 43(5):1051–1079

Turco E, Barchiesi E, Giorgio I, dell'Isola F (2020) A Lagrangian Hencky-type non-linear model suitable for metamaterials design of shearable and extensible slender deformable bodies alternative to Timoshenko theory. International Journal of Non-Linear Mechanics 123:103,481

Chapter 28
Tailoring 3D Buckling and Post Contact in Microlattice Metamaterials

Zacharias Vangelatos, Kyriakos Komvopoulos, and Costas P. Grigoropoulos

Abstract Mechanical metamaterials have been established as the paragons of enhanced mechanical performance, due to their properties inherited by their architected microstructure. The rapid progress in additive manufacturing has enabled the fabrication of complex geometries even at the microscale, with nanoscale features. For the case of ultra-light, ultra-stiff mechanical metamaterials, the high stiffness and increased energy dissipation is associated with the controlled buckling and subsequent post contact of the lattice members. In addition, architected defects, scaled up in the microscale, inspired by the structural defects in the crystal structure, have accomplished tailoring the plasticity mechanisms and localized deformation in a similar manner as slip planes through dislocation motion and vacancies in bulk crystalline materials. This chapter will focus on the design principles that must be addressed to effectively fabricate novel geometries possessing these intrinsic properties. These principles will be instantiated through the design of intertwined microlattices, inspired by crystal in-growth mechanisms, and architected 3D vacancies, inspired by crystal defects. All of the structures were fabricated through multiphoton lithography, the most suitable technique for fabrication of complex geometries at the micrometer-nanometer length scale. Through finite element analysis and nanoindentation experiments, we delineate how the mechanical behavior is manifested and can be tailored.

Keywords: Mechanical metamaterials · Controlled buckling · Architected defects · Intertwined structures

Z. Vangelatos, K. Komvopoulos, C. P. Grigoropoulos
University of California, Berkeley, 2521 Hearst Ave, 94720-1740, Berkeley, CA, USA
e-mail: zvaggelatos@gmail.com, kyriakos@me.berkeley.edu, cgrigoro@berkeley.edu

B. E. Abali and I. Giorgio (eds.), *Developments and Novel Approaches in Biomechanics and Metamaterials*, Advanced Structured Materials 132,
https://doi.org/10.1007/978-3-030-50464-9_28

28.1 Introduction

The advances in additive manufacturing utilizing techniques such as multiphoton lithography (MPL) (Sakellari et al, 2012) and numerical modeling (Yildizdag et al, 2018, 2019; Milton et al, 2017), has provided the inexorable progress in the design of architected mechanical metamaterials. Mechanical metamaterials possess properties bestowed by their engineered structure. Indisputable paradigms are the ultra-light, ultra-stiff metamaterials for extremely low weight and enhanced strength (Bauer et al, 2015; dell'Isola et al, 2019b), plate structures for stiffness reaching the theoretical limit of bulk materials (Berger et al, 2017), malleable structures for tailored deformations (dell'Isola et al, 2019b; Barchiesi et al, 2018; dell'Isola et al, 2015, 2018, 2019a) and controlled wave propagation (Berezovski et al, 2018). More specifically, microlattice structures are among the most studied architectures because of their simple design and scalability. Post-processing techniques, such as plasma etching, can provide hollow beams with tens of nanometers thickness with unique scale effects (Meza et al, 2014). The reason behind the objective to achieve nanoscale complex features is to imitate nature. Natural materials, through structural hierarchy can exhibit enhanced strength and resilience to large deformations, even if the initial bulk material is brittle (Gao et al, 2003). The primary mechanisms that must be fine-tuned, are the controlled buckling of the lattice members (dell'Isola et al, 2016; Misra et al, 2018), which will initiate large deformations and the subsequent post contact of them (Virgin, 2018). The contact of the lattice members through friction will enhance the energy dissipation and increase the stiffness and strength of the structure (Gibson and Ashby, 1999). Many different buckling and post contact mechanisms can commence during the deformation, rendering their design characterization crucial for a functional design (Giorgio et al, 2017; Vangelatos et al, 2019b).Furthermore, another design principle that has provided remarkable mechanical performance is the elevation of crystal structural defects. More specifically, by the tactical removal of lattice members from specific unit cells, we can architect defected unit cells which will localize the buckling and plasticity in the structure (Gross et al, 2019). It was proven that when FCC and BCC unit cells deform in macroscale, they substantiate mechanisms such as twin boundaries and slip planes with specific orientation, where localized failure can be isolated (Pham et al, 2019). For the first case, the challenge to imitate natural hierarchy is the constraint to create arbitrary complex features with high resolution, restrained by Abbe's diffraction limit (Vicidomini et al, 2018). Up to this point, functional complex geometries with nanometer length scale features have not been fabricated without post-processing techniques such as pyrolysis (Bauer et al, 2016). For the latter, there are myriad of possible orientations and designed defects that can be potentially utilized, and thus far they have only been tacitly studied. In this chapter we will present a novel approach to design hierarchically assembled unit cells at the same length scale to alleviate the fabrication challenge. By intertwining simple polyhedral structures, we create a complex unit cell with controlled buckling behaviour (Vangelatos et al, 2019a). Moreover, we assemble neighboring unit cells such that the lattice members of each one penetrate the others. Therefore, the buckled lattice members will come into contact with a large number of internal members, enhancing the energy dissipation through friction and the intrinsic geometric stiffness of the structures. This design principle is inspired by crystal twinning, where the crystals form in such a way that they penetrate each other, enhancing the strain hardening of the structure (Liang et al, 2015). The fundamental geometry is based on the first stellation of rhombic dodecahedron (Cromwell, 1999). To validate the new design, it is compared with one of the most thoroughly studied mechanical metamaterials, the octet-truss (Bonatti and Mohr, 2017). Finite element analysis (FEA) and nanoindentation experiments were performed to evaluate and measure the strain hardening and stiffness of the structures. In the next section, we present how 3D vacancies can be designed in octet truss unit cells to isolate large deformations at specific positions of the structure on demand, controlling where the structure will fail. Through nanoindentation experiments, we proved that test specimens designed on the basis of this principle have higher energy dissipation compared to the unblemished octet truss structures and lower buckling load (Vangelatos et al, 2019a). The experimental results were ratified by comparison with eigenvalue buckling analysis.

28.2 Design of Intertwinned Microlattice Structures

To elucidate the principles of controlled buckling and post contact, a novel design strategy was introduced. The design steps are delineated in Fig. 28.1 The fundamental unit cell is the first stellation of a rhombic dodecahedron. This geometry belongs to a vast group of 3D geometries, the stellation of regular polyhedral. Myriads of different geometries can be employed by this design space, as the regular icosahedron only can generate 472 stellated geometries (Shephard, 2000). The unit cells are assembled utilizing octahedra with a height-to-base ratio of 1:2. To control the buckling mechanism, the coalescence of the unit cells is organized such that the edges of each octahedron penetrate the neighboring ones. Figures 28.1A and 28.1B illustrate that the first stellation unit cell is comprised of three octahedra, rotated by 90° with respect to each other. To create a hierarchical structure, a second design step must be introduced. Hence, two unit cells are amalgamated by sharing their edges to form an intermediate structure (Fig. 28.1C). Following the same procedure to the other vertical direction, a hyper unit cell is constructed (Fig. 28.1D). This pattern can be utilized to form an array in the 2D space, and then expanded in the third dimension to form a multi-block structure (Fig. 28.1E). This design strategy can provide a substantial number of more plausible architected structures, regarding the spatial orientation of the intertwined unit cells with respect to the adjacent ones. The inspired in-grown mechanism of the lattice members can be produced by the perpetual intertwining of hyper unit cells. SEM micrographs of the hyper unit cell, with each octahedron colorized, are shown in Fig. 28.1F, and an SEM capture of a three-layer specimen is presented in Fig. 28.1G.

For the case of stretching dominated structures, the equivalent elastic modulus $\bar{E}$ and the equivalent yield strength σ_Y of the structure are functions of the relative density of the geometry as following (Ashby, 2005):

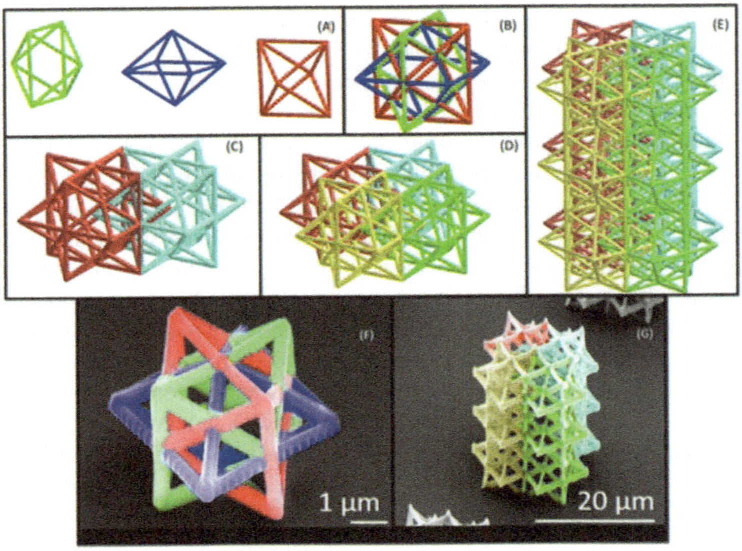

Fig. 28.1 (Vangelatos et al, 2019a). Conceptual design of a new mechanical metamaterial. (A-D) The conceptual process to design intertwined lattice structures, through the conglomeration of regular polyhedra. (E) The expanded metamaterial structure consisting of three layers of four-unit cell assemblies. (F) SEM micrograph of a fabricated unit cell. For improved legibility, each octahedron is distinguished by a different color in the whole assembly. (G) SEM micrograph of a fabricated three-layer hyper unit cell. Each subunit cell of the hyper unit cell is distinguished by a different color. The novelty of the design is the increase in strength of each unit cell through the mutual sharing of truss members between neighboring unit cells and the enhancement of strain hardening by introducing a deformation mechanism in the interior of each hyper unit cell to tailor the buckling behavior of the whole structure. Reprinted with permission from SAGE publishing.

$$\begin{aligned} \bar{E}/E &\cong \bar{\rho}\,, \\ \bar{\sigma}_Y/\sigma_Y &\cong \bar{\rho}\,. \end{aligned} \tag{28.1}$$

where E and σ_Y are the elastic modulus and yield strength of the bulk material, respectively. For the case of the octet truss, the relative density (defined as the volume ratio of the structure to the corresponding bulk material) (Deshpande et al, 2001) is given by,

$$\bar{\rho}_{OT} = \frac{6\pi\sqrt{2}R^2}{L^2}\,. \tag{28.2}$$

The relative density $\bar{\rho}$ can be used as a measure to compare the mechanical performance of metamaterials with others. The first stellation consists of 12 beams of length L and 24 beams of length equal to ($\sqrt{3}$/2)L. The volume of the bulk material occupying the same space as the first stellation is equal to $\sqrt{2}$ L× $\sqrt{2}$ L × L μm^3. Thus, the relative density of the first stellation $\bar{\rho}_{FS}$ is is given by

$$\bar{\rho}_{OT} = \frac{(12 + 24\sqrt{3}/2)L\pi R^2}{2L^3} = 6\pi(1 + \sqrt{3})\frac{R^2}{L^2} \,. \tag{28.3}$$

For the same parameters, $\bar{\rho}_{FS} \simeq 1.93\bar{\rho}_{OT}$. This indicates that the space occupied by the stiffer and stronger first stellation is two times larger than the octet truss. However, a novelty of this design is the dramatic decrease of the space occupied through unit cell intertwining. More specifically, if L = 10 μm the base area of the hyper unit cell of 4 × 4 unit cells is 14.5 × 14.5 μm^2, whereas the base area of an octet truss are with the same number of unit cells has a base area of 40 × 40 μm^2. The deformed stellated geometries can facilitate the buckling of proximal beams, accommodating large strains while the structural failure is limited. To evaluate the efficiency of the structure, it was compared with one of the primordial and most thoroughly studied mechanical metamaterials, the octet-truss. To study the buckling mechanism on each structure, an eigenvalue buckling analysis was performed. For the case of large deformations, the overall stiffness of the structure is the sum of the elastic stiffness $\boldsymbol{K}_e$ and the initial stiffness $\boldsymbol{K}_\sigma$. From the theory of elastic stability, buckling initiates when the displacement is $\boldsymbol{x} \neq \boldsymbol{0}$, even though the product of the stiffness with the displacement is equal to 0 (i.e. the force $\boldsymbol{f} = \boldsymbol{0}$), which implies that the stiffness matrix becomes semi positive definite, and instability commences. Palpably, for the case that the stiffness matrix is a scalar, buckling is initiated when the slope in the force displacement curve becomes 0, a notion which can be expanded in semi positive definiteness for higher degrees of freedom. Mathematically, this can be expressed utilizing the following equation (Krenk, 2009).

$$[\lambda \boldsymbol{K}_\sigma[\sigma_o] + \boldsymbol{K}_e]\boldsymbol{x} = \boldsymbol{f} = \boldsymbol{0} \,. \tag{28.4}$$

Since the vector $\boldsymbol{x}$ cannot be zero, Eq. (28.4) represents an eigenvalue-eigenvector problem. Therefore, the following eigenvalue equation must be solved

$$det[\boldsymbol{K}_\sigma^{-1}[\sigma_o]\boldsymbol{K}_e + \lambda \boldsymbol{I}] = 0 \,. \tag{28.5}$$

The physical interpretation of λ is that buckling occurs when the structure is loaded with force of magnitude equal to λ times a unit compressive load. The ratio of the load multipliers of the octet truss (OT) structure λ_{OT} to that of the first stellation (FS) structure λ_{FS}, obtained from an eigenvalue buckling FEA, was calculated to be $\lambda_{OT}/\lambda_{FS} = -0.55$. The reason for the negative sign is the negative load multiplier of the octet truss, indicating that buckling commences under tensile loading. Therefore, the first stellation structure will have the proclivity to buckle at lower loads compared to the octet truss. Hence, strain hardening, manifested by buckling events at relatively high loads, will have a predilection in the first stellation hyper unit cell. Furthermore, to investigate how buckling is evinced in the hyper unit cell, an elastic-plastic (large deformation) FEA was also performed. Fig. 28.2 depicts the plastic buckling of beams in one layer of the hyper unit cell deformed 40% in height. The plastic beam bending will increase the structure stiffness, as the bended beams will come in close proximity and simultaneously the plastic deformation will provide strain hardening.

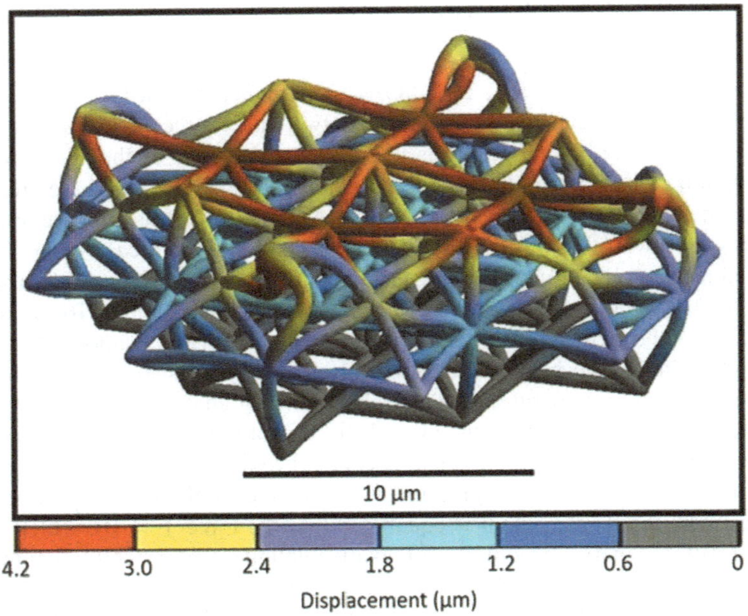

Fig. 28.2 (Vangelatos et al, 2019a). FEA results of an elastic-plastically compressed single-layer structure consisting of four first stellation unit cells. Compression by a distance of 4 μm promoted beam buckling in two of the three octahedra of the unit cells, a mechanism observed with octet truss structures at much larger deformations. The first stellation structure together with the specific arrangement of the unit cells are responsible for the significant enhancement of the mechanical behavior compared to the octet truss structure. Reprinted with permission from SAGE publishing.

To examine the validity of the simulation results, both of the structures, along with the bulk material, were fabricated by MPL. The fabricated structures are shown in Fig. 28.3A and 28.3B. To inquire the mechanical performance, nano-indentation experiments were performed utilizing the Hysitron TI 950 triboindenter. This indentation apparatus provides high precision, localized nano-mechanical testing with measurement errors up to 10-6 μN. Each structure was compressed by a distance of 4.5 μm (15% height decrease) three consecutive times and each force-displacement response is shown in Fig. 28.3C. Even though both structures demonstrated strain hardening after each loading cycle, due to the increase in the energy dissipation of each cycle, the first stellation structure has superior energy dissipation and strain hardening. More specifically, the octet-truss structure had 67% percentage increase in the energy dissipation, while the first stellation structure has 95%. Moreover, the strain energy per unit volume ratio for both structures was 12.16, one order of magnitude higher for the case of the first stellation. In addition, the first stellation sustained higher load before fracture and had a much steeper slope in the unloading section (i.e. ratio 1.99), despite the 30% higher volume and 4 times more unit cells of the octet truss structure. Furthermore, the residual strain (permanent height decrease) of the first stellation structure did not change significantly after the second loading cycle compared to the octet truss structure, indicating more pronounced strain hardening of the former structure. Finally, for the case of the bulk material the energy increase was 30%. Therefore, it can be concluded that the tactical coalescence of the lattice structures and the enhanced buckling mechanism and post contact, provided approximately 3 times higher strain hardening than the bulk material. Hence, the controlled assembly and spatial orientation of basic polyhedral structures can provide a new class of ultra-light, ultra-stiff mechanical metamaterials with enhanced strain hardening behavior. Random foams can provide these mechanical effects, because their random structure provides as massive number of fibre features conglomerated together (Ashby, 2005). Nevertheless, their inherent complexity renders their fabrication and design process challenging, whereas lattice structures, utilizing simple polyhedra, are easy to be fabricated, scalable and have a rigorous design methodology to be designed and improved.

28.3 Design of of Architected Vacancies in Metamaterial Structures

In this section we will introduce how to design three-dimensional (3D) metamaterial architectures containing vacancies in their lattice structures, using as geometry of reference the octet-truss structure. To facilitate localized buckling and plasticity in the structures, specific lattice members must be selectively removed from unit cells at locations of high deformation. Characteristic examples of the defected unit cells, along with the original octet truss, are presented in Fig. 28.4. From the first design (Fig. 28.4B) the internal octahedron is removed, while for the second design (Fig. 28.4C) the lattice members of two faces and the internal octahedron were removed by a set of two unit cells. The octet-truss is a convenient structure to examine this principle, as it consists of two parts. A regular octahedron at the core of the geometry and two lattice members forming an "X" on each face. Therefore, the lattice members can be removed from these two geometric groups, transcending the structure from stretching to bending dominated or even a hybrid design. As bending dominated structures sustain larger deformations and have higher energy dissipation, they can be positioned of locations of higher deformation in the structure to enhance their effects, while stretching dominated unit cells may be positioned in locations where high resilience to large deformations and high strength are required (i.e. stress concentration areas) (Khakalo et al, 2018).

Characteristic designs of octet truss arrays, including the distorted unit cells, defined as D1, D2, D3, and D4 respectively, are shown in Fig. 28.5. As it has been proven that these structures sustain failure in orientations that resemble slip planes for dislocation motion, the architected vacancies were positioned in a sloped arrangement in the structure. This non-orthonormal planar orientation is expected to sustain excessive shear stress during compression. The hollow unit cell shown in Fig.

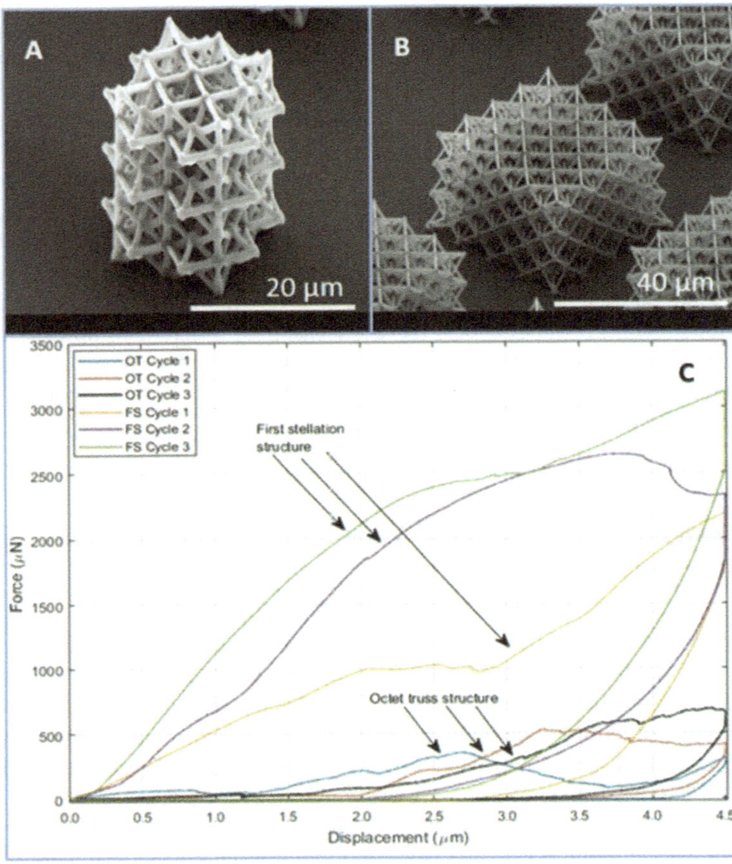

Fig. 28.3 (Vangelatos et al, 2019a). (A) SEM micrograph of the first stellation structure fabricated by MPL. (B) SEM micrograph of the octet truss structure. (C) Force-displacement responses of first stellation and octet truss structures. The three sequential load/unload cycles of each structure reveal a more significant increase in stiffness, strength, and strain hardening of the first stellation (FS) structure than the octet truss (OT) structure, despite the 30% higher volume and 4 times more unit cells of the octet truss structure. Reprinted with permission from SAGE publishing.

28.4B was used for the structures D1, D3, and D4, whereas the unit cell presented in Fig. 28.1C was employed to architect the D2 structure.

All of the structures were fabricated utilizing MPL, as in the previous section. Characteristic samples, such as the octet-truss and D4 are presented Fig. 28.6A and 28.6B. The height of the unit cells is 10 μm and the thickness of the lattice members is 0.5 μm. It must be noted that this design procedure is advantageous for the fabrication process, as it becomes faster and the non-polymerized material can be dissolved easier, as the structure has more voids).

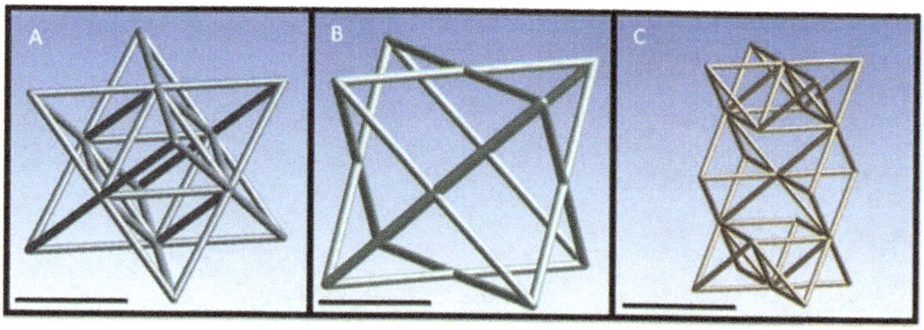

Fig. 28.4 (Vangelatos et al, 2019a). The unit cells used to design the metamaterial structures of this study. (A) The original octet-truss unit cell (scale bar = 10 μm). (B) A design consisting of two octet-truss unit cells where the lattice members of the top and bottom half of the two unit cells have been removed to increase the instability of the whole assembly (scale bar = 10 μm). (C) A design in which the internal octahedron of the octet-truss structure has been removed (scale bar = 5 μm). Reprinted with permission from SAGE publishing.

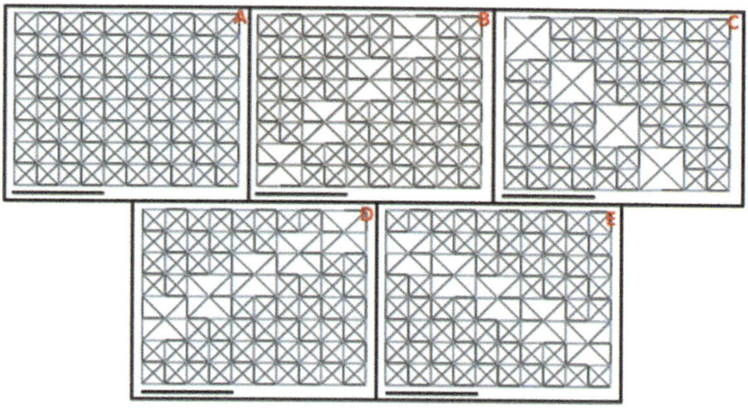

Fig. 28.5 The 5 × 5 × 4 array of unit cells used in each metamaterial structure: (A) octet-truss structure, (B) D1 structure, (C) D2 structure, (D) D3 structure, and (E) D4 structure. Even though the D1, D3, and D4 structures have the same 3D vacancy, the edges contacting the neighboring unit cells are aligned differently. The scale bar in each figure is 20 μm.

To validate the mechanical performance of the fabricated structures, quasi-static nanoindentation experiments were performed. The force-displacement curves after one loading cycle for each structure are presented in Fig. 28.7. For all of the structures with the vacancies, the maximum load is much higher for the same maximum deformation reached and the energy dissipation has been increased substantially. To provide the performance metrics, for each new design the energy dissipation ratio, with respect to the original octet truss structure, u/u_{OT} is equal to 2.35, 1.35, 1.24 and 1.61. Notably, for the design D1 the energy dissipation is two times larger than the original structures, a corollary of the localized buckling in the designated unit cells.

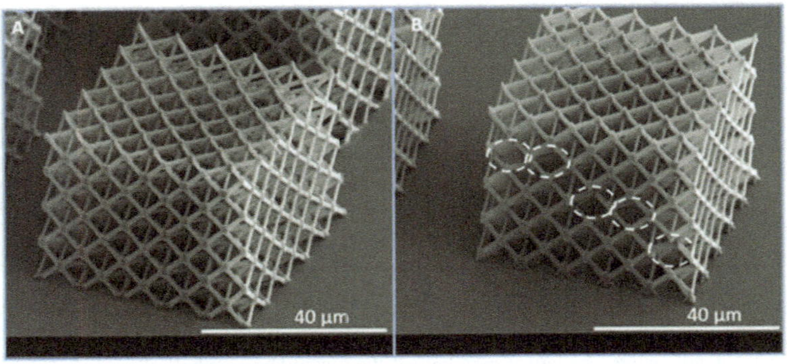

Fig. 28.6 (A) Isometric view of an octet-truss structure with 5 × 5 × 4 unit cells and (B) isometric view the structure D4, comprised of the same number of unit cells.

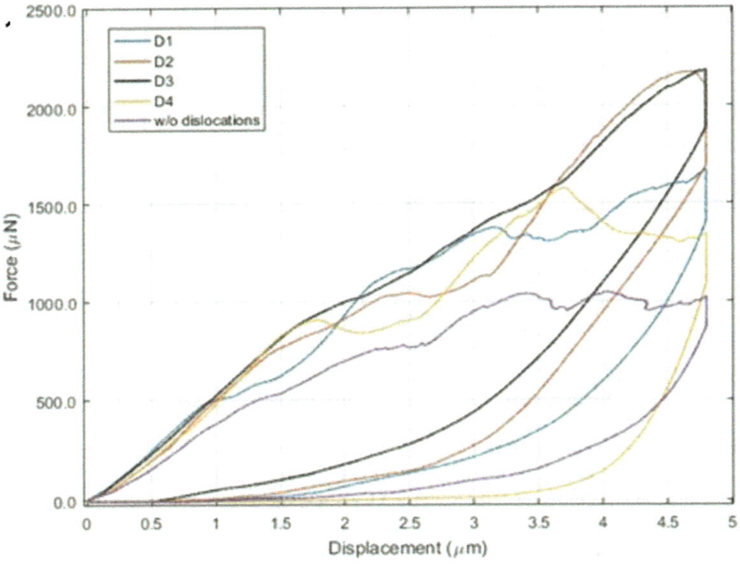

Fig. 28.7 (Vangelatos et al, 2019a). Indentation force versus displacement response of the various metamaterial structures. Reprinted with permission from SAGE publishing.

A possible explanation why D1 has higher energy dissipation than the rest of the structures is that it is comprised of fewer vacancies than the D2 and D3 and the fact that the array of vacancies must converge to the maximum shearing direction encountered during compression than any other structure. However, the D3 and D4 structures have vacancy architected unit cells at the top and the bottom of their respective structure. Therefore, the unit cell array with vacancies in the D1 structure has higher resilience to large deformation, as the neighboring unit cells function as escutcheons, protecting the vacancies from fracture while localized buckling occurs. This parameter ameliorates the mechanical response of this structure compared to the others. In addition, eigenvalue buckling FEA analysis provided an insight at which load buckling will commence for each structure, an indicator of localized large deformation. For each geometry, the load multiplier was calculated utilizing equation (28.2). The values of the load multiplier λ of each structure were normalized with respect to the load multiplier of the octet-truss structure. The ratios for each design were calculated equal to 0.780, 0.377, 0.698 and 0.782, respectively. The substantially smaller buckling load for the case of D1–D4 structures than that of the octet-truss designate for the case of the defected structures, buckling commenced earlier than the octet-truss. This is the evidence that the goal of tailoring localized buckling in the designed structures was achieved. A potential cause for the lowest load multiplier of the D2 structure is that its unit cell at the boundary edge of the structure is the least supported, with the smallest number of lattice members. Since this unit cell is not augmented by any bordering unit cells, it has higher predilection for buckling. The localized buckling in the sloped arrays of the structures can be observed in the large deformation FEA of Fig. 28.8. The localized buckling, attributed to the designed vacancies, is not observed in the original octet-truss array (Fig. 28.8A).

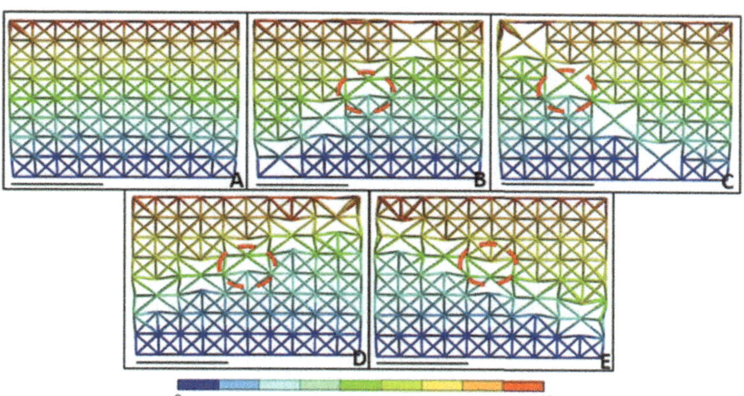

Fig. 28.8 Quasi-static FEA simulation results of compressed structures. Side view of (A) octet-truss, (B) D1, (C) D2, (D) D3, and (E) D4 metamaterial structures. The scale bar is 20 μm.

28.4 Conclusion

In this chapter we demonstrated two design paradigms to create arbitrary mechanical metamaterial lattice structures. By either assembling regular polyhedral structures to form a more complicated geometry or removing members of an already existing structure, we enhanced the energy dissipation and the strain hardening of the final design. The principal mechanisms that must be controlled are the localized buckling of the lattice members and their subsequent post contact. Through simulations, fabrication and indentation experiments, the validity of these design concepts was proven. This research paves the way for the expansion of the design space of ultra-light metamaterial structures. It also illuminated the importance of optimization techniques (Chen et al, 2018; Gu et al, 2018) to determine the optimum intertwining and the optimum aphaeresis of lattice members from a structure. This field is still incipient. Nevertheless, its advance can lead more efficient architected geometries that can be utilized for engineering applications.

Acknowledgements The authors acknowledge collaboration with M. Farsari, V. Melissinaki, and I. Sakellari (IESL/FORTH) on the MPL fabrication method. They also thank the UC Berkeley Professors A.M. Minor of Materials Science and Engineering and P. Hosemann of Nuclear Engineering for their help with the nanoindentation experiments. This research was partially supported by the National Science Foundation (NSF) under the Scalable Nanomanufacturing (SNM) Program, Grand No. 1449305.

References

Ashby MF (2005) The properties of foams and lattices. Philosophical Transactions of the Royal Society A: Mathematical, Physical and Engineering Sciences 364(1838):15–30

Barchiesi E, Spagnuolo M, Placidi L (2018) Mechanical metamaterials: a state of the art. Mathematics and Mechanics of Solids 24(1):212–234

Bauer J, Schroer A, Schwaiger R, Tesari I, Lange C, Valdevit L, Kraft O (2015) Push-to-pull tensile testing of ultra-strong nanoscale ceramic–polymer composites made by additive manufacturing. Extreme Mechanics Letters 3:105 – 112

Bauer J, Schroer A, Schwaiger R, Kraft O (2016) Approaching theoretical strength in glassy carbon nanolattices. Nature Materials 15(4):438–443

Berezovski A, Yildizdag ME, Scerrato D (2018) On the wave dispersion in microstructured solids. Continuum Mechanics and Thermodynamics

Berger JB, Wadley HNG, McMeeking RM (2017) Mechanical metamaterials at the theoretical limit of isotropic elastic stiffness. Nature 543(7646):533–537

Bonatti C, Mohr D (2017) Large deformation response of additively-manufactured fcc metamaterials: From octet truss lattices towards continuous shell mesostructures. International Journal of Plasticity 92:122 – 147

Chen D, Skouras M, Zhu B, Matusik W (2018) Computational discovery of extremal microstructure families. Science advances 4(1):eaao7005

Cromwell PR (1999) Polyhedra. Cambridge University Press

dell'Isola F, Giorgio I, Pawlikowski M, Rizzi N (2016) Large deformations of planar extensible beams and pantographic lattices: heuristic homogenization, experimental and numerical examples of equilibrium. Proceedings of the Royal Society A: Mathematical, Physical and Engineering Sciences 472(2185):20150,790

dell'Isola F, Lekszycki T, Pawlikowski M, Grygoruk R, Greco L (2015) Designing a light fabric metamaterial being highly macroscopically tough under directional extension: first experimental evidence. Zeitschrift für angewandte Mathematik und Physik 66(6):3473–3498 dell'Isola F, Seppecher P, Alibert JJ, et al (2018) Pantographic metamaterials: an example of mathematically driven design and of its technological challenges. Continuum Mechanics and Thermodynamics 31(4):851–884

dell'Isola F, Seppecher P, Spagnuolo M, et al (2019a) Advances in pantographic structures: design, manufacturing, models, experiments and image analyses. Continuum Mechanics and Thermodynamics 31(4):1231–1282

dell'Isola F, Turco E, Misra A, Vangelatos Z, Grigoropoulos C, Melissinaki V, Farsari M (2019b) Force–displacement relationship in micro-metric pantographs: Experiments and numerical simulations. Comptes Rendus Mécanique 347(5):397 – 405

Deshpande V, Fleck N, Ashby M (2001) Effective properties of the octet-truss lattice material. Journal of the Mechanics and Physics of Solids 49(8):1747 – 1769

Gao H, Ji B, Jäger I, Arzt E, Fratzl P (2003) Materials become insensitive to flaws at nanoscale: Lessons from nature. PNAS; Proceedings of the National Academy of Sciences 100(10):5597–5600

Gibson LJ, Ashby MF (1999) Cellular solids: structure and properties. Cambridge university press

Giorgio I, Rizzi N, Turco E (2017) Continuum modelling of pantographic sheets for out-of-plane bifurcation and vibrational analysis. Proceedings of the Royal Society A: Mathematical, Physical and Engineering Sciences 473(2207):20170,636

Gross A, Pantidis P, Bertoldi K, Gerasimidis S (2019) Correlation between topology and elastic properties of imperfect truss-lattice materials. Journal of the Mechanics and Physics of Solids 124:577 – 598

Gu GX, Chen CT, Richmond DJ, Buehler MJ (2018) Bioinspired hierarchical composite design using machine learning: simulation, additive manufacturing, and experiment. Materials Horizons 5(5):939–945

Khakalo S, Balobanov V, Niiranen J (2018) Modelling size-dependent bending, buckling and vibrations of 2d triangular lattices by strain gradient elasticity models: Applications to sandwich beams and auxetics. International Journal of Engineering Science 127:33 – 52

Krenk S (2009) Non-linear modeling and analysis of solids and structures. Cambridge University Press

Liang H, Chen W, Wang R, Qi Z, Mi J, Wang Z (2015) X-shaped hollow α-FeOOH penetration twins and their conversion to α-Fe2O3 nanocrystals bound by high-index facets with enhanced photocatalytic activity. Chemical Engineering Journal 274:224 – 230

Meza LR, Das S, Greer JR (2014) Strong, lightweight, and recoverable three-dimensional ceramic nanolattices. Science 345(6202):1322–1326

Milton G, Briane M, Harutyunyan D (2017) On the possible effective elasticity tensors of 2-dimensional and 3-dimensional printed materials. Mathematics and Mechanics of Complex Systems 5(1):41–94

Misra A, Lekszycki T, Giorgio I, Ganzosch G, Müller WH, dell'Isola F (2018) Pantographic metamaterials show atypical Poynting effect reversal. Mechanics Research Communications 89:6 – 10

Pham MS, Liu C, Todd I, Lertthanasarn J (2019) Damage-tolerant architected materials inspired by crystal microstructure. Nature 565(7739):305–311

Sakellari I, Kabouraki E, Gray D, Purlys V, Fotakis C, Pikulin A, Bityurin N, Vamvakaki M, Farsari M (2012) Diffusion-assisted high-resolution direct femtosecond laser writing. ACS Nano 6(3):2302–2311

Shephard G (2000) Isohedral deltahedra. Periodica Mathematica Hungarica 39(1-3):83–106

Vangelatos Z, Gu GX, Grigoropoulos CP (2019a) Architected metamaterials with tailored 3D buckling mechanisms at the microscale. Extreme Mechanics Letters 33:100,580

Vangelatos Z, Melissinaki V, Farsari M, Komvopoulos K, Grigoropoulos C (2019b) Intertwined microlattices greatly enhance the performance of mechanical metamaterials. Mathematics and Mechanics of Solids 24(8):2636–2648

Vicidomini G, Bianchini P, Diaspro A (2018) Sted super-resolved microscopy. Nature Methods 15(3):173–182

Virgin L (2018) Tailored buckling constrained by adjacent members. Structures 16:20 – 26

Yildizdag ME, Demirtas M, Ergin A (2018) Multipatch discontinuous Galerkin isogeometric analysis of composite laminates. Continuum Mechanics and Thermodynamics pp 1–14

Yildizdag ME, Ardic IT, Demirtas M, Ergin A (2019) Hydroelastic vibration analysis of plates partially submerged in fluid with an isogeometric fe-be approach. Ocean Engineering 172:316 – 329

The manufacturer's authorised representative in the EU is Springer Nature Customer Service Centre GmbH, Europaplatz 3, 69115 Heidelberg, Germany. If you have any concerns regarding our products, please contact ProductSafety@springernature.com

Printed and bound by CPI Group (UK) Ltd, Croydon, CR0 4YY
28/09/2026
02224057-0005